BASIC INSTRUMENTATION

FOURTH EDITION

by

Will L. McNair

published by

THE UNIVERSITY OF TEXAS AT AUSTIN

Division of Continuing & Innovative Education
PETROLEUM EXTENSION SERVICE

2002

Library of Congress Cataloging-in-Publication Data

McNair, Will L.
    Basic instrumentation / by Will L. McNair. — 4th ed.
        p. cm.
    ISBN 0-88698-197-2 (alk. paper)
    1. Measurement fundamentals.   I. Title.
TA165.M37 2002
620'.0044—dc21

2002002831

© 2002 by the University of Texas at Austin
All rights reserved.
First edition published 1943. Fourth edition 2002
Third impression 2013
Printed in the United States of America

This book or parts thereof may not be reproduced in any form without permission of Petroleum Extension Service, The University of Texas at Austin.

Brand names, company names, trademarks, or other identifying symbols appearing in illustrations or text are used for educational purposes only and do not constitute an endorsement by the author or publisher.

The University of Texas at Austin is an equal opportunity institution. No state tax funds were used to print or mail this publication.

Catalog No. 1.20040
ISBN 0-88698-197-2

Contents

Preface   xiii

Acknowledgments   xv

Chapter 1.   Introduction   1
- The Need for Measurment and Control   1
- Methods of Measurement   2
- Types of Control   4
- Methods or Modes of Control   9
- Summary   14
- Review Exercise   14

Chapter 2.   The Units of Measurement   15
- Comparison of Systems of Units   15
- Système International (SI) d'Unités   16
- Measuring Length   16
- Measuring Time   18
- Measuring Temperature   18
- Measuring Mass, Weight, and Force   19
- Measuring Work and Energy   22
- Measuring Dimensions of Various Quantities   24
- Summary   27
- Review Exercise   28

Chapter 3.   Final Control Elements   29
- Valves   29
- Sizing and Piping Arrangements   39
- Actuators   39
- Controlled-Volume Pumps   48
- Variable-Volume Pumps   50
- Other Final Control Elements   50
- Summary   50
- Review Exercise   50

Chapter 4.   Pneumatic Automatic Controls   51
- Pneumatic Controls   51
- Commercial Pneumatic Controllers   61
- Volume Booster Relays   64
- Valve Positioners   65
- Summary   69
- Review Exercise   69

Chapter 5.   Electronic Automatic Controls   71
- Analog Circuits and Equipment   71

        Modes of Control and Control Loops   73
        System Stability and Loop Tuning   78
        Programmable Logic Controllers (PLC) Control Systems   79
        Specialized Flow Computers   81
        Distributed Control Systems   81
        Human-Machine-Interface (HMI)   83
        Summary   84
        Review Exercise   84

Chapter 6.   Pressure Measurement and Control   85
        Units of Pressure Measurement   85
        Mechanical Pressure Elements   87
        Electronic Pressure Measurement   92
        Vacuum Measurements   95
        Pressure Control   97
        Summary   104
        Review Exercise   105

Chapter 7.   Temperature Measurement and Control   107
        Defining Temperature Measurement   108
        Mechanical Temperature Sensors   109
        Electronic Temperature Measurement   112
        Wheatstone Bridges   117
        Electronic Temperature Transmitters   117
        Temperature Control   121
        Summary   124
        Review Exercise   124

Chapter 8.   Liquid-Level Measurement and Control   125
        Defining Level Measurement   125
        Mechanical Level Sensors   125
        Electrical Level Measuring Devices   134
        Level Control   138
        Summary   140
        Review Exercise   141

Chapter 9.   Flow Measurement   143
        Defining Flow Measurement   143
        Mechanical Flow Sensors and Meters   144
        Electronic Flow Sensors and Meters   150
        Summary   156
        Review Exercise   156

Chapter 10.   Flow Control   157
        Mechanical Flow Control Elements   157
        Electronic Flow Controllers   159
        Integral Flow Controllers   162
        Summary   170
        Review Exercise   170

Chapter 11.   Gravity, Viscosity, Humidity, and pH   171
    Measuring Specific Gravity and Density   171
    Measuring Viscosity   177
    Measuring Humidity and Dew Point   180
    Measuring pH   185
    Summary   187
    Review Exercise   187

Chapter 12.   Programmable Logic Controllers   189
    PLC Operating Concepts   189
    PLC Brands   195
    PLC Applications and Loop Tuning   203
    Summary   205
    Review Exercise   205

Appendix A.   Numbering Systems and Codes   207

Appendix B.   Temperature Sensor Reference Tables   213

Glossary   297

Figures

1.1	Direct measurement examples of level and pressure	3
1.2	A method for inferring liquid level from hydrostatic pressure	3
1.3	Method of manual control of liquid level	4
1.4	Manual control of hot-water temperature	4
1.5	Automatic control of level	5
1.6	Automatic control of temperature in hot-water heater	5
1.7	Example of resistance in transfer of energy from one capacity to another	7
1.8	Reaction to a sudden change in opening of valve X in a liquid-level control system	8
1.9	Reaction to step change in control valve X in a water-temperature control system	9
1.10	Floating mode of liquid-level control	10
1.11	A proportional-speed floating control-level control system	10
1.12	A proportional control system with proportional band of 0 to 100 percent	11
1.13	A system using proportional plus-reset mode of control	13
2.1	Balance scale for determining mass	21
2.2	A spring scale determines mass by measuring the distance the spring is stretched.	22
2.3	A U-tube manometer is a common device for measuring pressure.	25
3.1	Double-ported valve	29
3.2	Two types of ported valves	30
3.3	A split-body valve	31
3.4	A typical angle-body valve	31
3.5	Three-way valves	31
3.6	Butterfly valve	32
3.7	A pipeline gate valve	33
3.8	Flow curves	33
3.9	Types of valve plugs	34
3.10	Equal percentage plug	35
3.11	Linear plug	35
3.12	Plugs for low-flow rates	36
3.13	Stuffing box	37
3.14	Self-lubricating stuffing box with Teflon packing	37
3.15	Valve bonnets and extensions	38
3.16	Throttling valve	39
3.17	Diaphragm actuator	40
3.18	A spring-loaded, reverse-acting diaphragm actuator	41
3.19	Air-loaded diaphragm actuator	41
3.20	Pneumatic piston actuators	42
3.21	Arrangement of a typical air supply for pneumatic actuators	43
3.22	Solenoid actuator and valve	44
3.23	Circuit diagram for floating control of reversible motor	45
3.24	Proportional control system	46
3.25	Balancing relay	47

3.26	A speed control system for a steam turbine	47
3.27	Controlled-volume pump as a final control element	48
3.28	A combination open-loop and closed-loop system with a controlled-volume pump	49
3.29	Gas-powered injection pump	49
3.30	Variable frequency, variable speed pump drive	50
3.31	An electric switch as the final control element in an electric hot-water system	50
4.1	A self-contained force-loaded pressure regulator	51
4.2	A self-contained spring-loaded regulator	52
4.3	A fixed orifice as part of a pneumatic controller	53
4.4	A simplified pneumatic controller	53
4.5	A two-position (on-off) controller in a liquid-heating system	54
4.6	Using an air relay to provide linear control	55
4.7	A proportional controller with a bellows-spring assembly	56
4.8	Proportional controller in a liquid-heating system	57
4.9	Performance of controller at different throttle range settings	58
4.10	Proportional controller with automatic reset	59
4.11	Performance curves for multi-mode controller action	60
4.12	Response curves of different modes following quick change in load	61
4.13	Continuous-bleed air relay	62
4.14	Nonbleed air relay	62
4.15	Pneumatic controller, Model 40	63
4.16	Flapper-nozzle assembly, Model 40	64
4.17	A booster relay, used to overcome response lag due to long connecting lines between controller and valve	65
4.18	Schematic of a simplified valve positioner	66
4.19	Valve positioner	67
4.20	A valve positioner for a pneumatic piston actuator	69
4.21	Valve positioner relay	69
5.1	Two- and four-wire transmitters	72
5.2	Electronic transmitter and other components in a control loop	73
5.3	Electronic proportional controller	74
5.4	Open loop and proportional control	76
5.5	Proportional-plus-integral control	77
5.6	Open loop and proportional-plus-integral control	77
5.7	Proportional-plus-integral-plus-derivative control (PID)	78
5.8	Open loop and proportional-plus-integral plus derivative control	78
5.9	PLC closed-loop control system	80
5.10	Flow controller	81
5.11	Panelview HMI	83
6.1	Pressure measurement comparisons	86
6.2	Three basic types of Bourdon tubes	88
6.3	Bourdon tubes arranged to measure differential pressure	88
6.4	A pressure element with single-shell metallic diaphragm	89
6.5	Pressure element with a capsule of two metallic diaphragms	89
6.6	A compound element with three diaphragm capsules	89

6.7	A nonmetallic diaphragm with calibrated spring	89
6.8	Using a metallic diaphragm-variable resistor to measure liquid level	90
6.9	Differential-pressure gauge using a spring-opposed bellows	90
6.10	A mercury-bell differential-pressure gauge	91
6.11	Protective devices for mechanical instruments	92
6.12	Typical four-wire voltage transmitter	92
6.13	A two-wire current transmitter	93
6.14	Electronic two-wire pressure transmitter with zero and span adjustments	93
6.15	Smart pressure transmitter and interface unit	94
6.16	A type of McLeod gauge	95
6.17	A Pirani gauge using a resistance element	95
6.18	A Pirani gauge using a Wheatstone bridge	96
6.19	Thermocouple vacuum gauge	97
6.20	Pressure-controlled air compressor	97
6.21	A system with hydrostatic pressure control	98
6.22	A system with air or gas pressure control	98
6.23	Relief valve with typical spring-loading	99
6.24	Relief valve that protects against overpressures and underpressures	99
6.25	A pressure relief valve	99
6.26	Location of instruments in a rural gas line	99
6.27	A relief valve used with a pump	100
6.28	An intermediate-pressure regulator	101
6.29	A low-pressure gas regulator	101
6.30	Performance curves for a 1-inch low-pressure regulator	102
6.31	A high-pressure regulator	103
6.32	A weight-loaded back-pressure regulator	103
6.33	Diagram of a proportional band pressure controller	104
6.34	PLC-controlled air compressor	105
7.1	Types of temperature-measuring devices and their ranges	107
7.2	Comparison of four temperature scales	108
7.3	A simple liquid-in-glass thermometer	109
7.4	A mercury thermometer used as a component of a temperature alarm system	110
7.5	Filled-system thermometer with no compensating elements	110
7.6	Bimetal elements used in filled systems as compensating components	111
7.7	Bimetal thermometer elements	112
7.8	Bimetal elements in wound forms	113
7.9	Thermometer with bimetal element	113
7.10	A butt-welded thermocouple	114
7.11	Thermocouple schematics	115
7.12	Schematic millivoltmeter with thermocouple	115
7.13	Optical pyrometers that infer temperature from color and brightness	116
7.14	Typical Wheatstone bridge circuit	117
7.15	A two-wire current transmitter for measuring temperature	118
7.16	Electronic temperature transmitter connections with RTD sensor	118
7.17	Electronic temperature transmitter connection diagram with thermocouple sensor	118
7.18	Electronic two-wire temperature transmitter with zero and span adjustments	119

7.19	Test calibration circuit of temperature transmitter	119
7.20	Smart temperature transmitter and interface unit	120
7.21	Block diagram of a typical smart temperature transmitter	121
7.22	Temperature safety control valves	122
7.23	Mixing valve	122
7.24	PLC temperature controller	123
8.1	Basic types of sight glasses	125
8.2	Sight glass for low-pressure, low-temperature systems	126
8.3	Sight glass for high-pressure, high-temperature systems	126
8.4	Buoyancy-type instrument with full-range indicator	127
8.5	Restricted range level controller	127
8.6	A float-operated liquid-level controller	127
8.7	Float-operated controller used with dump valve on oil and gas separators	127
8.8	A pilot-operated controller	128
8.9	Pilot valve for the controller in Figure 8.8	128
8.10	Displacer-float assembly of a torque tube displacer-type level indicator and controller	129
8.11	Schematic of a control system using displacer-type level indicator and controller	129
8.12	Displacer-type indicator and controller with flexure tube	130
8.13	Liquid-seal type of hydrostatic pressure measuring system	131
8.14	Diaphragm-box type of hydrostatic measuring system	132
8.15	Air trap method of level indication	133
8.16	Air bubble or air-purge level system	133
8.17	Pressurized tanks with differential pressure gauge	134
8.18	Electric liquid-level controller operating a switch with a float and flexure tube	135
8.19	Use of level switch in dry oil petroleum storage tank	135
8.20	A capacitor-type liquid-level gauge	136
8.21	Metritape liquid-level sensor	136
8.22	Electronic differential-pressure transmitter measuring liquid level	137
8.23	Open-tank level measurement with required elevation of zero	137
8.24	Open-tank level measurement with required suppression of zero	138
8.25	On-off level control with two outflow pumps	139
8.26	Level control using variable speed outflow pumps	140
9.1	Flowmeter restrictive elements	144
9.2	Typical venturi tube	145
9.3	Dall tube	145
9.4	Flow nozzles	146
9.5	Three views of installed orifice plate	147
9.6	Three types of orifice plate	147
9.7	Straightening vanes	148
9.8	Orifice meter with a modified U-tube manometer	149
9.9	Bellows assembly for a bellows-type orifice meter	150
9.10	Electronic differential-pressure transmitter for measuring flow	151
9.11	Exploded view of magnetic flowmeter	151
9.12	Mass flowmeter that measures torque	152
9.13	Mass flowmeter	152

9.14	Turbine flowmeter	153
9.15	Demonstrating the vortex principle of flow	153
9.16	A typical vortex flowmeter	154
9.17	Gas meter	154
9.18	Cutaway view of gas meter case	154
9.19	Operation of the gas meter	155
9.20	Phantom view of a gas meter	156
10.1	Fixed flow bean	157
10.2	Adjustable flow bean	157
10.3	Force-balance differential-pressure sensing instrument	158
10.4	Electronic flow control elements	159
10.5	Multivariable transmitter for measuring mass flow rate	160
10.6	Multivariable smart flow transmitter block diagram and sensor module	161
10.7	Flow control system using turbine flowmeter feedback and PLC	161
10.8	Flow controller	162
10.9	Installation for maintaining constant downstream pressure	163
10.10	Installation for maintaining constant upstream pressure	163
10.11	Downstream pressure sensing device for a steam-driven pump	164
10.12	Throttling valve in bypass line for a postive-displacement pump	165
10.13	Control valve installed downstream from a centrifugal pump	165
10.14	Feed-rate control for initial column of fractionating columns	166
10.15	Liquid-level controller for succeeding fractionating columns	166
10.16	Electronic/pneumatic controller	167
10.17	A system with simple flow-rate control	167
10.18	Liquid-level control for a system without a constant rate of feed	167
10.19	Liquid-level control system with accumulator tank	168
10.20	System using vapor pressure differential to establish set point of controller	169
11.1	Typical hydrometers used to measure specific gravity	173
11.2	An air-bubbler system with reference tank for measuring specific gravity	173
11.3	Air-bubbler system without reference tank	174
11.4	A displacer-float specific gravity meter	175
11.5	Schematic of system using displacer float	176
11.6	Lab equipment used over a century ago to determine viscosity of liquids	177
11.7	Direct-reading instruments for measuring viscosity	179
11.8	Control system for maintaining viscosity in a liquid	180
11.9	Chart of relative humidity versus dry- and wet-bulb temperatures	181
11.10	A laboratory method for determining vapor pressure	181
11.11	Principles involved in hair-type indicator used to measure relative humidity	182
11.12	A sling psychrometer	183
11.13	A recording psychrometer for measuring and recording humidity	183
11.14	A gold-leaf grid, lithium chloride element used for measuring humidity	184
11.15	A Bureau of Mines type of dew-point tester	184
11.16	Dew-point electrical measuring device	185
11.17	pH factor and ion concentration	186
11.18	Portable electric instrument that uses a glass electrode to measure pH factor	186

12.1	Typical relay ladder logic diagram	189
12.2	Five components of a PLC system	190
12.3	Ladder logic diagram of air compressor control	191
12.4	Pipeline pump operation	193
12.5	Allen-Bradley processor modules	196
12.6	Allen-Bradley I/O modules	197
12.7	Typical chassis arrangement of Allen-Bradley processor and I/O modules	197
12.8	Chassis arrangement with input and output field devices	198
12.9	Allen-Bradley analog input and output modules	199
12.10	Functional diagram of analog input modules	199
12.11	Process transmitter connection to a 12-bit analog input module	200
12.12	Analog output module of PLC	201
12.13	Frequency and RTD/thermocouple input PLC modules	202
12.14	Temperature control using PID module	203
12.15	Process variable responses to disturbances	204
12.16	Closed-loop tuning methods with software	205

Tables

2.1	Conventional and SI Units of Measurement	17
6.1	Types of Sensors Used to Measure Various Pressure Ranges	87
7.1	Relative Merits of Filled Systems	111
12.1	Binary and Octal Numbers	194
12.2	Binary and Hexadecimal Numbers	194
12.3	Binary, Octal, Hexadecimal, and Decimal Numbers	195
12.4	Decimal to Binary to BCD Conversions	195

Preface

Technology has continued to expand in many areas since the last edition of *Basic Instrumentation* was written in 1983. In particular, the measurement and control industry has evolved from the use of instruments that produce a visual metering process to one where the measured information is transmitted and processed through electrical signals without any human intervention.

This text has been prepared to remind us of what was used in earlier times and to acquaint us with new technology now in use in the process industries. The reader of this text should have had exposure to the basic high school technical courses of algebra, chemistry and physics, some knowledge of electrical basics, and interest or experience in the oil, gas, petrochemical, and related process industries.

Effort has been spent to reduce the complexity of the subject matter by omitting some formulas, equations, and other highly technical presentations and replacing them with descriptive words and language that make the text more readable and enjoyable. A number of components used in instrumentation currently in use may be foreign to some. Effort has been spent to describe the basic functions of these components in laymen's terms and how they are used in instrumentation.

The units used in process measurement can be described in conventional as well as metric units. When the process variables of pressure, temperature, flow, and level are being discussed, common industry units will be used with descriptive notes for other units when it is warranted.

Information for this book has been drawn from many sources, including instrument experts, publications, and practical experience. It is hoped that the reader will derive benefit from having read this book, by working sample problems, observing testing and calibration techniques, and understanding how various instruments function.

Acknowledgments

Behind the development of this book are over 55 years of contributions by various individuals and organizations. Earlier contributors include Bruce Whalen, who was PETEX's publications coordinator prior to his retirement in 1983; content consultants J. J. Sergesketter, Willis Finley, Jerry Campbell, and Louis Johnson; and several contributors employed by petrochemical companies throughout the world. Also, numerous instrumentation companies were a substantial resource for operational descriptions, photographs, diagrams, and charts. Moreover, I wish to recognize Shell Oil Company for giving me the opportunity to provide instrumentation instruction to various companies at the Robert Training Center in Louisiana.

Finally, I wish to recognize the work of the PETEX publications staff, without whom this book would not have been possible. Kathryn Roberts helped keep me on track while writing and editing was underway, and spent countless hours organizing and placing the many drawings and diagrams in the text. Debbie Caples laid out the book, and Doris Dickey proofread the manuscript for typos. Leslie Kell drew many of the diagrams and charts and made them easy to see and understand. Finally, PETEX director Ron Baker gave the book a final read. My heartfelt thanks to them all.

Will L. McNair, P.E.

Units of Measurement

Throughout the world, two systems of measurement dominate: the English system and the metric system. Today, the United States is almost the only country that employs the English system.

The English system uses the pound as the unit of weight, the foot as the unit of length, and the gallon as the unit of capacity. In the English system, for example, 1 foot equals 12 inches, 1 yard equals 36 inches, and 1 mile equals 5,280 feet or 1,760 yards.

The metric system uses the gram as the unit of weight, the metre as the unit of length, and the litre as the unit of capacity. In the metric system, for example, 1 metre equals 10 decimetres, 100 centimetres, or 1,000 millimetres. A kilometre equals 1,000 metres. The metric system, unlike the English system, uses a base of 10; thus, it is easy to convert from one unit to another. To convert from one unit to another in the English system, you must memorize or look up the values.

In the late 1970s, the Eleventh General Conference on Weights and Measures described and adopted the Système International (SI) d'Unités. Conference participants based the SI system on the metric system and designed it as an international standard of measurement.

Basic Instrumentation gives both English and SI units. And because the SI system employs the British spelling of many of the terms, the book follows those spelling rules as well. The unit of length, for example, is *metre*, not *meter*. (Note, however, that the unit of weight is *gram*, not *gramme*.)

To aid U.S. readers in making and understanding the conversion to the SI system, we include the following table.

English-Units-to-SI-Units Conversion Factors

Quantity or Property	English Units	Multiply English Units By	To Obtain These SI Units
Length, depth, or height	inches (in.)	25.4	millimetres (mm)
		2.54	centimetres (cm)
	feet (ft)	0.3048	metres (m)
	yards (yd)	0.9144	metres (m)
	miles (mi)	1609.344	metres (m)
		1.61	kilometres (km)
Hole and pipe diameters, bit size	inches (in.)	25.4	millimetres (mm)
Drilling rate	feet per hour (ft/h)	0.3048	metres per hour (m/h)
Weight on bit	pounds (lb)	0.445	decanewtons (dN)
Nozzle size	32nds of an inch	0.8	millimetres (mm)
Volume	barrels (bbl)	0.159	cubic metres (m^3)
		159	litres (L)
	gallons per stroke (gal/stroke)	0.00379	cubic metres per stroke (m^3/stroke)
	ounces (oz)	29.57	millilitres (mL)
	cubic inches (in.^{3})	16.387	cubic centimetres (cm^3)
	cubic feet (ft^3)	28.3169	litres (L)
		0.0283	cubic metres (m^3)
	quarts (qt)	0.9464	litres (L)
	gallons (gal)	3.7854	litres (L)
	gallons (gal)	0.00379	cubic metres (m^3)
	pounds per barrel (lb/bbl)	2.895	kilograms per cubic metre (kg/m^3)
	barrels per ton (bbl/tn)	0.175	cubic metres per tonne (m^3/t)
Pump output and flow rate	gallons per minute (gpm)	0.00379	cubic metres per minute (m^3/min)
	gallons per hour (gph)	0.00379	cubic metres per hour (m^3/h)
	barrels per stroke (bbl/stroke)	0.159	cubic metres per stroke (m^3/stroke)
	barrels per minute (bbl/min)	0.159	cubic metres per minute (m^3/min)
Pressure	pounds per square inch (psi)	6.895	kilopascals (kPa)
		0.006895	megapascals (MPa)
Temperature	degrees Fahrenheit (°F)	$\dfrac{°F - 32}{1.8}$	degrees Celsius (°C)
Thermal gradient	1°F per 60 feet	—	1°C per 33 metres
Mass (weight)	ounces (oz)	28.35	grams (g)
	pounds (lb)	453.59	grams (g)
		0.4536	kilograms (kg)
	tons (tn)	0.9072	tonnes (t)
	pounds per foot (lb/ft)	1.488	kilograms per metre (kg/m)
Mud weight	pounds per gallon (ppg)	119.82	kilograms per cubic metre (kg/m^3)
	pounds per cubic foot (lb/ft^3)	16.0	kilograms per cubic metre (kg/m^3)
Pressure gradient	pounds per square inch per foot (psi/ft)	22.621	kilopascals per metre (kPa/m)
Funnel viscosity	seconds per quart (s/qt)	1.057	seconds per litre (s/L)
Yield point	pounds per 100 square feet (lb/100 ft^2)	0.48	pascals (Pa)
Gel strength	pounds per 100 square feet (lb/100 ft^2)	0.48	pascals (Pa)
Filter cake thickness	32nds of an inch	0.8	millimetres (mm)
Power	horsepower (hp)	0.7	kilowatts (kW)
Area	square inches (in.^{2})	6.45	square centimetres (cm^2)
	square feet (ft^2)	0.0929	square metres (m^2)
	square yards (yd^2)	0.8361	square metres (m^2)
	square miles (mi^2)	2.59	square kilometres (km^2)
	acre (ac)	0.40	hectare (ha)
Drilling line wear	ton-miles (tn•mi)	14.317	megajoules (MJ)
		1.459	tonne-kilometres (t•km)
Torque	foot-pounds (ft•lb)	1.3558	newton metres (N•m)

I
Introduction

In broad terms, an *instrument* is a mechanical or electronic device that measures the present value of a quantity under observation. A *control* is a device that regulates and guides a process quantity against a previously selected standard or reference. A third term, *instrumentation*, suggests the measurement and control of a process.

This book uses many terms to describe the process of instrumentation. It is important to understand these terms so that you can understand the text. The terms are regularly used in the process industry and are commonly understood by those who work in it.

Instrumentation generally includes any arrangement of instruments used to measure, indicate, record, or control variable quantities that exist in a process. *Variable quantities* include such items as pressure, temperature, flow, and level. They are also referred to as *process variables*. A system of instrumentation may include transmitters, resistance temperature detectors, thermometers, pressure gauges, transducers, and control valves.

THE NEED FOR MEASUREMENT AND CONTROL

Early humans used crude devices, such as simple clubs, which were instruments of survival. Many centuries passed before people developed instruments that improved the environment and were not just for survival. They devised ways to observe the stars; measure distances, angles, and times; and to monitor natural phenomena more accurately.

Improvement in measurements also improved and adjusted human activity to an advantage. By obtaining measurement data, people could exert control over their basic needs and environment. In early Roman times, piping and aqueducts distributed water to homes and businesses in Rome from a central water supply. Customers were charged according to the size of the pipe or the channel that delivered the water. One consequence of developing such projects led humans to observe that they could improve products, conserve time, and produce better product quality through instrumentation.

Early process industries in Europe and Asia included brewing and winemaking, which used measurement and control to insure success. Measurement may have been as simple as visual observation of the fermenting process, and control as simple as locating the product in a cool cellar. Instruments as we know them today were crude and almost nonexistent.

In modern industrial processing, such as chemical manufacturing, the quality of the product may depend on the proper proportioning of ingredients by weight or volume, maintaining a constant pressure in a reaction vessel for a prescribed time, and adjusting the acidity (or pH) of the final product by adding a corrective agent. The economic gains achieved through proper measurement and control of processes are of primary importance in the instrumentation field.

Not only is instrumentation applied in manufacturing to increase savings in material and labor, but also it is used to improve the overall quality of the product. Even in the average modern home, instrumentation is applied in our heating and air conditioning systems, sprinkler irrigation, and security systems. This instrumentation provides us with basic needs and allows us to do a better job in a variety of environments.

One major benefit of instrumentation is to reduce the labor required to monitor and operate process equipment. However, officials of a Middle Eastern country contracted with an automation firm for the design of a modern refinery. When the plans were completed and submitted for approval, the officials were alarmed to find that the refinery would possess such a degree of instrumentation that only a small group of personnel would be needed for its operation. The officials considered such extensive automation undesirable because of the enormous supply of personnel available to operate the refinery. The officials requested that the engineering firm redesign the plan to eliminate all automatic controls and make provisions for the manual operation of valves and other equipment. After brief consideration, the firm withdrew its bid to equip the refinery, pointing out that a modern refinery simply cannot be manually controlled, since the coordination needed among the hundreds of individuals to maintain the close tolerances could never be achieved. Controls often do a better job than a person because they offer improved reliability, stamina, and speed inherent in properly designed systems.

Since the beginning of the industrial revolution in the late eighteenth century, the need for measurement and control has kept pace with the increase in the number of variable conditions and factors that have evolved in industrialization. Although the number of variables that could be measured and controlled is almost countless, it is heartening to know that for any given process only a few need to be considered.

Instrumentation—that is, measurement and control—has found wide use in the petroleum industry; in fact, the chemical and petroleum industries are characterized by the extensive degree to which they have adapted to automation. Of all the variables that might be considered, only a few are of primary importance. While pH, humidity, frost, and other conditions and properties are important, the basic process variables of temperature, pressure, flow, and level are of principal interest. When control is exerted on these process variables within very close limits, we call them *controlled variables*.

METHODS OF MEASUREMENT

Human responses are too insensitive, and vary from individual to individual, to be useful in most measurement activities. We cannot accurately judge temperature, distances, or other forms of measurement with acceptable precision needed in most processes.

Therefore, instruments are used to make measurements or to help a person make them. Instruments that augment a person's native faculties enable him or her to conduct measurements with any degree of accuracy desired, although excessive accuracy could be cost prohibitive.

When we measure a level of liquid, we do so by measuring the length, or height, of a column of liquid starting at a reference point in the liquid's container, which is usually at or near the bottom. The level measured is often recorded in units of feet (ft) and inches (in.) or metres (m) and centimetres (cm). The measuring units correspond to the nature of the variable. In the case of liquid level, we use units of length or height.

Pressure is defined as force per unit area and can be measured directly by its ability to lift a weight against the force of gravity. Thus, pressure units are usually expressed in pounds per square inch (psi) or kilograms per square centimetre (kg/cm^2).

Flow rate is usually expressed in terms of volume units per time, such as minute, hour, day, or some other period. Flow rate can be measured by noting the rate at which a tank is filled or emptied, or noting the passage of fluid in a pipe over time. The units can be gallons per minute (gpm), litres per minute (L/min), barrels per day (bpd), cubic metres per day (m^3/d), cubic feet per second (ft^3/sec), and so forth.

Traditional physical methods cannot be used to measure temperature because temperature is a result of the molecular activity in a gas, liquid, or solid. The laws of physics that establish relationships between temperature and other physical quantities allow for the measurement of temperature. Temperature causes physical changes in sensors that are exposed to its effects and these changes can be used to infer the temperature. When temperature or a similar variable is determined by indirect means, its value is *inferred*. All practical means of measuring temperature are inferential devices, as are most systems that measure variables. The following examples should aid in the understanding of direct and inferential measurements.

A gauge glass having a scale calibrated in inches shows the level of liquid in the tank above the reference level, which is the bottom of the tank (fig. 1.1A). The liquid level is read directly from the gauge glass; so, it is an example of direct measurement.

Introduction

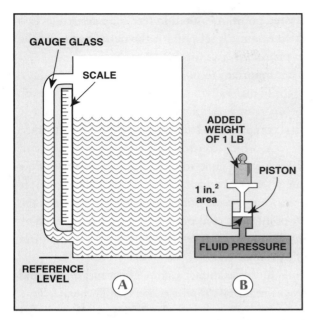

Figure 1.1   Direct measurement examples of level and pressure

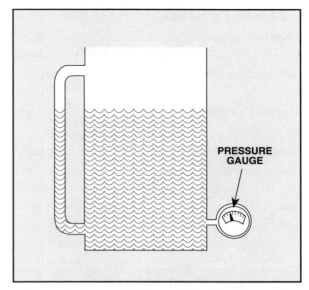

Figure 1.2   A method for inferring liquid level from hydrostatic pressure

A weight placed on a piston that is connected to a line containing fluid under pressure is a direct method of pressure measurement (fig. 1.1B). In this case, a balance exists between two forces: (1) the force of gravity acting downward on the piston and the added weight; and, (2) the force acting upward on the piston caused by fluid pressure in the line. Assume the piston weighs 1 pound (lb) and has a cross-sectional area of 1 square inch (in.^{2}). The downward pressure is the weight divided by the area, or 1 pound per square inch (psi). Therefore, for every 1 psi of pressure in the line, a force of 1 lb is exerted against the piston face, which is 1 in.^{2}. The mass of the piston and added weight exactly balance the force caused by the fluid pressure in the line. This setup directly measures the pressure.

Although liquid level is easy to measure directly with a gauge glass, it readily lends itself to inferential measurement. Bear in mind that a liquid in a container exerts *hydrostatic pressure*. Hydrostatic pressure is the pressure a column of liquid produces when it is not moving. (It is also called hydrostatic head or pressure head.) Hydrostatic pressure can be used to infer height. The liquid column produces the same pressure at a given height (h), regardless of the diameter of the tank, and this pressure (p) is determined by the specific gravity, or density (G), of the liquid. The more the liquid weighs—the higher its specific gravity—the higher is the hydrostatic pressure it exerts.

If we place a pressure gauge at the reference level in the tank (fig. 1.2), we can determine the fluid level indirectly by reading the pressure gauge and using an equation. The equation includes a fixed multiplier of 0.433 to account for converting each foot of liquid into hydrostatic pressure in psi.

$$h = \frac{p}{0.433 \times G} \quad \text{(Eq. 1.1)}$$

where
- h = liquid-level height, ft
- p = pressure gauge reading, psi
- G = specific gravity of liquid with respect to water.

For example, suppose the pressure gauge reads 15.2 psi and the liquid in the tank has a specific gravity of 0.97. In this case,

$$h = \frac{15.2}{0.433 \times G}$$
$$= \frac{15.2}{0.42}$$
$$h = 36.19 \text{ ft.}$$

The liquid-level height of 36.19 ft in this example is inferred from the physical effects of density and the force of gravity.

TYPES OF CONTROL

While we can measure process variable quantities with direct or inferential measurements, we can also control processes with either manual or automatic control.

Manual Control

When operators use a measuring instrument and a control device in process control, they are using *manual control*. The measuring device indicates the pressure, flow, level, or temperature of the process variable. The control device may be a valve or a temperature adjustment that is manually operated to bring about changes in the variable to be controlled.

Manual control is simple when only one variable requires control, but when operators must control several variables, they have to coordinate many adjustments; consequently, frustration and errors can occur. An example of complicated manual control is an aircraft flying without an autopilot. In this case, a human pilot must coordinate fuel, altitude, attitude, bank angle, ailerons, rudder, and other variables to control the airplane.

A water-filled tank with an inlet, an outlet, two valves (labeled *X* and *Y*), a float, and a level meter is another example of having to coordinate several variables (fig. 1.3). In this case, the operator needs to control the level of water in the tank at a desired point, measured in feet. The desired level is the set point. A set point is the desired value of the *controlled variable* (in this example, water level) and is an important term in instrumentation.

To maintain the desired set point of liquid level, the operator must manually balance valves *X* and *Y*. To accomplish this balance, the operator must adjust valve *X*, which controls the average rate of flow of water into the tank, to equal the average rate of flow out of the tank, which valve *Y* controls. This example points out a basic principle of instrumentation: the set-point value of a controlled variable is obtained by maintaining a balance between the input and output of a process.

In this example, a simple float measures liquid level. The float is the *primary element*. The float follows the water level, and a set of mechanical linkages operates the level meter. By noting the level measurement on the meter, the operator manually controls the inlet and outlet valves to maintain the set-point level.

A hot-water heater presents another example of manual control. It requires that the controlled variable be a specific temperature in degrees Fahrenheit, or °F (fig. 1.4). This hot-water heater has three valves: one (valve *X*) allows steam (the heat source) into the tank; another (valve *Y*) controls the entry

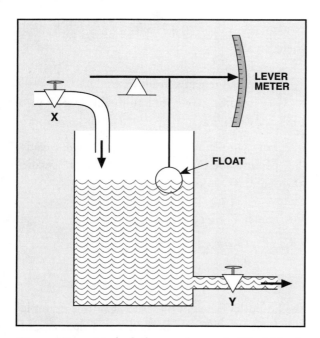

Figure 1.3   *Method of manual control of liquid level*

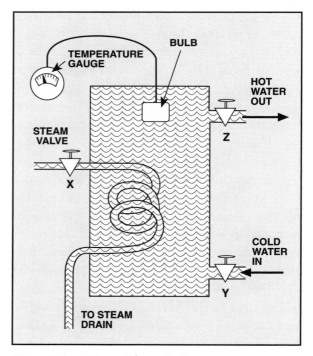

Figure 1.4   *Manual control of hot-water temperature*

Introduction

of cold water; and the third (valve *Z*) controls the exit of hot water. A temperature gauge shows water temperature. To maintain the set-point value of the temperature in the hot-water heater, steam valve *X*, which delivers the source of heat, must be manually balanced with coldwater valve *Y*. Steam flowing through the heating coils in the hot-water heater loses heat to cold water flowing into the tank. So, the operator has to adjust each valve to obtain a stable temperature at the desired set point.

An inferential system of pressure, temperature, and a temperature instrument (a bulb) achieves the temperature measurement in the hot-water heater. The primary element is a bulb that contains a fixed quantity of gas. As temperature changes, gas pressure in the bulb changes. When the temperature rises, gas pressure increases; when temperature falls, gas pressure decreases. This gas exerts force on a movable linkage that indicates temperature. The operator initiates manual control of the valves as the measured temperature varies from its set point.

Manual control can be accomplished when the variables are few. However, to achieve the desired set point accurately and within a reasonable amount of time requires some effort by the operator.

Simple Automatic Control

If we remove an operator from the manual control of a process, we can institute *automatic control*. For example, in level control, a float can operate not only the indicator, but also an inlet valve (*X*) when the flow through an outlet valve (*Y*) increases or decreases (fig. 1.5).

If discharge through valve *Y* increases and causes the water level in the tank to fall below the desired set point, then linkage between the float and valve *X* forces valve *X* to open wider and allow more water into the tank. Systems of this type are common. Note that a turnbuckle allows an operator to adjust the level set point by turning it to different lengths.

Control of hot-water heater temperature can be automated by using a bellows-actuated control valve (fig. 1.6). The bellows changes length by responding to pressure changes caused by changes in temperature. A turnbuckle or, as shown in the figure, a set-point screw and a spring can be used to vary the temperature's set point.

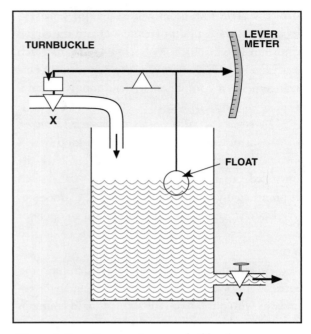

Figure 1.5   Automatic control of level

Feedback

Automatic control of a process requires *feedback*. Feedback is information that the process sends to the control system, which the system then compares to the desired set point. If deviation from the controlled variable's set point occurs, the control system takes

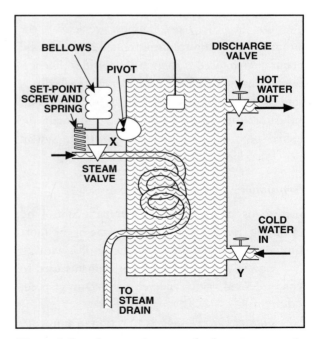

Figure 1.6   Automatic control of temperature in hot-water heater

corrective action. In other words, if a difference exists between the set point (reference) and the actual condition (feedback), the controlling means (such as a valve) activates to make the correction. This arrangement of components is commonly called a *closed-loop control system.*

In contrast, an *open-loop,* or *open-sequence, system* controls some processes. An open-loop system measures the process, but it does not feed back the controlled variable for comparison with the desired set point. Many influences may cause the process to vary, but changes do not take place until an operator readjusts the set point. Open-loop systems are limited in use.

An automobile is an example of open-loop and closed-loop control. Say that we are driving on the expressway with our foot on the accelerator to maintain 55 miles per hour (mph) or 90 kilometres per hour (kph). We are operating in an open-loop mode. That is, as we approach a hill, our speedometer shows that our car is slowing down. So, we apply more accelerator pressure to regain 55 mph (90 kph). Conversely, as we descend the hill, we let off on the accelerator to maintain 55 mph (90 kph). We are monitoring our speed with the accelerator pedal. No automatic system in the car takes corrective action to maintain our set-point speed of 55 mph (90 kph). Thus, we are exercising open-loop control of the car's speed.

On the other hand, suppose we activate the automatic speed control incorporated in the car and set our desired speed at 55 mph. As outside influences cause the car to decelerate or accelerate, the feedback from the car's speed sensor is compared to the set point and causes the throttle to change and maintain the desired speed. The car's built-in speed-control system uses closed-loop control to maintain the speed.

Terminology

Many terms describe and define automatic control and measuring devices. Following are some of the more common ones the reader should be familiar with.

Control agents are devices or elements used to achieve desired values of level, temperature, speed, and so forth.

Manipulated variable is an action in a process, such as the rate of flow, which changes control variables.

Controlling means are elements or components of a controller that produce corrective action.

Final control element is the part of the controlling means that directly changes the value of the manipulated variable.

Measuring means are elements of a controller that measure and communicate to the controlling means either the value of the controlled variable or its deviation.

Primary element is the part of the measuring means that first uses or transforms energy from the controlled medium to produce an effect in response to a value change in the controlled variable. The effect produced by the primary element may be a change in pressure, force, position, electrical resistance, and so forth.

Control variables are variations in processes or control agents that regulate such items as liquid level, water temperature, room temperature, and car speed.

Be aware that the control and measuring elements and devices may exist separately or together in a system. And, it can be difficult to determine where one ends and another begins. However, in most systems, devices exist as separate units in the form of a process sensor, instrument, set-point device, comparator, and actuator.

Automatic Control Systems
Characteristics

Several characteristics affect the performance of the controls in an automatic control system.

Response lag is the time lapse that occurs between the moment a set point sends a command to make a change in the controlled variable and when the change actually occurs. Lag exists whether the value is rising or falling. A number of factors in a process affects this time lag and each system has its own characteristic lags. Lags may be attributed to primary elements, measuring means, sensors, or other items in the system. In some systems, a large lag is acceptable, while in others it must be minimal.

Cycling is the periodic deviation above and below the set-point value. Cycling is also called *hunting.* When a change occurs in the controlled variable, such as level, temperature, or speed, an automatic closed-loop system takes corrective action with its final control element. A small deviation above and below the set point, coupled with system lag, may result in an oscillation, or cycling, about the set point. Any automatically controlled system or process must allow tolerance for deviations in its controlled variables. While some tolerance may have to be extremely fine, in others it may be quite broad. Small deviations are usually acceptable, but continued, or extensive, cycling is generally not acceptable.

Dead band is an area of movement where no output action occurs in a component in the system. Some systems (both mechanical and electronic) have components that do not transmit a continuous command from its input to its output. An example is a linkage with loose tolerances that require additional movement of its input shaft to produce an output response. The area of movement where no output action occurs is referred to as *dead band* or *dead zone*. The dead band in an automatic control system produces overshoots and undershoots—that is, cycling, or hunting, in the control. Other mechanical characteristics that produce a dead band include sticking friction, poor machine tolerances, and sloppiness in fit. Electronic devices with low gain or sensitivity may have a dead band, but it is typically small.

Capacity is the amount, or quantity, of liquid a tank holds. However, capacity also has another meaning in instrumentation. A storage tank stores energy in the form of the weight and the hydrostatic pressure of the liquid it contains. Instrument technicians think of this stored energy in terms of capacity. In two tanks of identical height with the same liquid level, we can measure and control the liquid level in each. However, if the diameters of the two tanks are different, and although the hydrostatic pressure is the same, the capacities, and thus the stored energy, of each tank is different. The differences in capacity affect the response of the automatic control that regulates liquid level in the two tanks. Where the only stored energy in a system is the liquid weight and height, or level, the system is referred to as a *single-capacity system*.

On the other hand, a hot-water system is a *multiple-capacity system*. A hot-water system stores heat energy in the bulk of the water within the heater tank, but so do the heating coils and the metal walls of the tank. Consequently, more than one capacity is present: one is the heat energy in the bulk of the water; the other is the thickness of the tank wall and heating coil wall. Both capacities affect control.

Resistance is the opposition to the free transfer of energy between two capacities. A control system or process encounters resistance when it is necessary to transfer energy from one capacity to another. An example of resistance occurs when transferring liquid from one tank to another through a pipe between two tanks (fig. 1.7). As the liquid flows through the pipe, the pipe resists the flow.

In a hot-water system, steam coils transfer heat energy to the water. However, this transfer is not

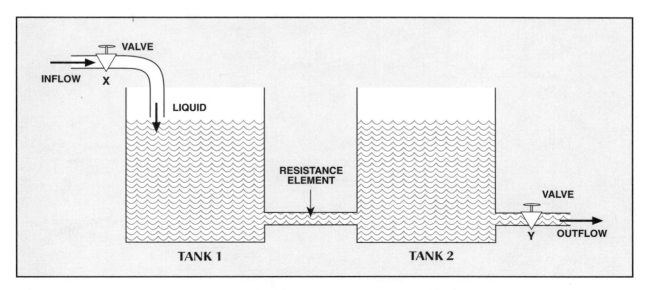

Figure 1.7   *Example of resistance in transfer of energy from one capacity to another*

instantaneous. One or more components in the system always resist the transfer to some degree—for example, the walls of the steam coils and the layers of steam and water on either side of the coils. Any part of a system that opposes the free transfer of energy between two capacities is resistance.

Process reaction rate is the rate of change in the controlled variable that takes place when a sudden, or *step, change* occurs. Although a change in the final control element (such as a valve) may occur suddenly, the controlled variable (such as liquid level) does not change immediately. Instead, it changes only after some time has passed. This time passage is the process reaction rate. Knowledge of the process reaction rate allows selection of equipment ratings that satisfactorily meet the needs of the process.

Process reaction rate can be shown in a liquid-level tank (fig. 1.8A). Assume that valve *X* undergoes a step change—that is, it is suddenly operated to a larger opening. This step change is shown graphically in figure 1.8B. At point t_0, note how the line representing flow rate through the valve suddenly increases in one step. But in figure 1.8C, note how the process reaction is a curve past point t_0; the curve shows that the process reaction rate is gradual, rather than sudden. In other words, it takes time for the process to react to the sudden change in the controlled variable (inlet valve *X* in this case).

In figure 1.8A, the liquid level in the tank immediately increases in height until a stable level is achieved between the additional fluid inflow and the liquid column pressure acting on valve *Y*. This system is relatively easy to control, largely because only a single capacity is involved and little or no resistance is encountered to complicate the problem.

Figure 1.9A shows a more complicated system in which temperature control is involved. When valve *X* is suddenly changed, the temperature indicator does not begin to move simultaneously with the valve opening. Note the graphs (figs. 1.9B and 1.9C), which show the step change in steam flow and the process reaction curve in the temperature rise. Once the controlled variable of temperature begins to change, it accelerates for a while and then, as it

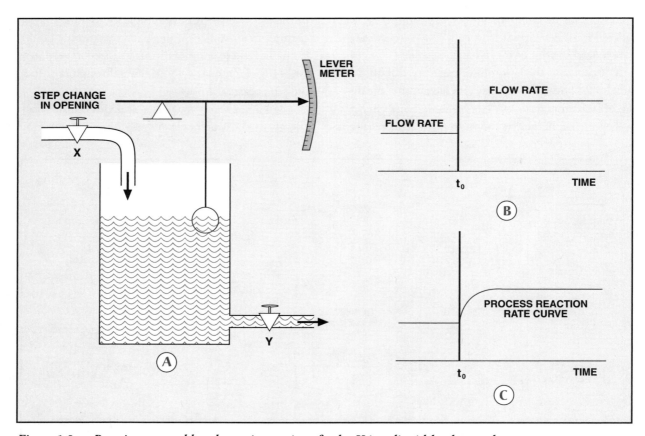

Figure 1.8   *Reaction to a sudden change in opening of valve X in a liquid-level control system*

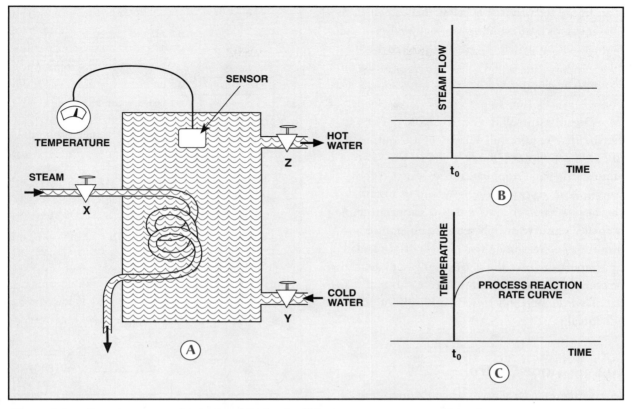

Figure 1.9 Reaction to step change in control valve X in a water-temperature control system

approaches the new stable value, its rate of increase declines.

Several factors cause the delay in response of the indicator to a change in the final control element setting. For one thing, a transfer of energy from one capacity to another occurs—in this case, from heating coils to water. This delay is *transfer lag*. The delay is caused by the fact that it takes a certain amount of time to carry the temperature change from the steam coils to the location of the primary element. The delay is also called *transportation lag*, or *dead time*. Another delay occurs as the primary element responds to the change in water temperature.

Process control systems can exist in many forms but the basic concepts outlined above can be found in most systems.

METHODS OR MODES OF CONTROL

A controller, or instrumentation system, that monitors and controls a process can respond in a number of ways when a sudden change occurs in the process. This systematic action of the controller in response to changes in the controlled variable is a method of response, or a *mode*.

We can put modes into several categories.
- on-off or two-position,
- floating,
- proportional,
- proportional plus reset, and
- proportional plus reset plus rate.

On-Off, or Two-Position, Mode Control

In many cases, an automatic control is made up of simple control elements. The on-off, or two-position, mode of control is an example. Processes that use this mode typically have appreciable lags in their controlled variables and therefore their set point does not require close control. An example of on-off mode control is a hot-water heater whose heat source is either gas or electricity. When the desired set point is established with a thermostat, the hot-water heater is either off (if the temperature is above the set point) or on (if the temperature is below the set point). A differential, or dead band, zone usually exists in the thermal sensing switch. The sensing switch is either open or closed to a signal that turns the process on or off. Because water heating takes time, system response is usually slow to prevent excessive cycling.

On-off control is also used on devices or systems that require full output or no output to function. Examples include a shutdown valve, a pump starting and stopping, and an alarm. Another example is an air conditioning system in a building where two set points control the air temperature. A thermally sensitive switch senses temperature and turns the air conditioning system on or off to regulate the temperature over the narrow range inherent in the switch. Electronic thermostats are also available that allow the operator to set the on-off set points individually.

Floating Mode Control

Floating mode control exists when the final control element is forward or increasing, reverse or decreasing, or off (its floating mode) as it regulates the controlled variable. An example of floating mode control is a liquid-level control system that uses a motor-operated valve on its inflow, a fixed valve on its outflow, and a level sensor that operates the motor-operated valve as well as indicating the level (fig. 1.10). In this system, the electric motor opens and closes a control valve through a worm-gear drive. An electrical switch, identified in the figure with electrical terminals labeled *2-1-3*, can reverse the motor. The motor rotates in one direction when voltage is applied across contact terminals *1* and *2* and rotates in the opposite direction when voltage is applied across terminals *1* and *3*. Contacts *2* and *3* are stationary; but the primary element (the float) and its connecting linkages are fitted to move contact *1* up or down.

When the liquid level in the tank changes enough to cause contact between contacts *1* and either *2* or *3*, the motor moves the valve stem open or closed to restore the liquid level to the set point. Once the float is restored to the set point, the contact breaks and the motor stops. As long as the controlled variable stays within its prescribed differential gap,

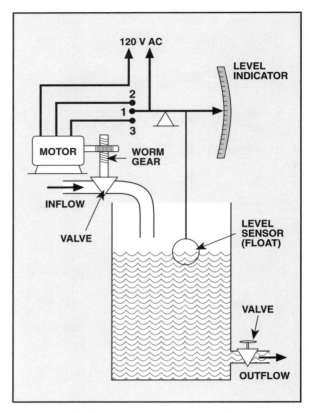

Figure 1.10   Floating mode of liquid-level control

no contact is made with terminal *1* and the motor is off. With the motor off, the valve is in a partially open, or float, position.

To create a proportional-speed floating-control mode, a variable speed controller is installed between the float sensor and the motor (fig. 1.11). In this mode, the speed at which the motor operates to

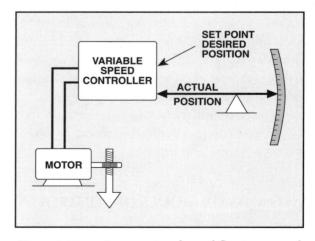

Figure 1.11   A proportional-speed floating control-level control system

Introduction

position the final control element is in proportion to the amount of deviation of the controlled variable beyond the differential gap. The motor is energized for each direction of rotation through the variable speed controller. As the deviation between desired and actual level decreases, the motor speed decreases proportionally. Unlike the single-speed mode, this system can respond to processes with fast reaction rates.

Proportional Control

Proportional control takes place when a linear relationship exists between the controlled variable and the final control element. In the tank containing liquid in figure 1.12A, control valve X is allowing water to flow into the tank at a rate of 50 percent of maximum flow. If discharge valve Y is open enough to require half again as much flow—that is, 75 percent—the level in the tank begins to fall as the tank liquid supplies the new demand to the outlet (fig. 1.12B). Control valve X opens wider, but the level declines until inflow equals outflow. If the system is controlling liquid-level, the point at which the system stabilizes will be below the original set point for 50 percent of maximum flow. A similar situation exists if the discharge rate falls somewhat below 50 percent of maximum. In this case, the liquid level will stabilize above the set point.

Let's assume that the tank in figure 1.12A and B is 100 in. deep, that the float can move up and down over this entire distance, and that control valve X is completely closed only when the tank is full and is completely open only when the water level in the tank falls to zero on the liquid-level scale. Also assume that the flow rate through the control valve is a straight-line function of its opening and that flow is proportional to valve opening. Therefore, for each change in flow from discharge valve Y, a proportional change in the opening of valve Y also occurs. A proportional controller is controlling the level in the tank. A continuous linear relationship exists between the value of the controlled variable and the full range of positions of the final control element.

The one-to-one relationship of a proportional controller yields another term: *proportional band*, or *throttling range*. The proportional band is the range of values of a controlled variable that corresponds to the full operating range of the final control element.

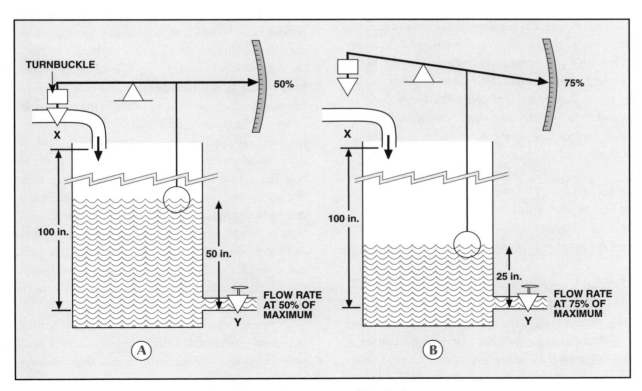

Figure 1.12   *A proportional control system with proportional band of 0 to 100 percent*

In the previous level-control example, the proportional band is 100 percent because a full range in a controlled variable produces a full range in the setting of the final control element.

By selecting the range of the controlled variable (level, in this case) that allows the control valve to be fully open to fully closed, the proportional band can be changed. For example, assume that we want to regulate the level to be near the middle of the tank and have the control valve go from fully open to fully closed when the level changes from 45 to 55 in. The change is 10 in. out of the full tank capacity of 100 in. or 10 percent proportional band. The proportional band can be adjusted to any percentage desired within reason, from a few to 100 percent.

Proportional band can be defined in terms of the range involved in sensing the controlled variable and the full range of the final control element. In mathematical terms, proportional band is—

$$P_{pb} = \frac{R_{cv}}{R_{fce}} \times 100 \; percent \quad (Eq. 1.2)$$

where

P_{pb} = percent proportional band
R_{cv} = range of controlled variable
R_{fce} = full range of final control element.

In electronic controllers containing amplifiers, the operator can adjust proportional band electronically with a potentiometer. The narrower the band, the tighter the control the system imposes on the set point. A proportional controller keeps the controlled variable within a range, but cannot keep the controlled variable at the set point as input conditions change.

A term used in lieu of band is its reciprocal: *gain*. Thus, a 5 percent proportional band can also be said to have a gain of 20, while a proportional band of 50 percent has a gain of 2. Mathematically, gain is related to proportional band as—

$$gain = \frac{1}{P_{pb}} = \left(\frac{R_{fce}}{R_{cv}}\right). \quad (Eq. 1.3)$$

Proportional controllers are suitable for use in systems having a slow reaction rate, load changes that are neither large nor rapid, and small transfer lag and dead time. A limit can be reached if appreciable dead time and lag are present in the system, which are coupled with small proportional bands and high gain. Exceeding this limit results in unceasing oscillations (hunting) about the set point, which is undesirable.

Proportional Plus–Reset Mode Control

In many cases, it is desirable to readjust the set point of a proportional controller to bring the controlled variable back to the desired position. This readjustment could be done to reestablish the set point of a proportional controller when the system has varying discharge rates or in systems where large load changes occur, such as in pressure control of turbine fuel gas or in flow control of a centrifugal compressor.

Refer back to figure 1.12. This simple system can be modified by adding the equivalent of a turnbuckle to the linkage system that is used to change the set point of the controller. Electronic controllers can also be adjusted to change the set point of the controller.

The system in figure 1.12 was assumed stable when the set point was 50 in. of water and the input-output rates were 50 percent of maximum flow. When the discharge rate reached 75 percent of maximum flow, the water level dropped until the final control element opened to 75 percent of maximum. The level can be readjusted to the 50-in. mark by manually adjusting the turnbuckle to open the control valve wider. This action causes the water level to rise in the tank because the rate of inflow exceeds the discharge rate. Eventually, the original set point will be reached, and the inflow and discharge rates will be equalized.

Another way to implement this concept is to replace the turnbuckle with a variable speed, reversible motor and control (fig. 1.13). This arrangement combines a proportional-speed floating mode with a proportional controller. The system is referred to as proportional plus-reset and combines the great stabilizing features of a proportional controller with the added advantage of the reset capability offered by the floating mode.

Proportional plus-reset controllers can be used to control processes that have a large amount of dead time or transfer lag. If these factors are unusually large, it is necessary to use a broad proportional band (low gain) and slow reset rate. The reset, or repeat,

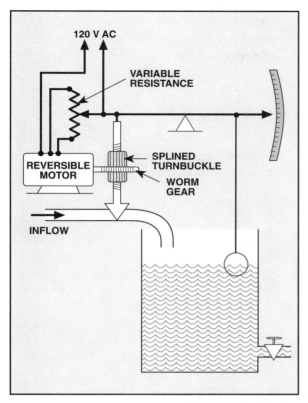

Figure 1.13 A system using proportional plus-reset mode of control

function is expressed in terms of reset or minutes per repeat and determines how often the final control element is changed. To obtain a feel for the numbers involved, a small reset rate of 0.2 minutes per repeat provides aggressive correction while a large reset rate of 10 minutes per repeat gives weak correction.

In summary, a proportional plus-reset controller constantly monitors the system for an error against the set point and performs corrections in discrete time intervals to maintain the set-point level.

Proportional Plus-Reset-Plus Rate

An additional feature is sometimes incorporated in a proportional plus-reset controller that largely offsets the effect of dead time and transfer lag. It is called *rate response*—a term derived from the fact that the control action produced by the feature is based on the rate at which the controlled variable changes value. This added rate response can be expressed in terms of a control that incorporates proportional plus-reset-plus rate.

The proportional plus-reset-plus rate control adds the necessary correction to the final control element in anticipation of the final value when the process variable is rapidly changing. This control is usually expressed in minutes. The larger the time factor in minutes, the more aggressive the controller exerts correction to rapidly changing process variables.

This function can only be added to systems that already have proportional and/or reset controls. By itself, it causes hunting since it constantly modifies the set point against process variable changes. Many processes do not require this function and others perform poorly if this feature is used.

A typical application for this system is in temperature control, where considerable dead time and transfer lag is present.

PID Controls

Because of the increasing use of electronics to perform process control functions, various modes of control incorporate PID control. PID refers to the common use of closed-loop controls utilizing feedback, which include proportional (P), integral (I), and derivative (D) functions. These functions are mathematical terms applied to the theory of automated control systems and relate to the more practical terms of proportional, proportional plus-reset, and proportional plus-reset-plus rate.

Assume we have a heat exchanger and we want to keep the system temperature at a certain point. The cooling water temperature will vary based on load conditions, but we have control over the output valve to regulate the temperature. The output valve is a proportional valve that allows us to go from fully open to fully closed and in between. Hotter cooling water temperature causes us to open the valve more to keep the temperature constant. Using this example, let's define the various controls we can apply in terms of PID.

Proportional control, P—comes into play when controlling temperature—that is, the higher the temperature, the more we want the valve to open. We open the valve in proportion to the amount of temperature difference that exists from the set point. The difference between the set point and the actual temperature represents the error.

The range of temperature over which the valve will be fully opened to fully closed is referred to as the *proportional band*. Proportional band can be expressed in percent or gain. Proportional gain is expressed as K_c. The lower the proportional band, or the higher the proportional gain, the more aggressive is the correction involved in the control. A 10 percent proportional band means that the output valve goes from fully open to fully closed as the temperature changes by 10 percent of the possible change level. The gain in this case is 10.

Integral control, I—reset comes into play when the control observes that the longer the time the temperature is too high, the valve needs to be opened more to gain control at a faster rate. In the control, the amount of error is multiplied by the amount of time the error has been detected, which determines how severe the problem is. Integral gain (K_i) is a multiplier that gives operators control over how aggressive they want this control to be in correcting the problem.

Derivative control, D—rate of change takes place when the control observes that the faster the rate of rising temperature, the valve should be opened more. The derivative control is monitoring the temperature for rate of change and adjusting the output valve. This system offers fast response to errors and improves response to system dead time and lags; but, it is more difficult to adjust because of a tendency to overshoot.

SUMMARY

This chapter covers the foundation of instrumentation—measuring and controlling a process. The terms used are discussed many times, not only in this book, but also in applications using the systems described in the chapter. Therefore, it is important for the reader to grasp their meaning.

REVIEW EXERCISE

- Define instrumentation.
- Name four important process variables.
- What is meant by direct and inferred measurements?
- What are the two broad categories of control?
- Define these terms:

  Controlled variable

  Set point

  Primary element

  Feedback

  Closed-loop

  Response lag

  Hunting, or cycling

  Dead band

  Capacity

  Resistance in process

  Process reaction rate

  Step change

  Transfer lag

  Proportional control

  Proportional plus-reset control

  Proportional band

  Proportional plus-reset-plus rate

  PID

2

The Units of Measurement

Instrumentation involves measuring relatively few quantities—for example, length, mass, time, and temperature. Such quantities are fundamental quantities because we cannot divide them into other quantities. By comparison, speed is not a fundamental quantity. We can measure it, of course, but also we can divide it into length and time.

All quantities have dimensions. Some dimensions are easy to see, such as length. Others, however, may be a little harder to make out. For example, mass and time also have dimensions, but we cannot physically measure them with a ruler or a yardstick. Instead, we have to apply a measuring tool, such as a clock or a scale, which, when set to a standard (is calibrated), makes the measurement and indicates it to us. Also, some quantities feature several dimensions. For example, as mentioned earlier, speed has dimensions of length and time; and force has dimensions of length, time, and mass.

A unit is a standard measure of a quantity. Laws establish some units of measurement while we adopt others by common usage. We use units to measure quantities of any size, and we always express the measurements in terms of the chosen unit.

COMPARISON OF SYSTEMS OF UNITS

Over the centuries, countries and regions initiated their own system of measurements. However, they rarely shared it with other countries. Moreover, many of these measurement systems were so crude and ill conceived that it was virtually impossible to convert one system to another.

As communication, transportation, and commerce expanded, measurement units evolved, merged, and became standardized. Today, the world is well on its way to adopting a single set of measurement standards common to all nations.

In instrumentation, it is important to use common units so they can be shared between companies, organizations, and countries. In most cases, measurements and readings are either in the English system of units (also called the conventional system) or in the Système International (SI) d'Unités (International System of Units), which is based on the metric system.

Conventional System of Measurement Units

The United States uses the *English*, or *conventional*, *system of measurement* for most of its trade and commercial dealings. People in the U.S. have used this system for a long time and are therefore comfortable with it. Unfortunately, it is ambiguous and it is difficult to convert from one unit to another that measures the same quantity. For example, the unit of mass in the conventional system is the pound, which in the U.S., surprisingly enough, is defined in terms of the kilogram, which is an SI unit. The pound is divided into ounces, drams, grains, and other units, each of which relates to the pound. To convert from one unit of weight to another, users have to remember such facts as 16 ounces make up a pound and that a ton weighs 2,000 pounds. The yard is the standard length in the system, and it is divided into feet and inches. To convert from one unit of length to another, users have to remember that 36 inches or 3 feet make up a yard. Also, 5,280 feet or 1,760 yards make up a statute mile. Interestingly, the yard, like the pound, is also defined in terms of the metric (SI) system.

Besides the difficulty of converting from one English unit to another, other shortcomings exist. For example, at least three kinds of miles are used in the system: nautical, geographical, and statute; only two (geographical and nautical) are the same length. The same is true for the gallon. An Imperial gallon, which the British used to employ, is larger than the U.S. gallon. (An Imperial gallon contains 277.42 cubic inches, while a U.S. gallon contains 231 cubic inches.

SYSTÈME INTERNATIONAL (SI) D'UNITÉS

The Système International (SI) d'Unités (International System of Units) is an outgrowth of the metric system. A group of French scientists developed the metric system in the 1790s. The metric system represented a remarkable and logical advance in the science of weights and measures, and nations rapidly adopted it.

SI retains the standards set up by the metric system but encompasses many improvements. Virtually unambiguous, calculations involving SI units are facilitated by the logical relationships existing among quantities and the powers-of-ten multiples used to measure a given quantity.

The metric system recognized only three fundamental quantities—mass, length, and time. SI added four others—temperature (Kelvin), electric current (ampere), amount of substance (mole), and luminous intensity (candela).

The use of SI requires strict attention to symbols and typographical style. The SI system uses symbols to abbreviate units, although most of the symbols are conventional letters of the alphabet. The SI unit of length is the metre, and its symbol is m, with no period. (For the most part, the SI system uses British spelling—for example, the unit of length is spelled *metre* and not *meter*. However, *gram* is not spelled *gramme*.) The unit of force is the newton, and its symbol is N, with no period. The SI temperature scale is Kelvin, and its symbol is K with no degree mark (°). The Celsius scale (C) is not strictly an SI entity, but it is used for most nonscientific needs. The appendix contains a more extensive discussion of SI.

Table 2.1 is a tabulation of conventional units and symbols and their relation to SI units and other quantities.

MEASURING LENGTH

Length is a fundamental quantity, and people express it in many ways. Instrumentation usually employs feet and inches in the conventional system, and metres and millimetres in SI. Further, we not only use length to measure distance, but also to measure volume and area.

Area Measurement

If two dimensions of length form a product—that is, if two dimensions are multiplied together—the result is a dimension of area. Area is expressed in square units, such as square feet (ft^2) or square metres (m^2). The equation is—

$$A = l \times w \quad \quad (Eq.\ 2.1)$$

where

A = area, ft^2 or m^2
l = length, ft or m
w = width, ft or m.

As an example, determine the area of a plot of land 90 ft (27 m) long and 40 ft (12 m) wide. The area of the land is—

A = 90 × 40 or
   27 × 12
A = 3,600 ft^2 or
A = 324 m^2.

Notice that ft times ft equals ft^2, just as x times $x = x^2$. Notice too that the names of the quantity units can be treated as mathematical terms—that is, area is A, length is l, and so on.

Be aware that an area measurement can turn up where we may not expect it. For example, the expression for kinetic energy is one-half mass times velocity squared, or $\frac{1}{2}Mv^2$. Since velocity is defined as length divided by time, or $L \div t$, velocity squared (v^2) is length squared (L^2) divided by time squared (t^2) or $L^2 \div t^2$; or area (A) divided by time squared ($A \div t^2$). While it may be surprising that an equation for kinetic energy involves an area measurement, the expression is solidly founded in physics and mathematics.

TABLE 2.1
Conventional and SI Units of Measurement

Quantity Name And Symbol	Dimension (Relation to Other Quantities)	Conventional Unit and Symbol	SI Unit and Symbol
Length (L)	Fundamental	foot (ft)	metre (m)
Mass (M)	Fundamental	pound (lb)	kilogram (kg)
Time (t)	Fundamental	second (sec)	second (s)
Temperature (T)	Fundamental	degree Fahrenheit (°F) degree Rankine (°R)	degree Celsius (°C) kelvin (K)
Electrical current (I)	Fundamental	ampere (A)	ampere (A)
Amount of substance (mol)	Fundamental	mole (mol)	mole (mol)
Luminous Intensity (l)	Fundamental	candela (cd)	candela (cd)
Acceleration (a)	L/t^2	feet per second per second (ft/sec²)	metre per second per second (m/s²)
Area (A)	L^2	square feet (ft²)	square metre (m²)
Density (δ)	M/L^3	pounds per cubic foot (lb/ft³)	kilogram per cubic metre (kg/m³)
Energy (E), Work (W), Heat (Q)	$M \cdot L^2/t^2$	foot-pound force (ft-lbf)	joule (J)
Force (F)	$M \cdot L/t^2$	pound force (lbf)	newton (N)
Frequency (f)	$1/t$	hertz (Hz)	hertz (Hz)
Power (P)	$M \cdot L/t^3$	watt (W) horsepower (hp)	watt (W)
Pressure (p)	$M/(L \cdot t^2)$	pound force per square inch (psi)	pascal (Pa)
Velocity (v)	L/t	foot per second (ft/sec)	metre per second (m/s)
Volume (V)	L^3	cubic foot (ft³)	cubic metre (m³)

Volume Measurement

Take three dimensions of length, multiply them together, and the result is a measurement of volume. The equation is—

$$V = l \times w \times h \quad \text{(Eq. 2.2)}$$

where
- V = volume, ft³ or m³
- l = length, ft or m
- w = width, ft or m
- h = height, ft or m.

As an example, determine the volume in ft³ of a box whose dimensions are 2 ft by 3 ft by 4 ft (0.6 m × 0.9 m × 1.2 m).

$V = 2 \times 3 \times 4$ or
$V = 0.6 \times 0.9 \times 1.2$
$\quad = 2 \times 12$ or
$\quad\quad 0.54 \times 1.2$
$V = 24$ ft³ or
$V = 0.65$ m³.

Notice again that the unit name foot (ft) is treated as a mathematical entity. Volume can be expressed in many units, but in instrumentation, cubic feet (ft³), cubic inches (in.³), cubic centimetres (cm³), barrels (bbl), gallons (gal), and litres (L) are the most common.

Another volume calculation occurs when area (A) is multiplied by velocity (v), as in $A \times v = L^2 \times$

$L \div t = L^3 \div t$. The result of this multiplication is volume per unit of time.

MEASURING TIME

Time is measured in seconds, minutes, hours, days, and so on, and these units have been almost universally adopted. In both conventional and SI measurements, the second is the standard unit.

The second was formerly defined in terms of rotation of the earth about its axis. Standard setters chose the year 1900 as the base for the invariable unit of time, which they called the *ephemeris second*. (Ephemeris is a word that describes a chart, table, or almanac related to astronomy—specifically, the orbital motions of the planets. The standard setters could have just as easily called it a universal, or standard, second, for that is what they meant.)

Today, the second is intended to be exactly equal to the ephemeris second, but modern standard makers did not define it in terms of planetary motion. Instead, they defined it in terms of the frequency of radiation from cesium 133 atoms. (Cesium is a chemical element. Cesium 133 is a radioactive isotope of elemental cesium.) Atomic clocks controlled by cesium radiation are extremely accurate and provide a time base adequate to meet the exacting demands of modern technology. Such clocks are expensive, but their time signals can be made available through radio signals to laboratories and other facilities.

MEASURING TEMPERATURE

Temperature is regarded as a dimensionless quantity—that is, we cannot directly measure it with something like a yardstick, which we can use to measure length. The presence of temperature in an object is clear indication that the object contains energy. As noted earlier, temperature is a fundamental quantity.

Three temperature scales are of importance in instrumentation: the Celsius scale (C), formerly called the centigrade scale, which is an SI unit; the Fahrenheit scale (F), which is the conventional unit; and the Kelvin scale (K), which is also an SI unit. The Celsius scale is based on 0° and 100° as the freezing and boiling points of pure water at standard atmospheric pressure. The Fahrenheit scale has equivalent points at 32° and 212°. One hundred degrees occurs between the freezing and boiling temperatures of water on the Celsius scale, and 180° between the same points on the Fahrenheit scale. The span of 1°C is exactly equal to the span of 1 K.

The Kelvin scale is used to measure thermodynamic temperatures and is an absolute scale. The significance of an absolute temperature scale is that gas properties change proportionally with absolute temperature. When the temperature of an ideal gas is increased by 10 percent, the volume increases by 10 percent if pressure is constant or the absolute pressure increases by 10 percent if the volume is constant. On the other hand, a 10 percent change in Celsius temperature does not cause a 10 percent change in the gas properties, but some other value dependent on the beginning and ending temperature.

The Kelvin scale hypothesizes the existence of an absolute zero temperature, meaning that as the temperature is lowered to ever-smaller values, a point is eventually reached at which it is impossible to go lower. At this point, all heat energy in an object has been removed, and all molecular motion within the object ceases. Absolute zero temperature has not been achieved, but its existence is accurately established. Temperatures to within a fraction of a Kelvin from absolute zero have been attained. Based on the Celsius scale, absolute zero temperature is –273.15°C. It is simple to convert from Celsius to Kelvin by adding 273.15° to the Celsius temperature. For example, 0°C is 273.15K.

An absolute temperature scale based on Fahrenheit measurement is the Rankine (R) scale. To convert from Fahrenheit to Rankine, simply add 459.6°. Thus, 32°F equals 491.6°R.

This book mostly uses the Fahrenheit scale. However, where appropriate, it also uses Celsius, Kelvin, and Rankine. For example, when discussing the behavior of gases under various conditions of temperature and pressure, it is easier to speak in terms of absolute pressure and absolute temperature; therefore, °K and °R are used in such cases.

Equations are required to convert °C to °F and vice versa. The equation for converting °C to °F is—

$$°F = \tfrac{9}{5}(°C) + 32°. \qquad \text{(Eq. 2.3)}$$

For example, if the current temperature is 20°C, what is the equivalent temperature in °F? The answer is 68°F because ⁹⁄₅ of 20 is 36 and 36 + 32 is 68.

The equation for converting °F to °C is—

$$°C = \tfrac{5}{9}(°F - 32°). \quad \text{(Eq. 2.4)}$$

For example, if the current temperature is –10°F, what is the equivalent temperature in °C? The answer is about –23°C because –10 minus –32 is –42 and ⁵⁄₉ of –42 is –23.3.

MEASURING MASS, WEIGHT, AND FORCE

The concepts of mass, weight, and force, and the relations between them, are often misunderstood, so it is worth discussing these units in detail.

Mass and Weight

Mass is sometimes used as another word for weight, but in the strictest sense, it is different. For most purposes, the *mass* of a given quantity of something remains constant regardless of its location in the universe. The *weight* of an object, in the ordinary sense of meaning, varies according to the physical conditions that form its environment. For example, astronauts become weightless in orbit. Weightlessness occurs because the force of attraction (gravity) the earth has for the astronaut is exactly counterbalanced by the centrifugal force resulting from the orbiting motion. Another example occurs on earth's moon. An astronaut who weighs 180 lb (82 kg) on earth weighs only 30 lb (14 kg) on the moon, because the mass of the moon is about ⅙ that of the earth. Consequently, the moon's gravity, because it exerts a force only ⅙ as strong as earth's gravity, causes an astronaut on the moon, or any object for that matter, to weigh less than on earth.

Mass is the measure of the inertial properties of a quantity of substance. Inertia is the resistance a substance shows to a change in its state of motion. If it is standing still, it resists an effort to move it. If it is moving, it resists an effort to slow it down, speed it up, or change its direction of travel.

The effort that is resisted by the inertia of the substance is a *force,* and the amount of force that is required to effect a given change in the state of motion is a measure of the mass of the substance. A simple way to define mass is to say that it is a measure of the amount of any given substance.

Force

Force is closely related to mass and weight. Force is the product of mass and acceleration. Therefore, force has dimensions of mass times length divided by time squared ($ML \div t^2$).

For many practical purposes, weight and force are synonymous. If, for example, an astronaut weighs 180 lb (82 kg), in effect, the earth's mass attracts his mass with a force of 180 lb (82 kg).

Gravitational Force

Two bodies of mass interact with one another. They are drawn toward one another, and the force of attraction between them can be used to measure their masses. The force acting between two bodies is a gravitational attraction. For pairs of small masses, the force is extremely small and difficult to detect without special instruments. However, where one or both of the masses is very large (for example, if one of the masses is the earth), the force of attraction is quite significant.

Every physical force has direction as well as magnitude. The force of attraction between two bodies of mass has definite direction. The force of attraction acts along an imaginary line that connects two centers of mass. Thus, when astronauts are launched into orbit and become weightless, forces still act on their mass. The two main forces, however, are acting in exactly opposite directions. The force of gravity that pulls the astronauts toward the earth neutralizes the centrifugal force that tends to drive the astronauts farther into space. The net force acting on astronauts in orbit is virtually zero, so their weight is also zero.

Magnetic Fields

Two magnets, whether permanent or electromagnetic, are strongly attracted to one another if their unlike poles are brought close together. This attraction is not gravitational, but it is a force with direction and magnitude. A magnetic field is a force field, meaning that, if a magnetized particle mass is placed in the field, the particle is accelerated in a definite direction.

Electric Fields

Electric fields are also force fields in which electrically charged particles of mass are accelerated with a definite magnitude and in a definite direction. Electric and magnetic fields are related and similar. Their directions of force, however, may have different effects on a given particle of mass. An electric field accelerates an electrically-charged particle in a given direction. When the same sort of particle is placed in a parallel magnetic field, the field accelerates the particle in a direction that is perpendicular to that observed for the electric field.

Springs

A very common object that produces force is a spring. Springs take many forms—coil springs, spirals, and leaf springs. Indeed, almost anything that returns to its original shape after it has been stretched, compressed, bent, or twisted is a form of spring.

Other Forces

Besides the forces already mentioned, contained forces exist that result in pressure. For example, when an airtight container is pumped up to a pressure somewhat greater than the surrounding outside air, forces act on the inner walls of the container, trying to push them out. These forces are resisted by the opposing stresses in the material of the container.

Mass and Force

In the original metric system, the basic unit of mass is the gram. It represents the mass of 1 cubic centimetre (cm^3) of water at a temperature of 4°C. This temperature is the temperature at which water is densest. The gram is expressed in various powers of ten to arrive at milligrams, kilograms, and so on. In SI units, the kilogram (kg) is the base unit of mass. In the United States, the pound-mass is defined in terms of the kilogram, wherein 1 lb = 0.453592427 kg.

As previously noted, force is the product of mass and acceleration ($F = Ma$). In conventional units, the unit of force is the pound-force (lb_f), equal to the weight of one pound-mass (lb_m), when acted upon by the average acceleration due to gravity at sea level, which is 32.174 ft/s^2. In SI units, force is expressed in newtons. A newton (N) is the amount of force needed to accelerate a mass of 1 kilogram at 1 metre per second per second. Note that 1 lb_m weighs 1 lb_f at sea level, but 1 kg weighs 9.8 N.

A conversion factor is necessary to relate mass, weight, and acceleration. The following equation, as well as a conversion factor, is required to show the relation.

$$F = \frac{M \times a}{g_c} \qquad \text{(Eq. 2.5)}$$

where
  F = force
  M = mass
  a = acceleration
  g_c = gravitational constant.

For conventional units, the conversion factor is—

$$g_c = 32.174 \times \frac{(lb_m \times ft)}{(lb_m \times sec^2)}. \qquad \text{(Eq. 2.6)}$$

For SI units, the conversion factor is—

$$g_c = 1.000 \times \frac{(kg \times m)}{(N \times sec^2)}. \qquad \text{(Eq. 2.7)}$$

The poor mathematical relation among units makes measurement of mass and force in the conventional systems of weights and measures difficult. For example, the unit for weight is the pound (lb). Unlike in the SI system, where weight units can be expressed in multiples of 10—for example, a thousandth of a gram is a milligram (mg) and 1,000 grams is a kilogram (kg)—each conventional unit of weight has its own name—for example, ounces, drams, grains, and tons. To further complicate the issue, if we wish to convert from one unit to another, such as ounces (oz) to pounds (lb), we must memorize the conversion unit, which, in this case, is 16 oz—that is, 16 oz are in a pound. But a ton equals 2,000 lb and 16 drams or 437.5 grains equal 1 oz. Moreover, we must exercise care to avoid confusing the fact that pound represents mass in some instances and force in others. Commonly, we indicate mass by pound-mass (lb_m) and force by pound-force (lb_f).

So, to measure something, we must first establish scales and fundamental units. Then, we can select the scales and units to represent the unknown quantities. Finally, we can determine the values of the quantities in terms of the chosen units.

Determining Mass By Balancing Gravitational Forces

Once a standard of mass is established, the best way to use it in determining unknown quantities of mass is to set up a balance of forces. On the one hand, a gravitational force between the earth's mass and the standard mass exists; on the other hand, a similar force exists between the earth and the unknown mass. A balance, or scale, can be used to determine mass by balancing an unknown mass (M_2) with a standard mass (M_2) (fig. 2.1A). When M_1 and M_2 are equal, gravitational forces F_1 and F_2 are also equal (fig. 2.1B). Since the gravitational forces existing between earth and standard mass on the one hand and earth and unknown mass on the other are equal, then the masses are equal. Force (F) is the product of mass (M) and acceleration (a), which equation 2.5 shows (see Eq. 2.5).

For a condition of equilibrium, the balance of forces for weighing is—

$$\frac{M_1 \times g}{g_c} = \frac{M_2 \times g}{g_c} \qquad \text{(Eq. 2.8)}$$

where
- M_1 = standard mass
- M_2 = unknown mass
- g = local acceleration due to gravity
- g_c = gravitational constant.

Equations can be simplified—that is, terms in an equation can be removed without altering it if the same mathematical operation is done to both sides of the equation. This operation is also known as canceling. In the case of equation 2.8, the factor g can be removed, or cancelled, by dividing it out of the equation. With g removed, the equation is simply—

$$M_1 = M_2. \qquad \text{(Eq. 2.9)}$$

In short, the unknown mass, M_2 equals the standard mass, M_1.

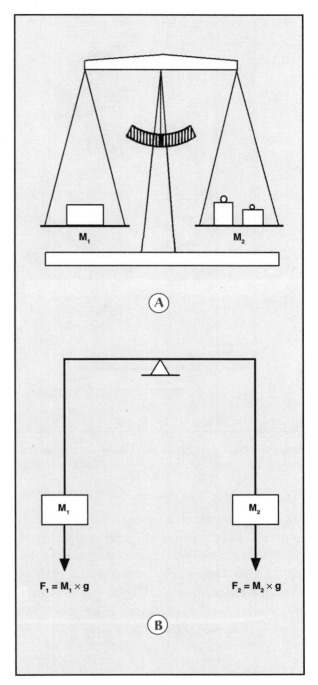

Figure 2.1 Balance scale for determining mass

Spring Scales for Measuring Mass

A gravity balance such as the one in figure 2.1, can determine mass without suffering an error because of deviation in the assumed value of g. However, a spring scale can also be used for weighing (fig. 2.2). The spring scale works on the principle of applying weight, or force, and measuring the distance the

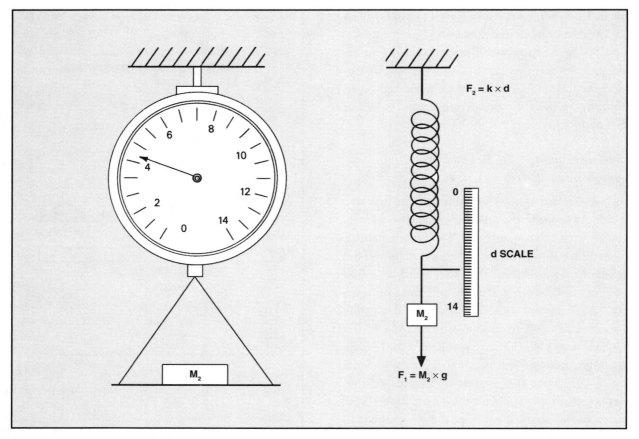

Figure 2.2   A spring scale determines mass by measuring the distance the spring is stretched. F_1 is the force exerted by the mass against the upward force F_2 of the spring.

weight, or force, causes the spring to stretch. Within limits, the spring obeys certain laws of physics and stretches an amount that is in linear relation to the applied force. That is, it stretches a certain amount if one unit of force is applied, twice that amount if two units are applied, and so on. The equation for weighing mass with a spring scale is—

$$F = k \times L \qquad \text{(Eq. 2.10)}$$

where
- F = force
- k = force constant for the spring
- L = distance spring stretches for the applied force.

The spring constant k is determined by the properties of a particular spring. A strong, stiff spring has a large k, while a limber spring has a small k. A spring scale, like a gravity scale, balances forces to weigh mass. The force $k \times L$ is balanced against the gravitational force that is attracting the mass $M \times g$. In equation form the balance of forces is—

$$k \times L = \frac{M_2 \times g}{g_c}. \qquad \text{(Eq. 2.11)}$$

The equation, and thus the scale itself, does not compensate for a change in the value of g. Therefore, an error can arise if g strays from the proper value. Consequently, spring scales are not usually legal for trade purposes. However, the error caused by deviation in the value of g is small. Spring scales can be up to 99 percent accurate.

MEASURING WORK AND ENERGY

Work and energy are sometimes used synonymously. Bear in mind, however, that the words have different meanings when used to discuss physical concepts on the one hand and nonscientific concepts on the other.

Work

In everyday use, the word work denotes any kind of mental or muscular effort. In science, however, the term work has a restricted and explicit meaning. In physical mechanics, when a force moves an object, work is accomplished. The distance the object is moved along a line parallel to the direction of the force, and the amount of force applied, determines the amount of work accomplished. A simple definition of work is that work, W, is the product of force, F, and length, L.

Remember: movement of the object must be parallel to the direction of the force. For example, if a force acts to push an object toward the east, and the object actually moves in a northeasterly direction, the work done is the product of the total force applied and the total distance the object moves in an easterly direction.

As noted earlier, force, in addition to having magnitude, has direction. An object that is being moved by force is moved along a line that has direction. However, although work is the product of two quantities—force and direction—both of which have direction, work has only magnitude. A quantity having both magnitude and direction is called a *vector* quantity, while one having only magnitude is called a *scalar* quantity.

Energy

Energy is another word that has different meanings in scientific and nonscientific usage. In everyday use, the word energy can denote a capacity for activity of a mental, physical, or even social nature. In physics, the term energy has an explicit and limited meaning. It denotes the capacity to do work. Energy, like work, has only magnitude and is measured in the same quantities—force and length.

Two kinds of energy are potential and kinetic. If an object or body is capable of doing work because it possesses a certain position, or because it exists in a stressed condition, it is a source of *potential energy*. A weight suspended above the floor has potential energy, and is an example of potential energy by virtue of position. A tightly wound clock spring also has potential energy, and is an example of energy by virtue of being in a stressed condition. If a massive object moves, it possesses *kinetic energy* because it moves and is capable of doing work. Spinning flywheels, projectiles in flight, and automobiles in motion are examples of kinetic energy. Both kinetic energy and potential energy are measured in the same quantities as work—that is, force and distance.

Equations of Work and Energy

Although work and energy are measured in the same quantities, the equations that represent them sometimes take forms that require interpretation to arrive at the basic product of force and distance. The equation for work is—

$$W = F \times L \qquad \text{(Eq. 2.12)}$$

where

W = work
F = force
L = the linear measurement between the beginning and ending points that represent the path over which the work is being computed.

Potential energy in the form of a suspended weight, or water trapped behind a dam, has the same equation as that for work. In this case, however, F is gravitational force, or weight, and L is the vertical distance between a reference point and the point at which the weight is located or the surface of the water behind the dam.

The equation for potential energy stored in a compressed or stretched spring is—

$$E_p = \tfrac{1}{2} k \times L^2 \qquad \text{(Eq. 2.13)}$$

where

E_p = potential energy
k = the force constant of the spring
L = the distance the spring is stretched or compressed from its reference position.

Because the force constant (k) times the distance the spring (L) is stretched ($k \times L$) is equal to force, equation 2.13 is easily resolved into—

$$E_p = \tfrac{1}{2} F \times L \qquad \text{(Eq. 2.14)}$$

where

$F = k \times L$.

The equation for kinetic energy, the capability for doing work by virtue of a moving mass, is—

$$E_k = \tfrac{1}{2} M \times v^2 \qquad \text{(Eq. 2.15)}$$

where

E_k = kinetic energy
M = mass
v = velocity.

If you recall, we can express the dimensions of force as $M \times \tfrac{L}{t^2}$ and velocity as $\tfrac{L}{t}$. Consequently, we can write the kinetic energy equation as—

$$E_k = M \times \frac{L^2}{t^2} \qquad \text{(Eq. 2.16)}$$

and as—

$$E_k = F \times L. \qquad \text{(Eq. 2.17)}$$

These two equations are the same as the equation for the compressed or stretched spring, but in a different form. Manipulation of the equations for potential and kinetic energy demonstrates that energy and work can be resolved into the two dimensions of force and distance.

Several other energy equations relate to rotating masses and spiral springs. These equations can be resolved into products that involve only force (F), distance (L), and perhaps a constant such as ½, as in the equations for kinetic energy (E_k) and potential energy (E_p).

When working with units of work and energy, take care to maintain consistent units. These equations mix units of force, mass, length, and time. Thus, we must use correct conversion factors to avoid improper results. The correct units for energy and for work are force pound (f-lb$_f$) in conventional units and newton metre (N-m) in SI units.

As an example, consider the equation for a stretched spring—

$$E_p = \frac{k \times L^2}{2}. \qquad \text{(Eq. 2.18)}$$

Since the force constant is in force per unit length ($k \times L^2$), no conversion factor is required because the product is already in the correct units of force times length.

However, in the case of kinetic energy, where the factors are mass and velocity, a conversion factor is needed. For example, the equation for the kinetic energy of a moving object is—

$$E_k = \frac{M \times v^2}{2}. \qquad \text{(Eq. 2.19)}$$

The result works out to $M \times \tfrac{L^2}{t^2}$. To obtain the desired result, we must divide by g_c. Thus, the equation becomes—

$$E_k = \frac{M \times v^2}{2 \times g_c}. \qquad \text{(Eq. 2.20)}$$

Relation of Power to Work and Energy

Work and energy are expressions for the product of force and distance. The time rate at which work is accomplished is defined as *power*. Power (P) is made up of force (F), length (L), and time (t). The equation is—

$$p = F\left(\frac{L}{t}\right). \qquad \text{(Eq. 2.21)}$$

In the conventional system of units, the work might be expressed in foot-pounds (ft•lb), and the time in seconds (sec) or minutes (min). A common unit of power in this system of measurement is horsepower (hp), which is defined as 550-ft•lb of force per sec or 33,000-ft•lb of force per minute.

Electric power is usually measured in watts (W) or kilowatts (kW). One hp equals 746 W, or 0.746 kW.

Utility companies commonly measure electric energy in kilowatt-hours (kW•hr). A kW•hr is equivalent to multiplying power times time (kW × t). Another way to express kW•hr is to say that it is the product of power and time. In any case, the result is energy, or work.

MEASURING DIMENSIONS OF VARIOUS QUANTITIES

In science and instrumentation, the word dimension means more than a measurement of length or distance. Instrumentation treats all fundamental quantities as dimensions. An interesting technique called dimensional analysis establishes certain physical quantities as fundamental, and from these quantities, all other compound quantities are derived in mathematical fashion. This sort of analysis is used to study cumbersome physical systems that do not lend themselves to easy, straightforward mathematical solution.

Pressure

Pressure (p) is defined as force (F) per unit area (A). The equation is—

$$p = \frac{F}{A}. \qquad \text{(Eq. 2.22)}$$

While we can use virtually any units of force and area to express pressure, the conventional system usually combines pounds force and square inches. That is, the conventional system expresses pressure in pounds per square inch (psi). The SI system uses newtons (N) and square metres (m²) to express pressure in pascals (Pa). In other words, 1 N/m² is equal to 1 Pa. However, instrumentation uses two other expressions for pressure. They deserve discussion because of their apparent departure from the basic dimensions of pressure of force and area. A U-tube manometer is a device commonly used to measure pressure (fig. 2.3).

The manometer tube contains mercury, which weighs about 0.5 lb per cubic inch (lb/in.³). With both ends of the U-tube open to the atmosphere, the surface of the left and right columns of mercury are at equal levels, and the gauge pressure reading is 0 (fig. 2.3A). If a pressure of x psi is applied to the left column (fig.2.3B), the surface of the mercury in the left part of the column falls and the surface of the mercury in the right part of the column rises. With the gauge scale graduated in inches, the left scale reads 2 in. below the zero mark, and the right scale reads 2 in. above the zero mark. The difference in level of the two surfaces is therefore 4 in.; thus, the column height is 4 in.

The diameter of the mercury columns in the U-tube has no effect on how high the mercury rises or falls in the two columns of the manometer for any applied pressure. In fact, the applied pressure, x psi, is simply the product of column height in in. and the weight in in.³ of the mercury. The calculation is—

$$x = 4 \text{ in.} \times 0.5 \text{ lb/in.}^3$$
$$x = 2 \text{ lb/in.}^2, \text{ or 2 psi.}$$

Since the applied pressure is directly proportional to the column height, it is a matter of convenience to express it in terms of inches of mercury. Measuring pressure in terms of length began in

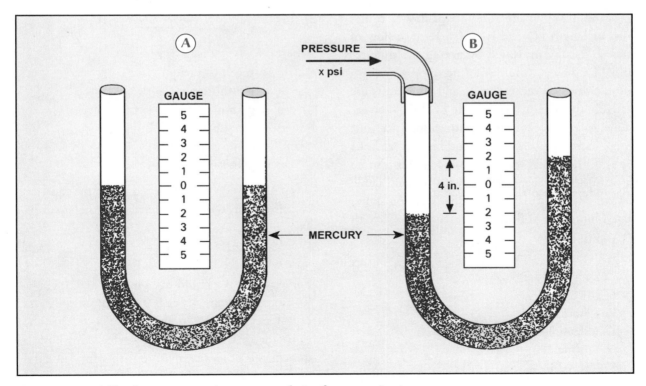

Figure 2.3   A U-tube manometer is a common device for measuring pressure.

laboratory work, but industry soon began using it to express low values of pressure. Meteorologists often express atmospheric pressure in inches of mercury, although they usually convert it to a barometer reading, which is not strictly a pressure reading.

A variation on the mercury manometer method of pressure measurement is to substitute water for the mercury. Imagine the U-tube in figure 2.3 using water instead of mercury. Since the density of water is only about $\frac{1}{13.6}$ as dense as mercury, a water manometer is much more sensitive. For example, pressure that causes mercury to rise 1 in. (25 mm) causes the water column to rise 13.6 in. (340 mm). The term inches or milli-metres of water is a popular expression for pressure in the gas industry where low values of gauge pressure are common.

In general, several popular expressions exist for pressure—pounds per square inch; newtons per square metre; and inches, centimetres, or millimetres of mercury or of water. Some of these are not true pressure dimensions, but from the standpoint of convenience, they are extremely useful and have a proper place in instrumentation.

Rate of Flow

A common way to measure rate of flow is to use units of length (L) and time (t). Rate of flow of fluids is usually measured by noting the time required for a given volume of fluid to pass a certain point. Scores of volume units and time units are available for measuring the rate of flow—for example, gallons per minute (gpm), gallons per hour (gph), barrels per day (bpd), and barrels per hour (bph). Other units include cubic feet per minute (ft³/min), cubic feet per hour (ft³/hr), and cubic feet per day (ft³/d). The following equation shows that volume rate of flow has dimensions of length (L) and time (t).

$$E_p = \frac{L^3}{t} \quad \text{(Eq. 2.23)}$$

where
R_f = rate of flow
L = length
t = time.

Rate of flow can also be measured in terms of mass per unit of time. Mass flowmeters, which utilize the mass of a flowing gas or liquid to determine rate of flow, enjoy considerable popularity today and will probably continue to do so. Converting from mass to volume is simple as long as the density of the gas or liquid is known. Many sources are available for obtaining such densities. The relation between volume (V) and mass (M) is—

$$V = \frac{M}{d} \quad \text{(Eq. 2.24)}$$

where
V = volume
M = mass
d = density.

Other Variables

Specific Gravity

Specific gravity may be defined in several ways. The most common is that specific gravity is the ratio of the mass of a given volume of substance to the mass of an equal volume of water. Let's assume that a grams (g) of substance has a volume of b cubic centimetres (cm³). The specific gravity (sg) of the substance can be determined from the equation—

$$sg = \frac{ag \times bcm^3}{cg \times bcm^3} \quad \text{(Eq. 2.25)}$$

where
sg = specific gravity
a = the substance's mass
g = grams
b = the substance's volume
cm^3 = cubic centimetres
c = grams of water.

All terms in the right-hand side of the equation cancel except a and c, which leaves the equation as—

$$sg = \frac{a}{c}. \quad \text{(Eq. 2.26)}$$

For example, if a liquid has a mass of 0.926 grams (g), its specific gravity is 0.926 because the mass of water is 1 and 0.926 divided by 1 is 0.926. The value for *sg* is a pure number—that is, it has no dimensions of length, mass, or other fundamental quantity. It is dimensionless. Incidentally, in the SI system, specific gravity is called *relative density*.

pH Factor

By virtue of its definition and use, *pH factor* is a pure number; it is dimensionless. The quantities from which it is derived, however, have a dimension of mass. The pH factor is arrived at by the equation—

$$pH = -\log_{10}[H^+] = -\log_{10}[10^{-7}]. \quad \text{(Eq. 2.27)}$$

The factor $[H^+]$ designates the hydrogen ion concentration in moles per litre of a solution. In pure water, which is neither acid nor base, the concentrations of hydrogen and hydroxyl ions are equal and amount to 10^{-7} mole per litre. So, the pH of pure water is—

$$pH = -\log_{10}[H^+] = -\log_{10}[10^{-7}]$$
$$= 7 \log_{10} 10$$
$$pH = 7.$$

If a substance such as pure water has a pH of 7, the substance is neutral. A substance that has a pH below 7 is acidic. A substance with a pH above 7 is basic. Note, too, that the pH scale is logarithmic and not linear. That is, a liquid with pH of 4 is ten times (not twice) as acidic as a liquid with a pH of 5. Consequently, pH is usually stated to one decimal place, such as 8.0, 7.2, or 5.3.

Viscosity

Mile is a common term; however, as mentioned earlier, at least three kinds of miles exist: statute, geographical, and nautical. Only two are the same length. Like the term mile, viscosity needs an adjective to specifically identify it. *Absolute viscosity* is expressed in a basic unit called a *poise*. The poise has dimensions of length, mass, and time. As a standard of viscosity, poise was established in the era when the centimetre-gram-second (*cgs*) formed one subsystem of the metric system. The poise is therefore best expressed in cgs units. The SI system, however, discourages the use of cgs units.

Consider the following situation. A viscous liquid that is 1 cm thick covers a solid smooth surface of indefinite length and width. Another solid surface with a unit area of 1 cm² is on the surface of the liquid. Now, if a force of 1 dyne (d), which is the unit of force in the cgs system, is required to move this unit-area surface at a velocity of 1 cm/sec, the absolute viscosity of the liquid is equal to 1 poise. Absolute viscosity in poises can be expressed as—

$$\text{poise} = \frac{d \times sec}{cm} \quad \text{(Eq. 2.28)}$$

where
- d = dimensions of mass and length
- sec = seconds
- cm = centimetre.

Equation 2.28 can also be stated as—

$$\text{poise} = \frac{M \times L \times t}{L^2 \times t^2} \quad \text{(Eq. 2.29)}$$

where
- M = mass
- L = length
- t = time.

Finally, equation 2.29 can be reduced to—

$$\text{poise} = \frac{g}{cm \times sec} \quad \text{(Eq. 2.30)}$$

where
- g = gram
- cm = centimetre
- sec = seconds.

What it all boils down to is that 1 poise equals 1 gram per centimetre per second (1 g/cm/sec).

Kinematic viscosity is obtained by dividing the absolute viscosity by the density of the fluid in grams per cubic centimetre (g/cm³). The equation is—

$$\text{kinematic viscosity} = \frac{g/cm^3 \times sec}{g/cm^3}. \quad \text{(Eq. 2.31)}$$

Equation 2.31 reduces to—

$$\text{kinematic viscosity} = (g/cm^3 \times sec) \times \frac{cm^2}{g}. \quad \text{(Eq. 2.32)}$$

$$\text{kinematic viscosity} = cm^2/sec.$$

The unit of kinematic viscosity is the stoke. Its dimensions are centimetres squared per second (cm²/sec). Several other forms of viscosity are in use and will appear in the chapter on measurement and control of other variables.

SUMMARY

Instrumentation deals with relatively few fundamental quantities. However, these quantities can be grouped in so many ways that problems sometimes arise.

An important concept to remember is that the names of the units of measurement are treated as mathematical entities. When a given unit appears in both the numerator and denominator of an expression, sometimes a cancellation step can be carried out, just as with equal numerical factors. Each variable measured and controlled is really a quantity made up of a particular grouping of one or more of the fundamental quantities.

REVIEW EXERCISE

1. Define dimensions.
2. What is a unit?
3. Convert 10 ft^2 into m^2.
4. 50°F is equivalent to how many degrees C?
5. A force of 100 lbs is acting on an area of 10 in.^{2} What is the pressure in psi?

3
Final Control Elements

In fluid flow processes, the final control element regulates the rate of flow. Most final control elements are valves; indeed, the two terms are almost synonymous. However, the petroleum industry also uses controlled-volume pumps, variable-speed pumping drives, and other devices as final control elements.

A final control element usually consists of a valve, an actuator, and piping. An actuator provides the force that operates the valve, which controls the rate of flow of a controlled variable through the valve. Mechanical, pneumatic, electrical, hydraulic, or a combination of these means operate the actuator.

A controlled-volume pump delivers a definite and predetermined volume of liquid with each stroke, or cycle. The petroleum industry widely uses controlled-volume pumps to force chemicals into lines and vessels.

In large-volume pumping systems, the final control element often includes variable-speed drives, which power variable-volume pumps. Many pipeline systems use variable-volume, variable-speed pumps to provide variable flow on a continuous basis, depending on demand. A signal from an electronic control, programmable logic controller, or similar electrical device operates the pumps. The magnitude of the signal determines the speed of the variable-speed drive, which, in turn, controls the volume rate of the pump.

VALVES

Valves have many parts (fig. 3.1). However, their function and use is straightforward and easy to understand. Important valve parts include the body, the plug, the guides, and the seats.

Also, keep in mind that this manual does not cover the extensive considerations involved

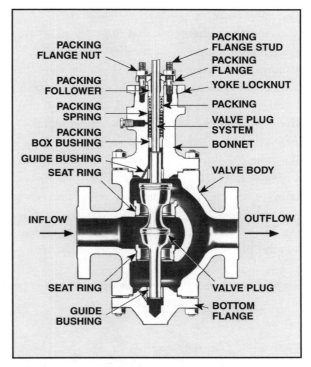

Figure 3.1. Double-ported valve (Courtesy Fisher Controls)

in the process of selecting valves for special applications. Valve selection is one of the many jobs engineers who design control systems do. But, readers should understand and appreciate the fact that control valve selection is based on the valve's having met many critical specifications to fulfill the exacting requirements of a particular control system.

Valve Bodies

Most control applications employ globe valve bodies. However, certain applications may use other types of body. The term globe comes from the round shape of the body.

Single-Ported Globe Bodies

Single-ported valves have a single path for passage of fluid (fig. 3.2A and B). The two single-ported valves in the figure are similar but have differences. One difference is the way the bonnets and bottom flanges are attached to the valve bodies. In figure 3.2A, studs and nuts attach the bonnet and bottom flange to the main valve body. In figure 3.2B, a clamp ring secures the bonnet and bottom flange to the main valve body.

Another difference is that the valve plugs are reversed with respect to each other. In figure 3.2A, pushing down on the valve stem closes the valve. In figure 3.2B, pushing down on the valve stem opens the valve. The valve in figure 3.2A is a direct-acting valve; the valve in figure 3.2B is a reverse-acting valve. Both are widely used in control applications.

The valve plugs in figure 3.2A and B can be reversed, which changes the valve from direct acting to reverse acting, and vice versa. The valve seats must also be reversed, but reversing them is easy. The greatest advantage of being able to reverse the valve plugs and seats is reduced manufacturing cost. The manufacturer can make only one set of parts to obtain either a direct- or reverse-acting valve.

Single-ported valves cost less, are easier to maintain, and are more resistant to leakage when fully closed than double-ported valves. However, single-ported valves have difficulty dealing with fluid pressure in the line, because pressure bears heavily against the plug in the closed position. But, where smooth valve operation is required—where it is not good for the plug to slam shut—single-ported valves can be installed in the line so that the fluid pressure tends to force the plug away from the seat and keep them from slamming shut.

Double-Ported Globe Bodies

Double-ported valves are widely used in automatic control applications. A double-ported valve requires less force than a single-ported valve to position its plug at any point between fully open and fully closed. Thus, double-ported valves are used wherever the valve needs to be sensitive to small pressure changes in the process.

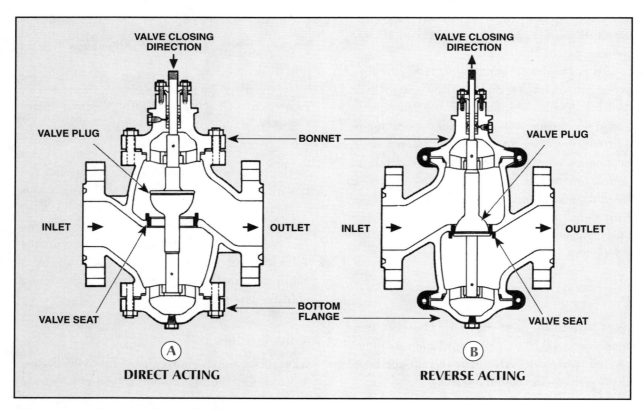

Figure 3.2 Two types of ported valves. A, direct-acting; B, reverse-acting

Final Control Elements

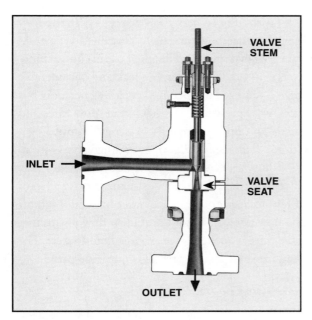

Figure 3.3 A split-body valve (Courtesy Fisher Controls)

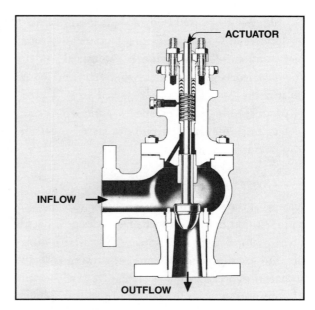

Figure 3.4 A typical angle-body valve

Split-Body Valves

A split-body valve has two major components that bolt together (fig. 3.3). The components are designed so that they can be connected to form either a straight-through or right-angle valve. The valve in figure 3.3 is configured as a right-angle valve. Split-body valves are easy to service. Also, sediment cannot readily build up inside them.

Angle-Body Valves

Angle-body valves are usually single-ported and are used where lines need regular draining (fig. 3.4). A typical right-angle valve is comparatively free from pockets and cavities, which gives it good drainage qualities. An angle valve is usually installed so that flow goes into the side port and out the bottom, because such an installation minimizes body erosion. One disadvantage is that angle-body valves tend to slam shut as the plug nears the seat.

Three-Way Valve Bodies

Three-way valves are available in two configurations: single-ported (fig. 3.5A) and double-ported (fig. 3.5B). Three-way valves are constant-flow devices—that is, the position of the plug does not affect the total flow through the valve.

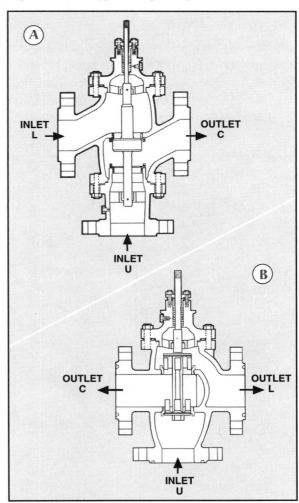

Figure 3.5 Three-way valves. A, single-ported; B, double-ported

The single-ported body style is used for mixing service, in which the *U* and *L* connections are inlets, and the *C* connection is the mixed outlet. The valve receives fluids from two sources and sends the mixture out. The double-ported style is used for proportioning service, wherein fluid enters one inlet *C* and exits through two outlets *U* and *L* in proportioned amounts.

Small Flow-Rate Valve Bodies

Because of the need for low-capacity control valves, manufacturers have developed several designs. These valves perform such functions as injecting inhibitors into lines or processes and performing tasks in laboratories and experiments. The valve-body design is less important than the design of the seats, plugs, and actuating mechanisms.

Butterfly Valve Bodies

Butterfly valve bodies are cylindrical and have a large capacity (fig. 3.6). When wide open, straight-through flow occurs without much impeding it. Only the small area of the edgewise surface of the disc or vane presents an obstruction.

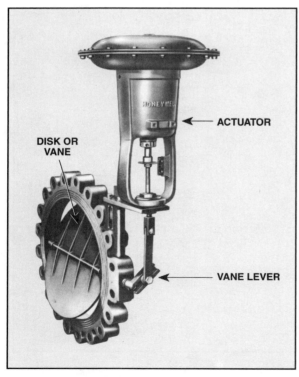

Figure 3.6   Butterfly valve (Courtesy Honeywell)

Although butterfly valves have a high capacity for a given size, and are usually less expensive to build than plug-type globe valves, their use in control functions is limited. They tend to leak and considerable force is required to operate the vane against any but the lowest pressures. Further, when flow rates and lift are low, flow through them is logarithmic. Logarithmic flow means that opening the valve twice as wide does not double the flow; instead, opening the valve twice as wide may increase the flow by a factor of ten, for example. On the other hand, at high lift and flow rates, flow is linear. Linear flow means that if the valve is opened twice as wide, flow doubles. The need for these characteristics is not widespread.

Gate Valve Bodies

Pipelines use a considerable number of gate valves, but, in process control, they are not used as much. In process control, flow through a valve usually needs to be *throttled*—that is, flow needs to be slowed or increased gradually between fully open and fully closed. Gate valves are either wide open or completely closed. Gate valves have full-flow capability in that they can open to the same inside diameter of the pipe in which they are installed (fig. 3.7). This full-flow feature allows scrapers and other full-diameter devices such as pigs to go readily through the pipeline. Also, gate valves have excellent shutoff capability.

Valve Characteristics

Various factors make one type of valve better suited for a given control problem than others. Of greatest importance are flow characteristics and flow capacity. Other considerations become important in specific applications—for example, ease and smoothness of operation, leakage characteristics, ease of maintenance, and self-cleaning features.

Flow Characteristics

The relation between the amount of fluid flow through the valve and the extent to which the valve is open is the flow characteristic of a control valve. Flow characteristic can be expressed on graphs (fig. 3.8A and B). In the figure, maximum flow is plotted on the horizontal axis of the graph (the *abscissa*), and maximum lift (opening) is plotted on the vertical

Final Control Elements

Figure 3.7 A pipeline gate valve

axis of the graph (the *ordinate*). The graph shows the characteristic curves of a control valve using five different plugs: equal percentage, linear, throttling, quick opening, and V-port. The curves are on linear graph paper in figure 3.8A and on linear-log, or semilogarithmic, paper in figure 3.8B. The graph on semilogarithmic paper (fig. 3.8B) gives a more detailed view of valve flow and lift at low flow rates—that is, at flow rates of 10 percent and lower.

As an example of what the graphs show, in figure 3.8A note the curve for an equal percentage plug. When the valve is 60 percent lifted, or open, only 20 percent flow occurs through the valve. The curves for the other plugs reveal their flow characteristics. As mentioned earlier, the curves in figure 3.8B show flow characteristics in detail at flows of 10 percent and below. Knowing flow versus lift helps system designers select the correct valve for a particular application.

Flow Capacity

Flow capacity describes the maximum volume of fluid that can go through a wide-open valve over a unit of time. With the valve wide open, flow is at a maximum; however, several factors are involved in measuring it. These factors include: (1) the pressure drop across the valve, (2) the energy difference between the inlet and outlet, and (3) the viscosity, density, and type of fluid.

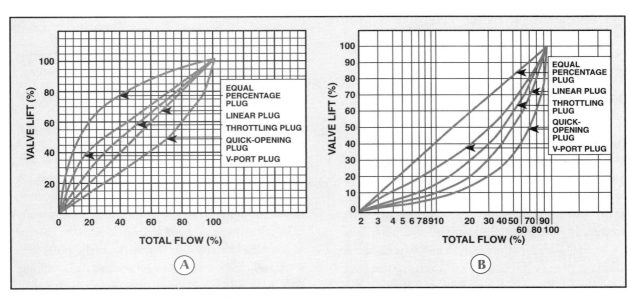

Figure 3.8 Flow curves (Courtesy Fisher Controls)

In the conventional (English) measurement system, using the term *flow coefficient* (C_v) has been a convenient way to express the flow characteristics of a control valve. C_v is the volume of water in gal that passes through a wide-open valve in 1 min with a pressure drop of 1 psi across the valve. This factor is widely used throughout the world. Currently, no equivalent is available in the SI system. However, a few valve manufacturers in Western Europe use a flow coefficient, K_v, which is based on flow per hr in m^3 and a pressure drop measured in bars. The bar is a pressure unit equal to 100 kPa (14.5 psi).

We can also use an equation to approximate the rate of flow through a valve—

$$Q = k\sqrt{\Delta_p} \qquad \text{(Eq. 3.1)}$$

where

Q = rate of flow
k = a constant
Δ_p = pressure drop across valve (or other restriction in the line).

This simple equation becomes complicated as more considerations are taken into account; however, for water the equation is—

$$Q = C_v\sqrt{\Delta_p}. \qquad \text{(Eq. 3.2)}$$

where

C_v = volume of water, gal.

Valve Plugs

Valve flow characteristics depend on the shape and performance of the valve plugs, or inner valves. Valve plugs (fig. 3.9A–C) can be designed to produce all forms of flow, from simple on-off service to any desired throttling action.

Quick-Opening Plugs

The intake and exhaust valves of most internal-combustion engines are quick-opening valves. Sometimes called *poppets*, they provide full flow with comparatively small stem movement. Quick-opening valves are used where on-off flow is required. They are not well suited to throttling action. Figure 3.9A is a double-ported, quick-opening valve plug. It is port-guided—that is, the wings that radiate from the axis of the plug ride against the seat rings to keep the plug aligned. Refer to figure 3.8A and B to see the flow characteristic of a quick-opening plug.

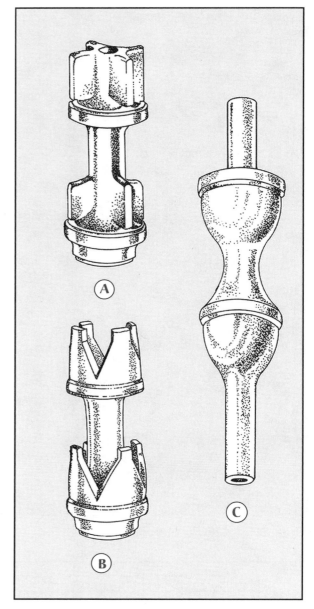

Figure 3.9 Types of valve plugs. A, quick-opening; B, V-port; C, throttling

Modified Linear, or V-Port, Plugs

Figure 3.9B shows a double-ported modified linear, or V-port, plug. This plug, like the quick-opening plug in figure 3.9A, is also port-guided. Refer to figure 3.8 for its flow characteristic. One company calls its version a *characterized V-port* because of V-shaped slots cut into the skirts of the plug. The term *modified linear* can be misleading, but the plug does have a linear characteristic for all high values of lift and flow.

Final Control Elements

Throttling Plugs

The throttling plug (fig. 3.9C) is a superior plug for handling slurries and other fluids having solids in them. Its flow pattern is similar to that of the V-port plug. It is stem-guided and lathe-turned. The stems ride in special bushings in the valve body, which hold the stems in alignment. Lathe-turned means the plug is machined from round stock in a lathe. Quick-opening and V-port plugs require more extensive manufacturing techniques.

Equal Percentage Plugs

Equal percentage plugs (fig. 3.10) have a logarithmic flow pattern. If the pressure drop across the valve is held constant, equal percentage plugs produce equal percentage changes in flow for equal changes in stem lift. That is, each time the stem lift is changed an equal percentage of the total lift, the flow through the valve changes 50 percent from the flow existing before the change in lift. For example, suppose flow through the valve changes 50 percent for each change of 10 percent in total lift. With the valve wide open—lift and flow at a maximum—the lift is reduced to 90 percent of maximum. This reduced flow produces a 10 percent change in lift and a 50 percent change in flow. Another reduction in lift to, say, 80 percent of maximum reduces flow to 25 percent of maximum, and so on until flow and lift reach 0.

Refer to figure 3.8A and B and find the curves for the equal percentage plug. At low values of flow, note that considerable stem movement is required for appreciable changes in flow. Also note that the opposite is true for large values of flow. Such flow characteristics are desirable for applications where a large portion of the pressure drop is normally absorbed by the system itself, with only a relatively small amount available at the control valve. These characteristics are also useful in systems where pressure drop is highly variable.

Linear Plugs

A linear plug (fig. 3.11) is made in lathe-turned and skirted types, as well as for single-ported and double-ported service. The lathe-turned types are

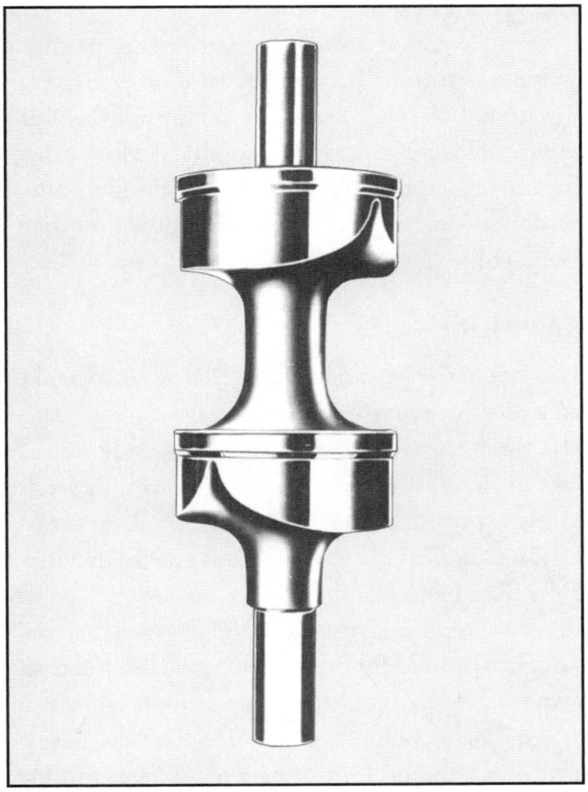

Figure 3.10   Equal percentage plug (Courtesy Fisher Controls)

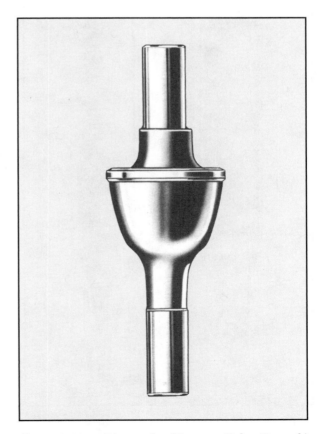

Figure 3.11   Linear plug (Courtesy Fisher Controls)

stem-guided, while the skirted plugs may be either stem- or port-guided. Port-guided plugs have two or more rectangular slots cut in the skirts. The flow characteristic of linear plugs is a straight line from zero to maximum flow and lift (see fig. 3.8A).

Plugs for Low Flows

Figure 3.12 shows two plugs designed to control low-flow rates. The flow characteristics of such plugs can be tailored to fit any pattern desired. The guide stems of low-flow plugs are larger than the controlling part of the plug. These plugs usually have a single port and top guides.

Valve Guides and Seats

Proper mating of the valve plug and seat requires a steady and durable means of maintaining correct alignment of the moving parts relative to the valve seat. Guides perform this function. Several forms are available.

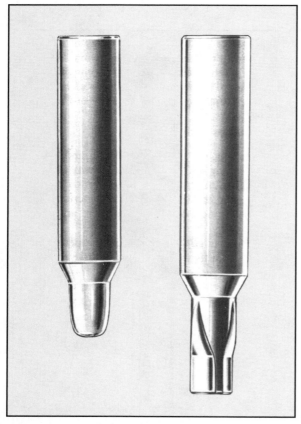

Figure 3.12   Plugs for low-flow rates (Courtesy Fisher Controls)

The quick-opening plug in figure 3.9A and the V-port plug in figure 3.9B are wing-guided and skirt-guided, respectively. Both plugs are port-guided. Valves with port-guided features are less expensive to build and work well if used in the service for which they are designed. They are not suitable in applications where wide ranges of throttling are needed. Also, at positions of low flow, any misalignment of the plug distorts the flow in the annular space between the seat ring and the plug, causing the plug to bind.

The stem-guided plug (fig. 3.9C) has top and bottom guides, which provide excellent stabilization for the plug. Stem-guided, double-ported units may have either top and bottom guides or top guides only. In either case, such guides prevent excessive valve vibration and binding.

Control valve seats are often made of metal, as are valve plugs; consequently, metal-to-metal contact with the valve plug occurs. However, if positive shutoff of long duration is required, metallic seats are not satisfactory. Thus, seats made of resilient compounds are available for use in systems where long-term positive shutoff is required. Where complete shutoff is not necessary, metal seats work well.

Single-seated valves give better shutoff than double-seated units because they are easier to keep aligned and their wear is more uniform. However, because of the difficulty in overcoming the forces that affect a single-ported valve near the closure point, the double-ported valve is favored where throttling is required.

Valve Trim

Trim refers to the kind of material used to make internal valve parts. Trim includes the plug, seats, valve stem, valve-stem guide bushings, and internal parts of the stuffing box. The type of trim depends mainly on two factors: (1) the pressure drop across the valve, and (2) the corrosive and erosive qualities of the fluid being controlled.

Stainless steel is the most popular trim material and it can be used for every part. Stainless steel trim is not only mechanically tough and durable, but also it resists corrosive fluids. Manufacturers have developed many special forms of stainless steel trim for valves, which bear registered trade names. Examples include Stellite, Colmony, Stoody, and Monel. Also,

Final Control Elements

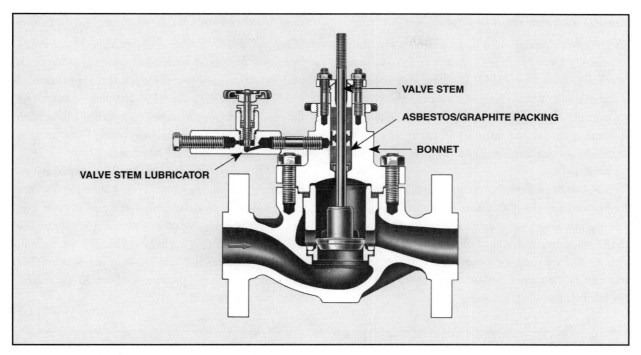

Figure 3.13   Stuffing box (Courtesy Fisher Controls)

chrome plated stainless steel is used as trim in large-size valves handling corrosive material.

Because of its low cost, bronze trim is used in a few instances in mild service, but the trend is away from using such soft materials.

Valve Design Details

Valves include many design details; the most important are stuffing boxes (valve-stem seals), bonnets, and end connections.

Stuffing Boxes

The valve stem transmits motion from an actuator to the plug through a seal capable of withstanding line operating pressure. Such seals are usually called *stuffing boxes*. Good seals must not only be leak resistant, but also they must not create too much friction; otherwise, the actuator cannot smoothly position the valve plug. One type of stuffing box uses several rings made of a composition material, such as asbestos and graphite, and a system for lubricating the stem (fig. 3.13). Composition material can also be mica with artificial rubber.

Teflon, a special plastic, is popular as a packing material because it is inert to most chemicals, has a wide temperature range, and is self-lubricating. It is often molded into V-ring, or chevron, packing (fig. 3.14). Valves with molded Teflon packing require careful maintenance, and procedures recommended by the valve manufacturer should be followed to avoid difficulties.

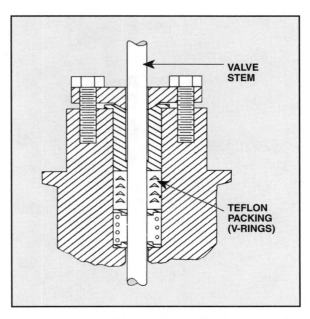

Figure 3.14   Self-lubricating stuffing box with Teflon packing

Bonnets and Bonnet Extensions

The bonnet of a valve can be fastened to the main valve body by: (1) direct mating of internal- or external-threaded joints, (2) the fastening together of flanges with studs and nuts, and (3) the clamping together of flanges by a clamp ring. A bonnet provides a mounting support between the main body and the actuator mechanism, and it contains the stuffing box.

Sometimes, bonnets are provided with radiating fins to carry away the heat of the process fluid, because this heat might cause serious deterioration of the packing material (fig. 3.15B). Extensions without fins are used for low-temperature service to help insulate packing and actuator parts from the low temperature (fig. 3.15A).

End Connections

Valves are usually attached to piping and equipment by mating flanges, threaded fittings, or welding. Each method has an advantage for a given application.

The valve in figure 3.15B has flanged ends. The flange on the valve attaches to a flanged fitting on the piping. Flanges are extensively employed because they are inexpensive and reliable.

Threaded fittings are used in applications requiring valves less than 2 in. (50 mm) in size, although high-pressure or low-temperature service, or unusually corrosive conditions, might dictate the use of flanged fittings. Threaded fittings are used in large-sized valves for low-pressure gas service, but threading valves and pipe beyond certain diameters becomes impractical.

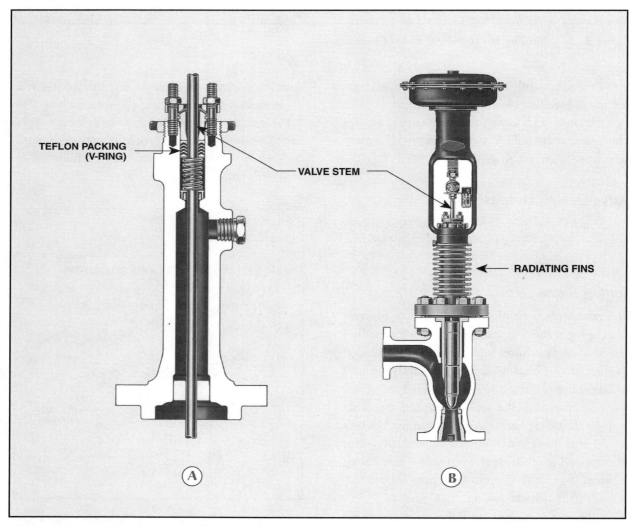

Figure 3.15   *Valve bonnets and extensions (Courtesy Fisher Controls)*

Final Control Elements

Welded ends are used in processes that handle dangerous or valuable products where leakage cannot be tolerated. If the installation is permanent, welded connections offer real advantages. They are capable of withstanding maximum pressure and are inexpensive.

SIZING AND PIPING ARRANGEMENTS

Selecting the right type and size of control valve for some processes and systems can be complicated. Selection entails expertly applying known process flow data and using assumptions that come from previous experience. This manual does not discuss valve sizing. However, keep in mind that throttling valves usually should be considerably smaller than the line in which they are installed—for example, half the size of the line. A rule of thumb for sizing throttling valves is that not less than one-third of the pressure drop in the system should occur across the valve during maximum flow conditions. (Exceptions exist, but this rule usually works.) If a large-size control valve is used—one that produces only a small part of the overall system pressure drop—then its ability to control flow near maximum flow rate is seriously impaired. Such a valve will probably operate at less than half-open positions for all conditions.

In spite of the relatively large pressure drop needed across a control valve, the valve must still be able to pass a quarter to a half more flow than the maximum rate ever demanded by the process or system.

In the typical piping arrangement for control valve installation shown in figure 3.16, block valves and a bypass line are provided so that repairs, cleaning, or complete removal of the control valve can be carried out without putting the line out of service. The control valve is smaller than the line in which it is installed, and the bypass line and valve have a similar relation to the main line.

ACTUATORS

An *actuator*, or *operator*, is a device that provides the force to vary the orifice area through which the control agent flows. It accomplishes this action by positioning the valve stem, or other driven element, in the valve or other control element.

Actuators are classified according to the form of input signal and output power used. Thus, actuators can be mechanical, pneumatic, electric, hydraulic, or a combination. For example, an electrohydraulic actuator is one that receives an electrical signal to

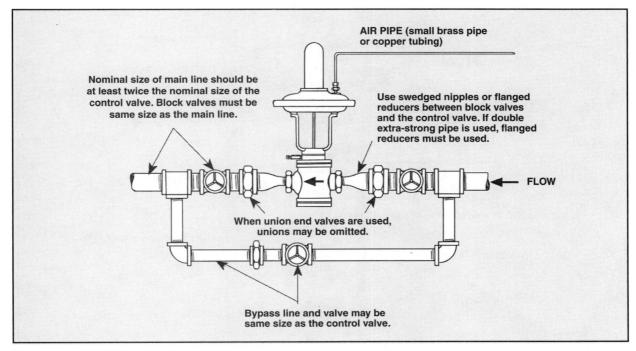

Figure 3.16   *Throttling valve (Courtesy Foxboro)*

control a hydraulic system, which performs the mechanical motion.

Mechanical Actuators

Mechanical actuators use fittings such as bolts, screws, and rods to transmit motion between a sensing device and the final control element. Liquid-level control systems extensively use mechanical actuators, although mechanical actuators are not sensitive enough for processes that require extremely accurate control.

Pneumatic Actuators

The petroleum and chemical industries widely use pneumatic actuators. Such factors as safety, simplicity, and reliability make pneumatic actuators popular. Two types of pneumatic actuators are the diaphragm and the piston.

A diaphragm actuator usually contains a spring that opposes the air pressure applied against the diaphragm, although springless types in which controlled air pressure can be applied to either side of the diaphragm are quite common. A piston actuator is usually springless.

Spring-Loaded Diaphragm Actuators

Figure 3.17 shows a spring-loaded, direct-acting diaphragm actuator. In a direct-acting actuator, air pressure is applied to the inlet at the top of the diaphragm case, and a heavy spring opposes downward motion of the diaphragm and actuator stem. Air pressure drives the actuator stem down to close a direct-acting valve.

A reverse-acting, spring-loaded diaphragm actuator works in a similar manner (fig. 3.18), but important differences exist. In a reverse-acting actuator, air pressure applied to the lower portion of the diaphragm case drives the actuator stem up. This action simultaneously compresses the actuator spring and opens a direct-acting valve.

Two concepts involving forces appear in the operation of the valve actuator in figure 3.17. The first is the force caused by air pressure bearing against the diaphragm, which is equal to the product of the area of the diaphragm and the pressure. The second is the force needed to compress the spring, which is equal to the product of the distance the spring is compressed times a spring constant.

If the spring rests against the diaphragm assembly with virtually no force, the actuator stem

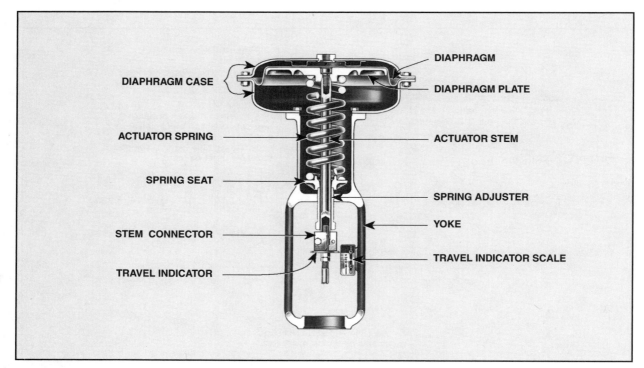

Figure 3.17   Diaphragm actuator (Courtesy Fisher Controls)

Final Control Elements

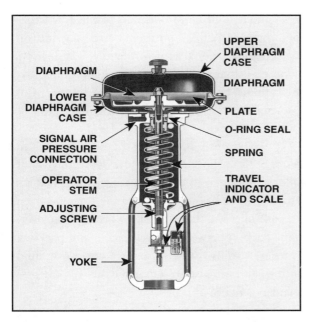

Figure 3.18   A spring-loaded, reverse-acting diaphragm actuator

will begin to move once gauge pressure exceeds zero. The springs in virtually all pneumatic actuators are compressed a small amount to hold the valve firmly open or closed with zero gauge pressure applied to the diaphragm. Therefore, a certain minimum pressure must be applied to the diaphragm before it can overcome the force caused by the slightly compressed spring.

Manufacturers and control system engineers have agreed on certain standards regarding the relationship between applied air pressure and actuator stem displacement. One standard concerns the range of air pressure values. The most popular range is from 3 to 15 psi (20 to 100 kPa), which means that when 3 psi (20 kPa) is applied to the diaphragm, the actuator stem just begins to move; when 15 psi (100 kPa) is applied, the stem is fully extended in its travel.

Pressure applied to a fixed area produces linear travel of the actuator stem. Pressure and travel are directly proportional to one another if no friction or other restriction exists. However, in practical actuator systems, friction occurs, which results in nonlinear travel with pressure. Where friction and nonlinear travel create a problem, designers use an auxiliary device, a *positioner*, with the actuator to ensure that the valve is positively positioned at the setting required by the process under control.

Air-Loaded Diaphragm Actuators

Replacing the spring in a spring-loaded diaphragm with air pressure creates an air-loaded diaphragm actuator (fig. 3.19). Air-loaded and spring-loaded diaphragm actuators are not directly interchangeable, and, unless it is used with a valve positioner, an air-loaded actuator is suitable only for on-off service. For on-off service, a constant air pressure of a few psi, depending on the particular type and use of the valve, is applied to one side or the other of the diaphragm, and air pressure from the controller is applied to the opposite side.

In many applications that require a valve positioner, the air-loaded diaphragm actuator is superior to the spring-loaded type because the force exerted by a spring is constant at any given degree of compression. Consequently, where the force required to position the valve stem varies widely for different positions, spring action is likely to be erratic. On the other hand, an air-loaded actuator with a proper positioner is capable of exerting a force

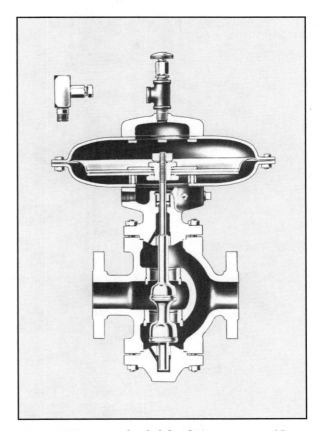

Figure 3.19   Air-loaded diaphragm actuator (Courtesy Fisher Controls)

in either direction of travel equal to the product of the effective area of the diaphragm and the applied air pressure.

Piston Actuators

Using pneumatically- or hydraulically-driven pistons to actuate valves has been around for a long time, and they are still in specialized use today. Piston actuators possess two advantages that make them desirable for some applications.

1. They are rugged and capable of handling high operating pressures. These features enable them to provide rapid response and enormous linear forces.
2. They can deliver large linear movement.

Figure 3.20A and B depicts two piston actuators. In figure 3.20A, a positioner assembly rides on top of a proportional actuator. Figure 3.20B shows an on-off actuator. The on-off actuator requires that the cylinder be loaded and unloaded by means of a solenoid valve, a pneumatic switching valve, or similar equipment.

A valve positioner fits on the valve actuator in figure 3.20A. It assures that valve stem movement follows the demands of the controller with great accuracy. The positioner amplifies the force between the controller and the actuator. Positioners help overcome friction between the valve stem and packing. They are also used when the force of flowing fluid on the valve plug causes unsatisfactory response by unaided actuators.

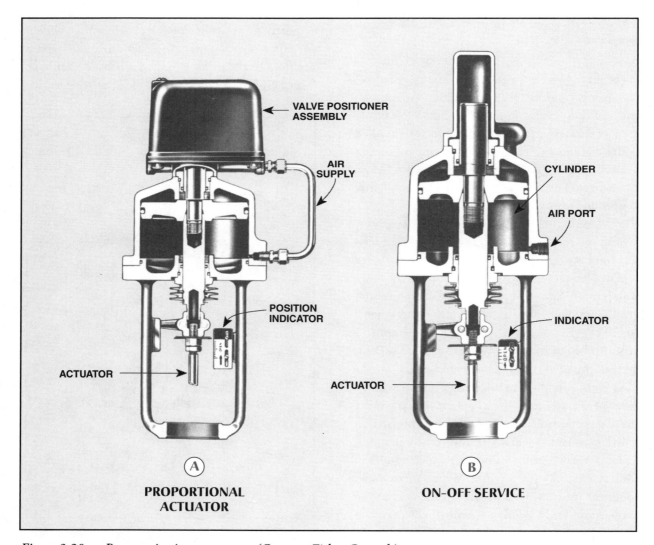

Figure 3.20   Pneumatic piston actuators (Courtesy Fisher Controls)

Final Control Elements

Air Supply to Pneumatic Actuators

Air for control purposes should be as free of moisture, oil, and contaminants as practicable. Sometimes, air to operate an actuator comes directly from the controller. In this case, a system of piping or tubing exists between the two components, and a filter is installed ahead of the controller.

Many control-valve actuators using positioners or volume boosters receive air from a source separate from that supplying the controller (fig. 3.21). In such instances, a filtering system is usually installed near the valve. A sump collects condensed moisture that can be drained out through a bottom petcock. The filter, which is often a part of a pressure-reducing regulator, prevents the passage of viscous oils and particles of dirt.

Electric Actuators

Two types of electrically powered devices that position final control elements are (1) solenoid-operated and (2) electric-motor-operated. Manufacturers usually provide the actuator and valve as an integrated unit.

Solenoid Actuators

Solenoid-operated valves are widely employed in automatic control systems where two-position, or on-off, flow control is required. They are often employed as fuel-flow control devices in automatic central heating systems and in industrial processes where on-off action is acceptable.

Although solenoid actuators are reliable and inexpensive to manufacture, they cannot operate large valves at even moderate pressures. Large valve operation requires a heavy-duty solenoid that consumes a considerable amount of power to obtain the required actuating force. Consequently, solenoid valves are not used to handle flow in 1-in. (25-mm) lines at 600 psi (4,000 kPa). On the other hand, solenoid valves are available for control in a 6-in. (150-mm) line at low pressures up to 10 psi (70 kPa).

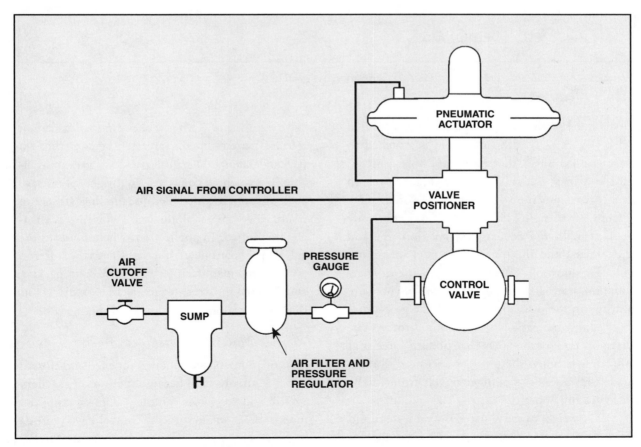

Figure 3.21   *Arrangement of a typical air supply for pneumatic actuators*

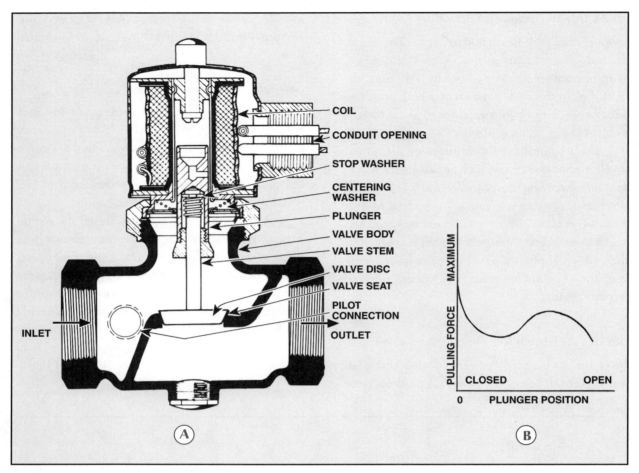

Figure 3.22   Solenoid actuator and valve (Courtesy Honeywell)

In the sectional view of a solenoid-operated valve (fig. 3.22A), the valve stem is attached to a soft iron plunger that rides in a hole centered on the axis of the solenoid coil. The valve is in deenergized position—that is, it is closed. Flow is from left to right. The valve disc is held firmly in its seat. In the deenergized position, the coil is energized and the plunger is under maximum pull. This position enables the actuator to overcome the seating force caused by the line pressure bearing on the disc. Figure 3.22B is a graph that plots plunger position versus pulling force. The curve on the graph shows that pulling force is at a maximum when the plunger is closed. Plunger forces are less as the plunger travels from totally closed to fully open.

The weight of the plunger, the valve stem, and the disc close the valve (see fig. 3.22A). Once the disc nears its seat, flow over the top of the disc snaps the valve tightly shut. This sort of closing action cannot tolerate friction from the valve stem's passing through packing material. Therefore, packless valves are available. The solenoid coil is in a fluid-tight cylinder, so packing is not required. Because this liner is made of nonmagnetic material, the unit's efficiency is virtually unaffected. In applications where fluids in the valve would corrode or otherwise harm the soft iron plunger, the manufacturer uses valve-stem packing and provides springs to overcome friction that resists the closing motion.

Electric-Motor-Operated Actuators

Electric motors have operated (opened and closed) valves for many years because they are an excellent way to control valves remotely. The exacting demands of automatic process control have resulted in the development of electric-motor actuators that are capable of any desired mode of control.

The motors used for powering actuators can be reversing or unidirectional types. They are available in a multitude of power and voltage ratings. A system of reduction gears usually connects the valve stems to the motors. Unfortunately, reduction gears decrease the speed with which a valve responds to a control signal. The time required for such actuators to move from one extreme position to another may be as little as 2 seconds or as much as 4 minutes.

Various devices can convert a reversible electric motor's rotary motion to linear motion for operating a valve. The main problem is controlling the action of the motor itself. One solution is to provide floating control of the motor (fig. 3.23). A mercury switch acts as a single-pole, double-throw switch capable of assuming a neutral or no-contact center position. A weak force, such as a Bourdon spring, can move the switch. If the instrument makes contact from C to L, the open winding of the motor is energized and the actuator continues to move toward its open position until the result of its control action or some other influence causes the contact between C and L to be broken. When contact is broken, the motor stops and the valve remains set until contact is made again from C to L or from C to H. If contact is made from C to H, a similar action occurs, except the actuator moves toward the closed position. As long as the mercury switch stays in its neutral position, the actuator does not move and the valve simply floats in its set position.

Electric-motor-operated actuators usually require a means for preventing overtravel of the valve stem or other device being positioned by the actuator. Limit switches that open the motor circuit at each extreme position of travel are in common use. Such switches may form a part of the actuator mechanism, or the valve stem may actuate a separate assembly. This type of floating control is not suitable for many applications because the amount of process deviation permitted by the floating band might be in excess of critical demands of the control.

A proportional control system permits specific positioning the valve in accord with the amount of deviation detected by the measuring means of the

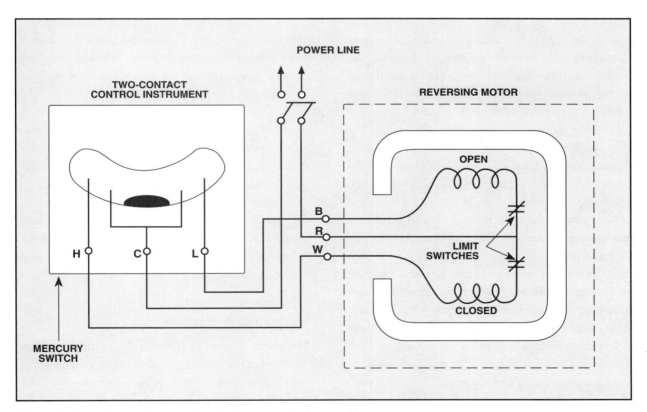

Figure 3.23   Circuit diagram for floating control of reversible motor

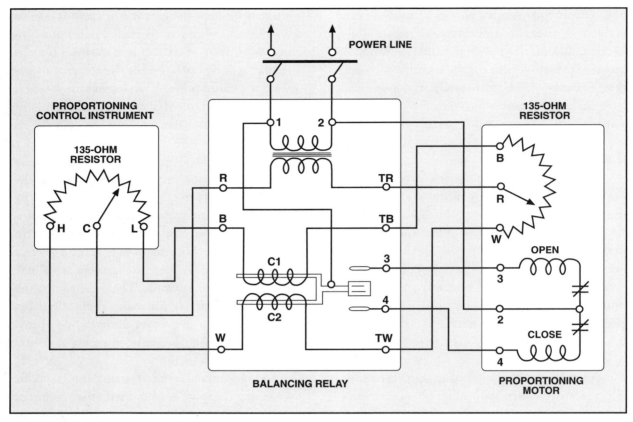

Figure 3.24 Proportional control system (Courtesy Honeywell)

system (fig. 3.24). This set up includes a reversible motor. The 135-ohm (Ω) resistor that forms a part of the proportional control instrument is positioned by the measuring means. The other 135-Ω resistor is directly connected to the actuator mechanism for positioning purposes. The position of the balancing relay controls the direction of the motor's motion. If more current flows in coil $C1$ than in $C2$, the center contact of the relay closes on relay contact 3, thus completing the motor circuit, and driving the actuator toward the open position. If coil $C2$ carries heavier current than $C1$, the relay operates to drive the actuator closed.

To better understand how a balancing relay works, note figure 3.25, which is a simplified version of the schematic shown in figure 3.24. With the sliding arms or movable contacts of the two resistors resting in their center positions, equal currents flow through coils $C1$ and $C2$, and the relay's center contact does not move. A decrease in the controlled variable (at left) drives the controller resistor's movable contact toward L. This movement adds resistance to the circuit containing $C2$ and subtracts a like amount from the circuit containing $C1$. Consequently, less current flows through $C2$ and more current flows through $C1$.

This unbalanced condition causes the relay to close the circuit to the motor and drive the actuator to the open position. The actuator-resistor movable contact begins traveling to a new position, simultaneously reducing the unbalanced condition caused by movement of the controller-resistor movable contact. Eventually, the currents in $C1$ and $C2$ become balanced and the motor stops. This action provides a linear relation between the controlled variable and the position of the actuator.

Hydraulic Actuators

Hydraulic systems in automatic process control are not used as much as electrical or pneumatic systems, but hydraulic systems have some advantages. The principal advantage lies in the fact that liquids are virtually incompressible. Because liquids are not

Final Control Elements

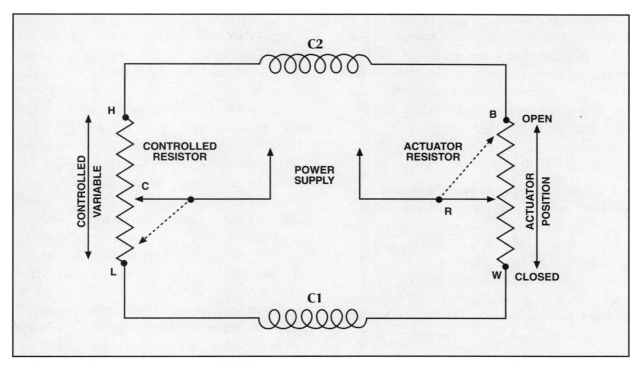

Figure 3.25   Balancing relay (Courtesy Honeywell)

compressible, they transmit pressure changes immediately. On the other hand, air is compressible. So, in a pneumatic system, a delay in transmission can result because it takes time for air to compress in the system.

A simple speed control for a steam turbine uses one type of hydraulic actuator (fig. 3.26). The system in the figure is in a balanced position. The steam valve is passing the correct quantity of steam to maintain the desired speed. The entire governor block (at left) is revolving. The two weights compress the spring as they revolve. The governor shaft protruding from the center of the spring moves up or down as the spring compresses or decompresses. The shaft's vertical motion is transmitted to the stationary shaft between *A* and *B* through the ball bearing at *A*. The rigid bar *BCD* pivots at *C*. The tension of the speed-adjustment spring keeps the rotating and stationary shafts firmly against the ball bearing.

If the turbine speeds up, the weights compress the governor spring and raise the rigid bar. The pilot valve piston moves upward and admits oil to the top of the actuator (servomotor) piston. At the same time, the drain opens below the actuator to allow oil to exit through the pilot valve. The actuator then

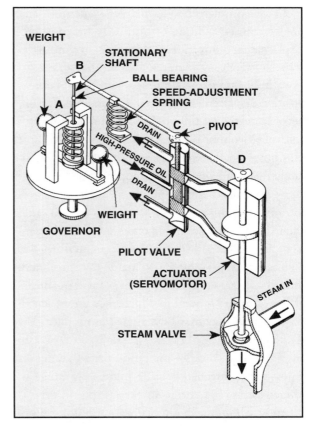

Figure 3.26   A speed control system for a steam turbine

moves in response to the hydraulic pressure and reduces the steam flow through the valve. As soon as the actuator piston begins its movement, the rigid bar repositions the pilot valve and shuts off the flow of oil to the actuator cylinder. If the turbine slows down, the actuator opens the steam valve to admit more steam to the turbine.

Combination Actuators

A hydraulic or pneumatic actuator becomes electrohydraulic or electropneumatic if electronic signals trigger the action of a hydraulic or pneumatic actuator. Electrohydraulic and electropneumatic actuators combine the advantages of using hydraulic or pneumatic pressure to drive the actuator mechanism with the ease of producing and transmitting electrical signals to activate the response of the actuator.

CONTROLLED-VOLUME PUMPS

A controlled-volume pump is one in which each cycle, or complete stroke, delivers a definite and predetermined volume of liquid. It is a positive-displacement pump. When used as a final control element, it usually contains refinements that set it apart from mud pumps, well pumps, and other familiar positive-displacement devices. The main difference between a regular positive-displacement pump and a controlled-volume pump is that a controlled-volume pump has smaller capacity and therefore can accurately meter the liquid it is pumping.

Controlled-volume pumps are used as injectors for forcing emulsion breakers, inhibitors, and other chemicals into process lines, tanks, and other vessels. They are capable of delivering enormous pressures—as high as 50,000 psi (350,000 kPa). On the other hand, they usually are employed to deliver small quantities of liquid to the process to which they are attached. For example, they may be designed to produce flow rates as low as 2 in.^{3} (33 cm^3) per day.

The liquid ends—the pump parts that deliver liquid—of controlled-volume pumps are made in several forms, but the reciprocating piston type is the most common. In a reciprocating piston pump, pistons move back and forth (reciprocate) in cylinders. Most controlled-volume pumps are provided with an adjustment that varies the length of the piston stroke to achieve a change in liquid volume delivered per stroke. Capacity is a function of displacement and speed, and displacement is the product of piston area and stroke.

Controlled-volume pumps are often used to inject chemicals into a pumping well (fig. 3.27). The injection pump has an adjustable stroke length and ratchet mechanism. An operating lever actuates the ratchet mechanism. The operating lever can be linked to the walking beam of the well pump. As each stroke of the well pump delivers a constant volume of oil to a flow line, the walking beam actuates the injection pump. Such a system makes it easy to vary the piston stroke of the injection pump to obtain the proper quantity of chemical additive. This setup is an open-loop system of control because the final product delivers no feedback. It is an example of one of the few practical applications of an open-loop system.

A combination open-and-closed-loop system (fig. 3.28) employs a controlled-volume pump. A variable-speed electric motor drives the pump, which two variables—rate of flow and pH—control. The rate of flow controls the motor speed within certain wide limits, allowing greater speeds with increasing flow values. The rate of flow also governs the injection of corrective chemical into the line, but no mutual adjustment exists between the correct amount of injection and the rate of flow. In other words, this part of the system is an open loop.

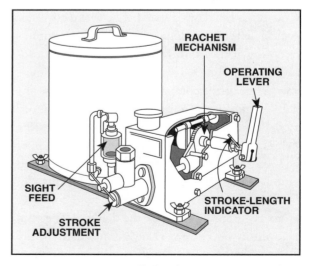

Figure 3.27   Controlled-volume pump as a final control element

Final Control Elements

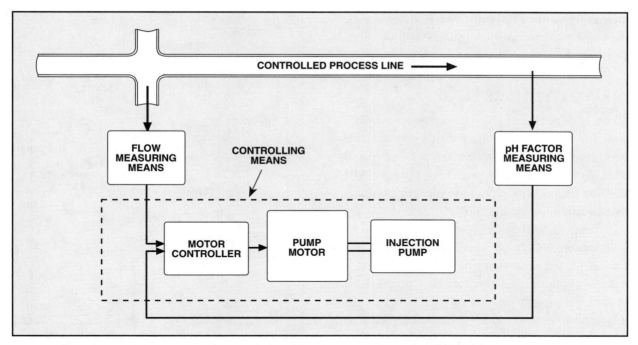

Figure 3.28   A combination open-loop and closed-loop system with a controlled-volume pump

A pH detector transmits corrective information to the controller. The controller, in turn, adjusts the speed of the pump, which injects chemical to acquire the proper pH. This part of the system is a closed loop.

In the system shown in figure 3.28, regulating the motor speed varies the capacity of the pump. Accurate injection control can be accomplished equally well by having the feedback loop actuate a mechanism that varies the piston-stroke length, leaving the open-loop portion of the system to control the motor speed.

An actuator powered by an air signal whose pressure is established by the measuring means can automatically accomplish stroke length variation. Some reciprocating pumps are air- or gas-operated (fig. 3.29). In such pumps, a pneumatic positioner that limits the piston travel partially controls the capacity.

Capacity regulation by pump-speed control takes on one of three forms.

1. The pump has a fixed stroke length and is driven by a variable-speed rotary motor.
2. The pump has a fixed stroke length and is driven by a reciprocating pneumatic or hydraulic actuator.
3. A constant-speed electric motor drives the pump whose speed is regulated by mechanical speed drives installed between the motor and the pump.

Each type of regulation is useful and practical for specific purposes. Bear in mind, however, that a variable-speed motor is not satisfactory for very low speed ranges. For example, its delivery would probably be erratic at low speed ranges if it was designed to

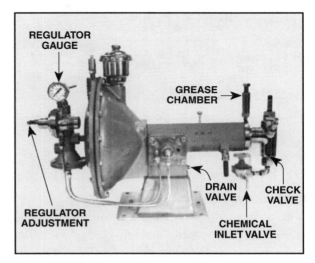

Figure 3.29   Gas-powered injection pump (Courtesy Texsteam)

cover a range of 1 to 100 strokes per min. Pneumatic and hydraulic drives are better suited to applications requiring about 1 stroke per min or less.

VARIABLE-VOLUME PUMPS

When an electric variable-speed drive controls a centrifugal pump, the pump can move variable amounts of liquids. Pump flow rate is proportional to the square of the input speed. Such pumps can be used as a final control element in pipeline pumping systems or similar applications.

Figure 3.30 shows a pipeline pump driven by an induction AC motor. The desired set point (the flow rate) is fed to an electronic flow controller that provides a control signal to a variable speed, variable frequency motor. A flow transmitter monitors the output flow rate and provides electrical feedback to the flow controller for comparison with the desired set point. If the flow rate is below the desired set point, the controller automatically adjusts the motor's speed to increase the flow rate until the flow transmitter's output corresponds to the desired set point.

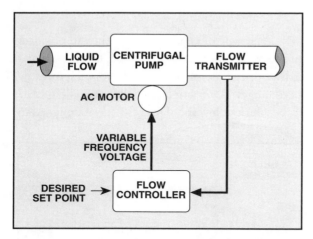

Figure 3.30   Variable frequency, variable speed pump drive

OTHER FINAL CONTROL ELEMENTS

In addition to valves and pumps, louvers and dampeners are sometimes used as final control elements in an automatic control system. Electric switches also serve as final control elements (fig. 3.31). In

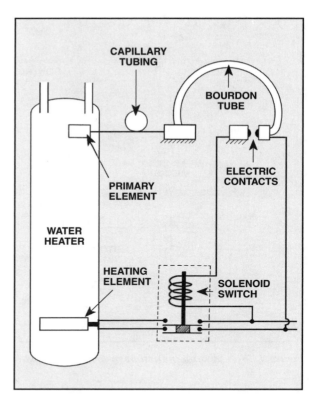

Figure 3.31   An electric switch as the final control element in an electric hot-water system

the figure, an automatic hot-water system's final control element is a simple, heavy-duty, and remote-controlled switch that controls current to the heating element of the hot-water heater.

SUMMARY

Final control elements are available in many forms in control systems. The type of final control element depends on the desired control and process. Most final control elements are valves, but controlled-volume pumps and variable-volume pumps are also employed.

REVIEW EXERCISE

Define the following:
1. final control element
2. positioner
3. actuator
4. variable speed drive

4
Pneumatic Automatic Controls

Automatic control of processes has evolved from simple control systems to the complex systems in today's plants and facilities. Electronic controls, sensors, and measuring devices are significant developments that have advanced automation. We now can set adjustments with dials and digital switches. We can push a start button and watch a system perform its function completely and automatically without the intervention of an operator. Microprocessors have not only put personal computers within reach of almost everyone, but also they have taken instrumentation processes to a new level.

This chapter reviews pneumatic concepts that many facilities still employ. Although many facilities use electronic automatic controls, learning about pneumatic controls leads to a better understanding of electronic controls. Because electronic controls form a significant part of process instrumentation and automatic control, they are covered in chapter 5.

PNEUMATIC CONTROLS

Automatic regulators and controls perform self-correcting functions—that is, once operators correctly set the automatic controls, they do not have to do anything further to control the system. Examples of automatic regulators include automobile speed controls, air conditioner thermostats, and oven temperature regulators—devices we use daily. We set them and forget them, as the saying goes. The system does the rest.

Regulating functions use devices that are hydraulically, pneumatically, or electrically controlled. This chapter covers pneumatic controls.

Pressure Regulators

Pneumatic devices depend on pressure from an air supply. For a pneumatic device to perform properly, the supply air pressure must be held steady at the required value. In short, the air pressure must be regulated. Thus, it is important to understand how a pressure regulator adjusts and holds pressure at a constant value. Let's say we have a source of air pressure delivering 100 psi (700 kPa). This pressure is too high for most control devices. So, a device is needed to reduce this pressure to an acceptable level. Moreover, once the device reduces the pressure, it must also regulate it—that is, maintain the reduced pressure at a constant value.

Weight-Loaded Regulators

A weight-loaded regulator (fig. 4.1) is a self-contained device that reduces and regulates pressure at its output. It is a double-ported valve with a poppet-type plug that a diaphragm actuates. The diaphragm also supports a weight, which is sized for the particular regulator. A flexible diaphragm isolates the weight

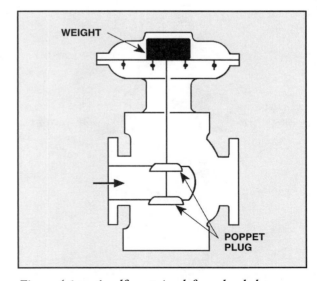

Figure 4.1 *A self-contained force-loaded pressure regulator*

from the pressure inside the valve assembly. The weight rests on the top of the diaphragm; the bottom of the diaphragm is open to the pressure inside the valve's outlet side. The weight tends to force the valve plug open by moving the diaphragm down. As pressure builds on the outlet side of the valve, it tends to lift the weight and close the valve. By choosing the correct value of mass to bear down on the diaphragm, the desired pressure can be obtained at the valve outlet. This type of regulator is self-contained because the controlling weight and the diaphragm pressure connections are internal features of the unit.

Spring-Loaded Regulators

A self-contained spring-loaded regulator (fig. 4.2) is similar to a weight-loaded regulator, but a spring under compression replaces the weight. Spring-loaded regulators are popular because they are simple, light in weight, require little maintenance, and are reliable. The compressed spring tends to push the valve plug open, while pressure on the outlet side of the unit exerts an opposing force against the lower side of the diaphragm. The valve plug settles at an opening determined by the amount of pressure required to balance the force of the compressed spring. The outlet pressure is adjusted by varying the tension of the spring with the spring-compressing screw.

Pneumatic Controllers

Closed-loop controllers have several functions in common. Functions include maintaining the desired set point, sensing the process variable with a primary element, establishing proper signal levels with a converter, monitoring the process with a controller, and performing the actual process control with a final control element.

Pneumatic controllers consist of a constant and regulated air supply, a fixed orifice, and a variable orifice. Designers of pneumatic controllers must determine the variable orifice function to control the process variable. The following are examples of how this control is achieved.

Fixed and Variable Orifices

A basic pneumatic controller uses a regulator, a fixed orifice, a pressure gauge, and a variable orifice in the form of a needle valve (fig. 4.3). In this setup, a regulator reduces 100-psi (700-kPa) air pressure to 20 psi (140 kPa) and maintains it to the left of the fixed orifice. If the variable orifice (the needle valve) is tightly closed, the pressure to the right of the fixed orifice is also 20 psi (140 kPa), as indicated by the pressure gauge.

If the variable orifice is opened two to three times larger than the fixed orifice, air escapes so fast from the larger opening that no pressure builds up between the two orifices. Consequently, the pressure gauge indicates zero, or near zero.

If a diaphragm actuator is substituted for the pressure gauge, changing the opening of the variable orifice causes the actuator to follow the pressure changes. Unless the fixed orifice is quite small, such a system wastes a large volume of air. Therefore, a small orifice, about 0.01 in. (0.3 mm) in diameter, is installed, which institutes a variable, or modulating, method of control. Because a variable orifice in the form of a needle valve is not readily adaptable to the requirements of automatic control, another means is usually employed.

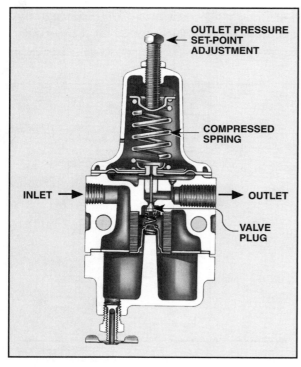

Figure 4.2   *A self-contained spring-loaded regulator*

Pneumatic Automatic Controls

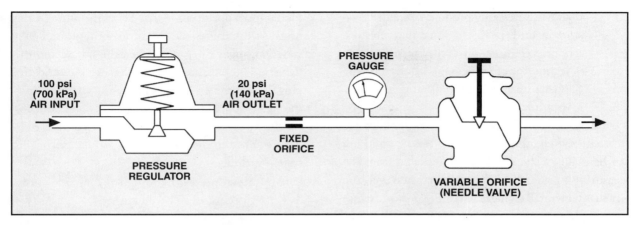

Figure 4.3 A fixed orifice as part of a pneumatic controller

Nozzle and Flapper

A better system utilizes a 0.01-in. (0.3-mm) fixed orifice and a 0.025-in. (0.6-mm) nozzle for the variable orifice (fig. 4.4). As the regulator feeds constant air pressure to the fixed orifice and to the unrestricted nozzle, air escapes through the nozzle at a rate that prevents the buildup of any appreciable pressure between the fixed orifice and the nozzle.

A light strip of metal on a pivot serves as a flapper, or baffle, near the nozzle opening. The flapper, or baffle, restricts the nozzle opening, which increases the pressure between the orifice and the nozzle. This pressure increase operates the valve actuator and the pressure gauge indicates a pressure near the maximum available from the regulator. If the flapper is retracted a short distance from the nozzle, the pressure decreases, which the gauge shows.

Keep in mind two major points about the flapper and its effect on the controlling pressure.

1. A small force is required to position the flapper and produce a full range of pressure values for control. However, the force required is not a linear function, largely because of the opposing force of the jet of air as the flapper approaches the nozzle. In other words, applying twice the force to the flapper

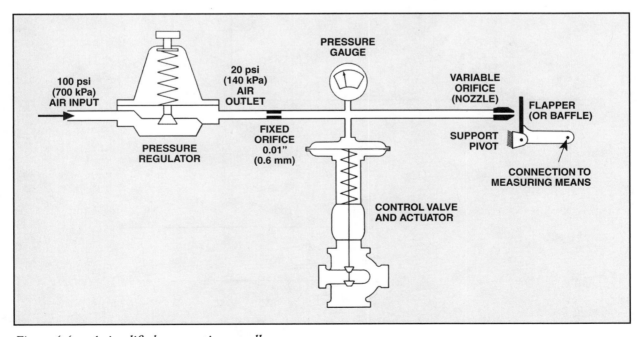

Figure 4.4 A simplified pneumatic controller

does not necessarily produce twice the pressure; instead, twice the force may produce ten, or even more, times the pressure.
2. The amount of movement required by the flapper to produce the full range of values is very small.

Because applying only a small positioning force creates broad changes in control pressure, good use can be made of the small forces available from the measuring means. At the same time, however, the fact that very small changes in flapper position cause large pressure changes also presents problems that must be overcome. For example, only 0.01 inch (0.3 mm) of flapper movement produces enough change in nozzle back-pressure to drive the actuator from one extreme to the other.

Controller Improvements

A pneumatic controller using a fixed and a variable orifice can be improved by installing a set-point device. For example, in a liquid-heating system (fig. 4.5), a set-point adjusting screw and a Bourdon spring refine the flapper's operation. A rise in temperature in the heating tank causes the Bourdon spring to flex in a way that opens the gap between the nozzle and the flapper. The pressure to the valve declines and the plug slowly begins to close. The closing valve decreases the steam supply to the heating tank. The device is essentially an on-off controller, but it permits wide departures from the set point. Disadvantages of this system include—

- Slow bleed-off through the nozzle causes the control to be sluggish.
- Severe departures from the set point occur.
- A nonlinear relationship exists between the nozzle-flapper clearance and nozzle back-pressure. This nonlinear characteristic can make control of the system difficult over a wide operating range.

Adding an air relay valve to the nozzle-and-flapper system significantly improves the system (fig. 4.6). A small volume of air actuates the air relay valve, which, in turn, controls large-volume air

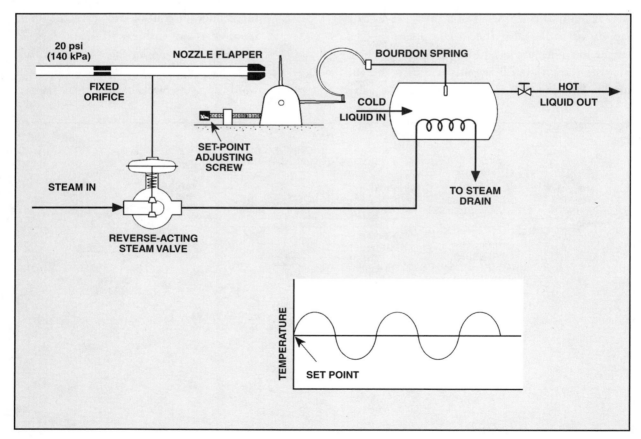

Figure 4.5   A two-position (on-off) controller in a liquid-heating system

Pneumatic Automatic Controls

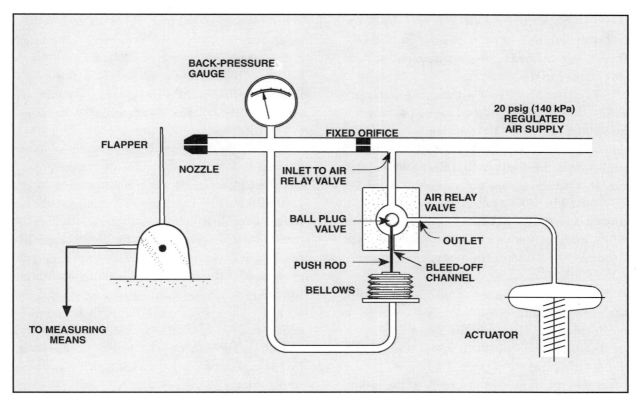

Figure 4.6 Using an air relay to provide linear control

impulses to drive the actuator. To reduce the effects of nonlinearity, a nearly linear portion of the operating curve for nozzle-flapper clearance and nozzle back-pressure is selected. For example, the range represented by 2- and 4-psi (15- and 30-kPa) back-pressure is an acceptable linearity and any problem of flapper displacement caused by jet action at high back-pressure is eliminated.

The air relay must be capable of 2- to 4-psi (15- and 30-kPa) impulses of back-pressure to provide linear control of the 3 to 15 psi (20 to 100 kPa) needed to operate the actuator. Instead of a diaphragm case, the air relay uses a small-volume bellows. Nozzle back-pressure actuates the bellows. A push rod, which is attached to the free end of the bellows, operates a special ball-plug valve. This special valve has two seats, and the small bellows is capable of driving the ball plug from one seat to the other.

Full 20-psi (140-kPa) regulated air pressure is applied to the inlet of the air relay valve. The relay valve's outlet is connected to the actuator diaphragm case. With the ball plug positioned between its two seats, some air is applied to the push rod. Careful design of the valve achieves a proportional relationship between nozzle back-pressure and the 3- to 15-psi (20- to 100-kPa) pressure needed for operating the actuator. In summary, this relay is crude but effective. Other and better relays exist, but substituting this relay device into the control shown in figure 4.5 results in a fast-acting on-off controller.

Proportional Controllers

A proportional controller incorporating an air relay valve improves control. However, because only 0.002 in. (0.05 mm) of flapper movement is present to create a full range of pressure values from the air relay in figure 4.6, mechanical linkages are ruled out unless ¼-in. (6-mm) movement in the linkage can be achieved.

To achieve this increase in linkage movement and actuate the flapper, two methods may be considered: (1) the action of the measuring means, and (2) the related action of the air relay output pressure. The measuring means displaces the flapper and a bellows-spring arrangement, which operates the output pressure from the air relay and drives the flapper back to near its previous position (fig. 4.7). This action provides a form of negative feedback and adds stability to the system.

Keep in mind that all moving parts are restricted to motion in the plane of the diagram shown in figure 4.7. The levers and links are designed to prevent any component of motion perpendicular to the diagram. Rod B is a stiff member capable of vertical motion only. Proportioning lever P is a curved bar whose radius of curvature is equal to the length of link S. Lever P is pivoted at three points, but the location of the pivot point formed with link S is adjustable from one end of lever P to the other.

The bellows-spring assembly consists of a pair of bellows arranged concentrically and fitted to common end plates, thus forming a single unit. The bellows assembly is fitted into a sturdy metal cylinder and compresses a special spring. A setscrew is provided for adjusting the tension of the spring.

When assembled, the unit comprises two airtight compartments: one is the bellows and the other is the metal cylinder surrounding the spring and bellows. The cylinder is vented. Rod B moves a little more than ¼ in. (6 mm) when the pressure applied to the bellows is varied from 3 to 15 psi (20 to 100 kPa).

With the flapper midway between its limits of travel and positioned well away from the nozzle, nozzle back-pressure is at a minimum when 20-psi (140-kPa) regulated air pressure is applied to the air relay. Under these conditions, the air relay rapidly transmits air to the bellows. As the bellows expands, rod B pulls down on lever P, causing the flapper to move toward the nozzle. The resulting back-pressure ultimately positions the ball-plug valve of the air relay. The system quickly reaches equilibrium with a nozzle-flapper clearance of about 0.004 in (0.1 mm). The system reacts quickly to changes in this clearance and repositions the actuator to a new equilibrium position.

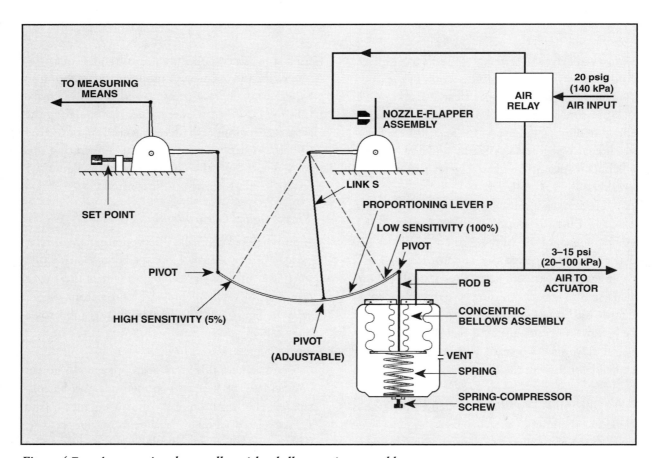

Figure 4.7 A proportional controller with a bellows-spring assembly

Link *S* can be positioned at any point along lever *P*, and the action produced at several points can be analyzed. With link *S* attached to the left end of lever *P*, movement of the measuring means linkage produces a practically equal movement of the flapper, while movement of rod *B* has little or no effect on flapper position. The controller is extremely sensitive to changes in the controlled variable under these conditions, because only 0.002 in. (0.05 mm) of net flapper movement can cause a full range of output from the air relay. This effect produces an on-off controller action.

If lever *P* is positioned to obtain full movement of rod *B*, ¼ in. (6 mm) of flapper travel occurs. The result is that the full range of the measuring means is required to produce a full range of output from the air relay. Such an arrangement provides 100 percent throttling range. Placing link *S* at other positions along lever *P* produces various throttling results.

Controller Operation

We can use the controller in figure 4.7 to maintain the temperature of a liquid at 150°C or 302°F (fig. 4.8). The steam supply and energy requirements of the process are such that the steam control valve operates at 50 percent of maximum flow. The measuring means indicates a scale range of 100°C to 200°C (212°F to 392°F), and this range produces ¼ in. (6 mm) of linear motion in the measuring means linkage. Link *S* is set at full throttling range (see fig. 4.7).

Starting with a controller set point of 150°C (302°F), the system is operating in a stable mode with normal load. A sudden change in demand for a larger quantity of hot liquid upsets the energy balance and the amount of steam currently flowing cannot maintain the desired temperature with the increased outflow of liquid. Consequently, the temperature falls below 150°C (302°F). The measuring means detects this fall and causes the flapper to move. Nozzle back-pressure falls, which increases the air relay output and forces the reverse-acting steam valve to open wider. Simultaneously, the controller bellows is actuated to reposition the flapper. The system soon settles to a new temperature a little below the desired value of 150°C (302°F). It remains stable at this new set of conditions as long as the load does not change. This action emphasizes a characteristic of all proportional controllers: to correct for a change in load on the process, a proportional shift in the value of the controlled variable must take place.

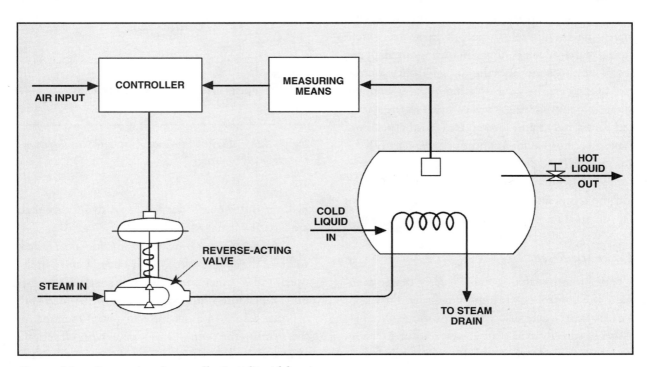

Figure 4.8   Proportional controller in a liquid-heating system

If it is important to maintain the controlled variable at a set point—it usually is—further improvements must be made in the controller.

If we shift link S to the 50 percent throttling range point on lever P (see fig. 4.7), only half the full range of the measuring means is required to throttle the steam valve in figure 4.8 from fully open to fully closed. With the same set of conditions, the controller is now capable of maintaining the controlled variable within a band only half as broad as the 100 percent throttling range. Thus, the offset from the set point is only half as great as before. However, offset is undesirable and needs to be corrected. Further adjustment of link S, pushing it ever closer to its limit at the left side, is not the answer. Adjusting link S to the left creates an on-off controller because of the tight operating band. Figure 4.9 shows the various offsets created by different positions of link S. These offsets are often referred to as an error in the control.

Mechanical Reset Adjustments

To correct the situation in the example temperature controller—to reduce the error to zero—first note that the problem is caused by the control valve's not passing enough steam to counteract the increased volume of fluid in the tank. The solution lies in increasing the air relay's output pressure, which increases the valve opening. Increasing the flapper-nozzle clearance increases the air relay's output pressure to increase the valve opening. Also, tightening the spring-compressing screw compresses the spring and forces rod B and lever P back to the position they had at normal load conditions (see fig. 4.7). As this change is made, the air relay's output pressure immediately increases, the steam valve opens wider, and the temperature rises to the desired set point. This manual reset device solves the problem.

Automatic Reset

A better solution than a manual reset device is to make the controller work automatically. First step is to charge the cylindrical bellows-spring container with pressure from the air relay's output pressure (fig. 4.10). This pressure adds force in the direction of the spring's force. Applying equal pressure to the

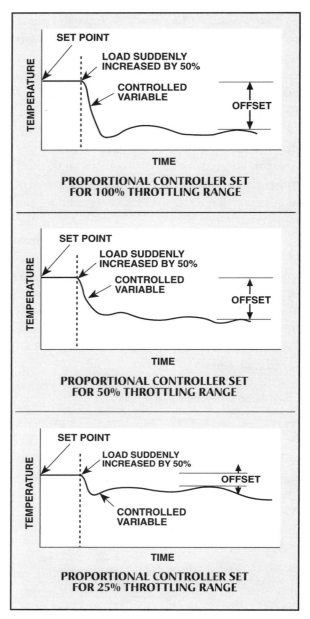

Figure 4.9   Performance of controller at different throttle range settings

bellows and the cylinder results in two forces that are virtually equal and opposite.

Valves X and Y provide the adjustments needed to make the system work. With valves X and Y fully open, rod B remains virtually stationary at all relay output pressures. Equal pressure in the bellows and cylinder results in a net force of zero. Operated in this manner, the controller produces an on-off action. However, if valve Y is closed until the charging rate of the cylinder lags behind the charging rate of the

Pneumatic Automatic Controls

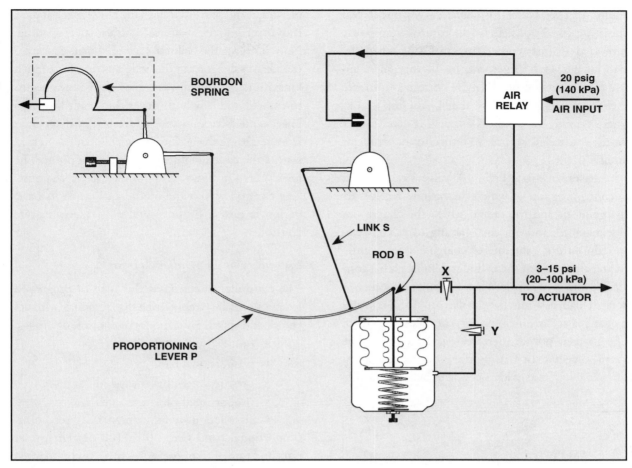

Figure 4.10 Proportional controller with automatic reset

bellows, rod B moves link S to move the flapper to a position that corrects the system to the temperature set point.

By adjusting valve Y to slow the charging rate of the cylinder, the system maintains the temperature set point. First, assume that the system is running smoothly at normal load and temperature. Suddenly, the load (liquid outflow) increases by 50 percent. The measuring means detects the temperature change and causes the flapper-nozzle clearance to broaden. Wider nozzle clearance opens the control valve wider, and the bellows of the controller expands rapidly, causing lever P to move and narrow nozzle-flapper clearance. The cylinder slowly charges with air and drives the bellows and rod B toward their midpoint of travel, which causes nozzle-flapper clearance to broaden again.

The system eventually settles, meets the new load requirements, and maintains liquid temperature at the set point. The action caused by the slow charging of the cylinder tends to increase the flapper-nozzle clearance and thus keep the steam valve sufficiently open. At the same time, the heating effect of the steam is reflected in increased liquid temperature and in the efforts of the measuring means to reduce the flapper-nozzle clearance. So, the system can only settle at the set point. This modification to the controller creates a proportional-plus-reset controller. The combined proportional and integral (PI) elements in a control system, such as the reset or integral function, return the controlled variable to the desired set point at regular intervals to eliminate error in the system.

Rate of Response Adjustment

Some processes require that corrective actions to reduce error from the set point be performed rapidly.

Figure 4.11 shows performance curves under two operating conditions. In both, controller response is plotted against temperature and time when the load increases by 50 percent. In the top graph, no rate correction is applied. In the bottom graph, rate correction is applied. Note that in the top graph, where no rate correction is applied, it takes longer for the controlled variable to return to the temperature set point.

Refer back to figure 4.10. Valve *X* can be used to control the rate at which the bellows receives its charge of air and the rate at which the charge can be exhausted. Indeed, the opening of valve *X* can be reduced until the rate of charge is only slightly greater than that of the cylinder. Under such a condition, and with the process running smoothly at normal load, assume that the demand for hot liquid is changed to require a 50-percent increase in load. The air-relay output increases rapidly, opening the steam valve widely. During the previous operation, where valve *X* was wide open, the bellows quickly expanded and narrowed the flapper-nozzle clearance, thus throttling the steam valve to a smaller opening. Now, however, the bellows is retarded in its action, so the steam valve stays at the wide position for a much longer time. Eventually, of course, the bellows begins to expand and slowly throttles the steam flow to a lower value. Meantime, the added energy that comes from retarding the action of the bellows, has helped restore the controlled variable to the set point. The lower curve in figure 4.11 shows a faster response. The controller with the modifi-cations to its basic design is now a proportional-plus-reset-plus-rate controller.

Summary of Controller Action

The controller discussed in this manual progressed from a sluggish system to one that responds with accuracy and speed. Four modes of controlling are—
- on-off (two position),
- proportional band,
- proportional-plus-reset, and
- proportional-plus-reset-plus-rate.

Controllers have adjustments for set point, proportional band (from 0 to 100 percent), reset rate, and rate of response. Also, they can be adapted to control other process variables such as rate of flow, pressure, and liquid level. The various modes available in the controller can be combined as needed to achieve the desired system results. A review of these modes follows. The modes apply to the liquid heating system discussed earlier.

Proportional Action

Proportional action positions the valve in proportion to change in the controlled variable. If a 100-percent proportional band is used—that is, full movement of the measuring means linkage produces exactly full movement of the valve plug—the motion of the valve plug is directly proportional to the deviation of the controlled variable. If a 50-percent proportional band is used, a direct proportion between deviation of the controlled variable and the motion of the valve plug does not exist, because, in a 50-percent proportional band, half the range of the measuring means linkage can produce full operation of the valve. The relationship between controlled-variable and valve-stem motion is always a linear, or straight-line, function. The

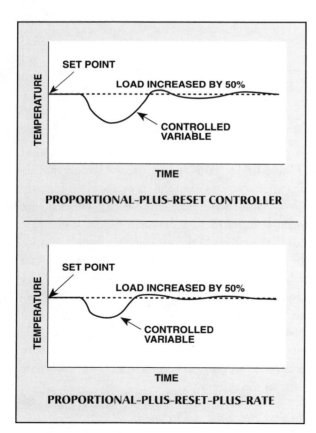

Figure 4.11   Performance curves for multi-mode controller action

change in process input energy caused by proportional action is sustained only as long as the controlled variable is displaced from the set point.

Reset Action

The energy change that results from proportional action is only a temporary change when it is followed by reset action. The reset action positions the valve at a rate that is proportional to the change in the controlled variable. The input energy change caused by reset action, unlike that in proportional action, is a permanent one.

Note that once the controlled variable is restored to the set point by reset action, the left end of proportioning lever P (see fig. 4.10) is again in its normal position. Normal position is the position lever P had before the load change occurred. This action eliminates the initial input energy change caused by proportional action.

Rate of Change

The positioning of the valve by rate response is an action that is proportional to the rate of change in the controlled variable. When the load suddenly increases, for example, the controlled variable begins a rapid departure from the set point. Its rate of change is at a maximum at this time because as yet no corrective measures are active in the process.

The rapid rate of change in the controlled variable causes the valve to open wide, because needle valve X (see fig. 4.10) prevents the controller bellows from being rapidly charged. Slow charging of the controller bellows tends to narrow nozzle-flapper clearance. Eventually, as the controller bellows begins to expand, the rate action narrows nozzle-flapper clearance.

Once the controlled variable has reached its maximum deviation, a moment occurs when the rate of change is zero. At this same moment, the valve-stem displacement caused by the rate response is also zero, because it is proportional to the rate of change in the controlled variable. After this pause, the controlled variable begins changing again, seeking the set point. In this case, the rate of change is reversed. As a result, the component of valve-stem displacement due to rate response actually becomes less than that existing at the set point.

The curves in figure 4.12 show several modes of operation when a sudden load change occurs.

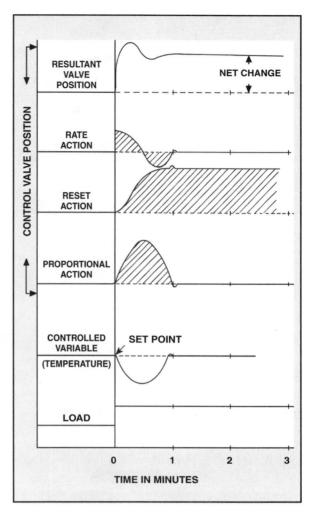

Figure 4.12 Response curves of different modes following quick change in load

Although proportional and rate action resist change and contribute to the system's stability, neither is responsible for any final correction. Reset action accounts for the change in energy input to compensate for load change. The shaded areas in the figure represent the input energy change caused by each mode of the controller.

COMMERCIAL PNEUMATIC CONTROLLERS

Pneumatic controllers for commercial applications usually have proportional-plus-reset-plus-rate capability. The controllers have air relays, bellows, orifices, baffles, and components similar to those previously discussed.

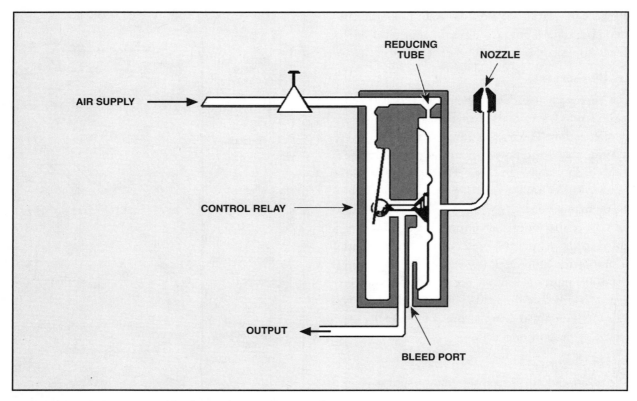

Figure 4.13   Continuous-bleed air relay (Courtesy Foxboro)

Air relays are generally one of two types: continuous bleed and nonbleed. In a continuous bleed relay (fig. 4.13), air is constantly lost from the relay unless nozzle back-pressure is at either a maximum or minimum. A nonbleed relay (fig. 4.14) allows a loss of air only when it is necessary to bleed off the pressure applied to the valve actuator.

Figure 4.15A shows a Model 40 pneumatic controller manufactured by Foxboro. Figure 4.15B is a schematic of the controller. The controller provides

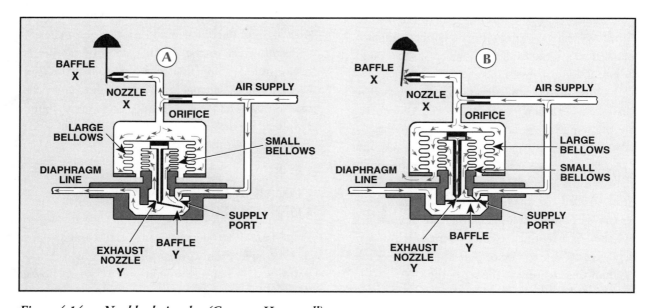

Figure 4.14   Nonbleed air relay (Courtesy Honeywell)

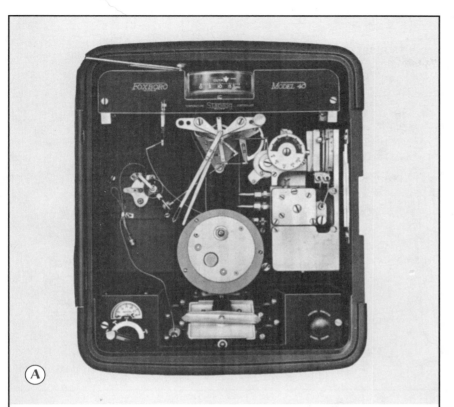

Figure 4.15 Pneumatic controller, Model 40 (Courtesy Foxboro)

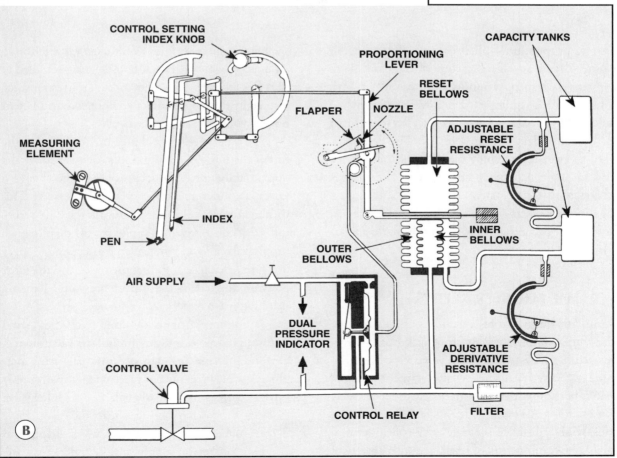

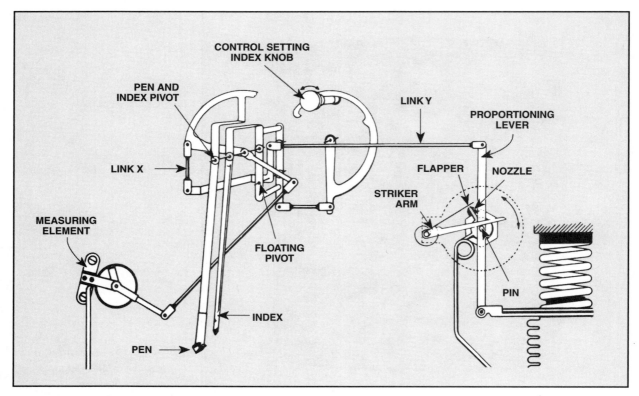

Figure 4.16 Flapper-nozzle assembly, Model 40 (Courtesy Foxboro)

on-off, proportional-plus-reset-plus-rate action. Figure 4.16 is an enlarged view of the controller's flapper-nozzle proportioning-lever assembly. This controller allows unimpeded air flow to and from the inner bellows (see fig. 4.15A), which is a refinement to the controllers discussed earlier. The bellows is sized to exert a force only one-fifth of the exterior unit for a given pressure application. The bellows adds stability to the system. It keeps the valve actuator from responding violently should a mechanical disturbance occur that could upset the nozzle-flapper relation.

VOLUME BOOSTER RELAYS

Pneumatic output from controllers usually operate diaphragm motor valves. An air relay is usually incorporated in the system to speed valve response. Also, some systems require a faster valve-actuator response than is possible with an ordinary controller-air-relay combination. An example is an installation where a considerable distance separates the controller and valve. The combination of resistance and capacity of the long connecting lines would slow valve action to an unacceptable degree. A booster relay installed at the valve location overcomes most of the response lag created by the long connecting lines and insufficient output volume.

In figure 4.17, signal pressure is applied at the top of the booster relay and fills cavity A above the upper diaphragm B. Equal pressure applied beneath lower diaphragm C opposes signal pressure. This combination is the output pressure to the valve actuator. With equal pressures applied to the diaphragms, supply valve D and relief valve E are seated. Supply pressure tends to seat the supply valve, and the same pressure applied to the top of pressure-balancing diaphragm F balances this tendency.

Output pressure is admitted to cavities G and H through hole I, midway on hollow valve stem J. Output pressure beneath pressure-balancing diaphragm F balances output pressure on top of supply valve D. Forces are similarly balanced on relief valve E. Thus, the force of spring K seats the valves.

A signal pressure increase causes diaphragm assembly L to unseat supply valve D and increase the

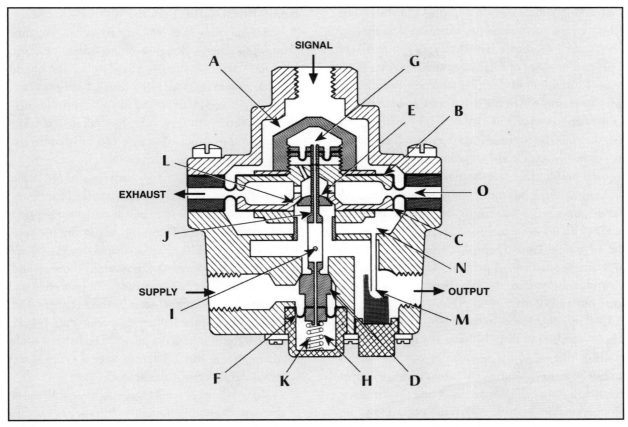

Figure 4.17   A booster relay, used to overcome response lag due to long connecting lines between controller and valve

pressure in the output line. The air bleeds through pressure-balancing tube M in the output line into lower cavity N. Pressure in lower cavity N brings diaphragm assembly L to its balanced position and seats supply valve D.

A decrease in signal pressure causes the pressure in lower cavity N to raise diaphragm assembly L and unseat relief valve E. Relief valve E and port O vent excess pressure until pressure balance is regained. Diaphragm assembly L then resumes its balanced position and seats relief valve E.

VALVE POSITIONERS

Volume booster relays improve the response speed of valve actuators, but they do not improve valve positioning accuracy. Thus, manufacturers offer valve positioners. Valve positioners accurately position valves. What is more, they can force valves with sticky stems or other problems to assume a precise position.

Manufacturers have developed controllers capable of responding to the smallest variations in the controlled variable. The immediate manifestation of this response is a change in the air relay output pressure, and this pressure change repositions the control valve plug. It is important to realize that a controller cannot be arbitrarily applied to solve every control problem. The controller may fail to stabilize the process within the narrow band of values needed for satisfactory performance.

Although the controller is working properly, a stubborn control valve can fail to respond to pneumatic nudges that are below a certain level of force. Further, valve characteristics can upset the control pattern—for example, the effects of flow through the valve and static and sliding friction between the valve stem and packing material. The coefficient of static friction is usually significantly greater than the coefficient of sliding friction. Thus, a greater force is required to unstick the valve stem, but once it begins moving, this same force is capable of accelerating it.

Such action can cause overshooting and instability. However, a more serious problem is that static friction causes a dead spot. The dead spot can represent 5 percent or more of total valve stem travel.

To understand this phenomenon better, consider an example. Assume that a control valve having a total stem travel of 1 in. (25 mm) is being used, and that the objective is to maintain the controlled variable to within 1 percent of the set point. This requirement requires positioning the valve plug to within 0.01 in. (0.25 mm) of the optimum position. Since changing the diaphragm pressure from 3 to 15 psi (20 to 100 kPa), which is a net change of 12 psi or 80 kPa, moves the valve stem 1 in. (25 mm), a 1 percent change in pressure represents 0.12 psi (0.83 kPa). To meet the control requirements, the valve must respond to this small amount of diaphragm pressure.

If inherent forces in the valve—friction for example—are greater than the force tending to properly position the valve, the fineness of control needed cannot be achieved. A ready solution is to use a larger diaphragm, since the force exerted by a diaphragm is proportional to the square of its diameter. However, because of the increased size of such a unit and the greater volume of air needed for operating it, other methods have been developed.

Basic Pneumatic Positioner

A better solution is to amplify the small pressure changes to a high enough value to overcome static friction. A basic system uses a fixed orifice and nozzle and diaphragm pressure using nozzle back-pressure. This system is similar to the on-off controller discussed earlier. Although it is slow because it takes a long time to build up and bleed off diaphragm pressure, it does accurately position the valve.

The basic positioning system consists of a baffle plate (a form of flapper), which is pivoted for rotary motion at the actuator stem and at the bellows-spring assembly (fig. 4.18). The pivot pin at the bellows-spring assembly slides in an elongated slot. As the actuator stem and bellows-spring assembly move up and down, the distance between the pivot pins changes. The elongated slot compensates for these changes. The curved portion of the baffle helps maintain a proper tangential relation with the nozzle. The baffle is resilient enough to withstand flexing when it occasionally is driven tightly against the nozzle.

To see how it works, assume that full travel of the stem causes nozzle-baffle clearance to change 0.5 in. (12 mm). Also, bellows-spring action causes a similar change in response to the full range of pressure values from the controller. The controller's full

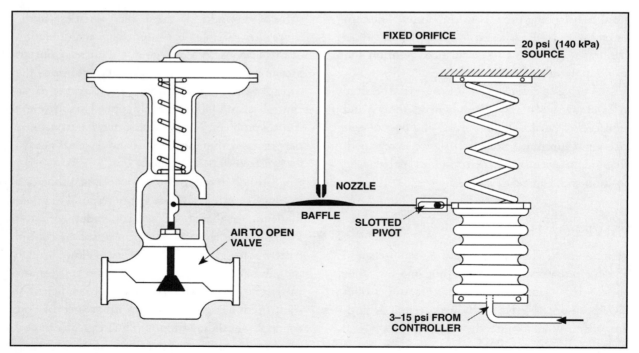

Figure 4.18   Schematic of a simplified valve positioner

Pneumatic Automatic Controls

range is from 3 to 15 psi (20 to 100 kPa). Assume that the controller's output changes to 12 psi (80 kPa). Also, assume that this 12-psi (80-kPa) change causes a 0.5-in. (12-mm) change in nozzle-baffle clearance. With such an arrangement, a change in controller output of 0.12 psi or 0.8 kPa (a change of 0.01×12 psi or 0.01×80 kPa), causes a nozzle-baffle clearance change of 0.005 in. or 0.12 mm ($0.5 \times 0.01 = 0.005$, or $12 \times 0.01 = 0.12$ mm). Because a nozzle-flapper arrangement is sensitive, this change in nozzle-baffle clearance brings about a change in nozzle back-pressure so drastic that diaphragm pressure either bleeds off toward zero pressure or builds up toward maximum. Whether diaphragm pressure bleeds or builds depends on the direction of the clearance change.

With an air-to-open, or reverse-acting, valve in use, and by noting that the controller's input increases when a decrease in the controlled variable occurs, the positioner's reaction to a fall of slightly less than 1 percent in the value of the controlled variable can be seen. The response of the controller to this change is an increase in output pressure of less than 0.12 psi (0.8 kPa)—say, 0.10 psi (0.7 kPa). The valve-and-actuator combination does not respond to such a change applied directly to the diaphragm, but this change is probably enough to close the nozzle tightly.

The pressure on the diaphragm builds up as quickly as the fixed orifice permits and, if necessary, ultimately attains the full 20 psi (140 kPa) to release the valve stem. Of course, in any properly operating valve and actuator, the response is rather quick, and the correct relation between nozzle and baffle is reestablished rapidly. This system shows that a minute deviation of the controlled variable can be relayed and amplified into a very powerful force for positioning the valve.

Commercial Valve Positioner

The basic positioner just described works, but adding an air relay and other refinements to the device make it adaptable for use with a wide variety of valves and actuators. For example, a positioner that adjusts to the differences that may be encountered in valve stem stroke uses an air relay to good advantage.

In the positioner shown in figure 4.19A and B, a system of linkages and pivots allows adjustment for valve stem strokes ranging from 0.2 to 30 in. (0.5 mm to 75 cm). The baffle (fig. 4.19B) is actuated by the motion of a bar whose left end is attached to the

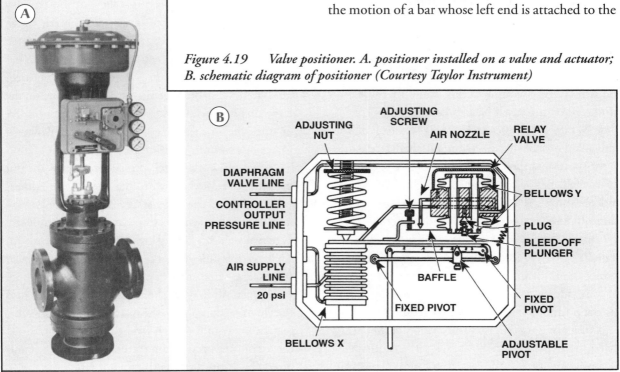

Figure 4.19 Valve positioner. A. positioner installed on a valve and actuator; B. schematic diagram of positioner (Courtesy Taylor Instrument)

free end of bellows *X* and whose right end is moved by action of the valve stem. Downward movement of the valve stem pulls the baffle away from the nozzle, while an increase in controller output causes the bellows to expand and counteract this action.

The air relay valve used in this positioner is unusual. The 20-psi (140-kPa) air supply is applied to the relay and enters the small cavity just above the cone-shaped, spring-loaded plug. The air supply also passes through a fixed orifice, which is a part of the relay, and then takes two paths—one to the nozzle and the other to an airtight chamber fitted over the upper bellows of the relay. A free passage for air exists between the interior portions of the upper and lower bellows, because considerable air space around each of the rods connects the upper and lower free ends of the bellows. Also, the air line leading to the valve actuator comes directly from the air space formed by the interior of these bellows.

Assume that a decrease in controlled variable has just caused the controller to produce an increased output pressure. This increase, when applied to the main bellows of the positioner, narrows the nozzle-baffle clearance. The back-pressure in the relay rises, causing the upper bellows to compress, and this motion is transmitted through the two rods to the free end of bellows *X*. This action pulls the cone-shaped plug away from the seat, allowing a rapid inflow of air to the interior of the bellows and the actuator. At the same time, the baffle, which has one end attached to the free end of bellows *X*, is moved away from the nozzle, thus counteracting some of the effect produced by increased back-pressure.

When a significant rise in the value of the controlled variable occurs, the nozzle-baffle clearance in the positioner increases, causing nozzle back-pressure to fall. Bellows *Y* expands, which pulls the free end of bellows *X* upward. This action unseats the bleedoff plunger and allows air from the actuator and the interior space of the bellows to bleed off until action of the valve stem and the system of linkages restores equilibrium.

In addition to the adjustable pivot for adapting the positioner to various stroke requirements, an adjusting nut is used to vary the tension of the spring that opposes the main bellows. A light tension makes the positioner provide more action for small pressure variations. The adjusting screw that depresses the baffle is useful for making initial settings to adapt the positioner to a new process and valve combination. Once this setting is made, further adjustment should not be necessary.

Positioner for Piston Pneumatic Actuators

A piston pneumatic actuator uses another type of positioner (fig. 4.20A and B). In this positioner, the air signal from a pneumatic controller is applied to a reversible bellows, which is attached to a beam that pivots about a fixed point. Special air relays are located on opposite sides of the pivot point, and the beam acts as a flapper in relation to nozzles that form part of the relays.

An increase in controller output pressure causes the bellows to expand, and this action tends to close off the nozzle of relay *X* and open up the nozzle of relay *Y*. An increase in nozzle back-pressure in these relays causes them to pass more air into the cylinder, while reduced back-pressure causes the relay to vent the cylinder until a correct proportion between nozzle back-pressure and cylinder pressure is achieved.

Figure 4.21 is a sectional view of the air relay used in the positioner in figure 4.20. Nozzle back-pressure is applied to the top of the diaphragm, while output pressure is applied to the small piston that acts as a seat for the exhaust valve. A balance of forces tends to develop between nozzle back-pressure on the large diaphragm, output pressure, and the pressure on the smaller area of the piston. This balance of forces represents a position of equilibrium.

Should nozzle back-pressure decrease, output pressure acts on the piston-like seat of the exhaust valve and drives the exhaust valve open. The open exhaust valve allows air to escape from the exhaust vent. When nozzle back-pressure increases, the force on the large diaphragm forces the piston-like seat hard against the exhaust valve plug and drives the supply valve open. Air flows to the actuator cylinder and into the chamber containing the piston-like exhaust valve seat. Soon, equilibrium is established between forces due to output pressure on the exhaust valve seat and nozzle back-pressure on the main diaphragm.

Pneumatic Automatic Controls

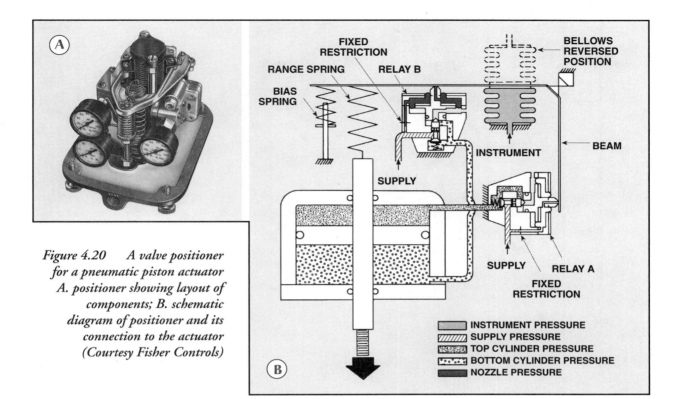

Figure 4.20 A valve positioner for a pneumatic piston actuator A. positioner showing layout of components; B. schematic diagram of positioner and its connection to the actuator (Courtesy Fisher Controls)

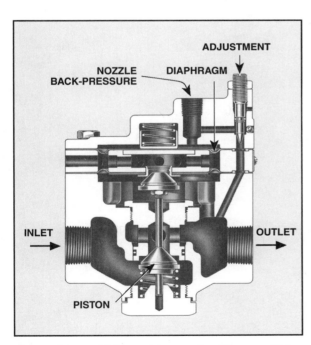

Figure 4.21 Valve positioner relay (Courtesy Fisher Controls)

SUMMARY

Using pneumatics to monitor and control processes has evolved into a well-established and mature science. Pneumatic components include pressure regulators, pneumatic controllers, air relays, volume boosters, and valve positioners. Pneumatic controllers have overcome mechanical restrictions and the limitations of friction, low and high pressure, vibration, poor-quality air supply, and other limitations to produce reliable and efficient controls.

REVIEW EXERCISE

1. Name two types of pressure regulators.
2. State the other name for a flapper.
3. Identify three modes of control.
4. List the functions that an air relay provides.
5. What is the function of a volume booster relay?

5

Electronic Automatic Controls

Electronic devices can duplicate all pneumatic control effects and they can do it with less maintenance, greater flexibility, and easier adjustment. In addition, electronic controls provide virtually immediate response, transmit control signals over long distances, and are easily modified when using devices incorporating microprocessors.

This chapter assumes that readers have a basic knowledge of electricity and of such electrical components as resistors, capacitors, potentiometers, rheostats, and switches. This chapter also explains the fundamental differences and similarities between analog and digital equipment.

ANALOG CIRCUITS AND EQUIPMENT

The word *analog* refers to a signal that is continuous and has an infinite number of points between its beginning and ending values. For example, an analog pressure signal of 3 to 15 psi (20 to 100 kPa) varies between 3 and 15 psi (20 and 100 kPa), but it has an infinite number of points, or values, in between. Similarly, an electrical analog signal of 4 to 20 milliamperes (mA) varies between 4 and 20 mA and has an infinite number of values in between.

Essentially, analog signals are an analogy, or a representation, of a process. For example, an electronic pressure transmitter can sense a pressure range of 0 to 200 psi (0 to 1,500 kPa) and produce an electrical signal of 4 to 20 mA that corresponds to this range of pressure. Zero psi corresponds to 4 mA and 200 psi (1,500 kPa) corresponds to 20 mA. A pressure between these two limits produces a corresponding electrical signal output—for example, 100 psi (750 kPa) produces a signal of 12 mA.

A signal range of 4 to 20 mA is a standard value in process systems. Other less frequently used signals from process transmitters include values such as 0 to 5 volts direct current (VDC), 1 to 5 VDC, or 10 to 50 mA. Typically, most electronic process transmitters produce a 4-to-20 mA signal that is converted to 1 to 5 V when the signal loop is terminated to a programmable logic controller (PLC), recorder, metering device, or other indicator. A simple but accurate 250-Ω resistor converts current to voltage in accordance with Ohm's law. Ohm's law is stated mathematically as—

$$V = I \times R \qquad \text{(Eq. 5.1)}$$

where
- V = voltage drop across resistor, volts (V)
- I = current in signal loop, amperes (A)
- R = resistance of signal terminating resistor, ohms (Ω).

Thus, if I is 0.004 A (4 mA), and R is 250 Ω, then—

$V = 0.004 \times 250$

$V = 1.$

Electronic signals in the form of current in mA are preferred over voltage for several reasons. For one thing, if a long length of wire is used from the transmitter terminals to the signal interface point, which may be a PLC, a recorder, or the like, resistance in the wire reduces the signal's voltage. Voltage can, however, represent a process input accurately if the signal is near the transmitter's signal terminals. In any case, if a signal reduction occurs, it represents a measurement error and is undesirable. On the other hand, if a 4 to 20 mA current signal range is used, resistance in the wire does not affect its mA value even if the wire is miles in length. When the current signal reaches the measuring point, an electronic device then converts it to an accurate voltage of 1 to 5 V.

Voltage transmitters are also sensitive to interference from external current and voltage sources.

Because a voltage transmitter's output impedance is low, power circuits can induce voltages in the transmitter's signal, which causes electrical interference. Electrical interference can, in turn, cause inaccurate measurements. On the other hand, current transmitters have infinite output impedance and are less susceptible to radio frequency interference.

Voltage transmitters are *four-wire transmitters* because two wires carry power (the power supply wires) and two wires carry the signal (the signal wires). Current transmitters are referred to as *two-wire transmitters* because the power supply is in series with the signal wires (fig. 5.1).

Electronic components and circuitry in a current transmitter process a sensor signal that the process variable creates from the primary element. The current transmitter then produces an mA signal output. Within this circuitry are *operational amplifiers* (*op amps*) that perform functions such as amplifi-cation; comparison; addition and subtraction; integration, or summing; and differentiation, or rate. Op amps in a current transmitter receive low-level signals from the sensors, amplify and convert these signals into usable voltages, and then convert the voltage into a current (mA) signal output.

Figure 5.2 shows a typical control loop using a two-wire electronic transmitter, power supply, 250-Ω resistor, and PLC. Two-wire transmitters are installed where the process variable range of operation is known. Once the process variable range is known, the transmitter is adjusted to allow accurate measurements over this operating process range. A technician can manually adjust a transmitter's zero and span, or *range*, to allow for a lower range than the full capability of the transmitter.

For example, suppose we have a pressure transmitter and wish to monitor pressure over a range of from 0 to 200 psi (0 to 1,500 kPa). To obtain accurate results, we calibrate, or *rerange*, the transmitter over the same range—that is, from 0 to 200 psi (1,500 kPa). Because the lower range value (LRV) of the pressure is 0 psi (0 kPa), we make the zero adjustment produce 4.0 mA at zero pressure. We then elevate the pressure to its upper range value (URV) of 200 psi (1,500 kPa), and adjust the span, or range, to produce 20.0 mA at 200 psi (1,500 kPa).

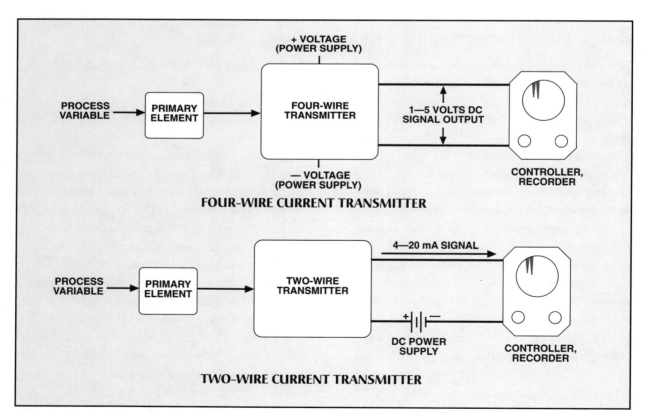

Figure 5.1   Two- and four-wire transmitters

Electronic Automatic Controls

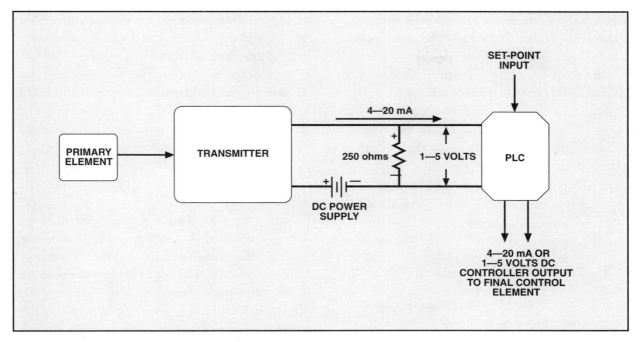

Figure 5.2 Electronic transmitter and other components in a control loop

Finally, we verify *linearity*—that is, we ensure that our adjustment is accurate—by checking the mid-span of 100 psi (750 kPa) and making sure it corresponds to 12.0 mA. Once verified, the transmitter is now calibrated for operation at 0 to 200 psi (0 to 1,500 kPa). Four mA corresponds to 0 psi and 20.0 mA corresponds to 200 psi (1,500 kPa). Intermediate mA readings correspond to pressures between 0 and 200 psi (1,500 kPa).

In the control loop using the two-wire transmitter (see figure 5.2), we can determine the process variable, or signal current, with simple equations since the relationship between the two is linear. The general equation for the process variable when the current in mA is known, is—

$$PV = [(mA - 4) \div 16] \times (URV - LRV) + LRV \quad \text{(Eq. 5.2)}$$

where

PV = process variable
mA = milliamps
URV = upper range value
LRV = lower range value.

For example, if we measured 8.6 mA and wish to know the corresponding pressure in psi, we use psi for PV in the general equation.

$$\begin{aligned} psi \ (PV) &= [(8.6 - 4) \div 16] \times (200 - 0) + 0 \\ &= (4.6 \div 16) \times 200 \\ &= 0.2875 \times 200 \\ psi &= 57.5. \end{aligned}$$

A similar relationship exists if we want to find the signal loop current when the process variable is known. The equation is:

$$mA = \{[(PV - LRV) \div (URV - LRV)] \times 16\} + 4. \quad \text{(Eq. 5.3)}$$

For example, if the measured pressure is 150 psi and we wish to know the corresponding current in mA, we use equation 5.3.

$$\begin{aligned} mA &= \{[(150 - 0) \div (200 - 0)] \times 16\} + 4 \\ &= [(150 \div 200) \times 16] + 4 \\ &= [(0.75) \times 16] + 4 \\ &= 12 + 4 \\ mA &= 16. \end{aligned}$$

MODES OF CONTROL AND CONTROL LOOPS

With pneumatic controllers, various modes of control can be incorporated to optimize processes. These modes include proportional, reset, and rate. Electronic controls also offer such modes, but with different names. Adjustments, or modes, in electronic controls can (1) vary the width of the proportional band (the gain), (2) provide for returning a variable to its set point under varying load conditions, and (3) adjust to the rate at which loads are changing. These three modes have a collective name, which is *proportional-integral-derivative* (*PID*) *loop control*. Additional terms under each main term describe control adjustments.

- Proportional, or proportional control
  + Gain adjustment, or proportional band
  + Set-point adjustment
- Integral, or reset
  + Timing, or repeats-per-minute adjustment
  + Set-point adjustment
- Derivative, or rate
  + Rate adjustment
  + Set-point adjustment

Proportional (P) Control Mode

One advantage of electronics is its flexibility. Electronic equipment and components can provide the proportional mode of control using many different methods.

The desired set point and proportional band (the gain) are usually adjusted with a *potentiometer*, which is a variable resistor used for voltage and resistance adjustments. By adjusting voltage and resistance, a technician can set the control loop for any desired set point of operation and provide the desired proportional band being sought.

A proportional control uses the set point to achieve the desired process level (reference) and compares its signal with the information provided by the primary element or process variable sensor (feedback). Any difference between these two signals creates an error (E), which the proportional control conditions and multiplies to operate the final control element.

A proportional control usually employs the error represented by the difference between the desired process level and that which is actually occurring (fig. 5.3). The bigger the error, the larger the signal is to the final control element. Because the strength of the signal is proportional to the size of the error, the controller is called proportional. A proportional controller aids in maintaining the desired process level, but it cannot keep it at the exact set point when input and output conditions change. The closeness of agreement between the desired set point and the actual process level is determined by the gain, or proportional band, of the controller. The proportional band determines how much change in the controlled variable is required to operate the final control element (the valve) from fully open to fully closed.

In the proportional mode of control, if the initial measured variable is equal to the set point, the

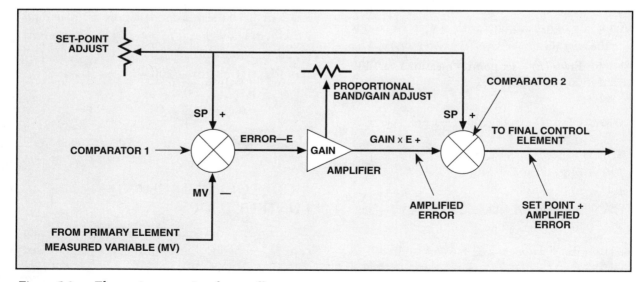

Figure 5.3   Electronic proportional controller

Electronic Automatic Controls

system is in equilibrium and the error is zero. If the measured variable changes, the controller provides a signal to the final control element by an amount equal to the change, as long as the proportional band is 100 percent or gain is 1.0. In figure 5.3, electronic comparator 1 compares the set-point adjustment signal (SP) and the inverted measured variable (MV) signal. If a difference exists in these two signals, an error (E) is produced and sent to the amplifier for further processing.

As an example, assume that the proportional band is set at 100 percent (or amplifier gain is 1.0). Also, the set-point adjustment (SP), whose range is from 1 to 5 V, is set at 3.0 V at comparator 1's positive (+) input. This 3.0 V is 50 percent of the range. The measured variable (MV) feedback signal is also set at 3.0 V and is fed to the comparator 1's inverting negative (−) input. Since the signals are identical, the difference is zero and thus the amplifier's output is zero. At the same time, set-point signal (SP) is also provided to comparator 2, which adds the amplified error (which is zero) to the set-point value of 3.0 V. The resulting output signal is 3.0 V and this signal ultimately determines the position of the final control element. The system is, at this point, in equilibrium.

In summary—
1. Gain of the amplifier is set at 1.0.
2. Set point is set at 50 percent, which is 3.0 V.
3. Initial final control element signal is 50 percent, or 3.0 V.
4. Measured variable is also at 50 percent, or 3.0 V.
5. Error (E) = set point − measured variable
   E = 50% − 50%, or = 3.0 − 3.0
   E = 0.
6. Final control element signal = set-point signal + (gain × error)
   = 50 percent + (1 × 0 percent) = 50 percent,
   or
   = 3.0 + (1 × 0)
   E = 3.0 V.

If a change in the process changes the measured variable to 2.0 V, the difference between the set point and measured variable becomes 3.0 − 2.0 = 1.0 V.

The amplifier multiplies this 1.0-V error by 1, which produces an output to comparator 2 of 1.0 V. This voltage is added to the set-point signal of 3.0 V, producing a signal of 4.0 V to the final control element. Some correction is now achieved, but an error still exists between the set point and controlled variable.

In summary—
1. Amplifier gain is set at 1.0.
2. Set point is set at 50 percent, or 3.0 V.
3. Initial final control element signal is 50 percent, or 3.0 V.
4. Measured variable is at 25 percent or 2.0 V.
5. Error (E) = set point − measured variable
   = 50 percent − 25 percent
   = 25%
   = 3.0 − 2.0
   E = 1 V.
6. Final control element signal = set-point signal + (gain × error)
   = 50 percent + (1 × 25 percent)
   = 75 percent
   or
   Final control element signal = 3.0 + (1 × 1)
   = 4.0 V.

By setting the proportional band at 50 percent, which is an amplifier gain of 2, an initial measured variable signal change from 3.0 to 2.0 V, or 1.0 V, results in an amplified output from the amplifier of 2.0 V. The 2.0 V is added to the SP voltage of 3.0 V, which produces a signal to the final control element of 5.0 V, or 100 percent. A stronger process correction results because the final control element correcting voltage is higher than it is with a gain of 1.0. An error still exists, but it is less than it is with a gain of 1.0. In summary—

1. Amplifier gain is set at 2.0.
2. Set point is set at 50 percent, or 3.0 V.
3. Initial final control element signal is 50 percent, or 3.0 V.
4. Measured variable changes and reduces to 25 percent, or 2.0 V.
5. Error (E) = set point − measured variable
   = 50 percent − 25 percent
   = 25 percent
   = 3.0 − 2.0
   E = 1 V.

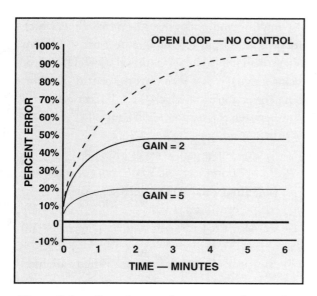

Figure 5.4   Open loop and proportional control

6. Final control element signal = set-point signal + (gain × error)
   = 50 percent + (2 × 25 percent).
   Final control element signal = 100 percent, or
   Final control element signal = 3.0 + (2 × 1) = 5.0 V.

Increasing the proportional gain reduces the error between the set point and measured variable. For example, setting the gain at 2 has an initial percentage of error of a little more than 10 percent. Setting the gain at 5 has an initial percentage of error of a little more than 5 percent. However, over time, a gain of 2 has a percentage of error of over 40 percent, while a gain of 5 has a percentage of error of only about 15 percent (fig. 5.4). Bear in mind, however, that real systems have time lags in their response, so the system is unstable and oscillates. Another method can be used to reduce the error with relatively low proportional gains.

Proportional-Plus-Integral Control (PI), or Proportional-Plus-Reset Mode

While proportional controllers make some corrections to return the controlled variable to the initial set point, they are not entirely satisfactory. The problem is that an offset, an error, or a difference in the set-point setting and the value of controlled variable always occurs. Adding additional gain reduces the error, or difference, but a lack of agreement between the set point and controlled variable still exists. To eliminate the error, an integral control, or reset, can be added to an electronic controller. The integral control, or reset, returns the controlled variable to the set-point value at appropriate intervals, which eliminates the error.

Think of integral control in terms of how fast and how aggressive the controlled variable returns to the set-point position when process conditions change. If the error that exists between the set point and controlled variable is multiplied by the time the error has existed, we can exert relative amounts of influence on how fast we want the recovery to occur. Adding time as a correction factor is referred to as *integral*, or *reset*.

Integral is the time it takes the controller to adjust itself back to its set point. Integral is expressed in repeats per minute or, in some cases, minutes per repeat. Integral time is a function of a reset circuit. A circuit called an *integrator* accomplishes the reset. An integrator consists of a resistor and a capacitor, whose product is expressed in seconds. The equation is—

$$T_i = R \times C \qquad (Eq.\ 5.4)$$

where
   T_i = integrating time, sec (time constant)
   R = resistor value, Ω
   C = capacitor value, farads.

If a fixed input voltage is applied to an integrator for an extended length of time, the integrator's output saturates to an appropriate positive or negative voltage value. The speed with which this action occurs depends on the time constant of RC.

Figure 5.5 is a schematic of a proportional-plus-integral control. It is similar to the controller in figure 5.3, but has an integrator circuit. In figure 5.5, note that the amplified error signal, indicated by $K_p \times E$, which is to the right of the gain amplifier, is also applied to the integrator circuit through its integral adjustment, R. (R is an adjustable resistor.) The value of R, in conjunction with capacitor C, determines the integral time constant in seconds. Therefore, any change in the amplified error signal ($K_p \times E$) is sent to comparator 2, which adds it to the set point and amplified error signal before proceeding to the final control element.

Electronic Automatic Controls

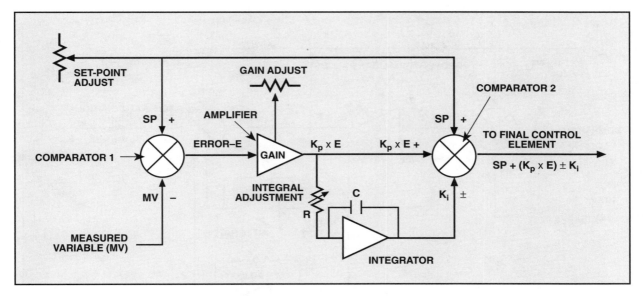

Figure 5.5   Proportional-plus-integral control

The integral control circuit detects any error existing between the set point and controlled variable and sends it to the final control element to force an agreement between the two. Until the controlled variable is returned to the set point, the integrator stage continues to produce a voltage of appropriate polarity—that is, either positive or negative (±)—to drive the final control element toward a position of stability. Once the controlled variable is returned to the set point, the reset action ceases. No steady-state error exists in this type of control.

Figure 5.6 compares open loop (no control) with proportional-plus-integral (PI) control. Note that over a 2-min period, PI control corrects the error to zero.

The integral time constant or its reciprocal, reset rate, is set to equal the response of the proportional mode of the controller. It is expressed in repeats per minute. If we assume that the output of the proportional amplifier is 1 V and the output of the integrator matches this voltage with 1 V in 30 sec, we say that the integrator responds in 30 sec and the reset rate is 2 repeats per minute. The output of the integrator will be two 30-sec intervals in 1 min to produce 2 V of reset action. In reality, the controlled variable will probably be well on its way toward the set-point value before the reset action is completed.

The PI method of control is appropriate for many applications, including control of flow, gas

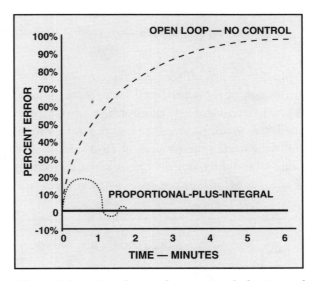

Figure 5.6   Open loop and proportional-plus-integral control

and liquid pressure, temperature, and level. Use of such a system allows close control of the controlled variable without sacrificing stability.

Proportional-Plus-Integral-Plus-Derivative (PID) Control, or Proportional-Plus-Reset-Plus-Rate Mode

A relatively small modification to PI control introduces the concept of derivative, or rate, control (PID), which improves response of the system.

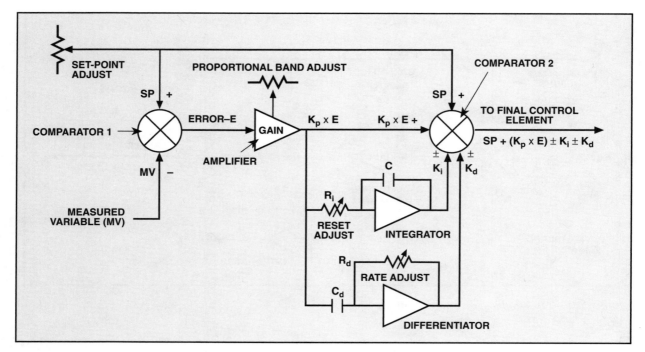

Figure 5.7   Proportional-plus-integral-plus-derivative control (PID)

Figure 5.7 shows a PID control. The circuit responds to the rate of error produced from the amplifier. When it senses this rate of error, the output from the differentiator produces a rapid response, which quickly corrects the process. A PID circuit also reduces instability and overshoot when correction or changes occur. The response curves in figure 5.8 compare a PID control's performance with an open loop, or no control. The PID curve shows how quickly the control responds to changes.

Besides set-point changes, other sources of change occur in a control system. For example, the outflow in a level-control system can change, or the input pressure in a flow-control system can change. Also, ambient or external temperatures in a controller may change. In any case, the automatic control detects the process variable (PV) changes in these systems and makes adjustments to satisfy the set-point requirements. The result is that the set point equals the measured variable (MV).

PID systems can be used in all systems but perform especially well in systems that have long delays in their responses—for example, temperature control.

SYSTEM STABILITY AND LOOP TUNING

The primary goal of all types of control is to have the set point and process variable in agreement at all times. The automatic control must be able to correct for slow system responses, environmental conditions, and nonlinearities in processes that cause disagreement in the set point and process variable.

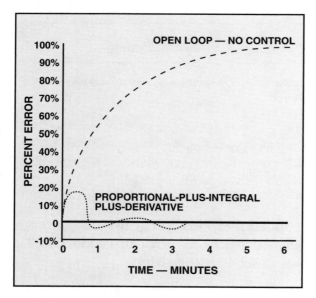

Figure 5.8   Open loop and proportional-plus-integral plus derivative control

In the process of correction, the system may use all or part of its energy and resources in the effort to regain zero error. Some systems respond slowly (they are *overdamped*). Some respond fast and oscillate (they are *underdamped*). And some systems are just right (they are *critically damped*). Still other systems become unstable and oscillate without control from the system. How the system is configured with appropriate gain, K, integration time, T_i, and differentiation time, T_d, influences the system's recovery performance. Most system designers recommend that system response be slightly underdamped so that each successive overshoot is reduced by 25 percent until stable operation resumes.

A common approach to establishing the optimum performance is the Ziegler-Nichols method. This procedure applies an experimental method to an actual system to determine the gain and time constant settings. When performing these tests, technicians must be sure that they can operate the system in an unstable mode without damaging equipment, spoiling a process, or injuring personnel.

The system is initially set up with the integrator time constant, T_i, set for maximum (infinity) and the differentiator time constant, T_d, set for minimum (zero). By increasing the gain, K, as the system set point is changed in steps (a step change), the system eventually becomes unstable. A critical value of gain, K_c, causes constant oscillation. The technician then measures the frequency of these oscillations. The period of the oscillations (one cycle in time) is determined and identified as the critical time constant, T_c. From this information, the technician can determine the controller settings for optimum performance. The equations for each system follow.

For proportional (P) systems—

$$K = 0.5 \times K_c. \qquad \text{(Eq. 5.4)}$$

For proportional-plus-integral (PI) systems—

$$K = 0.45 \times K_c \qquad \text{(Eq. 5.5)}$$

$$T_i = 0.8 \times T_c. \qquad \text{(Eq. 5.6)}$$

For proportional-plus-integral-plus-derivative (PID) systems—

$$K = 0.6 \times K_c \qquad \text{(Eq. 5.7)}$$

$$T_i = 0.5 \times T_c \qquad \text{(Eq. 5.8)}$$

$$T_d = 0.12 \times T_c \qquad \text{(Eq. 5.9)}$$

where

K = gain
K_c = critical gain value
T_i = integrator time constant
T_c = critical time constant
T_d = differential time constant.

Because every system varies and has its own personality, technicians use the equations to determine initial settings. Then, they optimize the system by fine tuning the initial values. When derivative action is used, a good rule of thumb is to make $T_d = T_i \div 4$.

Typical gains, K, for control of temperature, level, flow, gas pressure, and liquid pressure is in the range of 0.5 to 2.0. Conversely, the proportional band is in the range of 2.0 to 0.5. The integral time constant (T_i) for flow control is approximately 10 repeats per min or, conversely, 0.1 min per repeat. For gas-and-liquid pressure control, T_i is approximately 0.2 repeats per min or 5 min per repeat. For level control, T_i is in the range of 0.02 to 0.2 repeats per min or 0.5 to 5 min per repeat. And, temperature control has a T_i in the range of 0.1 repeats per min or 10 min per repeat because of its slow response.

PROGRAMMABLE LOGIC CONTROLLERS (PLC) CONTROL SYSTEMS

Programmable logic controllers (PLCs) are electronic control devices that perform many logic functions. Incorporated in PLC processors are microprocessor functions, memory storage, and mathematical functions. They accept a variety of input signals, including discrete (on-off), analog (4 to 20 mA or 1 to 5 VDC), frequency (flow), resistance (RTD), voltage (thermocouple), and digital (BCD) signal sources used in instrumentation. Because PLCs are so versatile and can accept analog signals, they can perform in the closed-loop control modes of P, PI, and PID.

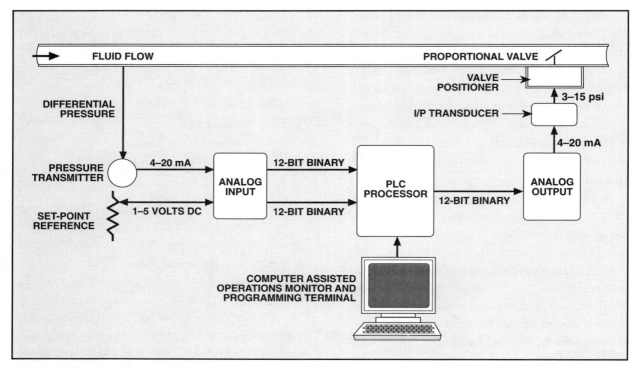

Figure 5.9 PLC closed-loop control system

Figure 5.9 is a diagram of a closed-loop system that uses a flow measuring process transmitter, an analog input card, a PLC processor, an analog output card, a current-to-pressure (I/P) transducer, and a proportional control valve in the flow line. These elements constitute the required components involved in controlling flow of a fluid.

The following discussion focuses on how the PLC performs the functions of accepting an analog input set point (1 to 5 VDC) and feedback analog signals (4 to 20 mA). The PLC performs addition, subtraction, multiplication, square root, and division functions for the PID functions and then converts the result into an analog output signal to operate the final control element, which is a flow control valve.

A system programmer writes the PLC software program, and a programming terminal connected to the processor downloads the program into the processor. The objective of the program and system is to set a desired flow rate and to maintain it automatically. The program determines how the input data is processed, calculated, and scaled in order to deliver an appropriate signal to the analog output card. The analog input card accepts the set-point voltage and transmitter current and converts them into 12-bit binary numbers. The analog output card converts the processor's binary information into a 4-to-20 mA signal that a current-to-pressure (I/P) transducer receives and produces a corresponding 3-to-15 psig signal. The I/P transducer sends this pressure signal to a valve positioner that sets the desired flow through the valve.

The differential-pressure transmitter monitors fluid flow through an orifice plate that produces a differential pressure that goes to the pressure transmitter. The pressure transmitter produces a 4-to-20 mA analog signal to the analog input, which converts the signal to 1-to-5 VDC. The analog input also converts and properly scales the 1-to-5 VDC signal into a 12-bit binary number and sends it to the PLC processor. Because flow is not linear with pressure, but is proportional to its square root, the 4-to-20 mA signal produced by the transmitter has to be processed with a square root function. Square root processing can be achieved either within the transmitter, if it has an optional square root function card, or in the PLC processor program. Proper scaling of the signals into appropriate engineering units is also required. Just as a drafting technician draws the plans for a building to scale rather than actual size—that is,

Electronic Automatic Controls

a certain number of in. (cm) on the plans represents a certain number of ft (m) in reality—so must a processor program scale the signals to units that the system can properly read and interpret.

In addition, the software program must be structured to perform the required mathematical calculations in the right sequence. Required calculations include—

- performing the square root function on the pressure signal from the transmitter, if needed;
- subtracting the flow signal from the set point to determine the error;
- multiplying the error with the gain;
- introducing the integrating time constant, T_i, into the system equations; and
- introducing the differentiating time constant, T_d, into the system equations, if required.

Also, the system must perform scaling that is consistent with the range of the I/P transducer, valve positioner, and valve. The result of these calculations is to produce a signal that can be used by the output functions and equipment. Consideration for fail-safe operation in the event of a malfunction along with such output limits as rate limit, hold limit, and suicide limit, also need to be considered in writing the program.

The ability of a technician to change the parameters of an automatic control from a keyboard or programming terminal is one the major benefits of PLCs. Moreover, the technician can adjust gain and time constants from the programming terminal to optimize the system without changing components or wiring.

SPECIALIZED FLOW COMPUTERS

A PLC can operate and control closed-loop systems quite well. However, specialized flow computers are also available that allow a technician to view and adjust such functions as set point, gain, integrating time constant, and differentiating time constant (fig. 5.10). Analog controllers can perform the three terminal functions of a PID system. Also, human-machine-interface (HMI) push buttons or a mouse allow technicians to program such controllers. Selected information and the program are usually

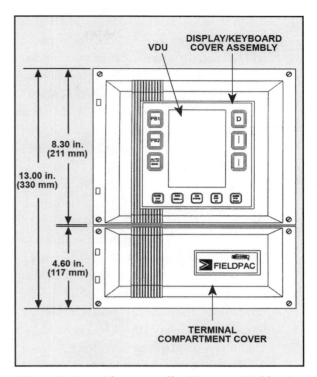

Figure 5.10   *Flow controller (Courtesy Fieldpac)*

provided on a visual display unit (VDU) associated with the controller.

DISTRIBUTED CONTROL SYSTEMS

In many facilities, PLCs and specialized controllers perform several functions. In some cases, it may also be desirable to exchange data and other information between controllers and computers. Communication between controllers and computers allows data to be gathered, the control of adjustments, or the operation of master-slave devices. Such systems are referred to as being part of a distributed control system (DCS).

To get systems to communicate with one another, technicians must consider several factors to minimize interference and errors. Considerations include—

- What standards and protocol are to be used? Among the choices for the set of rules involved in the communication link are voltage levels, connection and control of the interface between units, and the methods and procedures involved in the content and control of digital messages.

In most cases, serial transmission is used to connect the data terminal equipment (DTE) and the data communication equipment (DCE). The Electrical Industries Association (EIA) publishes *Standard RS232* that most technicians use as a guide when connecting serial transmission systems. *RS232* defines the voltages to be used in data signals and specifies that a 25-pin D connector be used for termination. Pin numbers in this connector are designated for certain functions and uses, such as ground.

However, in long distance communications (greater than 25 ft or 8 m) occurring at high speeds between DTEs and DCEs, *RS232* does not adequately define the requirements involved for good signal transmission. Therefore, the EIA created two additional standards to cover long distance applications: *RS422* and *RS423*. These two standards allow use of differential receivers to overcome the common problem of mode ground noise.

A set of rules, often referred to as *protocols*, cover the mechanics involved in data trans-mission. Protocols define (1) the message form in bits, (2) message initiation, (3) message termination, (4) error detection, and (5) what happens if a link is broken during a message. The American Standard Code for Information Interchange (ASCII) adopted *ISO 646* to use as a protocol, as well as IBM's BISYNC standard. Error detection uses a form of even or odd parity checks in the transmission between the sender and the receiver.

- Use of parallel or serial cabling?
  Serial cables are used to communicate between stations that are greater than 15 ft (4.5 m) apart. Transmission of serial data takes longer than parallel connections, but most systems can tolerate the slower speeds. Serial cables are usually satisfactory with high-data-transmission rates and improved noise-rejection techniques. For systems that are relatively close, such as a printer-PLC connection, parallel cables are typically used.

- Data transmission rate?
  The transmission rate is usually expressed in *bauds*, which indicates the number of binary bits per second. For a PLC communicating with an instrument, 1,200 baud is usually sufficient. On the other hand, high-quality cables, which can transfer data at a rate as high as 115 kilobauds, are used to communicate between PLCs or between remote input/output (I/O) links.

- Synchronous or asynchronous transmission?
  For a sender and receiver to communicate in both directions when transmitting on a serial cable, synchronization is required so that the data can be satisfactorily interpreted. Some systems use a single clock for both the transmitter and receiver. One clock ensures that the data stream is always in alignment. When one clock is used, the system is using the synchronous transmission of data. However, most systems use separate clocks; but, they are in close timing, which allows the transmission of data using asynchronous methods.

  The common method of asynchronous transmission is to break down the digital message into characters that are 5 to 8 bits in length. The two clocks become synchronized at the start of each character. Since the idle state of transmission is a logical 1, the character starts with a logical 0 that initiates transmission of the character and recognition by the receiver. A last character—an eighth bit—is sometimes added in the transmission to check for errors. Such a character is referred to as even or odd parity.

- Data code to be used?
  ASCII is generally accepted as the binary code for character transmission. It is a seven-bit code with 128 combinations of predetermined arrangement of 1s and 0s. The combinations cover full upper and lower case alphanumeric characters along with punctuation and 32 control characters.

- Type of error control to be used?
  Typically, parity checking is the method used to detect errors in the serial transmission of

data. Errors can result in serial transmission when electrical noise is coupled into the cables or poor connections corrupt the signal. Using even or odd parity checking can confirm that the character sent is received correctly.

For example, if even parity checking is used as the protocol, the 7-bit character is sent in the form of 1s and 0s. If the number of 1s in the character is 3 and the remaining 4 characters are 0s, the number of 1s is an odd count of 3. An additional logic 1 is added to the character code as the eighth bit to make the count an even number of 1s.

The same procedure with a character code of 4 1s and 3 0s is an even number of 1s. The parity code would be a 0 as the eighth bit to be given an even count of binary 1s.

- Network sharing?

Distributed control systems can have point-to-point connections, or links, for communications between the PLC and other equipment, or they can have PLC-to-PLC-to-computer communication on a common link. Communication between all members of the system needs to be done openly and freely for optimum functioning.

One method shares a common network, prevents time-wasting data transmissions, and uses an address system for direct communication. Local area networks (LANs) and wide area networks (WANs) are used, depending on the area involved and the number of stations that are connected.

PLC manufacturers typically have their own networking protocol and system. For example, Allen-Bradley uses Data Highway Plus in their PLC-5 product line, Gould/Modicon uses a Modbus system, and General Electric uses a GENET system. These proprietary systems use a protocol that is not common to each other. Therefore, mixing of various manufacturers' products cannot be achieved without some form of interpretative hardware and software interface.

Efforts have been made by several organizations to arrive at a common communication bus system that could be used by all manufacturers; but, it has not occurred for a number of reasons. Systems such as Fieldbus and Profibus are attractive to many manufacturers, but large-scale acceptance has not been obtained.

HUMAN-MACHINE-INTERFACE (HMI)

A PLC system generally operates without intervention from a technician or maintenance personnel. In some cases, however, adjustments and commands from the technician need to be made and status of process conditions needs to be understood. Equipment that allows the operator to interface with the system is *human-machine-interface* (*HMI*).

Formerly referred to as man-machine-interface (MMI), the new designation is better because it has a broader meaning. *Ergonomics* is the science that deals with the methods, procedures, and environment involved between humans and machines. Ergonomics works to assure that operators perform their assignments efficiently, with ease, and with a minimum of error.

One example of an HMI is the Allen-Bradley Panelview (fig. 5.11). This panel allows the operator to view information on a digital screen as well as perform limited control functions with a keypad or touch screen. A two-wire serial cable connects the panel to a PLC and allows communications in both directions—that is, it receives and sends data. The Panelview provides a number of functions that allow operators considerable flexibility with their process.

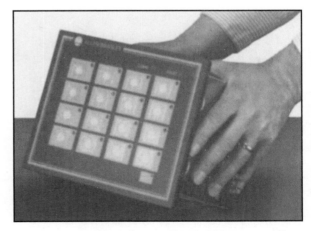

Figure 5.11   Panelview HMI (Courtesy Allen-Bradley Co.)

1. It provides an operator display that includes—
   a. a status indicator of functions—that is, the display shows whether the system is stopped, running, or off;
   b. numeric displays that can indicate such things as tank level and temperature;
   c. bar graphs;
   d. alarm messages; and
   e. message displays.
2. Data inputs from the operator include—
   a. push buttons that the operator uses to regulate functions of the PLC controls;
   b. numeric input that sets alarm levels, and
   c. a list selector, which includes such control options as calibrate, test, and run.

SUMMARY

Electronic control represents an improvement in instrumentation that we could not have envisioned 50 years ago. Within the last 15 years, substantial improvements have been made with the introduction of microprocessors in personal computers, programmable logic controllers, and smart process transmitters.

This chapter pointed out that electronic controllers improve a system's performance because it makes controlling a process easier and more flexible. Signals can be transmitted over long distances, processed, manipulated, diagnosed, and used in control. PLCs are important tools because they can interface with analog instruments, on-off controls, and other devices. Adjusting the PLC to perform PID functions adds to its importance as a controller in point-to-point control as well as in distributed control systems.

REVIEW EXERCISE

1. When working with process instruments, what do zero and span mean?
2. Calculate the pressure applied to an electronic transmitter calibrated for 0 to 300 psi if the measured signal current is 10 mA.
3. When an analog process transmitter sends a 12-mA signal to a PLC input card, in what form is the current converted and how much?
4. Define the following terms—
   a. integral,
   b. derivative, and
   c. proportional.
5. What are the benefits of adding integral control to a proportional control?
6. What is the benefit of having parity checking in a LAN network using PLCs?
7. If a control system becomes unstable, what adjustments can be made to stabilize it?

6

Pressure Measurement and Control

Over the past 20 years, technology in pressure measurement has advanced considerably, progressing from mechanical techniques to electronic methods. Although temperature measurement rivals pressure measurement in automatic control, pressure measurement is also vital. Pressure measurement can serve as an indicator and can control other process variables in the system. The measurement and control of pressure occurs in tanks, pipes, vessels, and other components in a process system. Pressure is also used in measuring such variables as temperature, level, and rate of flow.

In this chapter, pressure is discussed in its use to control process variables, as well as to provide a reference in checking other measurement methods. Mechanical methods of pressure measuring are covered first; electronic methods follow.

UNITS OF PRESSURE MEASUREMENT

When the word measure is used, it is typically meant in a broad sense because, in some instances, pressure is not literally measured. For example, pressure may actuate a measuring means that is not an indicator. A Bourdon tube may be attached directly to the flapper of a pneumatic controller and the controlled pressure applied to flex the tube. In this case, the Bourdon tube is the primary element and it measures the controlled variable although no graduated scale is present.

Another example is an electronic pressure transmitter where the pressure actuates a capacitor whose change results in a control signal from the transmitter. In this case, the primary element is the capacitor and it measures the controlled variable.

Pressure Scales

Pressure is defined as force per unit area. As pointed out earlier, in the U.S., pressure is usually stated in pounds per square inch, or psi. The SI system uses kilopascals (kPa), which are derived from newtons per metre (N/m). In the atmosphere, a uniform pressure of about 14.7 psi (101.4 kPa) exists all around us, although we are usually not aware of it. Some pressure measurements ignore atmospheric pressure and begin the pressure measurement at zero. We refer to measurements that ignore atmospheric pressure as *gauge pressure*. In the conventional measurement system, it is often abbreviated as psig, which stands for pounds per square inch gauge. Most pressure gauges indicate gauge pressure, which is the pressure above ambient atmospheric pressure. Pressure below atmospheric pressure is referred to as *vacuum pressure*.

If we change our reference pressure from atmospheric to that of space where no pressure exists, *absolute pressure* is obtained. In the conventional measurement system, absolute pressure is abbreviated as psia, which stands for pounds per square inch absolute. Using mechanical methods on earth, it is almost, but not quite, possible to attain a pure vacuum, which is the vacuum of space, or the complete absence of pressure. Gauges on an absolute scale indicate about 14.7 psi for atmospheric pressure, while a gauge pressure scale indicates zero for atmospheric pressure. Gauge pressure measurements are often referred to as *GP* while absolute pressure measurement is referred to as *AP*.

Another form of pressure measurement is *differential pressure*. Differential pressure is the difference between a low pressure and a high pressure at some point in a system. Gauges that measure pressure differences are differential-pressure, or *dP*, gauges.

Definitions

Industry has adopted several terms that relate to pressure measurement and typically coincide with each other. Terms include—

- *Absolute pressure*: a single pressure measurement referenced to a full, or perfect, vacuum. Absolute pressure is the measurement of the process pressure in excess of a full vacuum, or 0 psia. Zero pressure, absolute (0 psia) represents a total lack of pressure. Outer space is considered a full vacuum.
- *Atmospheric pressure*: the force exerted by the earth's atmosphere. Atmospheric pressure at sea level is equivalent to 14.7 psia (101.4 kPa). The value of atmospheric pressure decreases with increasing altitude.
- *Barometric pressure*: atmospheric pressure as measured by a barometer. In conventional units, it is often given in in. of mercury; in SI units, mm of mercury is used.
- *Differential pressure*: a difference in two pressures. One application of differential pressure is to monitor flow rate by measuring the drop in pressure across an orifice plate. That is, when fluid flows through an orifice (a hole) that is mounted in a pipe, or line, and the orifice is smaller in diameter than the pipe, a drop in pressure occurs. Special equipment can use this pressure drop to determine the amount of fluid flowing through the line (the flow rate).
- *Gauge pressure*: a single pressure measurement that indicates the pressure above that caused by the atmosphere. Gauge pressure represents the positive difference between measured pressure and existing atmospheric pressure. We can convert gauge pressure to absolute pressure by adding the actual atmospheric pressure to the gauge pressure reading. For example, at sea level, where atmospheric pressure is 14.7 psi (101.4 kPa), 10.0 psig (68.95 kPa gauge) is equivalent to 24.7 psia (170.35 kPa absolute) and 0 psig is equivalent to 14.7 psia (101.4 kPa absolute).
- *Hydrostatic pressure*: pressure exerted by a fluid that is at rest in a container. Hydrostatic pressure is encountered in liquid-level applications where pressure exerted by the liquid below the surface level is important to measurement.
- *Line pressure*: pressure exerted on a pipe wall by a fluid flowing parallel to the wall.
- *Static pressure*: line pressure. Static pressure in this context is confusing because although the fluid is flowing, or moving—it is not static, or at rest—instrument technicians nevertheless use the term static pressure.
- *Working pressure*: line, or static, pressure. Also, the pressure existing at a given time; the maximum operating rating of a vessel.

Figure 6.1 charts the relationship between absolute pressure, atmospheric pressure, gauge pressure, barometric pressure, and a vacuum. Note that in this case, vacuum refers to any pressure below atmospheric pressure. Be aware that the scientific definition of vacuum is the total absence of pressure.

Range of Pressure Values

Pressure values can range from miniscule amounts such as 1 micrometre (μm) of mercury (0.00002 psi) to large amounts such as 10,000 psi (68,950 kPa). Pressures below atmospheric pressure are considered negative values, or vacuum. The methods and components used to measure the range of pressures vary.

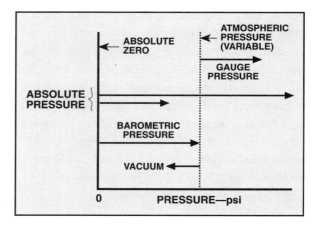

Figure 6.1   Pressure measurement comparisons

When dealing with very small values of pressure, scientists and technicians often express pressure in millimetres (mm) of mercury. The measurement uses the actual length of a column of mercury and can be expressed in metres (m), millimetres (mm), or micrometres (μm). A McLeod gauge, which is a special column of mercury, measures the smallest pressures—as low as 1 μm of mercury. The practice of measuring small pressures in mm of mercury is likely to continue for some time.

Another term used in pressure measurement is *head*. Head is the pressure that results from gravitational forces on liquids. It is measured in linear distance, such as ft (m), and thus deviates from true pressure units, just as mm of mercury does. Thus, a certain liquid may have a hydrostatic head (or just head) of 300 ft (90 m). If the liquid is water, this amount of head is equivalent to 130 psi (881 kPa). To convert head in ft to psi, remember that head pressure is based on a column of water that exerts 0.433 psi for each vertical foot of height. So, to convert a head of water to psi simply multiply the head height by the constant of 0.433. Thus, 300 ft × 0.433 = 129.9 or 130 psi. In SI units, a column of water exerts about 9.79 kPa/m, so 90 × 9.79 = 881.1 or 881 kPa.

Each incremental range of pressure can be measured with different types of sensors that lend themselves to these ranges. Table 6.1 shows several pressure ranges and the type of sensor used to measure them.

Note that intermediate pressures—for example, 5 to 30 psi (30 to 200 kPa)—can use a direct means of measurement, such as a spring or bellows. That is, the pressure acts directly on the spring or bellows to indicate its amount. Other ranges use an *inferential*, or an *inferred*, means of measurement, such as a mercury U-tube or a capacitor. In such cases, pressure is indicated as a pressure byproduct, such as a change in electrical capacitance or resistance.

MECHANICAL PRESSURE ELEMENTS

While electronic measurement is the trend in pressure instrumentation, many mechanical devices either already exist or are specified for particular applications in the process industry. Following is a summary of the more popular methods to measure pressure using mechanical technology.

Mechanical pressure elements use direct measurement—that is, fluid pressure causes mechanical motion to some part of the measuring means. Such devices are used extensively to measure intermediate pressure ranges, typically from less than 3 ounces (oz) per square in. to thousands of psi (1 to thousands of kPa). The variety of available mechanical pressure elements is almost limitless. The more popular ones include Bourdon springs or tubes, diaphragms, bellows, liquid manometers, and bell-type gauges.

Bourdon Tube Elements

The design of a Bourdon tube or spring utilizes concepts that have been developed from mathematical analysis as well as test findings. The free end of a Bourdon tube, within its elastic limits, experiences movement that is proportional to the fluid pressure applied to it. Also, fluid pressure applied uniformly

Table 6.1
Types of Sensors Used to Measure Various Pressure Ranges

Pressure Range	Sensor Type
Below 1 psi (7 kPa)	Pirani gauge, Televac gauge, McLeod gauge, electronic capacitor
0–7 psi (0 to 50 kPa)	Hg U-tube (differential), electronic capacitor
8–17 psi (50 to 120 kPa)	Low-pressure spring and bellows, electronic capacitor
0–17 psi (0 to 120 kPa)	Compound spring and bellows, electronic capacitor
5–30 psi (30 to 200 kPa)	Intermediate spring and bellows, electronic capacitor
30–4,000 psi (200 to 30,000 kPa)	Spiral spring, electronic capacitor
4,000–6,000 psi (30,000 to 40,000 kPa)	Electronic capacitor, electrical resistance strain gauges
6,000–100,000 psi (40,000 to 700,000 kPa)	Electrical resistance strain gauges

to the tube's exterior surface tends to cause motion opposite to that caused by pressure applied to the tube's interior.

Bourdon tubes are used in three forms: (1) the simple C-tube, (2) the spiral, and (3) the helical. Each has an important place in measurement and each possesses an advantage that makes it suited to a particular application.

The C-tube is the most commonly used Bourdon element. It is not only relatively easy to manufacture in large quantities of uniform quality, but also it can be made small and shallow enough for the pressure measuring instrument to have legible dials. A single C-tube has limited movement, but a spiral and a helical Bourdon tube amplifies movement (fig. 6.2A–C). Using appropriate gearing, expanded motion is achieved to position the pointer of an indicator gauge.

Differential pressure can be measured with Bourdon tubes by arranging them to measure pressure from two sources, P_1 and P_2 (fig. 6.3). With equal pressure, the net movement of the dial is zero, but any difference in pressure results in dial movement.

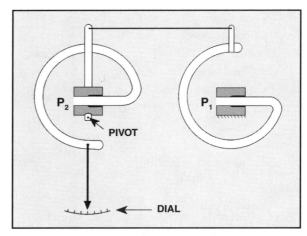

Figure 6.3   Bourdon tubes arranged to measure differential pressure

Diaphragm Pressure Elements

Metallic and nonmetallic, or *slack*, diaphragms are two broad types of diaphragm pressure element. Both respond to small pressure changes and are used extensively for differential pressure movement in low-pressure and low-vacuum ranges.

The material used for metallic diaphragms includes metals such as brass, phosphor bronze,

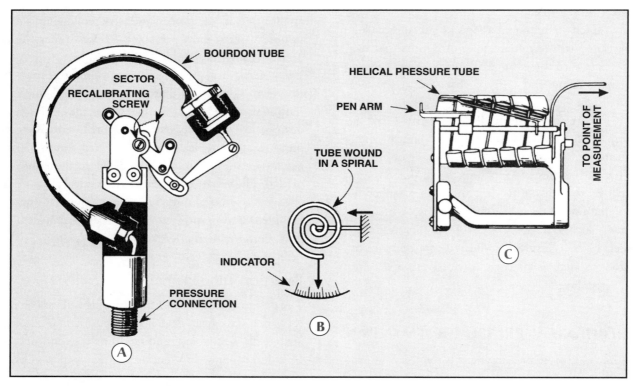

Figure 6.2   Three basic types of Bourdon tubes. A, simple C-tube; B, spiral; C, helical

Pressure Measurement and Control

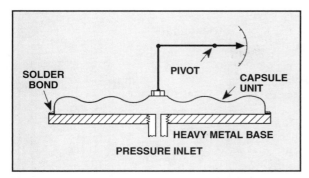

Figure 6.4   A pressure element with single-shell metallic diaphragm

beryllium copper, and stainless steel. Figure 6.4 shows a single metallic diaphragm that has a heavy metal base to which a lightweight metal unit called a *capsule* is soldered. Pressure causes the capsule to flex and move the pivoted indicator. Figure 6.5 shows a diaphragm that is made from two capsules placed back to back and mounted on a base. A C-spring strengthens the unit. Figure 6.6 shows three capsules

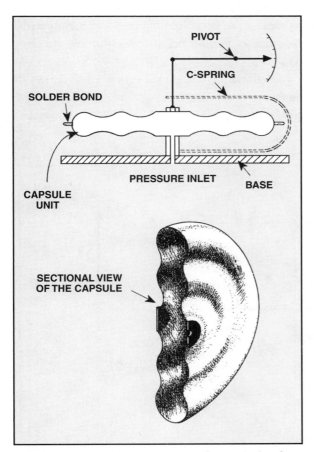

Figure 6.5   Pressure element with a capsule of two metallic diaphragms

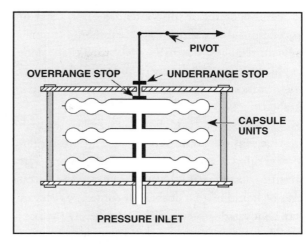

Figure 6.6   A compound element with three diaphragm capsules

making up a compound element. The unit also has underrange and overrange stops.

Nonmetallic, or slack, diaphragms are often used to measure very low pressure, where the pressure on one side of the diaphragm differs by only a small amount from that on the other side. Slack diaphragms are made of a variety of materials, such as leather, impregnated silk, and several plastic substances having such trade names as Teflon, Neoprene, and Koroseal. They are formed from circular sheets of the material and sometimes contain one or more corrugations. Some also feature a calibrated spring to provide additional tension on the diaphragm (fig. 6.7).

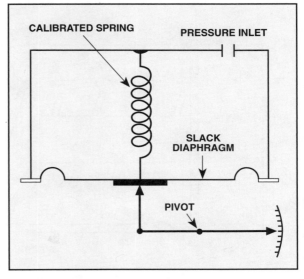

Figure 6.7   A nonmetallic diaphragm with calibrated spring

Nonmetallic diaphragms are often used in applications that require accurate response to small pressure changes, down to a few psi (kPa). Slack diaphragm elements also are commonly used in *draft gauges*, which measure the pressure of air within the furnace of a boiler.

On the other hand, liquid-level gauges, which must be sensitive to minute changes in pressure to be accurate, often use metallic diaphragms. The pressure-sensing unit is located as near to the lowest level of liquid as practicable. In some instances, it may be located somewhat above or below the bottom of the tank or vessel whose liquid level is being monitored, but the unit must be located no higher than the minimum level that is to be measured.

Figure 6.8 shows a hybrid unit for measuring liquid level. It has a metallic diaphragm element connected to a variable electrical resistance. As the pressure changes, so does the resistance value, which is used to electrically operate a remote dial.

Bellows Pressure-Sensing Elements

A bellows element is similar to a diaphragm element that is composed of a series of metallic capsules, although some characteristics distinguish bellows elements.

The materials used in a bellows are similar to those used in a metallic diaphragm, and the selec-

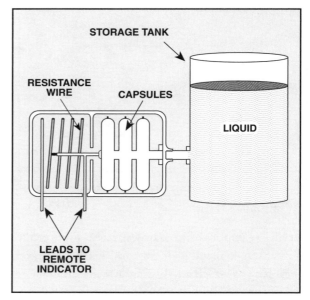

Figure 6.8   Using a metallic diaphragm-variable resistor to measure liquid level

tion of a particular metal depends largely on the corrosive conditions to which the unit is exposed. Stainless steel, beryllium copper, and metals with trade names like Monel and Everdur are in common use.

Bellows are made in a variety of sizes, ranging from ⅜ to 12 in. (9.5 to 305 mm) in diameter and from ¾ to 30 in. (19 to 762 mm) in axial lengths. Figure 6.9 shows a bellows being used as a

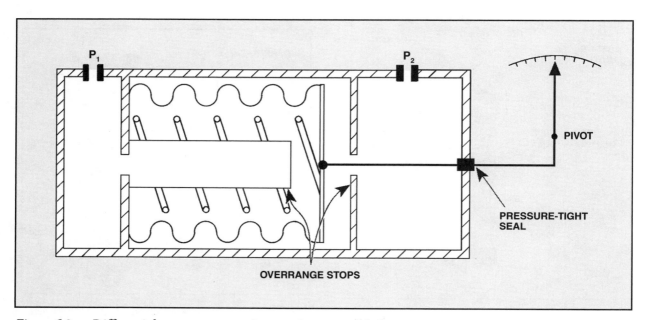

Figure 6.9   Differential-pressure gauge using a spring-opposed bellows

Pressure Measurement and Control

differential-pressure gauge. In the drawing, P_1 is on one side of the bellows and spring and P_2 is on the other. The spring opposes the bellows movement.

Bellows can also be used for purposes other than pressure measurement. For example, they frequently convert hydraulic pressure to linear motion to operate mechanical devices and pressure indicators. Bellows are also used extensively in pneumatic controllers, air relays, and valve positioners. In such applications, the bellows not only performs a measurement function, but also other functions. For instance, in air relays, the bellows closes and opens valves; in valve positioners and controllers, it positions flappers. However, in each of these actions, the bellows is measuring some value of air pressure to position a device.

Liquid Manometers

Manometers are frequently used in applications that use a liquid's buoyant effect, height, or both to perform a measurement or control. A *manometer* is a pressure gauge that responds to small pressure changes. It uses mercury or another liquid to measure low differential pressures.

Bell-Type Gauges

A bell-type gauge measures differential pressure by floating a relatively heavy bell-shaped device in a pool of mercury. The mercury-bell contact area tightly seals the cavity formed between the bell interior and the surface of the mercury (fig. 6.10). If the pressure in cavity V_1 equals that in V_2, as is the case when P_1 and P_2 are open, the weight of the bell pulling it down and the buoyant effect of the mercury pushing it up, determine the bell's vertical position. The bell settles into a position of balance. A bell-type gauge is sensitive, accurate, and stable. A dead band may exist if the shaft sticks in the gas-tight seal, but its heavy construction makes it able to withstand high static pressures.

Protection of Mechanical Pressure Instruments

Environmental factors can damage instruments unless precautions are taken in their design and manufacture. For example, if the pressure output of reciprocating pumps or compressors creates pulsating pressure, unnecessary wear and abuse of pressure gauges can occur. Pulsation dampeners in the form

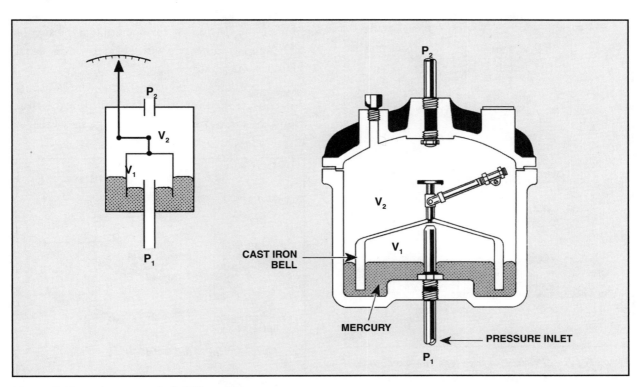

Figure 6.10   A mercury-bell differential-pressure gauge

of capillary tubes may be used to reduce pulsations while continuing to monitor pressure (fig. 6.11A–B). One type of dampener uses a felt filter to prevent the introduction of contaminants that might clog a valve (fig. 6.11A).

Another form of protection uses a rubber bulb, which provides a closed system for meter, small passage, and filters, thus assuring freedom from contaminants (fig. 6.11B). The rubber bulb is filled with glycerin or other relatively inert liquid, and process fluid acts on the exterior surface of the bulb. Process pressure is transmitted through the rubber wall to the glycerin, which then actuates the measuring device.

Figure 6.11C shows a simple loop—a pigtail—that protects the gauge by forming an isolation seal downstream from the gauge.

ELECTRONIC PRESSURE MEASUREMENT

New technology has broadened pressure measurement with electronic sensors and transmitters. Pressure can be measured and corresponding signals transmitted to remote locations for monitoring, measuring, processing, and manipulation. Calibration of these electronic instruments is relatively easy using a small screwdriver or a smart interface unit.

Electronic pressure transmitters are constructed in two basic configurations: two-wire current or four-wire voltage. Each type offers advantages, but the two-wire current transmitter is most commonly used.

Four-Wire Voltage Transmitters

Four-wire transmitters get their name from the number of electrical wires used to power and operate them (fig. 6.12). Two of the wires to the transmitter provide the DC power for operation (typically 24 VDC) and the other two wires are used for the signal loop or voltage output.

Electronic components in four-wire transmitters convert the proportional signal from the electronic sensor into an electrical voltage signal that is

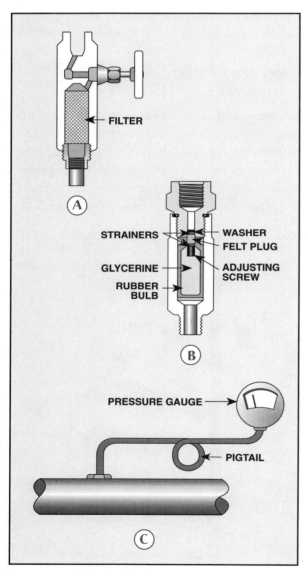

Figure 6.11 Protective devices for mechanical instruments. A, pulsation dampener; B, pulsation dampener; C, isolation seal.

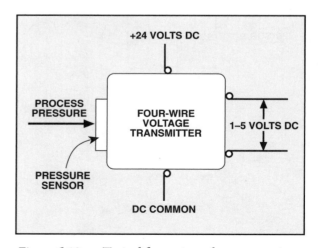

Figure 6.12 Typical four-wire voltage transmitter

Pressure Measurement and Control

proportional to pressure. For example, if the input pressure calibration range is 0 to 200 psi (0 to 1,400 kPa), the transmitter produces a proportional voltage signal of 1 to 5 VDC, or a similar voltage range. At 0 psi (0 kPa), which is the lower range value (LRV), the output voltage is 1 V. At 200 psi (1,400 kPa), which is the upper range value (URV), the output signal is 5 V. A linear relationship exists throughout the range between the lower range value and the upper range value.

If the equipment that uses the output signal of 1 to 5 V is close to the transmitter, its signal accuracy is not significantly impaired. However, as the signal is transmitted to a location over 50 ft (15 m) away, signal wiring resistance and coupling of electrical noise impairs the quality and accuracy of the signal. Because of this distance limitation, four-wire transmitters are infrequently used.

Two-Wire Current Transmitters

Eliminating signal inaccuracy caused by wire resistance and electrical noise is achieved with a two-wire transmitter (fig. 6.13). This type of transmitter is universally used in process applications where accurate measurement of pressure is needed.

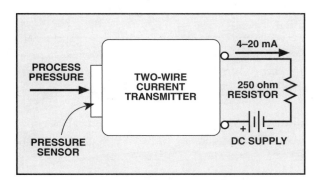

Figure 6.13   A two-wire current transmitter

The transmitter receives its pressure signal from the process source and operates a capacitive pressure sensor, which is a two-wire pressure transmitter (fig. 6.14). The transmitter converts a change in capacitance to an output signal current of 4 to 20 mA and sends it to a signal loop in the transmitter, which delivers it to the device performing the indication, recording, or reception. The signal-terminating

Figure 6.14   Electronic two-wire pressure transmitter with zero and span adjustments (Courtesy Taylor Instruments)

device typically contains a 250-Ω resistor, which converts the current range to 1 to 5 VDC. In many cases, the signal terminating device is a meter, which indicates process pressure either directly or as a percentage of full range.

Two-wire current transmitters may be reranged, or calibrated, within their operating range either manually with a screwdriver and a pressure source or with a hand-held interface unit, depending on their manufacture. Linearity exists between the process variable, which is pressure, and the output signal current whose lower range value begins at 4.0 mA. Because of this linearity, it is a simple procedure to relate the input pressure to the output current with a straight line.

The electrical current signal output of a two-wire transmitter whose LRV begins with 0 is related to the input pressure through the following equations—

$$mA = \left(\frac{P}{PSI} \times 16\right) + 4 \qquad \text{(Eq. 6.1)}$$

or

$$P = \left(\frac{mA - 4}{16}\right) \times PSI \qquad \text{(Eq. 6.2)}$$

where

mA = signal current at pressure P
P = actual pressure, psi
PSI = pressure span, psi.

These mathematical relationships apply to all units of pressure whose lower range value begins with 0. Instrument technicians can manually calibrate pressure transmitters that have screwdriver adjustments for adjusting zero and span, or range. To calibrate these units in the field, the following procedure is suggested.

1. Determine the desired calibration range—for example, assume the range is 0 to 100 in. of water for a low-pressure gauge or differential-pressure transmitter.
2. Confirm that the transmitter has an operating range greater than 100 in. of water and calibration over the range of 0 to 100 in. of water is permitted.
3. Place the transmitter in bypass to avoid upsets in the process.
4. Connect an adjustable pressure source that can operate over the desired calibration range to the transmitter's input port.
5. Connect a DC mA meter to the transmitter's test terminals to measure the signal current.
6. Set the input pressure to 0 in. of water; then, adjust the zero adjustment until the signal current on the meter is 4.00 mA.
7. Set the input pressure to 100 in. of water; then, adjust the span adjustment until the signal current on the meter is 20.00 mA.
8. Repeat this procedure until the zero and span set points remain unchanged. Check the midpoint of 50 in. of water to confirm linearity at 12.00 mA.
9. Return the transmitter to service.

Smart Transmitters

In many applications using transmitters, it may be economically advantageous to use smart pressure transmitters. Smart transmitters reduce the time involved in reranging or confirming transmitter set points. Such transmitters contain electronic microprocessors that process information from a digital communicator, or interface unit; store this information, and allow reranging without the need for a separate pressure source (fig. 6.15).

Smart transmitters can perform multiple functions, including reranging, store tag identification, messages, and other data. They can be trimmed to adjust the transmitter to specific applications for maximum accuracy.

Figure 6.15   Smart pressure transmitter and interface unit (Courtesy Fisher Rosemount)

Electronic Transmitter Configurations

Electronic transmitters used for pressure measurement can be obtained in a variety of configurations for various applications. Among these are gauge pressure, absolute pressure, differential pressure, and gauge pressures up to 6,000 psi (40,000 kPa).

The materials used to interface with the process can be stainless steel or other corrosion-resistant metals. Options include local metering, push-button set points, square-root extractors for flow measurement, and the like.

Pressure Measurement and Control

VACUUM MEASUREMENTS

When it is necessary to measure pressure that exerts less than 1 mm of mercury—in other words, pressures that are near zero absolute pressure—special measurement equipment is required. The range of pressures from 1 micrometre (μm) of mercury to 1 mm of mercury is referred to as *high-vacuum range*. Pressures below 1 μm are referred to as *ultra-high vacuum range*.

Bourdon gauges can measure down to 10 mm of mercury, while bellows and diaphragm gauges can measure values as low as 0.1 mm of mercury. The McLeod gauge, the Pirani gauge, and the thermocouple vacuum gauge measure pressures below 0.1 mm of mercury. To measure ultra-high vacuum, ionization gauges are used almost exclusively.

McLeod Gauges

A McLeod gauge is a column instrument capable of measuring pressures down to 1 μm of mercury (fig. 6.16). However, it cannot be used in automatic control because it cannot transmit information to a remote point. Thus, a McLeod gauge is primarily used as a standard of measurement to check the accuracy of other instruments.

Pirani Gauges

The Pirani gauge measures pressure by inferential means—that is, it does not directly measure pressure. Instead, a Pirani gauge measures electrical effects that occur in an environment of rarefied air or another gas. These measurements can be related to pressure values.

The Pirani gauge shown in figure 6.17 consists of a battery, ammeter, and a pressure-controlled glass bulb containing a quantity of resistance wire whose resistance depends on temperature. Current flowing in the resistance element causes it to heat up, which changes its resistance. Some of the heat, which is energy being dissipated at the same rate that the battery supplies it, is carried away by the supporting electrical leads, some by radiation, some by gas thermal conduction, and an insignificant amount by convection.

Only gas thermal conduction relates to pressure. Heat loss by gas thermal conduction reaches a high and relatively stable value for pressures greater than about 10 mm of mercury. For pressures above about 1 mm, heat loss by gas thermal conduction declines with pressure, becoming approximately linear at pressures below about 0.1 mm of mercury.

With a given current flowing in the resistance element and a given low-gas pressure existing in the bulb, the temperature and the resistance of the element reach stable values. If the pressure within the bulb is reduced to values within the range for which gas thermal conduction and pressure are definitely

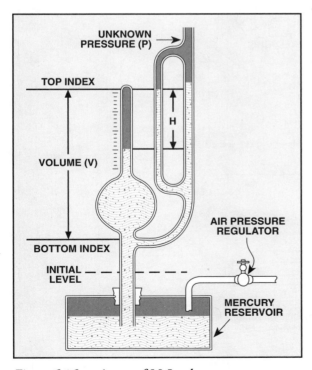

Figure 6.16   A type of McLeod gauge

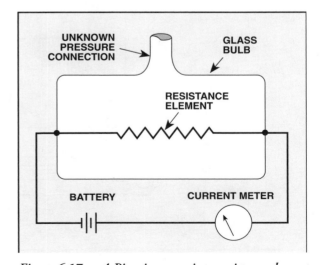

Figure 6.17   A Pirani gauge using a resistance element

related, heat loss from the element decreases and its temperature increases to a higher value, thus increasing its resistance and lowering the current flow.

A practical Pirani gauge is shown in figure 6.18. At top is a schematic of the gauge; at bottom is a drawing of the actual gauge. Referring to the schematic, note that the gauge has two similar resistance elements, one of which is evacuated and sealed off, and the other of which is fitted with an unknown pressure connection. These elements form arms of a *Wheatstone bridge circuit*. (A Wheatstone bridge is a circuit that measures the electrical resistance of an unknown resistor by comparing it with a known resistance. Typically, it consists of four resistors interconnected by a conductor to form a bridge.) When the unknown pressure is equal to that in the evacuated element, current flow and resistance values of the two elements are equal. Consequently, no current flows through the pressure-indicator meter. Once the unknown pressure rises in value, the ability of its associated resistance element to dissipate heat increases, lowering its resistance and upsetting the balance of current flow. This imbalance is reflected as a pressure reading on the meter.

The Pirani gauge is popular, not because it is a precision instrument, but because of its simplicity, low cost, facility for remote indication, ruggedness, and application to a very important pressure range.

Thermocouple Vacuum Gauges

The thermocouple vacuum gauge is similar to a Pirani instrument in that both depend on heat loss by gas thermal conduction for their operation.

In a simple thermocouple gauge (fig. 6.19), the thermocouple element is welded to the midpoint of a heater filament, and this joint represents the hot junction of the thermocouple. Its cold junction is one of the connections to the base terminals.

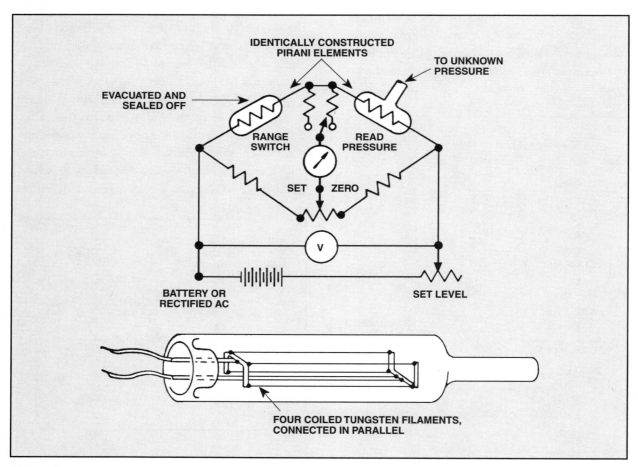

Figure 6.18 *A Pirani gauge using a Wheatstone bridge*

Pressure Measurement and Control

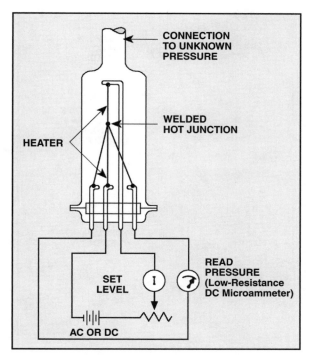

Figure 6.19   Thermocouple vacuum gauge

Current flow in the filament heats the hot junction and filament. The rate at which this heat is dissipated at low pressure is a function of pressure, just as in the Pirani gauge. The thermocouple, which produces a voltage proportional to the differences in temperature between its hot junction and cold junction, is capable of actuating a sensitive current-measuring device, such as a microammeter. The microammeter can be calibrated to read in pressure units.

Thermocouple gauges can be made to have usable ranges of a few to several hundred µm of mercury, although their accuracy is no better than a Pirani gauge. For the type of unit described, manufacturers provide a standard calibration curve; then, to compensate for production variations among units, a particular and appropriate heater (or filament) current is specified to bring individual units as near as practicable to the standard.

PRESSURE CONTROL

Pressure can be obtained from many sources and it can be used and controlled for many purposes. Pressure naturally exists in downhole formations penetrated by oilwells, and it exists in the form of liquid column pressure, which the liquid's weight and height causes. Compressors, pumps, and boilers also create pressure. Regardless of its source, to be useful, pressure must be controlled.

Safety may be the reason for controlling pressure. For example, steam boilers, petroleum stock tanks, and air compressors are equipped with pressure-relief valves to prevent excessive pressure from damaging equipment, which, in turn, can harm personnel.

Pressure control may be necessary to regulate specific pressure values. Regulation allows pressure to be applied to control and actuating devices to make them operate efficiently and effectively within their design limits.

In addition, pressure must be accurately controlled to produce a desired effect. For example, the decomposition of raw material supplied to a process might occur properly only when the temperature and pressure to which it is exposed is controlled within narrow limits.

In summary, pressure is controlled for safety, for efficient operations, and for material and chemical processing.

Pressure Control Devices

Control of pressure generally involves the flow of liquids or gases in a process. If a boiler or air compressor produces pressure, regulating the rate of pressure production controls it. Adjusting the flow of fuel to a boiler furnace controls the pressure. An on-off pressure-operated switch that stops and starts a motor-driven compressor can control the pressure in a compressed air system (fig. 6.20).

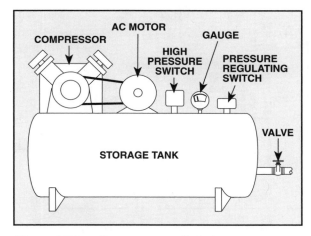

Figure 6.20   Pressure-controlled air compressor

The hydrostatic head developed by a liquid in an open or vented vessel can control the pressure in a piping system (fig. 6.21). With inflow and outflow adjusted to maintain a fixed liquid level in the tank, a particular pressure exists at the pressure tap. If inflow increases, the liquid level and pressure rise; the reverse is true if inflow decreases. Liquid level and pressure can also be manipulated by varying the outflow rate. In any case, varying the height of liquid and thus its hydrostatic pressure achieves practical pressure control.

When hydrostatic head alone cannot provide enough line pressure, a closed vessel with an additional pressure source can be used (fig. 6.22). By regulating the gas and air pressure above the liquid in the vessel, outflow pressure can be made to equal hydrostatic pressure plus the pressure on the surface of the liquid.

The main concern with pressure control is achieving and controlling reduced pressures. In most cases, high pressure flows to a control valve or a pressure regulator, which then lowers the pressure to a required value and keeps it there. Commonly, the high-pressure side is called *upstream* and the low-pressure side, *downstream*.

Although most pressure-control systems regulate the low-pressure, or downstream, side of flow, important exceptions occur. For example, boiler safety valves and vapor relief valves for petroleum stock tanks prevent the buildup of dangerous pressures on the high-pressure, or upstream, side. They also act as a seal against loss of steam or valuable

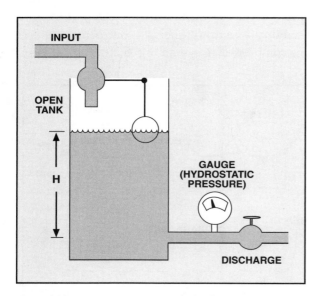

Figure 6.21   A system with hydrostatic pressure control

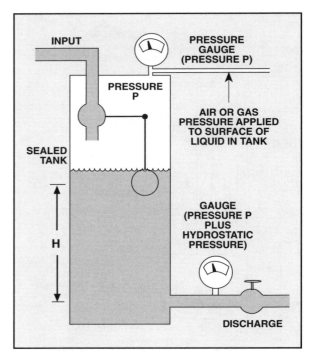

Figure 6.22   A system with air or gas pressure control

vapors at pressures below the danger point. Another exception is a regulator that maintains fixed pressure in an oil and gas separator by releasing fluid into a line on the low-pressure side. Such controllers are commonly called back-pressure regulators.

Pressure Relief Valves

Another name for a pressure relief valve is safety valve because it prevents damage to equipment and possible injury to personnel. The most common relief valve releases pressure above a set-point value, which can be varied by adjusting the loading on the valve stem. A spring-loaded relief valve (fig. 6.23) is often used to protect against overpressure in hot water and compressed air systems.

Another type of pressure relief valve protects against both overpressure and underpressure (fig. 6.24). Valves of this type are used on petroleum stock tanks that cannot withstand wide variations in pressure. For example, an oil tank interior is normally kept under a small amount of pressure to prevent the loss of volatile compounds through vaporization. Overpressure protection in the relief valve vents the tank should the pressure rise too high. At the same time, such a tank may undergo high temperatures during the day. Then, at night, it may cool off so

Pressure Measurement and Control

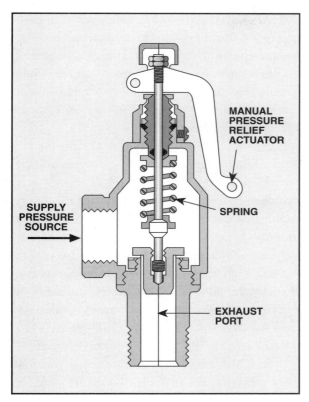

Figure 6.23   Relief valve with typical spring-loading

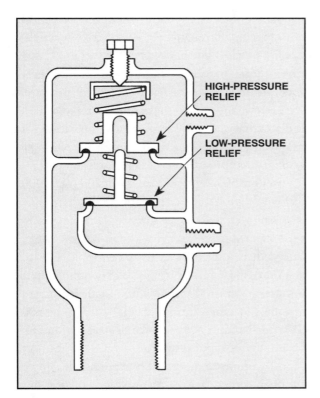

Figure 6.24   Relief valve that protects against overpressures and underpressures

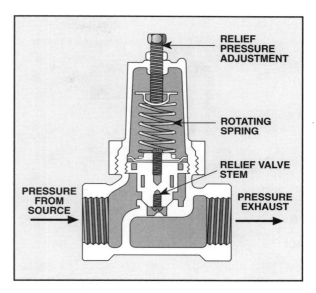

Figure 6.25   A pressure relief valve (Courtesy Fisher Controls)

much that the pressure can drop to a dangerously low level. That is, a partial vacuum occurs, which could collapse the tank. But, the relief valve opens to relieve the vacuum and protect the tank.

A third type of relief valve (fig. 6.25) protects a second-stage regulator, meter, and additional attachments in the event of failure in a first-stage regulator. A typical installation is in a rural gas line between the first and second stages of pressure regulation (fig. 6.26). Should the farm tap regulator fail, the pressure

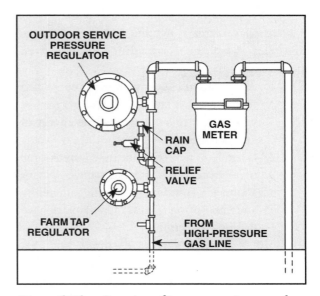

Figure 6.26   Location of instruments in a rural gas line (Courtesy Fisher Controls)

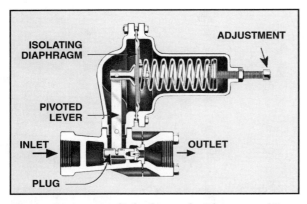

Figure 6.27   A relief valve used with a pump (Courtesy Fisher Controls)

relief valve protects the service regulator. (The regulator is called a farm tap because it frequently supplies low-pressure gas to a farm or ranch in rural areas.)

Another type of relief valve operates between the suction and discharge ports of a pump (fig. 6.27). Pressure applied to the inlet side of the valve encounters the normally closed plug. The diaphragm actuates a pivoted lever attached to the valve plug. Motion of the diaphragm caused by pump discharge pressure is opposed by the adjustable spring tension. This type of relief valve prevents the buildup of dangerously high pressures on the discharge side of a pump if a valve in the line is inadvertently closed or some other condition prevents sufficient release of the pump discharge.

Pressure Regulators

A pressure regulator is not the same as a pressure controller. Two characteristics distinguish one from the other.

1. A regulator uses energy from the controlled medium to actuate the controlling means, while a controller generally uses energy of another source.
2. A regulator may combine functions of some of its elements so that it has a single element action as a component of both controlling and measuring means. Controllers have distinctive sections.

Also, regulators, particularly simple ones, are not usually capable of the fine degree of control that characterizes pressure controllers. Thus, regulators are usually employed in service that requires only reasonably good regulation and a limited variation of flow rates. On the other hand, pressure controllers are usually employed where close control, pressure-recording capabilities, and transmission of measurement are required. Controllers are frequently closely attended, while regulators are often in isolated locations where long periods of service pass without their being attended.

Simple regulators consist of three elements: (1) a measuring element, (2) a loading element, and (3) a variable-area port. A more complex regulator may include a pilot relay. Measuring elements are usually diaphragms that act against the load element, which can be a spring or a weight. The variable-area port may be a simple plug and seat, but it is usually a disc with a soft material such as Neoprene or rubber, which assures positive shutoff. Simple units do not have large-volume capacity because the variable port, when fully opened, is still quite small—from 0.125 to 0.625 in. (3.175 to 15.875 mm) in diameter.

As a fuel, natural gas is best used at very low gauge pressures. For example, the gas pressure for ordinary domestic service is only about 5 to 6 ounces per square inch (2 to 2.6 kPa). However, gas is usually transmitted through pipelines at high pressures. Regulators, therefore, provide the several stages of pressure reduction that is necessary between a high-pressure pipeline and a low-pressure fuel line.

Intermediate-pressure regulators. Regulators usually work within a relatively narrow range of pressure. The narrow range allows the manufacturer to make a regulator that can provide close regulation within the pressures it is designed for. In other words, a single regulator that produced an output pressure of 0.75 to 7 psi (5 to 50 kPa) from a source of 700 psi (5,000 kPa) would probably not perform well.

An intermediate-pressure regulator (fig. 6.28) is used where pressures fall within a narrow range. Designed for pipe sizes of ¾ to 1½ in. (19.05 to 38.1 mm), the regulator is capable of handling input pressures of up to 425 psi (3,000 kPa). Its output pressure can range from 5 to 100 psi (35 to 700 kPa), depending on the choice of load spring and the amount of input pressure.

An intermediate-pressure regulator is simple and operates similar to a pneumatic regulator, but regulators of this type have certain properties that make them unique. For example, assume that the

Pressure Measurement and Control

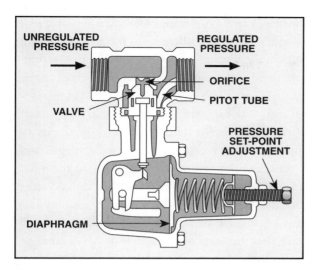

Figure 6.28   An intermediate-pressure regulator (Courtesy Fisher Controls)

regulator is set to deliver 14.5 psi (100 kPa) and that the setting is made under pressure but with no flow in the system. Once flow begins, a sharp drop in outlet pressure occurs—perhaps to 12 or 13 psi (83 to 90 kPa). However, as flow increases to higher values, outlet pressure rises and possibly exceeds the set value by a small amount. A Pitot tube that extends into the outlet stream causes the outlet pressure to rise because it has a *venturi effect* for any appreciable flow. (The venturi effect increases the velocity of flow of a fluid flowing through it and a corresponding reduction in pressure.) The venturi effect reduces pressure on the left side of the diaphragm, causing the valve to open wider and raise the pressure slightly.

The pressure drop is at its highest just as flow begins. Two factors cause this phenomenon.
1. An increase in the effective area of the diaphragm occurs as the valve opens.
2. The force of the spring wanes as the valve opens.

These factors achieve a lower balancing force and consequently a lower outlet pressure holds the valve in equilibrium.

Orifice sizes for this regulator range in diameter from 0.125 to 0.5 in. (3.175 to 12.7 mm) in increments of 0.125 in. (3.175 mm). Five color-coded load springs provide the 5 to 100 psi (35 to 700 kPa) range of outlet pressures. This regulator has no internal relief valve, and if it is used in a system such as that in figure 6.22, a follow-up relief valve is needed to protect against excessive pressures at the outlet.

Low-Pressure Regulators

The operation of a low-pressure gas regulator (fig. 6.29) is similar to a farm tap (see fig. 6.27). However, a low-pressure regulator incorporates an internal relief valve. The center of the diaphragm is connected to a linkage-shaft-pivot assembly that operates the disc-valve plug. This center connection forms a relief valve that is normally held closed by the tension of a small inner spring (not visible in the drawing).

Under normal conditions, outlet pressure pushes on the diaphragm to close flow through the regulator. However, if the disc valve fails to sufficiently restrict flow—say because of improper seating—pressure continues to build on the right side of the diaphragm. Eventually, pressure on the diaphragm compresses the inner spring, which releases gas into the chamber to the left of the diaphragm. This pressure exits the regulator through a screened, bug-proof vent (not visible in the drawing). The built-in relief valve opens when outlet pressure exceeds the desired setting by about 0.5 psi (about 3.5 kPa).

Low-pressure regulators are available in pipe sizes of ¾ to 1¼ in. (19.05 to 31.75 mm). Six orifices are available in sizes from 0.125 to 0.5 in. (3.175 to 12.7 mm), and inlet pressures range from about

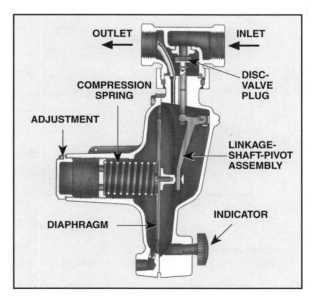

Figure 6.29   A low-pressure gas regulator (Courtesy Fisher Controls)

0.3 to 150 psi (2 to 1,000 kPa). Standard springs and orifices permit a range of outlet pressures from 0.1 to 1 psi (0.7 to 7 kPa), and a special model with high-pressure diaphragm assembly permits outlet pressures to about 7 psi (about 50 kPa).

Figure 6.30 shows performance curves for a 1-in. low-pressure regulator in which a gas of 0.6 specific gravity at a temperature of 60°F is flowing under 14.7 psia pressure. These curves show flow volume and regulation characteristics. Each curve is associated with a particular inlet pressure and orifice size—for example, 150–0.187 represents an inlet pressure of 150 psig and an orifice diameter of 0.187 in. Note, too, that the load spring in each case was set to deliver an outlet pressure of 7 in. of water for a gas flow of 50 cubic feet per hour (ft^3/hr). At a flow rate about 50 ft^3/hr, a sharp rise to 7½ in. of water occurs, shown on the left side of the graph. The performance of this regulator from shutoff to maximum flow is remarkably good.

High-Pressure Regulators

Diaphragms are good positioning members because they are broad and flexible, which makes them sensitive and able to travel linearly over a good distance. Unfortunately, these qualities make them unsatisfactory where large differential pressures exist—large differential pressures quickly wear out the diaphragm. (However, a method is available that provides for a small difference of pressure on the two sides of the diaphragm. This small differential pressure permits the use of broad and light diaphragms that have good sensitivity.)

At any rate, the regulators described earlier possess diaphragms that are spring loaded. Atmospheric pressure is on one side of the diaphragm, and downstream, or controlled, pressure is on the other side. A disadvantage is that, as the desired outlet pressure increases, the diaphragm must be made heavier, smaller, or both, to cope with the overall force that increases as the square of the diaphragm radius for a given pressure. Making the diaphragm heavier or smaller adds to its stiffness and lessens its sensitivity.

Regulators of large capacity and simple operation are available, and they differ in minor aspects from simpler types with limited capacity. Some large units are weight-loaded, and most of them use valve plugs and seats that resemble those found in control valves.

One high-pressure regulator is pilot-operated and piston-actuated (fig. 6.31). Although the device is a high-pressure regulator, it uses a single-seated valve. An external pilot valve and a piston actuator positions the valve plug in the presence of high pressure. Piston actuators are very small, yet high

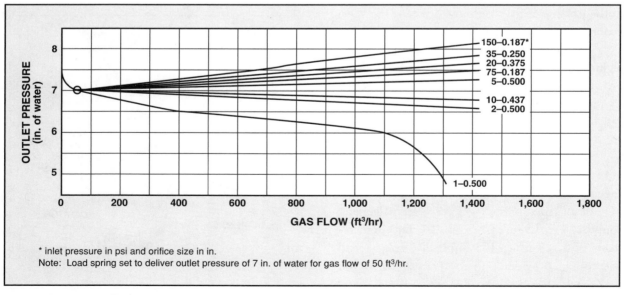

Figure 6.30 Performance curves for a 1-inch low-pressure regulator. Data is for flowing gas at 0.6 specific gravity, 14.7 psia pressure, and temperature of 60°F. (Courtesy Fisher Controls)

Pressure Measurement and Control

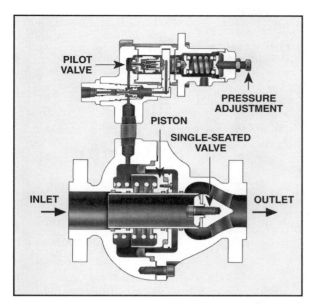

Figure 6.31   A high-pressure regulator (Courtesy Fisher Controls)

pressure causes them to develop great force. This high-pressure regulator is available in five pipe sizes ranging from 1 to 6 in. (25.4 to 152.4 mm), accepts inlet pressures up to 1,000 psi (7,000 kPa), and produces outlet pressures of 40 to 450 psi (about 300 to 3,000 kPa). The 6-in. (152.4-mm) size can handle as much as 16 million ft³/hr (450,000 m³) of gas.

Back-Pressure Regulators

Be aware that pressure control not only involves the control of downstream pressure, but also the control of upstream pressure. Regulating the rate at which fluid flows downstream controls upstream pressure. The control of upstream pressure is commonly called *back-pressure control*, and controlling back-pressure involves maintaining a fixed pressure in a vessel such as a fractionating tower or an oil and gas separator.

Regulators similar to those used for downstream pressure control are used for back-pressure control. One back-pressure regulator is a weight-loaded, double-ported type (fig. 6.32.) In many cases, spring-and-pressure loaded regulators have replaced weight-loaded regulators because weight-loaded regulators have disadvantages.

1. Weights have considerable mass to be moved and, once in motion, their momentum tends to make them overshoot. This action sometimes causes serious oscillations in the controlled variable.
2. Weight loading takes up more space than other types of regulator action. Also, the regulator must be positioned accurately so that weights and levers move without pivotal binding.
3. Weights and levers are readily accessible to tampering, thus ruling out their use in some critical applications.

Weight-loaded regulators do, however, have a few advantages.

1. Their design and manufacture is simple.
2. They can be adjusted easily for an almost unlimited range of loading forces.
3. Some designs provide a way to effectively reverse the action of the weights on the valve plug.

The operation of the weight-loaded regulator in figure 6.32 is simple, and its double-ported valve and the form of plug used enable it to handle relatively large volumes of fluid. One pivot (the one at right in the figure) is attached to the outside of the diaphragm case and supports the lever carrying the weight. The other pivot (the one at left) is attached to a shaft that passes through the upper cover of the diaphragm case and is connected to diaphragm and valve stem.

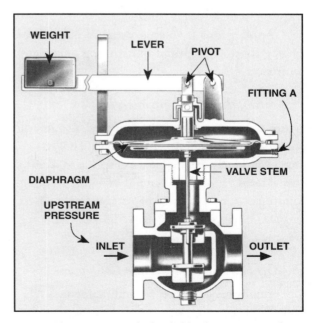

Figure 6.32   A weight-loaded back-pressure regulator

When used for back-pressure control, upstream pressure goes to the lower diaphragm case through fitting A by means of a short piece of small-diameter pipe or tubing. (Some back-pressure regulators have built-in channels that introduce upstream pressure into the diaphragm case without external piping.) When back-pressure increases to the point that its force on the diaphragm exceeds the opposing force of the weight loading, the valve plug opens. If back-pressure flow is sustained, as from an oil and gas separator, the plug soon settles at a reasonably stable opening, and back-pressure remains quite close to the desired value.

Pressure Controllers

Regulators, pneumatically-actuated valves, and pressure controllers are similar. But a pressure regulator uses energy from the controlled medium to operate, which limits its control ability, while a pressure controller relies on energy from other sources to accomplish its mission. Pressure controllers—

1. allow rapid and easy changes to pressure control set points.
2. are adaptable for remote indication, recording, and setting of pressure values.
3. offer a ready choice of on-off, proportional band, and reset and rate capabilities.
4. provide superior performance for use in very high-pressure applications.

Applications involving pressure controllers include control valve positioning using pneumatic controllers.

Proportional Band Controllers

Figure 6.33 is a diagram of a typical pneumatic proportional band pressure controller. This system uses a Bourdon tube, flapper, and spring to control downstream pressure. It contains a control set-point adjustment, a proportional band adjustment, and a separate 40-to-250-psi (275-to-1,700-kPa) operating pressure medium.

On–Off PLC Air Compressor Controllers

Air compressors compress air and maintain a relatively fixed pressure in a storage tank. Further, the pressure must be regulated. Several components

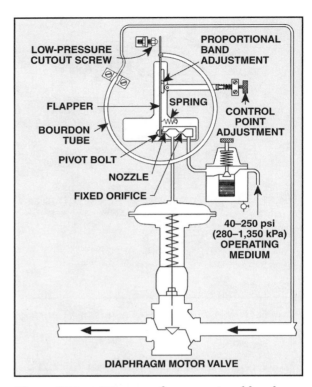

Figure 6.33   Diagram of a proportional band pressure controller

accomplish these tasks. One is a pressure-sensitive switch with a narrow dead band, which stops and starts the air compressor motor within the desired pressure range. Another is a PLC that reads and processes the pressure switch's opening and closing, along with other controlling inputs, to operate a motor starter (fig. 6.34).

The system allows flexibility in that the pressure switch can automatically control the pressure, or an HOA selector switch, which bypasses the automatic function, allows for manual pressure control. The system also shuts down in the event of high air temperature, low oil pressure, and high-high pressure conditions in the storage tank.

SUMMARY

Different units can measure pressure in several different ways. In all cases, pressure is measured against a reference, such as absolute zero, atmospheric pressure, or pressure in the system. Measurement against absolute zero is referred to as absolute pressure, against atmospheric pressure as gauge pressure, and against system pressure as differential pressure.

Pressure Measurement and Control

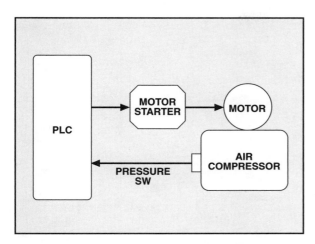

Figure 6.34 *PLC-controlled air compressor*

Bourdon tubes and gauges, diaphragm pressure elements, bellows sensing elements, and liquid manometers are mechanical devices that measure pressure. Transmitters that convert an electronic capacitor change into current over the range of 4 to 20 mA are electronic sensing devices.

Pressure control usually involves either the flow of a fluid in a pipe or vessel, such as a boiler or air compressor, or the hydrostatic head of a fluid to maintain a fixed level in a tank by adjusting inflow and outflow.

PLCs can be used with electronic transmitters to measure and control pressure in systems accurately and with fast response. Many systems are being converted to this method of control because of its ability to transmit signals remote from the pressure source and initiate control as needed.

REVIEW EXERCISE

1. Identify three pressure units used in measuring pressure.
2. A manometer works on the principle of _____.
3. What changes in an electronic capacitor to cause it to transmit a pressure signal to the transmitter?

7
Temperature Measurement and Control

Modern technology has vastly improved temperature measurement and control. While many mechanical, pneumatic, and hydraulic techniques for temperature measurement are still in use, electronic measuring devices have made significant inroads. Indeed, electronic measurement is now considered the standard method of temperature measurement and control.

Temperature is the most important variable encountered in automatic control, yet its quantitative value cannot be readily determined by direct means. Regardless of how it is determined, temperature has a profound effect on almost every process. (Its effect on personal comfort alone shows that it can bring about some spectacular events.) Temperature frequently acts with other variables to produce inter-related effects. Well known physical laws establish dependency between temperature and pressure and between temperature and volume. Also important, but not as obvious as temperature-pressure and temperature-volume relationships, is the relation between humidity and temperature. Humidity is a measure of air's ability to contain moisture at different temperature levels.

Inferential temperature measurement takes many forms, including the expansion and contraction of metals (bimetallic thermometers), changes in volume and pressure of liquids and gases (filled-system thermometers), change in electrical properties (resistance and thermocouples), and radiation energy that produces color and brightness (pyrometers). Figure 7.1 charts the devices and the temperature

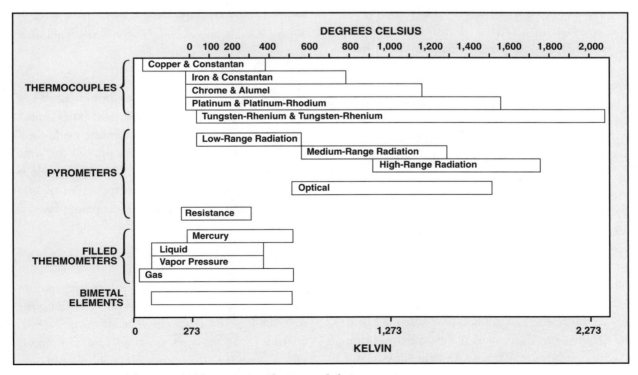

Figure 7.1   Types of temperature-measuring devices and their ranges

ranges they measure. The chart also divides each of the broad kinds of temperature measuring devices into different types. For example, thermocouple devices are divided into copper and constantan, which is a copper-nickel alloy; iron and constantan; chrome and Alumel, which is the trade name of a nickel, aluminum, manganese, and silicon alloy; platinum and platinum-rhodium; and two tungsten-rhenium alloys.

Electronic sensors display and transmit temperature signals different from the way in which mechanical sensors display and transmit temperature. Electronic sensors create proportional signals from temperature sources, and then transmit electrical current signals to remote locations.

DEFINING TEMPERATURE MEASUREMENT

Over the years, scientists who study the behavior of solids and gases have created and developed temperature measurement techniques. Temperature is involved in the behavior of solids and gases because it is a measure of molecular activity in a substance. Molecules move randomly within substances. Their movement is an irregular darting about of minute quantities of mass that possess kinetic energy. As heat is added to a substance, molecular motion increases. This motion represents the stored heat in a substance. Molecular motion is measured as temperature.

If all heat is removed from a substance, molecular activity ceases and it has no temperature. This fact suggests the existence of an absolute zero temperature, which is akin to absolute pressure, or a perfect vacuum. In each instance, a theoretically attainable zero value occurs from which there is only one way to go—up, or toward positive values. The Kelvin (K) temperature scale is based on absolute zero temperature.

Scientists involved in the study of gases and solids have created various scales to monitor temperature. The scales include the Fahrenheit scale (°F), the Celsius scale (°C), the Rankine scale (°R), and the Kelvin scale (K). Note that Kelvin temperature is written simply as the letter K and does not use the degree symbol, ° (fig. 7.2).

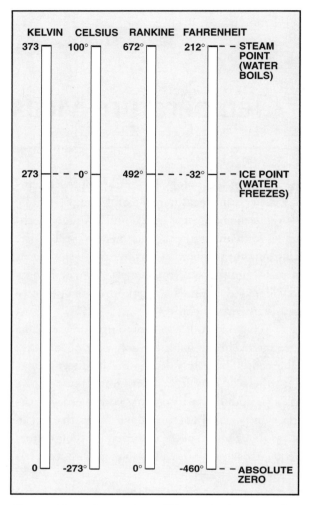

Figure 7.2   Comparison of four temperature scales

Fahrenheit Scale

The Fahrenheit scale is the most common domestic and commercial temperature scale used in the United States. It is not used much in scientific work. German scientist Gabriel Fahrenheit devised the scale in the eighteenth century. The Fahrenheit scale is usually employed to measure temperatures slightly below freezing and above the boiling point of water; however, its scale goes beyond these points.

Rankine Scale

The Rankine scale converts Fahrenheit temperatures to absolute temperatures, whereby 460°F is equivalent to 0° Rankine (R). It was developed in the nineteenth century by Scottish physicist William Thompson (also known as Lord Kelvin), who named it after the Scottish engineer William J.M. Rankine.

Celsius Scale

The Swedish astronomer, Anders Celsius, devised the Celsius scale in the mid-1800s. It is based on the freezing and boiling points of water, whereby 0°C is the freezing point and 100°C is the boiling point. Because the Celsius scale covers the span of 0 to 100, it used to be called the centigrade scale. The scale represents two criteria, boiling and freezing, the values of which can be reproduced easily and with acceptable accuracy.

Kelvin Scale

The Kelvin scale is an international thermodynamic scale. It possesses a convenient relation to the Celsius scale: a change of 1°C equals a change of 1K. Thus, a change from 40°C to 20°C results in a new temperature of 20°C—the same as a change of 20K.

The Celsius scale's 0° (freezing point) equals 273.16K. Conversely, the Kelvin scale's absolute zero value of –273°C is 0K. The Kelvin scale is named for its developer Lord Kelvin (William Thompson). The Kelvin scale better represents an interval of temperature rather than an everyday temperature.

Work performed by Jacques Charles, a French scientist who studied gas behavior under various conditions of temperature and pressure, led to Lord Kelvin's scale. Lord Kelvin's work, in turn, pointed to the existence of absolute zero temperature near –273°C.

MECHANICAL TEMPERATURE SENSORS

Measurement of temperature with mechanical means includes liquid-in-glass, liquid-filled thermometers, and bimetal thermometers. Each of these provides local indication of temperature near the site where measurement takes place.

Liquid-in-Glass Thermometers

Several mechanical temperature measuring devices use the expansion of liquid to create values on a graduated scale. Various liquids expand at different rates under various temperatures. Scientists say that liquids vary in their *coefficient of expansion* with temperature as well as other factors. The selection of the liquid and how it is contained is important in the development and construction of liquid-in-glass measuring devices.

The liquid is placed in a glass container, and the liquid moves up and down in an internal channel or capillary tube according to the temperature. The level of the liquid in the glass container indicates the temperature. A scale on the thermometer gives a direct reading. A glass thermometer is typical of liquid-in-glass thermometers (fig. 7.3). In this simple device, a thin column of fluid, usually mercury, is channeled in a thin stem with an indicating scale. A bulb contains a large percentage of the fluid. Thermometers with large bulbs generally cover a narrow range of temperatures, while small bulb thermometers cover a wide range of temperatures.

The glass not only channels the fluid as the fluid level moves with temperature changes, but also magnifies the fluid and its scale for better readability. An expansion chamber at the top of the glass allows upward movement of the fluid.

Many fluids can be used to fill the tube, but the most popular are mercury, alcohol, toluene, or pentane. Mercury can measure temperatures over the range of –30 to 925°F (–35 to 500°C). A mixture of fill fluids expands the temperature range of the thermometer. For example, combining mercury with thallium can

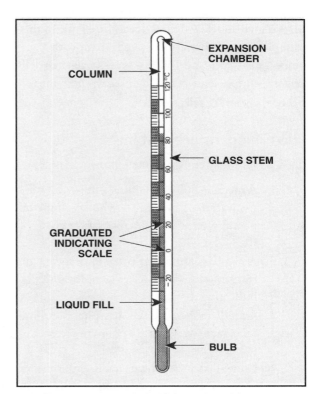

Figure 7.3   A simple liquid-in-glass thermometer

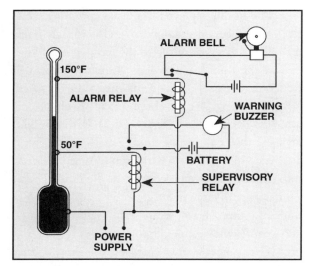

Figure 7.4 A mercury thermometer used as a component of a temperature alarm system

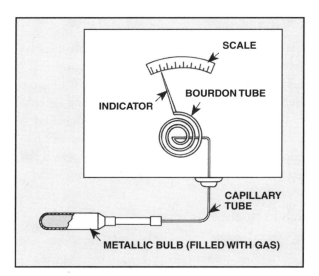

Figure 7.5 Filled-system thermometer with no compensating elements

register temperatures down to –75°F (–60°C). If an inert gas such as argon, nitrogen, or carbon dioxide is placed above the mercury column, the thermometer can measure temperatures up to 1,200°F (650°C).

One limitation of liquid-in-glass thermometers is that they cannot be used in automatic control applications. However, mercury's ability to conduct current can be used to trigger alarms if temperatures range above or below desired levels (fig. 7.4). In this example, a simple battery-powered circuit with leads connected at 50°F and 150°F sets off an alarm bell if the temperature rises to 150°F, and a warning buzzer if the temperature falls to 50°F.

Filled Temperature Systems

Filled systems may consist of the following elements:

1. A metallic bulb containing a fluid—liquid, gas, vapor, or mercury—whose volume or pressure responds to temperature changes (fig. 7.5).
2. A capillary tube that provides a means of transmission between the bulb and the indicating device.
3. An indicator connected to a spiral Bourdon tube, which may drive a pointer or a recording pen.
4. Compensating elements to offset the effects of a varying ambient temperature (not shown in the figure).

Filled systems are relatively trouble-free and accurate for most work, although failure of one component may require complete replacement of most of the mechanism. Table 7.1 compares the various types of fills and their performance.

Filled-system thermometers are relatively simple in design and inexpensive to manufacture. In process control, they can remotely indicate or record temperatures. Filled-system thermometers can be used as direct-reading devices, or they can be used with compensating elements to offset the effects of ambient temperature. Ambient temperature—the temperature of the air that surrounds the thermometer where it is mounted—can cause an erroneous temperature reading of the process being measured by adding to the reading. Consequently, in many cases, ambient temperature effects must be compensated for.

Types of filled systems include—

1. *Liquid- and mercury-filled thermometers* that operate on the principle of volumetric expansion.
2. *Vapor pressure thermometers* that operate on the principle of pressure expansion. They are partially filled with volatile substances such as methyl chloride, butane, propane, ether, or hexane. They must be used in applications where the boiling point of the volatile liquid is lower than the lowest temperature to be measured.

Table 7.1
Relative Merits of Filled Systems

System	Minimum Temperature Capability	Maximum Temperature Capability	Advantages	Disadvantages
Liquid, Class I	−300°F (−185°C)	700°F (370°C)	Linear scale; small bulb	Requires case and capillary compensation; elevation error occurs
Vapor Pressure, Class II	−300°F (−185°C)	660°F (350°C)	Low-cost; fast response; does not require compensation	Nonlinear scale; elevation error occurs; erratic response is possible at ambient temperature
Gas, Class III	−435°F (−260°C)	1,400°F (760°C)	Has low- and high-temperature capabilities; linear scale; withstands overranging	Has large bulb
Mercury, Class IV	−60°F (−50°C)	1,200°F (650°C)	Linear scale; small bulb	Requires case and capillary compensation; elevation error occurs; no low-temperature capability.

3. *Gas-filled systems* that contain nitrogen or helium under pressure. Gas-filled thermometers are capable of measuring extremely low temperatures as well as reasonably high ones.

Of the types of filled systems, gas-filled thermometers (see fig. 7.5) have the greatest measurement capabilities. They operate over wide temperature ranges and have fast response to temperature changes.

Bimetal elements can be used as compensating elements to provide correction in filled-system thermometers (fig. 7.6A and B). Figure 7.6A shows a bimetallic metal strip as part of a measuring spiral.

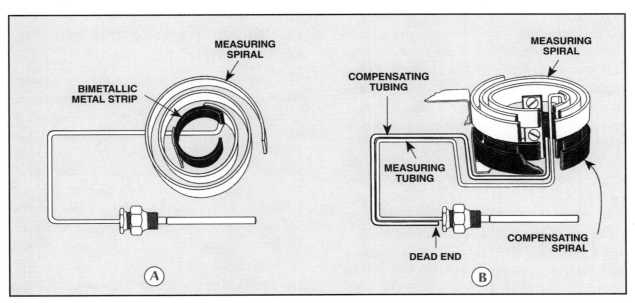

Figure 7.6   Bimetal elements used in filled systems as compensating components. A, for case compensation; B, for full compensation

The bimetallic strip compensates for ambient temperature in the case that holds the instrument. Figure 7.6B uses tubing and a spiral to fully compensate for ambient temperature.

Bimetal Thermometers

The principle behind bimetal thermometers has been known for more than a century, which is that two strips of metal with widely different coefficients of expansion—they expand at different rates at various temperatures—can be used to measure temperature. However, it was the development of alloys with widely different thermal expansion coefficients that provided the means for making bimetal elements extremely useful components in temperature-sensitive devices.

A bimetal thermometer utilizes two strips of metal, one having an extremely low coefficient of thermal expansion and the other a rather high coefficient (fig. 7.7A–C). A nickel-iron alloy containing about 36 percent nickel, called invar, has a thermal expansion coefficient that is very small, while certain other alloys containing nickel, chromium, and iron have comparatively large coefficients. The two strips can be welded, brazed, or riveted together. In figure 7.7B, note that at 0°C, the two strips are 10 centimetres (cm) long. At 100°C, however, one strip expands to 10.001 cm and the other to 10.01 cm (fig. 7.7A).

At a particular temperature, the two strips lie flat, but as the temperature changes, the two strips bend into an arc (fig. 7.7C). The outer strip forms a longer arc than the inner strip. Although the actual change in length of each strip is relatively small, because they are bonded, the free-end movement is quite impressive.

Bimetallic elements can be formed in several shapes, including spiral elements or helical elements (fig. 7.8A–C). Spiral elements (fig. 7.8A) are used in household temperature controls (thermostats) and in automatic chokes of carbureted engines. Helical elements are typically used in instruments and industrial applications, including temperature displays, temperature alarm switches, and other devices. Figure 7.9 shows a thermometer that uses a helical bimetal element in its stem.

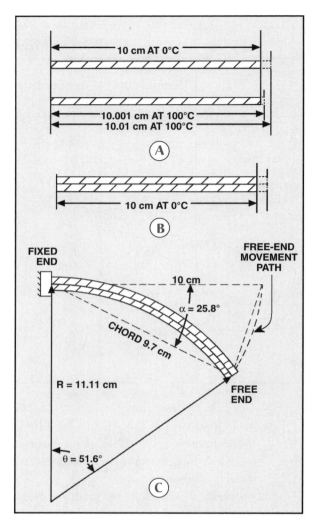

Figure 7.7   Bimetal thermometer elements

ELECTRONIC TEMPERATURE MEASUREMENT

Electronic temperature measurement systems overcome many disadvantages of other temperature systems. While not perfect, they are capable of temperature measurement up to 5,000°F (2,760°C). Electronic systems are relatively small, extremely accurate, respond quickly, and can transmit data to remote locations. The lowest temperature they can measure is about –200°F (–130°C). Filled systems can measure temperatures as low as –435°F or –260°C (see table 7.1).

Electronic Temperature Sensors

Electronic temperature sensors fall into several categories.

Temperature Measurement and Control

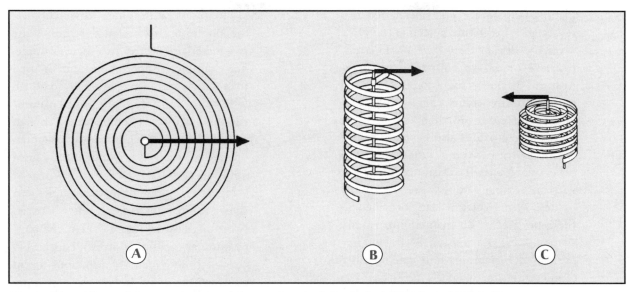

Figure 7.8   Bimetal elements in wound forms. A, spiral; B, single-helix; C, multiple-helix

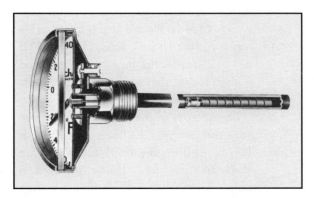

Figure 7.9   Thermometer with bimetal element (Courtesy Rochester Manufacturing)

1. *Thermoresistive elements.* These elements change their resistance value with temperature. Thermoresistive elements include silver, copper, and platinum. They produce linear resistance changes with temperature. Consequently, they are referred to as positive-temperature-resistance coefficient resistances.

   Platinum is often used in temperature sensors. It has a temperature coefficient of 0.00392Ω/Ω/°C over the range of 0°C to 100°C. This temperature coefficient means that for any temperature in the 0 to 100°C range, each ohm (Ω) of resistance a platinum wire contains becomes 1.00392 Ω for each °C increase, or 0.99608 Ω for each °C decrease.

Assume that a coil of platinum wire has exactly 100 Ω resistance at 0°C. In boiling ethyl alcohol the resistance increases to 130.58 Ω. The resistance increase is 130.58 minus 100, or 30.58 Ω. Each degree of temperature increase from the reference value of 0°C causes an increase in resistance of 0.00392 Ω for each Ω of resistance as measured at the reference temperature. Thus, the new temperature—that of boiling ethyl alcohol—can be calculated with the following equation—

$$T = \frac{R_t - R_o}{R_o \times a} \quad \text{(Eq. 7.1)}$$

where

T = new temperature in °C
R_t = resistance at temperature T in Ω
R_o = resistance at 0°C in Ω
a = resistance temperature coefficient of platinum.

Substituting values in the equation, the new temperature is—

$$T = \frac{103.58 - 100}{100 \times 0.00392}$$

$$T = \frac{30.58}{0.392}$$

$$T = 78°C .$$

Many applications use platinum temperature resistances to measure temperature.

Platinum devices are generally referred to as resistance temperature detectors (RTDs).

2. *Oxides of thermoresistive metals.* Thermoresistive oxides of copper, iron, magnesium, nickel, tin, and zinc have a temperature-resistance characteristic that is more pronounced than that of pure metal. These oxides also have a negative temperature-resistance coefficient. In other words, their resistance decreases with increasing temperature. Powdered oxides, with electrical leads attached to their body, can be formed into many shapes, sizes, and resistance values. They are called *thermistors* and can measure temperatures below 500°F (260°C).

3. *Thermocouples.* When certain two dissimilar materials are joined together by brazing, soldering, twisting, or welding, they produce a small voltage. Joining the two creates a *thermocouple junction* (fig. 7.10). Current flows from a junction at one temperature to another junction at a different temperature. One thermocouple combination is platinum-platinum/rhodium, which the American National Standards Institute (ANSI) designates as a type B thermocouple junction. Other thermocouple junctions include chromel-constantan (type E), iron-constantan (type J), chromel-alumel (type K), and copper-constantan (type T).

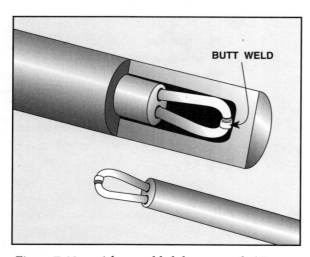

Figure 7.10   A butt-welded thermocouple (Courtesy Honeywell)

The temperature range to which thermocouples are subjected determines the two materials selected for a thermocouple. For example, copper-constantan is used for a temperature range of –300° to 600°F (–185° to 315°C). Platinum-platinum/rhodium is used at temperatures as high as 3,000°F (1,650°C). Iron and constantan operate favorably in a temperature range of –300° to 1,500°F (–185° to 815°C).

Figure 7.11 presents three examples that show the direction of current flow between two junctions. In figure 7.11A, the temperature at each junction is identical. No current flows when the junctions are at equal temperatures. If, however, one junction is heated, as the one at right in figure 7.11B or the one at left in figure 7.11C, current flows. Current flow indicates that a difference in electrical potential exists between the two junctions. Figure 7.11B and C identifies the junctions as hot and cold; in reality, they are better identified as measuring and reference junctions.

The polarity of the thermocouple junctions and therefore the direction of current flow depend on the relative temperature of the two junctions. If one is heated to a higher temperature than the other, current flows in a certain direction. If junction heating is reversed, current flow also reverses.

Thermocouples produce a small amount of voltage. For example, a copper-constantan thermocouple whose reference temperature is 32°F (0.55°C) produces an output voltage of 9.52 millivolts DC at 400°F (204.44°C). An iron-constantan thermocouple produces 11.03 millivolts DC. Bear in mind that a millivolt is one-thousandth of a volt. A typical flashlight battery produces 1.5 volts, or 1,500 millivolts; thus, thermocouple voltages are indeed small. Although their output voltage is small, the voltage relates directly and proportionately to temperature. This proportionality makes thermocouples very useful in temperature measurement (fig. 7.12).

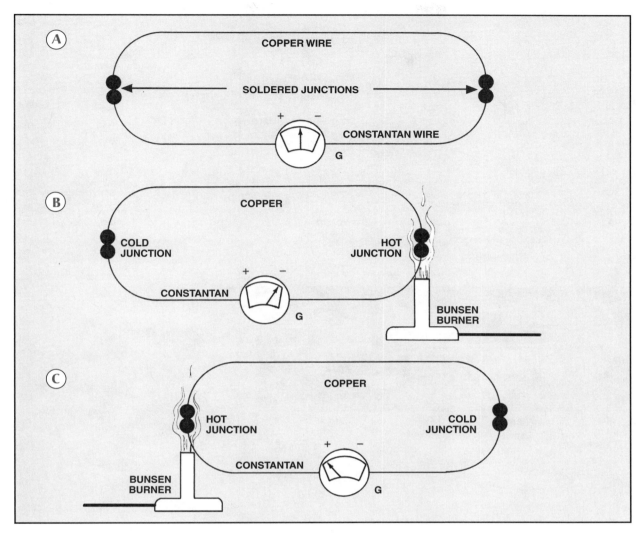

Figure 7.11   Thermocouple schematics

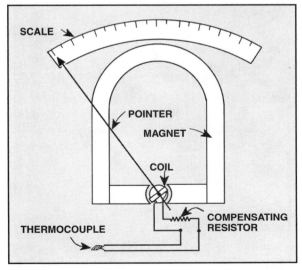

Figure 7.12   Schematic millivoltmeter with thermocouple (Courtesy Honeywell)

4. *Optical pyrometers.* The inferential temperature measuring devices discussed earlier are in direct contact with the temperature they are measuring. The maximum temperature such devices can measure is about 3,200°F (1,800°C) because most materials are unable to remain stable at higher temperatures. Thus, to measure temperatures above 3,200°F (1,800°C), an inferential device must be employed that is not in direct contact with the temperature it is measuring.

We can detect the heat of a hot substance, particularly a very hot solid such as a bar of steel, at some distance. Heat radiates from the object to us. Further, metals heated to very high temperatures emit light in the visible spectrum, beginning with dull red

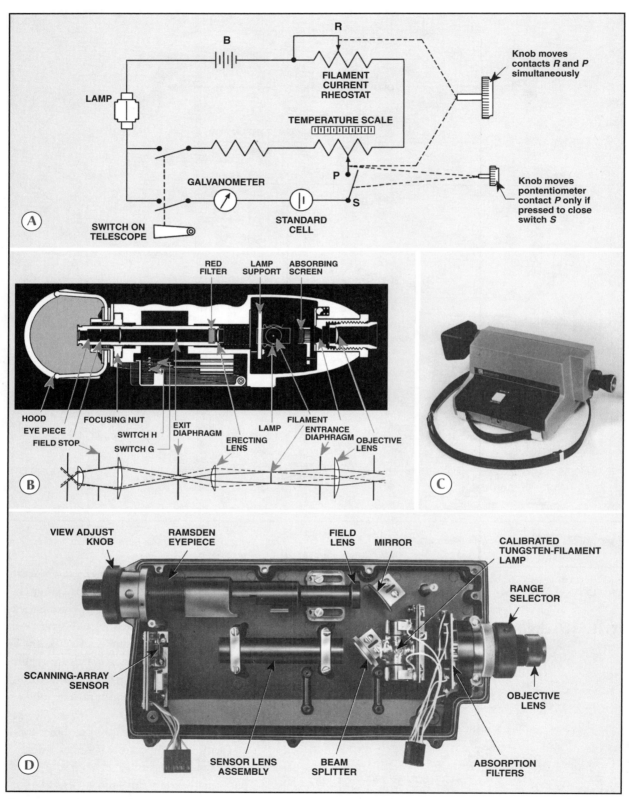

Figure 7.13 Optical pyrometers that infer temperature from color and brightness. A, simple schematic diagram of a basic idea; B, cross-sectional view and ray diagram of a commercial unit; C, automatic optical pyrometer; D, automatic pyrometer with eye shield and electronics removed to show optical components. (*Courtesy Leeds and Northrup Instruments*)

and progressing through brighter reds to an almost white incandescence. Engineers developed optical pyrometers that correlate radiated energy, color, and brightness with temperature. Figure 7.13A–D shows a schematic of an optical pyrometer (fig. 7.13A) and three actual units (fig. 7.13B–D). A typical optical pyrometer enables a technician to compare the color and brightness of an object with the color and brightness of a controlled tungsten filament to determine temperature.

Viewing the hot object through the pyrometer's eyepiece enables an observer to focus on the filament of a bulb and the hot source at the same time. The observer adjusts the filament current until the bulb and the hot object have the same brightness. At this point, the temperature of the hot source is the same as the adjusted filament. The observer then refers to a chart that plots temperature versus filament current to determine hot source temperature.

The latest pyrometers utilize scanners, electronics, and microprocessors to establish references and output temperature signals. Digital displays convey the temperature of the measured hot source.

Industrial uses of the optical pyrometer include gauging molten metal temperatures, determining the temperature of furnace interiors, and measuring the temperature of hot spots on equipment.

WHEATSTONE BRIDGES

In thermocouples, temperature variations create very small resistance element changes. Indeed, the low resistance coefficient extends to three decimal places. When current flows through the elements, it generates internal heat that may heat the resistance element that, in turn, creates a measurement error.

One circuit that accurately measures resistance with a minimum of current flowing through it is a Wheatstone bridge (fig. 7.14). It consists of four resistors connected in a manner to minimize the current flow through the sensing resistor, R_x, while producing a voltage output proportional to

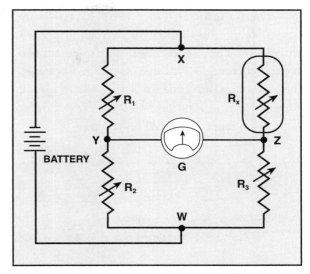

Figure 7.14   Typical Wheatstone bridge circuit

the change in resistance of the sensing resistor. The three precision resistors, R_1, R_2, and R_3, are equal in resistance value. If R_x is the same as R_3, no voltage is applied across meter G because the voltage drop from X to Y is equal to that from X to Z. However, as sensing resistor R_x changes with temperature, the bridge becomes unbalanced and a voltage difference appears across voltmeter G. This unbalanced voltage increases in value as the sensing resistor value changes. The unbalanced voltage is also proportional to the change. Devices such as temperature gauges, pressure gauges, mass flowmeters, gas detectors, and other instruments that use variable resistors as the sensing elements employ Wheatstone bridges.

A Wheatstone bridge provides a means of converting resistance change to a voltage change from zero, although the sensing resistor value does not begin at zero. And the voltmeter can be scaled to read temperature in whatever scale is desired, such as °F.

ELECTRONIC TEMPERATURE TRANSMITTERS

Just as pressure transmitters convert the process variable of pressure to a proportional electrical current, temperature transmitters also convert the process variable of temperature to a proportional electrical current. While four-wire, or voltage, transmitters are used in some instances, the more popular is the two-wire, or current, transmitter.

The transmitter uses resistive temperature sensors, or thermocouples, to detect the process temperature. The sensors convert the temperature to an electrical resistance change or a millivolt change. The transmitter then conditions the small signals and converts them to a proportional milliamp current over a given range of 4 to 20 milliamps.

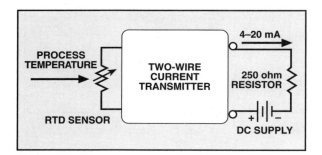

Figure 7.15 A two-wire current transmitter for measuring temperature

Two-Wire Current Transmitters

A two-wire transmitter eliminates signal inaccuracy caused by wire resistance and electrical noise. This transmitter is widely used in process applications where accurate measurement of temperature and remote transmission of its signal is needed (fig. 7.15). The transmitter receives its temperature signal from the process source and operates either an RTD (fig. 7.16) or thermocouple sensor (fig. 7.17). The change in RTD resistance or thermocouple

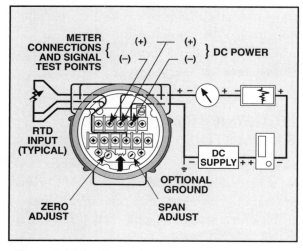

Figure 7.16 Electronic temperature transmitter connections with RTD sensor

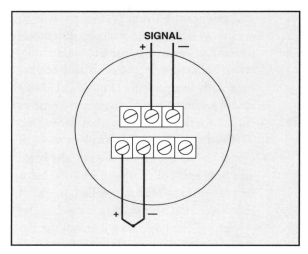

Figure 7.17 Electronic temperature transmitter connection diagram with thermocouple sensor

voltage is converted to an output signal current of 4 to 20 mA to the signal loop of the transmitter and delivered to the device performing the indication, recording, or reception. The signal-terminating device is typically a 250-Ω resistor where the current range is converted into a voltage range of 1 to 5 VDC. In some cases, the signal-terminating device is a meter that indicates in percentage of full range or process pressure.

RTD and thermocouple units may be re-ranged, or calibrated, within their operating range either with a screwdriver and pressure source or with a hand-held interface unit, depending on the manufacturer. Linearity exists between the process variable, which is temperature in this case, and the output signal current whose LRV begins at 4.0 mA and ends with a URV of 20.0 mA. Because of this linearity, it is a simple procedure to relate the input-sensed temperature to the output current with a straight line.

The electrical current signal output of a two-wire transmitter whose lower range temperature value begins with 0 is related to the input temperature through the following equations.

$$mA = \left(\frac{t}{T} \times 16\right) + 4 \qquad \text{(Eq. 7.2)}$$

or

$$t = \left(\frac{mA - 4}{16}\right) \times T \qquad \text{(Eq. 7.3)}$$

Temperature Measurement and Control

where

mA = signal current level at temperature, t
t = actual temperature, °F, °C, or K
T = temperature span, °F, °C, or K.

These equations apply to all units of temperature whose LRV begins with 0.

An instrument technician manually calibrates two-wire temperature transmitters that have a zero-and-span (or range) screwdriver adjustment as part of their assembly (fig. 7.18). To calibrate these units in the field, refer to figure 7.19 while reading the following suggested procedure.

1. Determine the desired calibration range; as an example, assume the range is 0 to 100°F.
2. Confirm that the transmitter has an operating range greater than 100°F and calibration over the range of 0 to 100°F is permitted.
3. Place the transmitter in bypass to avoid upsets in the process.
4. Connect an adjustable RTD simulator source to the transmitter in place of the actual RTD sensor that can operate over the desired calibration range.

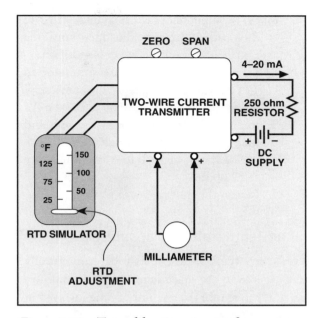

Figure 7.19 Test calibration circuit of temperature transmitter

5. Connect a DC milliameter in the transmitter's signal loop to measure the signal current.

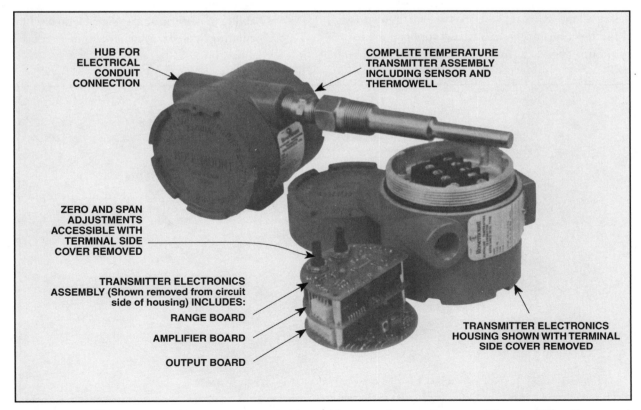

Figure 7.18 Electronic two-wire temperature transmitter with zero and span adjustments (Courtesy Fisher-Rosemount)

6. Set the input temperature simulator to 0°F; then, adjust the zero adjustment until the signal current on the meter is 4.00 mA.
7. Set the input temperature simulator to 100°F; then, adjust the span adjustment until the signal current in the loop is 20.00 mA.
8. Repeat this procedure until the zero and span set points remain unchanged. Check the midpoint of 50°F to confirm linearity at 12.00 mA.
9. Return the transmitter to service.

Smart Temperature Transmitters

With many transmitters, it may be economically advantageous to use smart temperature transmitters (fig. 7.20) to reduce the amount of time involved in reranging or confirming transmitter set points. These types of transmitters contain electronic microprocessors that allow them to process information from a digital communicator, or interface unit, store this information, and allow reranging without the need for a separate temperature simulator or source.

Smart transmitters can perform multiple functions, including reranging, store-tag identification, messages, and other data. They can be trimmed to adjust the transmitter to specific applications for maximum accuracy. In short, smart temperature transmitters can form an integral part of a temperature control system and lend flexibility to accurate and convenient measurements.

Electronic Smart Transmitter Configurations

Electronic transmitters used for temperature measurement can be obtained that allow a variety of configurations for various applications without changing transmitters. Configuration options include—

1. sensors that include RTDs and various types of thermocouple;
2. engineering unit selections, including °F, °C, °R, and K;
3. damping, or response time, of transmitter in seconds;
4. power line frequency filtering for 50 or 60 hertz (Hz) interference;
5. security jumpers to prevent unwanted tampering with calibration, or reranging;
6. upscale-downscale selection for fault conditions;
7. ability to electronically tag and identify units, and store reranging data and other information that can be written to memory;
8. ability to perform analog signal loop test by entering desired current level from interface unit; and

Figure 7.20 *Smart temperature transmitter and interface unit (Courtesy Fisher-Rosemount)*

Temperature Measurement and Control

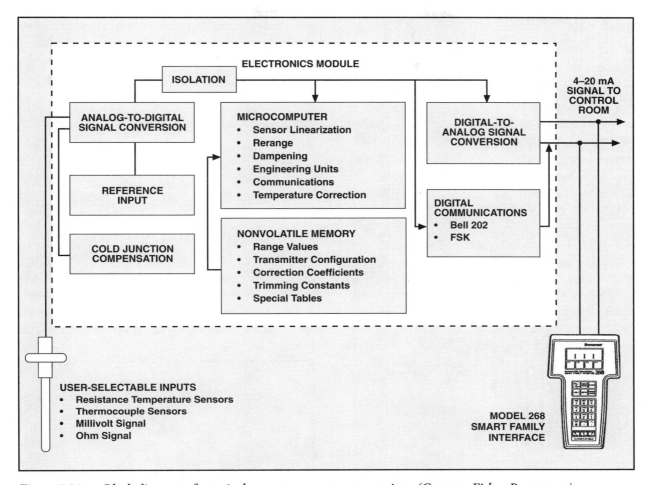

Figure 7.21 Block diagram of a typical smart temperature transmitter (Courtesy Fisher-Rosemount)

9. diagnostic testing of unit to determine its integrity.

Figure 7.21 is a block diagram that shows the major functions of a typical smart temperature transmitter. Note that the input temperature signal from an RTD or thermocouple is an analog signal. Thus, an analog-to-digital converter is needed to change the analog signal to a digital signal. After processing the data in digital format, the digital-to-analog converter transmits the result to the signal terminals as an analog 4-to-20-mA signal.

The transmitter stores reranging data in a special type of computer memory called nonvolatile EEPROM memory. Storing data in EEPROM memory simply means that a power loss or other interruption does not destroy or disrupt the information. Consequently, reranging data is immediately available as soon as power is recovered.

Another advantage of smart transmitter systems is that they allow a technician to calibrate the transmitter anywhere along the signal loop. Being able to calibrate any place in the signal loop means, for example, that if multiple transmitters are terminated to a central PLC system, all these transmitters can be reranged or tested from this central location without having to visit each transmitter.

TEMPERATURE CONTROL

Temperature is an important variable in industrial, commercial, and residential processes. For example, temperature controls in homes maintain constant, comfortable temperatures, and regulate hot-water systems. In industrial and commercial applications, temperature controls manage the flow of a fluid, such as natural gas, to the burner of a hot-water heater, or control the amount of electrical power to a heating element.

Relief Valves

Two types of relief valve are used as safety controls in liquid systems involving temperature. One contains a fusible plug that melts at a specified temperature (fig. 7.22A). Another has a bellows and sensitive bulb that is filled with an expansion fluid (fig. 7.22B). Excessive temperature near the fusible plug melts it, which releases fluid from the vessel or pipe to which the relief valve is attached. Excessive temperature near the bellows unit extends the bellows, which forces open a soft disk plug. Once temperature returns to normal, the valve closes.

Regulated Mixing Valves

Mixing a hot liquid with varying amounts of a cooler liquid can control temperature. An example is a hot-water heater in which heated water at 190°F (88°C) is mixed with cool water from the regular cold-water supply. A popular self-operated mixing valve is a three-way, piston-type proportioning unit (fig. 7.23).

Another method of fluid temperature control is transferring heat from one fluid to another through a metal wall that divides a vessel or pipe. The cool water on one side of the wall carries the heat away from the hot water on the other side of the wall.

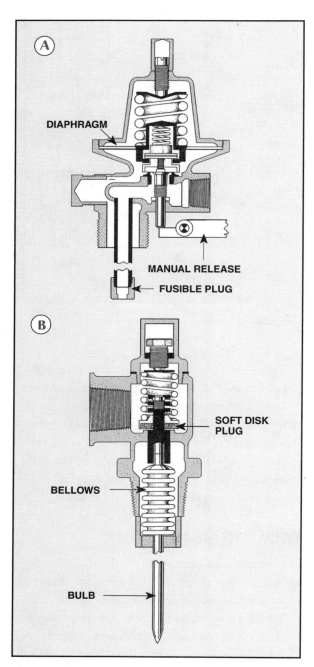

Figure 7.22 Temperature safety control valves. A, with fusible plug; B, with bellows unit and sensitive bulb

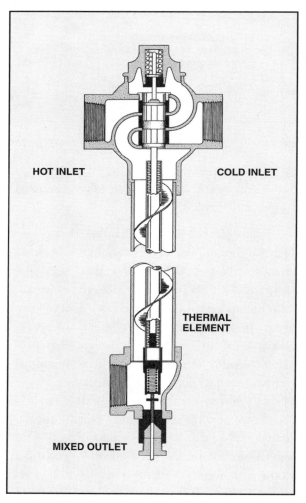

Figure 7.23   Mixing valve

Temperature Measurement and Control

Flow Regulators

Temperature regulators control the flow of a fluid control agent, such as steam, hot water, or natural gas, through a system. A common form of temperature regulator is used in systems that employ steam as a control agent. In this type of regulator, the actuator is a vapor-pressure system consisting of a sensitive bulb, capillary tubing, bellows unit, and compression spring. The valve is a double-ported poppet globe valve that operates smoothly and provides a throttling action. The unit is normally open and is therefore not a fail-safe device. Should the capillary tube, bulb, or bellows leak, the regulator provides no safety shutoff of the control agent.

PLC Temperature Control

PLCs are used extensively for many types of control, including process temperatures. Used in conjunction with electronic temperature transmitters, PLCs afford precise control of processes. PID controllers, such as PLCs, employ proportional, reset (integral), and rate (derivative) forms of control. PID control is best suited for processes such as controlling temperature in a system with large delay times and lags in system response.

Figure 7.24 shows a typical PLC temperature controller. This control scheme keeps the temperature of the liquid in the tank at a desired value. The set-point input device enters the value

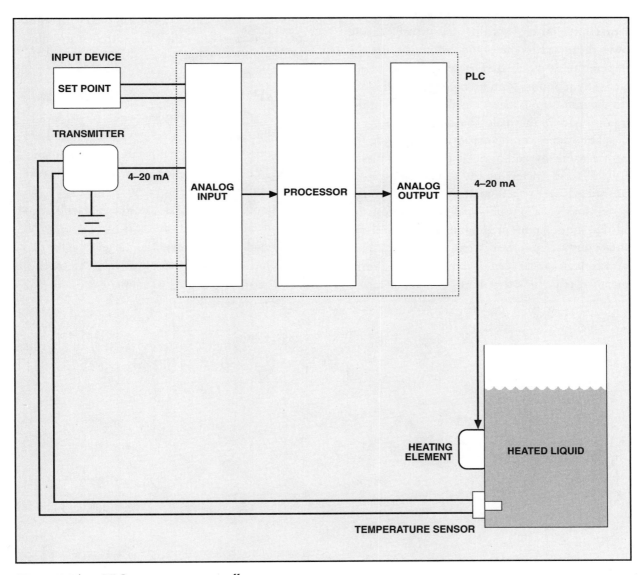

Figure 7.24 *PLC temperature controller*

of temperature to the PLC's analog input card. The analog input card converts the set-point signal into a 12-bit binary number and enters it (in block format) into the processor. A similar process occurs with the transmitter whose temperature sensor measures the actual liquid temperature. The electrical signal produced by the transmitter is also sent to the processor via the analog input card. The PID controller, following instructions from the processor's program, combines both the set-point signal and the transmitter's signal. The error between the set-point input and the feedback from the transmitter becomes a command signal, which is sent to the analog output card and distributed to the heating element controller at the liquid-filled tank.

The 4-to-20 mA signal to the heating element controller adjusts the amount of heating power disbursed to the liquid, which adjusts the temperature as required. The temperature sensor constantly monitors liquid temperature and sends it to the transmitter, where any deviation from the set point goes to the analog output card.

The system can be expanded to provide for alarm sensing by adding a discrete output card to the PLC. By detecting the actual temperature transmitted by the temperature sensor through the transmitter, a "greater-than" function can be added to the processor program to provide an alarm contact output from the discrete output card. This alarm contact can be used to activate a lamp, horn, shutdown relay, or other high-level temperature function.

SUMMARY

Temperature measurement and control exists in various forms, from mercury-filled thermometers to sophisticated electronic transmitters and controllers. Temperature has been, and continues to be, one of the most important forms of measurement because it is an integral part of the environment.

Mechanical measuring devices include liquid-in-glass thermometers, filled-system thermometers, and bimetal thermometers. Electronic devices include transmitters with resistance or thermocouple sensors, and pyrometers that measure light wavelength, or color, from temperature sources.

Temperature, unlike pressure, can vary widely, and protection against extreme limits is necessary. Control of temperature can be achieved with on-off controllers as well as with PID controllers for a finer control of its set point.

REVIEW EXERCISE

1. Name four temperature scales.
2. What produces a reaction when bimetal elements are used for temperature measurement?
3. In liquid-filled systems, what actuates the temperature display?
4. The output range of electrical current with a two-wire electronic transmitter is _____.
5. Two adjustments are available for reranging an electronic temperature transmitter. What are they?
6. What device is used to calibrate a smart temperature transmitter?

8
Liquid-Level Measurement and Control

Liquid level is a process measurement that can be achieved directly and is therefore easy to understand. In simple terms, level is a length measurement. However, its value can also be inferred by using various techniques and devices.

Many processes that deal with liquid products include level measurement. In flow processes, for example, level is often measured and controlled to keep enough fluid in a tank to equalize inflow and outflow. Also, accurate level measurement and control is very important to companies that sell products. The amount of revenue a liquid product generates is usually based on how much of it is in a sales tank, or a container. Consequently, accurate measurement of liquid level is vital.

DEFINING LEVEL MEASUREMENT

Liquid level is usually measured in length units such as in., ft, m, cm, and yards. The length, or height, of the liquid is based on a reference point located at or near the bottom of its container and the top surface of the liquid. Measuring actual liquid height is a direct measurement—that is, nothing is inferred by indirect means.

On the other hand, level measurement can be made by inference. For example, level can be determined from the weight, or *head pressure*, that a liquid exerts in a tank. In this case, the specific gravity of the liquid must be known so it can be related to a standard reference, which is the specific gravity of water. Since water has a specific gravity of 1.0, other liquids are either heavier or lighter and their head pressures vary accordingly. Since water is the reference when measuring level by means of head pressure, liquid level is usually stated in terms of in. of water (H_2O) in the conventional system. In the SI system, kPa is the preferred term, but millimetres (mm) of water can also be used.

MECHANICAL LEVEL SENSORS
Direct-Reading Instruments

People probably first measured liquid levels with a stick or rod. The stick determined the depth of a pond or a stream. In many instances, we still use graduated sticks and rods. For example, we use dipsticks to check oil level in an engine and gauge, or sounding, rods to measure fuel in buried storage tanks. Chains, or lead lines, fitted with weights on their ends, gauge the depth of water off the bow of a ship. And, personnel unwind steel tapes fitted with plumb bobs to determine, or *gauge*, liquid level in petroleum storage tanks. These methods are reasonably accurate when correlated to a specific temperature. Such measurements must be correlated with temperature because liquids in a vessel or tank expand and contract with temperature changes. Expansion and contraction alter the level of the liquid in the tank.

A *gauge cock* is a valve mounted on the side of a storage tank. When opened, liquid flows from it if the liquid in the tank is at least as high as the gauge cock. Several gauge cocks installed on the side of a tank can give an approximate measure of liquid level.

A *sight*, or *gauge*, *glass* mounted on the side of a liquid tank gives a visual indication of level (fig. 8.1A and B). Open-ended sight glasses are used on

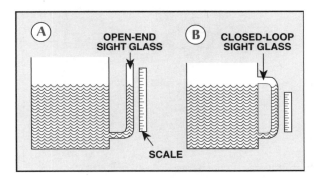

Figure 8.1 Basic types of sight glasses. A, open or vented vessel; B, pressurized vessel

open, or vented, vessels (fig. 8.1A). Closed-loop sight glasses are used on pressurized vessels (fig. 8.1B). On both tanks, a scale next to the sight glass indicates liquid level. Today, operators seldom use open-ended sight glasses, but they still employ closed-loop sight glasses on pressurized and open vessels. Gauge glasses provide a continuous visual indication of liquid level and are therefore better than dipsticks, tapes, or rods, because a person must correctly use such devices to obtain accurate level measurement.

Another type of sight glass consists of a glass tube between two valves (fig. 8.2). Adapters fit the gauge to the vessel in which liquid level is to be monitored. The metal balls in the valves block fluid flow should the sight tube rupture. Ordinarily, the balls allow free passage of fluid, but if flow velocity increases, as it does if the glass ruptures, the increased velocity drives the balls tightly against their seats to stop fluid loss. Tubular glass for this gauge is available in lengths up to about 6 ft (2 m) and for pressures up to 600 psi (4,200 kPa). Safe operating pressure

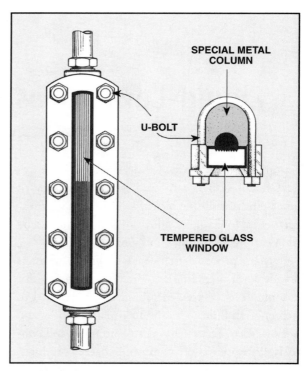

Figure 8.3 Sight glass for high-pressure, high-temperature systems

decreases with increasing length. Operators extensively use tubular glass gauges on low-temperature vessels and low-pressure steam boilers.

Operators can use gauges made of heavy brass and tempered glass windows (fig. 8.3) in tanks that store liquids under high temperature or high pressure. Grooves etched in one side of the window make it easier to see the liquid level, especially when the liquid is colorless like water.

Buoyancy Instruments

Objects lighter in weight than the liquid being measured can be used for level measurement. A common method is to float an object on the liquid and note position changes with liquid-level changes (fig. 8.4). A weight counterbalances and stabilizes the float.

Floats can also control liquid level in a tank or vessel (fig. 8.5). A valve connected to the float on a pivot restricts the float's range of travel. The valve stem limits the float's motion to about a fourth of the overall possible liquid-level change. Such level controllers often control water level in livestock troughs and toilet tanks.

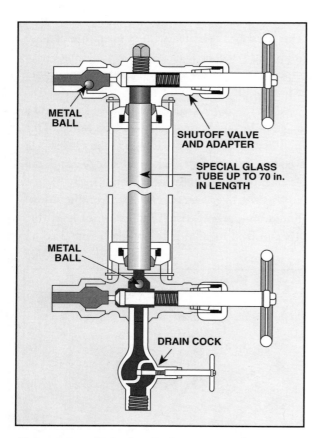

Figure 8.2 Sight glass for low-pressure, low-temperature systems

Liquid-Level Measurement and Control

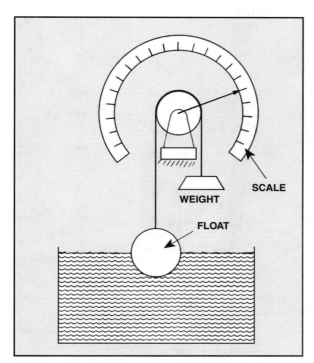

Figure 8.4 Buoyancy-type instrument with full-range indicator

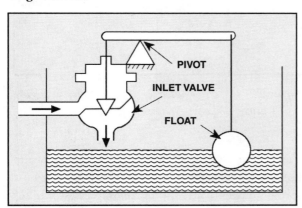

Figure 8.5 Restricted range level controller

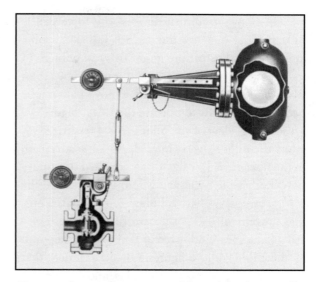

Figure 8.6 A float-operated liquid-level controller (Courtesy Fisher Controls)

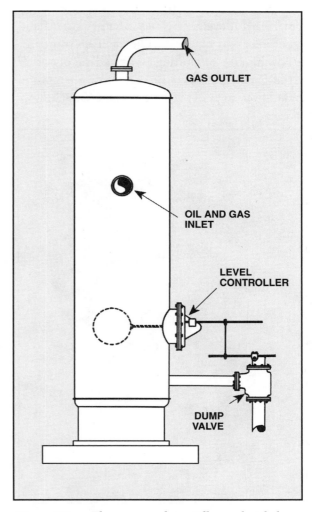

Figure 8.7 Float-operated controller used with dump valve on oil and gas separators

Operators can mount an industrial float-operated liquid-level controller (fig. 8.6) in a flanged hole in a pressurized vessel or in a float chamber. The length of the float rod and the angle of arc through which it is free to swing, limits vertical travel of this controller. Motion of the float turns a shaft that protrudes from the flanged housing through a stuffing box. Linkage attached to the rotary shaft can be arranged either to close or open the valve with a rise in liquid level. This type of liquid-level controller can control dump valves on oil and gas separators (fig. 8.7).

In buoyant control devices, the force of the buoyant float operates the control valves. In some applications, this force may not be great enough to operate the valve. To overcome this problem, a pilot valve (fig. 8.8) can be used to remove part of the load the float needs to operate the valve. Figure 8.9 is a sectional view of the pilot valve. Pressurized air enters the pilot valve at inlet A. An adjustable cam on the float's case (see fig. 8.8) is attached to a rotary shaft (not visible because it is in the case behind the cam). The rotary shaft is linked to the float. Thus, float movement drives the cam through the rotary shaft. (The cam is adjusted at point D, as shown in fig. 8.8.) Referring to figures 8.8 and 8.9, note that roller B rides against the cam and is linked to a pedal that actuates pilot stem C.

In figure 8.9, downward movement of C moves shaft N down against the tension of steel spring E. Downward movement closes exhaust port M and opens inlet port L. The open inlet allows supply air to flow into the supporting cavity and out of port H to the diaphragm. As pressure on the diaphragm and in the lower cavity of the pilot valve increases, bellows

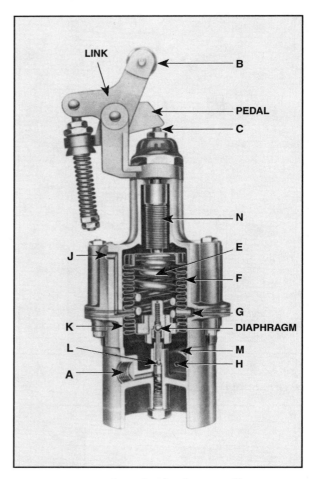

Figure 8.9   Pilot valve for the controller in Figure 8.8 (Courtesy Mason-Neilan)

F and K (whose exteriors are under the same pressure because of equalizing port G), are compressed against spring E. Eventually, spring E compresses until, at equilibrium, inlet valve L closes. With inlet valve L closed, pressure on the diaphragm is maintained at a fixed value. If the cam moves up, exhaust port M opens, which bleeds pressure on the actuator to reestablish equilibrium.

Pilot-operated liquid-level controllers give excellent control and can withstand considerable pressure. However, friction and changes in the specific gravity of the liquid under control can cause errors in the system.

Displacer Instruments

Place an object in a liquid that is heavier than the liquid. This object displaces liquid as it moves

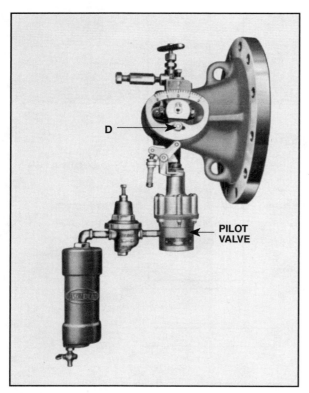

Figure 8.8   A pilot-operated controller (Courtesy Mason-Neilan)

Liquid-Level Measurement and Control

through it. Consequently, such objects are called displacer floats. They are usually used in combination with torque tubes or flexure tubes. Torque and flexure tubes do away with stuffing boxes and the friction such boxes cause. So, displacer floats can be accurately controlled and they can control a wide range of levels.

Displacer elements are also used to determine and control the level of the interface between two immiscible liquids such as gasoline and water. To use displacers to determine and control interface levels, the liquids must have different specific gravities, and the displacer float must be completely submerged. A totally submerged displacer can also measure the specific gravities of the two liquids.

One displacer device consists of a displacer float in the float cage with a connection to a float rod (fig. 8.10). The float rod is connected to a torque tube.

A typical displacer float produces a force of about 5 lb or 2.3 kilograms (kg) in air and a force of about 1 lb (0.45 kg) when fully submerged in water. Thus, the downward force the displacer float puts on the float rod varies from a maximum of about 5 lb (2.3 kg) with no water in the cage to about 1 lb (0.45 kg) when water fills the cage to a point above the displacer float. The amount of downward force that exists when the cage contains liquid is an inverse function of the density of the liquid—that is, force increases as liquid density decreases.

Figure 8.11 shows a complete control system using a displacer float and torque tube. The system controls the amount of liquid that flows into the vessel to which it is attached and thus controls the

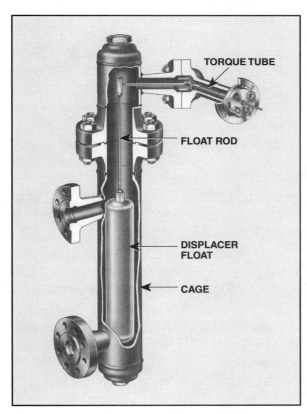

Figure 8.10 Displacer-float assembly of a torque tube displacer-type level indicator and controller (Courtesy Fisher Controls)

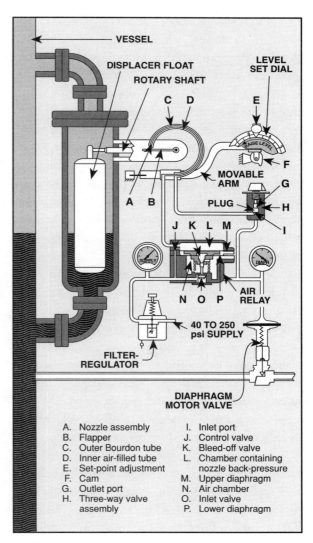

Figure 8.11 Schematic of a control system using displacer-type level indicator and controller (Courtesy Fisher Controls)

liquid level in the vessel. Flapper *B* is attached to the end of a rotary shaft. Inner air-filled tube *D*, which is fitted inside outer Bourdon tube *C*, sends air to nozzle assembly *A*. Air from the supply line goes through a filter-regulator and to the air relay. The relay contains a fixed orifice through which air flows to chamber *L*. From chamber *L*, air flows through a length of tubing to inner air-filled tube *D* and the nozzle. Chamber *L* contains nozzle back-pressure, and this pressure is exerted on diaphragm *M*. Pressure on diaphragm *M* opens inlet valve *O*, which admits air to chamber *N* and control valve *J*. To establish equilibrium, the pressure in chamber *N* works against diaphragm *P*, moving it up until inlet valve *O* is closed.

An increase in nozzle-flapper clearance causes a drop in back-pressure, which upsets the equilibrium. Thus, the pressure in air chamber *N* pushes up diaphragm *P*, which opens bleed-off valve *K*. Bleed-off valve *K* allows air to escape to the atmosphere until equilibrium is reestablished.

The same air pressure applied to the air relay is also applied to three-way valve assembly *H*. A manual adjustment of this valve positions the plug between inlet port *I* and outlet port *G*. When *G* is shut off completely, full diaphragm pressure in the relay valve is applied to Bourdon tube *C*. Pressure in the Bourdon tube pulls the nozzle away from the flapper.

If the three-way valve is positioned to shut off inlet *I* completely, the Bourdon tube remains stationary regardless of any pressure applied to the relay valve. Proper adjustment of the three-way valve gives a very sensitive control action. The three-way valve is used to adjust the proportional band from 0 to 100 percent. The proportional band adjustment varies considerably with the specific gravity of the liquid whose level is to be controlled. Manufacturers supply charts for obtaining correct adjustment.

The level-set dial operates cam *F*, which changes the position of the nozzle with respect to the flapper. The liquid level can be controlled to any value that lies between the top and the bottom of the displacer float. Floats are made in lengths of about 1 to 15 ft (30 cm to 4.5 m).

As mentioned earlier, the system in figure 8.11 controls the flow into the vessel to maintain desired liquid level. If desired, the block on which the Bourdon tube nozzle assembly is mounted can be easily repositioned to provide reverse action. Reverse action means that the system controls discharge from, rather than flow into, the vessel to maintain liquid level.

Another form of displacer-float device is shown in figure 8.12. It uses a flexure tube instead of a torque tube. The shaft in the flexure tube serves a purpose similar to the rotary shaft in the torque-tube device, but it does not rotate. Up or down movement of the displacer float positions the free end of the flexure tube. Free-end movement, in turn, can move a flapper or operate a pilot valve. The bottom drawing in figure 8.12 shows it moving a flapper. This simple system gives excellent results when an on-off action is acceptable.

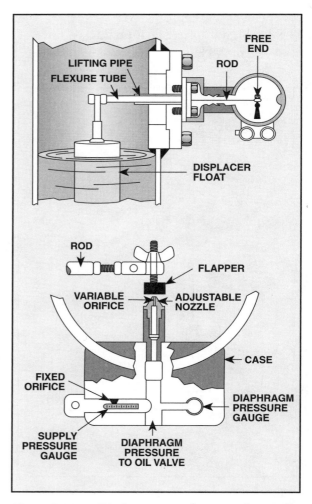

Figure 8.12   Displacer-type indicator and controller with flexure tube

Hydrostatic Pressure Instruments

Pressure caused by the height of a liquid column—that is, hydrostatic head—can be used to measure and control liquid level. For each foot of vertical height, a column of water produces a pressure of approximately 0.433 psi. (For each metre of vertical height, a column of water produces a pressure of about 9.795 kPa.) Liquids besides water also produce pressure, and the amount of pressure varies with the specific gravity of the particular liquid. A simple formula for determining hydrostatic pressure in psi is—

$$p = 0.433 \times G \times H \quad \text{(Eq. 8.1)}$$

where

p = hydrostatic pressure, psi
G = specific gravity of the liquid
H = height above the measuring point, ft.

For example, if a liquid's height is 28.3 ft from a reference point near a tank's bottom and the liquid's specific gravity is 0.9829, the hydrostatic pressure in psi is—

$p = 0.433 \times 0.9829 \times 28.3$
$p = 0.426 \times 28.3$
$p = 12.06$ psi.

For determining hydrostatic pressure in kPa/m, the equation is—

$$p = 9.795 \times G \times H \quad \text{(Eq. 8.2)}$$

where

p = hydrostatic pressure, kPa/m
G = specific gravity of the liquid
H = height above measuring point, m.

For example, if a liquid's height is 8.62 m from a reference point near a tank's bottom and the liquid's specific gravity is 0.9829, its hydrostatic pressure in kPa/m is—

$p = 9.795 \times 0.9829 \times 8.62$
$p = 9.628 \times 8.62$
$p = 82.99$ kPa/m.

Because specific gravity is a factor in the pressure value, temperature changes, or any change that affects specific gravity, also affects the accuracy of the measurement.

When using hydrostatic pressure to measure liquid heights in closed and pressurized vessels, positive gauge pressure that bears on the surface of the liquid must be compensated for. In a pressurized tank, pressure on top of the liquid can be significant. Compensation is merely a matter of adding the gauge pressure that bears on the liquid's surface to hydrostatic head pressure.

Level Measurement in Open Tanks

Figure 8.13 shows a liquid-level-measurement system in a tank open to the atmosphere. (Tanks open to the atmosphere are also called vented tanks.) A liquid

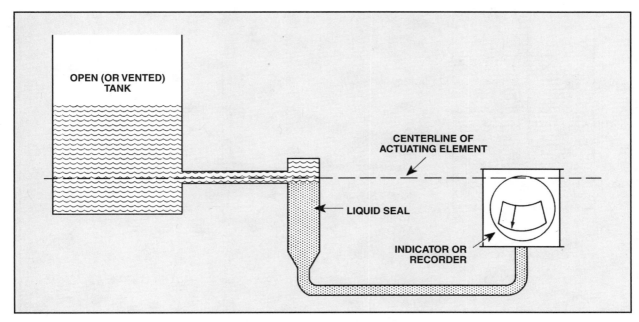

Figure 8.13   *Liquid-seal type of hydrostatic pressure measuring system*

seal in the line from the tank to the level indicator or recorder prevents tank liquid from reaching the indicator or recorder. This seal is necessary if the tank liquid is corrosive, viscous, or likely to clog the lines leading to the instrument. The centerline of the actuating element, which could be a bellows- or diaphragm-pressure unit calibrated in liquid-level units and is not shown in the figure, is located at the minimum level to be measured. The indicator can, however, be mounted lower if additional head needs to be accounted for. And, if the actuating element of the indicator or recorder is set up to position a flapper, this system can control liquid level by regulating flow in or out of the tank. In a similar way, the actuating element can operate a mercury switch and electric control system to achieve the same purpose.

Another open-tank method of liquid-level indication and control uses a diaphragm box to detect hydrostatic pressure (fig. 8.14A and B). One is an open type (fig. 8.14A); another is a closed type (fig. 8.14B). The open-type diaphragm box is either entirely open or has several vent holes. The closed-type diaphragm box mounts outside the vessel and has no openings. A pipe or a tube connects the box to the vessel. In either box, hydrostatic pressure presses against a soft neoprene diaphragm. Piping or tubing leads from the diaphragm to a sensitive pressure indicator, recorder, or controller, which is calibrated to indicate liquid level.

To place the diaphragm-box system into operation, the operator lowers the liquid level to the height of the diaphragm, which is zero liquid level. The operator then opens a vent valve for a few seconds to allow air that is trapped above the diaphragm and in the tubing to reach atmospheric pressure. After the air in the system equalizes with the atmosphere, the operator tightly closes the vent valve to maintain a constant quantity of air in the system.

As the liquid level in the vessel rises above the zero reference line, hydrostatic head exerts pressure on the diaphragm in direct proportion to the rise in level. As noted in equations 8.1 and 8.2, pressure equals either 0.433 psi/ft $\times G \times H$ or 9.795 kPa/m $\times G \times H$. Because the neoprene diaphragm offers only negligible resistance to hydrostatic pressure, hydrostatic pressure

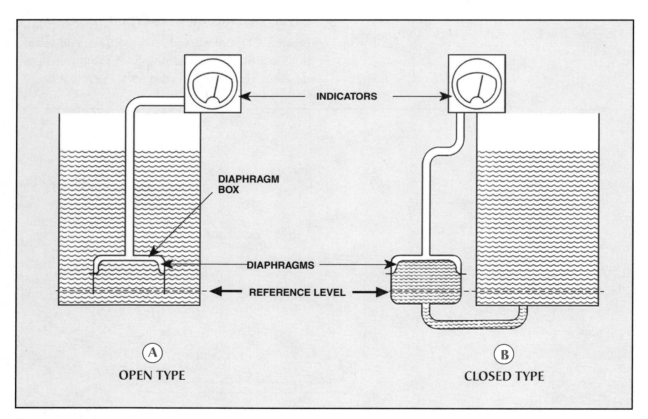

Figure 8.14 Diaphragm-box type of hydrostatic measuring system

deflects the diaphragm until air pressure above the diaphragm equals hydrostatic pressure under the diaphragm. The indicator responds to the air pressure and indicates the liquid level.

The weight of the air in the system is insignificant and thus presents no appreciable head. Thus, the elevation of the indicator does not affect the reading of the liquid level in the tank.

The open type of diaphragm box cannot be used in corrosive liquids, but the closed type is satisfactory if a liquid seal isolates it from contact with the corrosive liquid. The temperature of the liquid in contact with the diaphragm is also critical because temperatures above 150°F (65°C) can damage the diaphragm.

A special open diaphragm box, called an open air trap, can overcome the problems of corrosive liquids and high temperatures (fig. 8.15). The device traps air under it and transmits liquid hydrostatic pressure to the indicator-recorder.

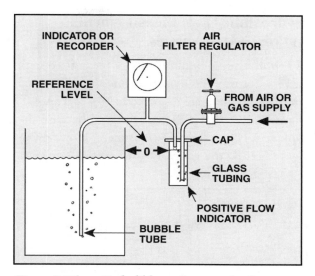

Figure 8.16   Air bubble or air-purge level system

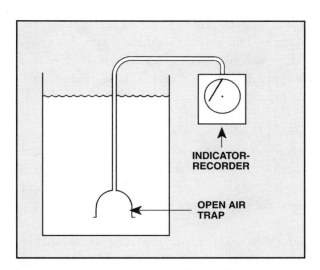

Figure 8.15   Air trap method of level indication

Air-Bubble, or Air-Purge, System

Another hydrostatic-pressure method of measuring and controlling liquid level in an open tank is the air-bubble, or air-purge, system (fig. 8.16). In this system, a pipe called a bubble tube is placed in the tank liquid. Its outlet is connected above the vessel where it represents zero reference level. Regulated air pressure is applied to the bubble tube through a positive-flow indicator.

The positive-flow indicator consists of a short length of glass tubing, which a cap seals tightly at the upper end. Liquid partially fills the tubing. Air or gas from a regulated supply goes through another tube that extends nearly to the bottom of the glass tubing. The air or gas bubbles to the top of the liquid. Accumulated air or gas at the surface then flows to the bubble tube and to the indicator or controller.

The bubble tube is large enough to pass air or gas freely if its open end is not obstructed. As liquid rises in the vessel above the bubble tube's outlet, more pressure is required to overcome hydrostatic head and allow the air or gas to escape at the open end of the tube. The positive-flow indicator shows the pressure needed to equal hydrostatic head.

The positive-flow indicator reveals problems in the system. For example, if hydrostatic head exceeds regulated pressure applied to the tube, air or gas flow ceases and the indicator-recorder indicates the pressure of the regulated air or gas supply. Hydrostatic pressure higher than supply air or gas is not detected by the indicator-recorder. It gives false readings until the liquid level returns to the point at which hydrostatic pressure is equal to or less than the air or gas supply pressure. Bubbles appear in the indicator when air or gas freely flows through the indicator. If no bubbles appear, then an interruption has probably occurred in the regulated air or gas supply, and the readings on the indicator-recorder are not necessarily correct.

Hydrostatic Level Measurements in Pressurized Vessels

When measuring liquid level in a pressurized tank or vessel, it is necessary to measure the differential pressure of the system correctly (fig. 8.17A–C). Figure 8.17A shows the correct level because the tank vent is open to the atmosphere. However, in figure 8.17B, the gauge is not accurate—it gives a higher reading than the hydrostatic head of the liquid in the tank because internal fluid pressure, p, also acts on the gauge. Figure 8.17C shows the correct way to set up the gauge. A line from the closed tank runs to the gauge and accounts for internal fluid pressure.

ELECTRICAL LEVEL MEASURING DEVICES

On-Off Level Control

Most mechanical level devices can produce an electrical signal by operating a sensitive electrical switch in the instrument. Because such switches are either normally open or closed, they are on-off devices. On-off switches are most often used in simple level-control systems.

Contact elements in such switches are either a conductive hard metal or mercury. The switches are usually in a liquid-level control assembly that is located away from the liquid being measured. Many on-off level-sensing devices have floats (fig. 8.18) that move upward as the level rises and downward as the level falls. This motion operates the switch.

Electric liquid-level controllers can be used for low- and high-level alarms and discrete level indicators. They can also operate multiple outflow pumps. For example, two outflow pumps and one inflow pump from a liquid source, such as a pipeline, can move the liquid in and out of the tank (fig. 8.19). In this case, three switches control tank level. One is installed near the bottom of the tank, another in the middle, and the third near the top. If the level drops below the bottom-level switch, the switch turns off both outflow pumps. The inflow pump continues to run and raises the level. As the level rises to the mid-level switch, this switch turns on an outflow pump. If the level continues to rise to the top-level switch, this switch turns on a second outflow pump. The mid- and top-level switches continue to turn the outflow pumps on and off to regulate the level between the two switches. If the liquid level rises above the top-level switch, the switch not only turns off, but also locks out the inflow pump until the problem is corrected.

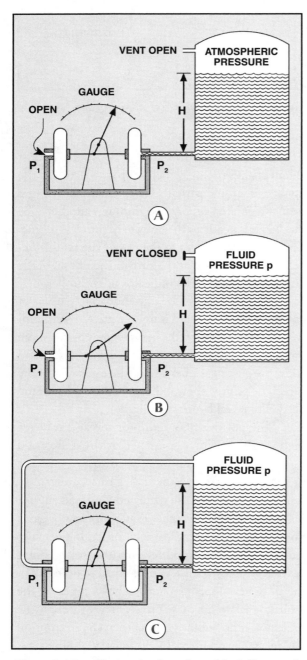

Figure 8.17 Pressurized tanks with differential-pressure gauge

Liquid-Level Measurement and Control

Figure 8.18   Electric liquid-level controller operating a switch with a float and flexure tube (Courtesy Instruments, Inc.)

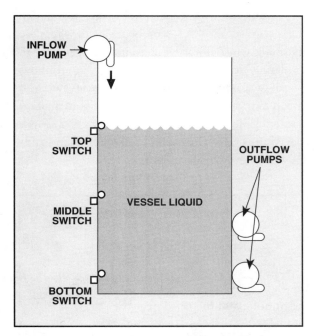

Figure 8.19   Use of level switch in dry oil petroleum storage tank

Proportional Level Measurement and Controls

Because of technological advances, many electronic devices are available for level measurement and control. The three basic electrical attributes of capacitance, resistance, and inductance form the basis for electronic level-measuring sensors.

Capacitance Level Measurement and Controls

Capacitance is the property of an electrical component that contains a conductor separated by a nonconductor, which permits the storage of energy. Such components are called *capacitors*. Air or another nonconductive material separates the conductive material in a capacitor. The conductive materials are *capacitor plates* and the nonconductive, or insulating, material is the *dielectric*. In its simplest form, the equation for capacitance is—

$$C = k \times A \div d \qquad \text{(Eq. 8.3)}$$

where

C = capacitance, farads

k = dielectric constant of insulating material (for air, $k = 1.0$)

A = plate cross-sectional area, in.^{2}

d = plate separation, in.

A simple capacitor consists of two tubes of the same length but of different diameters. One tube is inside the other and insulating material is between the tubes. The tube-within-a-tube forms a capacitance probe. The two tubes represent the plates, or *electrodes*, and the material between the tubes, which can be air or a liquid, is the dielectric. Place this simple capacitance probe in a liquid whose level is to be measured and an electrical component called a capacitance bridge can measure the resulting capacitance, which, when

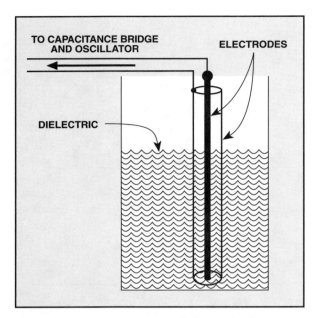

Figure 8.20   A capacitor-type liquid-level gauge

connected to an electronic oscillator, can determine liquid level (fig. 8.20). (An oscillator produces alternating current, AC, at various frequencies. The AC current operates the level readout.)

First, a zero level or lower range value (LRV) is established. With no liquid in the tube, the capacitor's dielectric is air. Air as the dielectric has a certain capacitance value. This value is established as zero, or the LRV. As liquid rises in the tank and thus the tube, a dielectric combination of air and liquid occurs. This combination dielectric has a higher capacitance than air alone. As more liquid fills the tube and the tank, the capacitance continues to increase. When the liquid level reaches its maximum, the upper range value (URV) is achieved. By calibrating the capacitance change to that of the liquid level, a means of accurately measuring the liquid level is obtained.

An electronic oscillator, whose frequency is typically in the range of 10 to 30 kilohertz (kHz), detects the capacitance change. Measuring the resulting current flow allows accurate calibration of level with capacitance. Moreover, measuring the current flow associated with capacitance changes allows continuous measurement of liquid level. Also, if the vessel is not cylindrical but of complex shape, the capacitance probe can be profiled to the vessel's irregular shape. For example, a capacitance probe measuring fuel quantity in an airplane must conform to the shape of the aircraft's fuel tanks, which are often located in the wings.

Resistance-Level Measurement

Other level measuring techniques use electrical resistance. One uses resistance strips in a probe, which is placed in the liquid along with magnetically attractive switches. The float, which moves with the liquid level, contains a magnet. The strip with its magnetic switches is placed next to the float magnet. Thus, as the float magnet rises and falls with the liquid level, it moves up and down the resistance strip. At whatever level the float magnet attains, it shorts out a portion of the magnetic switches on the resistance strip. This shorting out changes the resistance of the strip. The resistance changes in the strip are then correlated to liquid level.

Metritape is another level-measuring device that uses resistance. It is a probe inside of which is a resistance wire, which is wound in a helix around an insulating, or shorting, strip (fig. 8.21). A continuous length of conductor on the strip creates an electrical circuit to measure resistance. The probe's outer jacket is sealed. The probe containing the helix and strip is placed in the vessel containing the liquid and runs the full length of the vessel.

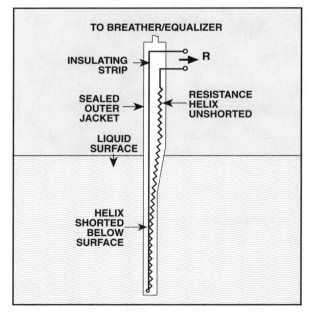

Figure 8.21   Metritape liquid-level sensor (Courtesy Metritape Corp.)

Liquid-Level Measurement and Control

The liquid's hydrostatic head pressure below the surface of the liquid causes the helix resistance wire to touch the shorting strip and the resistance change indicates liquid level. A dead band exists between the liquid surface and the point on the strip where the helix contacts the strip. This dead band is taken into account to obtain final measurement values.

Electronic Level Sensors and Transmitters

A popular method of liquid level measurement uses a differential pressure transmitter (fig. 8.22), which generates a 4-to-20 mA electrical current output signal in relation to the level of liquid in a vessel. The transmitter input tap, which is the high side of the transmitter, is placed at the lowest measured level. The low side of the transmitter is vented to the atmosphere. Such placement allows hydrostatic pressure in the tank to be measured. The system measures hydrostatic pressure in terms of the hydrostatic pressure of water (H_2O) in inches. Once the specific gravity of the liquid in the vessel is known, it is related to H_2O's hydrostatic pressure and converted to level.

Depending on whether the vessel is open or closed, and on the location of the transmitter, certain adjustments may need to be made with the resulting 4-to-20 mA output signal. If the transmitter's high side is located at the same level as the zero level point, and the low side is vented to the atmosphere, the measurement begins with zero differential pressure acting on the transmitter with an open vessel.

However, if the transmitter is located above the zero-liquid-level point, a vacuum exists on the transmitter high side although the vessel liquid level is at zero. The transmitter, in this case, produces an electrical signal output below 4.0 mA, which is undesirable. To compensate for this low output signal, the zero point of the transmitter is elevated electronically to produce a current of 4.0 mA when the vessel's liquid level is at its zero point. Put another way, bias is introduced into the transmitter to compensate for the vacuum that occurs when the vessel's liquid is actually at zero level. Thus, to *elevate zero* means to raise the signal current to 4.0 mA so that it corresponds to the actual zero level in the vessel.

The pressure, for example, may begin on the transmitter at –20 in. of H_2O at an output current signal of 3.20 mA. To begin at zero level, which is the level's LRV, the transmitter is biased with an elevation adjustment. That is, the electrical signal output is elevated to 4.0 mA when the vessel liquid level is at its zero level, or LRV. The elevation adjustment eliminates the existing pressure created by the vacuum pulled on the high side of the transmitter when it is mounted above the actual zero level of liquid (fig. 8.23).

When the transmitter is mounted below the zero level point of liquid in a vessel, the high side transmitter begins with higher-than-atmospheric pressure and a signal output above 4.0 mA. In this

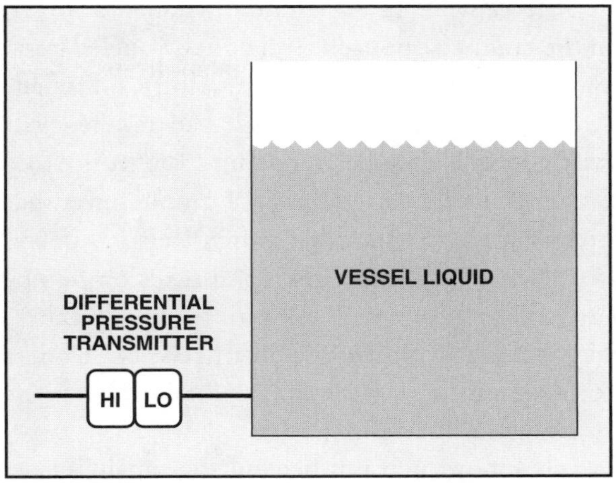

Figure 8.22 Electronic differential-pressure transmitter measuring liquid level

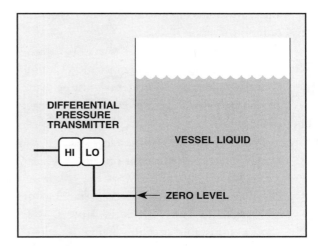

Figure 8.23 Open-tank level measurement with required elevation of zero

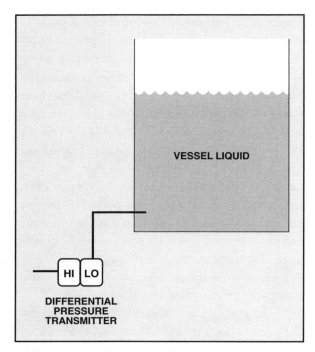

Figure 8.24   Open-tank level measurement with required suppression of zero

3. At the zero level point (LRV), adjust the zero adjustment of the transmitter for 4.0 mA. (In some cases, elevation or suppression of zero may be required.)
4. Raise the pressure (and level) to the URV. Adjust the span (or range) adjustment of the transmitter for 20.0 mA.
5. Verify the zero setting again after fine-tuning the zero and span adjustments.
6. Verify linearity of the transmitter by testing the mid-span point for 12.0 mA.

When using smart differential pressure transmitters, calibration, or reranging, can be accomplished either with a hand-held interface unit or actual process values. Select the measuring units that the process uses, such as in. of H_2O, in. of mercury (Hg), or kPa. Also, when calibrating any unit, follow the manufacturer's instructions.

LEVEL CONTROL

On-off control or proportional-plus-reset control can maintain tank levels within specified limits. Level switches that control either the inflow or outflow of the liquid in the tank can achieve on-off control. In many cases, the inflow may vary, requiring that the outflow be controlled with multiple pumps that are turned on at specified levels within the tank (fig. 8.25). For low inflow rates, only one pump may be required for the desired outflow while high inflow rates may require two, three, or four pumps to maintain the desired level. In figure 8.25, level switches are located in four positions on the tank. The bottom level switch (labeled bottom *LS*) assures that both pumps turn off when the liquid level reaches this level. The middle-level switch (middle *LS*) operates pump *P*1 to reduce the level if the level rises. If pump *P*1 cannot lower the level by itself, the top-level switch (top *LS*) operates when the level reaches it and turns on pump *P*2. Both pumps remain on until the level reaches middle *LS*, which turns off *P*2. *P*1 stays on. Under normal conditions, the level is maintained between the middle *LS* and the top *LS*.

case, the vessel liquid provides pressure through the tubing that goes to the transmitter (fig. 8.24). The pressure, for example, may begin on the transmitter at 30 in. of H_2O at an output current signal of 4.6 mA. To begin at zero level, which is the level's LRV, the transmitter is *suppressed*. That is, the electrical signal output is reduced, or suppressed, to 4.0 mA when the vessel liquid level is at its zero level, or LRV. Suppression eliminates the existing pressure created by the liquid acting on the high side of the transmitter when it is mounted below the actual zero level of liquid.

Elevation or suppression of zero to the transmitter must also be applied under other level conditions. For example, bias adjustments must be made in closed tank systems and under *wet leg measurement* conditions, which happens when filled-tubing pressure is applied on the transmitter's low side.

A calibration procedure for transmitters measuring level is—

1. Place the system in bypass so that reranging, or calibration, does not disrupt the process.
2. Use an adjustable pressure source or actual level conditions for setting the zero and span set points.

If inflow increases beyond the capability of the two outflow pumps (or if one pump fails), the level continues to rise until it operates the high

Liquid-Level Measurement and Control

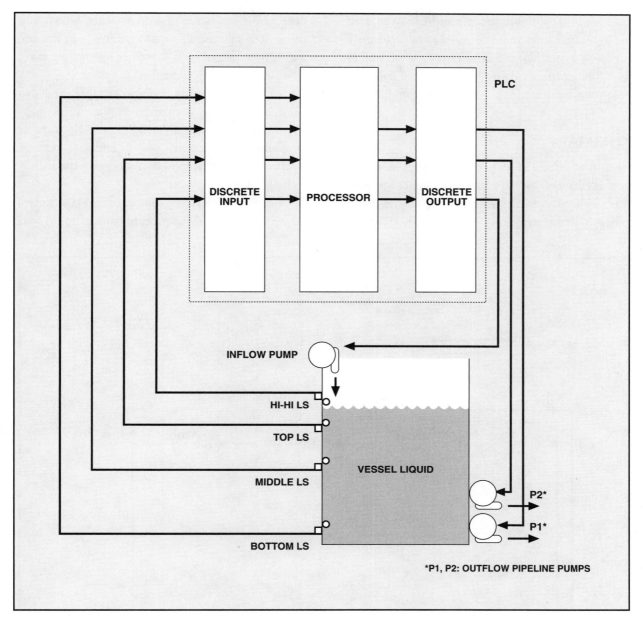

Figure 8.25   On-off level control with two outflow pumps

high-level switch (hi-hi *LS*). This switch turns off the inflow pump and allows *P*1 and *P*2 to pump the liquid level down to bottom *LS*, which then stops both pumps. A PLC controller that uses on-off, or discrete, input and output cards turns the pumps on or off.

Newer, more sophisticated, systems use variable-speed pump drives to control the flow and maintain the tank level within tight limits. A typical system includes an electronic differential pressure transmitter to monitor head pressure in the tank. The electrical signal current is delivered to an analog PLC card for processing. The PLC's processor compares the desired level set point to the level transmitter's signal. If a difference exists, an error signal goes to an analog output card. This card delivers a proportional 4-to-20 mA command signal to a variable frequency, variable speed, motor drive on the pump. (More than one motor drive and pump may be installed.) The signal adjusts motor speed to regulate the outflow and maintain the desired tank level.

Multiple variable-speed motors may be used and operated in parallel to relieve the load on a pumping system (fig. 8.26). This type of system also allows for redundancy in the event of failure of one pumping drive.

SUMMARY

Measuring liquid levels is relatively simple, and it employs devices that have been in existence for some time. Sight glasses, dipsticks, and measuring-gauge rods are direct-measuring devices. Other devices depend on buoyancy to actuate dials, levers, and other mechanisms. These mechanisms can provide a direct reading of level as well as operate electrical switches for further control.

Liquids create hydrostatic pressure, and this force can operate devices to provide level measurements. Diaphragm instruments; air-trap, or air-bubble, instruments; and differential pressure instruments are examples of devices that use hydrostatic pressure to measure level.

A popular electrical control uses level switches, which buoyant floats operate, to control liquid level

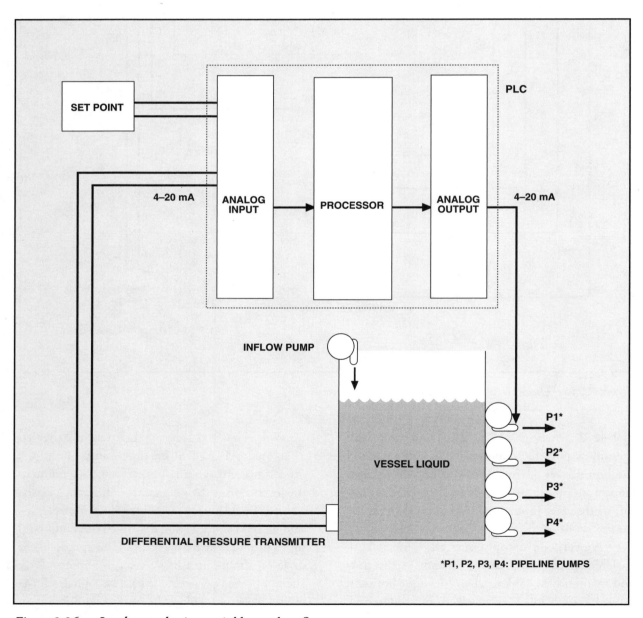

Figure 8.26   Level control using variable speed outflow pumps

and flow in on-off systems. When analog measurements are desired, variable resistance strips, variable capacitance tubes, and electronic differential pressure transmitters can be employed to provide level display information or to control pumping systems.

REVIEW EXERCISE

1. Describe the principle hydrostatic pressure instruments used to measure level.
2. Review and list the advantages of the air-bubble level measuring method.
3. What causes level indication when a float-operated switch is activated?
4. What changes in a capacitor to allow it to measure level?
5. What changes in a Metritape level sensor to allow it to measure level?
6. How can an electronic differential pressure transmitter measure level?

9
Flow Measurement

Fluid flow must be controlled if the flow regulates such variables as temperature, pressure, or liquid level. Controlling fluid flow to regulate variables requires that the flow itself be a manipulated variable. The fluid is the control agent, and temperature, pressure, or liquid level is the controlled variable. In flow measurement, fluid flow is treated as a controlled variable because it is measured and controlled to determine the quantity of fluid used or produced in a system or process.

DEFINING FLOW MEASUREMENT

Units

Flow measurement is the process of determining the quantity of fluid that passes a particular point in a given interval of time. Thus, gallons (gal) or litres (L) of water per minute (min), cubic feet (ft^3) or cubic metres (m^3) of gas per hour (hr), and barrels (bbl) or m^3 of oil per day are measurements of flow.

A quantity of fluid can be expressed as a volume or as a mass. Expression as volume is often flawed because of temperature effects. For example, a gal or L of gasoline at 40°F (4.4°C) becomes more than a gal or L at 100°F (37.8°C). Automobile owners of earlier days sometimes experienced an example of fluid expansion with temperature increases. If they filled their fuel tanks to the very top with cool gasoline and parked the car in the sun, they shortly noticed that gasoline ran out the vent hole of the filler cap. The warmth caused the gasoline to expand in volume. (Modern environmental practices prohibit gasoline or its vapors from being vented to the atmosphere.)

Wide variations in volume that accompany temperature changes in a liquid present a problem so troublesome that volume measurement has, in some cases, been abandoned. In many cases, operators and organizations use mass measurement to determine fluid volumes because the mass of a quantity of liquid or gas does not change with temperature. For example, military and commercial aviation express the quantity of gasoline, or other fuel that an aircraft carries, in terms of mass, usually in pounds (lb). Mass measurement is a much more accurate indication of the energy available from a fuel than volume measurement.

In many areas, however, volume measurement of fluids still prevails, despite its deficiencies. We still buy gasoline by the gal (L) and natural gas by the ft^3 (m^3). But, when companies transport and sell large quantities of fluid, they often state the conditions of temperature and pressure, which provide a way to determine the mass of the fluid.

Dimensions

Liquid-level measurement has one simple dimension: length. Flow measurement is more complex because it has two dimensions: volume and time, or mass and time. We can determine fluid mass if we know its density and volume because the mass of a fluid equals its density times its volume.

Sometimes only the total quantity of fluid transported, produced, or used is important. In this case, time is not a factor or dimension because quantity is more important than the speed with which it is transported or used. Many meters, such as those used for measuring the quantity of natural gas, register only the amount of fluid that passes, and not the time-rate of its passage. For example, a meter may indicate only that 25,000 ft^3 (700 m^3) of gas passed through the meter. Operators call such devices *quantity meters*. On the other hand, some meters measure quantity and time-rate. Such a meter registers, for example, 25,000 ft^3/hr (700 m^3/hr) of gas. Operators call meters that measure flow in terms quantity per unit of time *rate meters*.

MECHANICAL FLOW SENSORS AND METERS

Many types of meters and instruments measure flow. The type used depends on the kind of fluid involved, the required accuracy, and the kind of application.

Differential-Pressure Flowmeters

Differential-pressure measurement is the most popular way to determine the amount of gas that flows through a line. To create a pressure difference, or *differential pressure*, a restriction is placed in the gas flow line. When the gas flows through the restriction in the flow line, an energy change occurs. Because of certain physical laws, this energy change can be expressed in terms of the pressure change, or the differential pressure, that occurs across the restriction. The industry calls this differential-pressure *head*, and they call the meters either *head meters* or *differential-pressure meters*.

Restrictive Elements

Three devices frequently used in head meters to create a restriction in the flow line are a venturi section (fig. 9.1A), a flow nozzle (fig. 9.2B), and a plate with a hole, or an *orifice*, in it (fig. 9.1C). In each device, an indicating tube or tubes shows the differential pressure, or head, that occurs as the fluid flows through the restriction. For example, in figure 9.1C, a tube is installed on either side of the orifice plate. Note that the pressure in the line is high before the gas flows through the orifice and that it drops dramatically after it leaves the orifice. This pressure differential is noted by the height h in figure 9.1A–C. The industry employs each of these restrictive elements, but it mostly uses orifice plates, because of their simple design, low cost, and ease of installation and replacement.

Factors that enter into the choice of head meters include—

1. their accuracy and stability over long periods of use,
2. their initial cost and the cost of maintenance or replacement,
3. their ease of replacement (a factor when replacement is required to meet changes in the fluids or flow conditions),
4. their performance over wide variations in flow rate and in fluid density and viscosity,

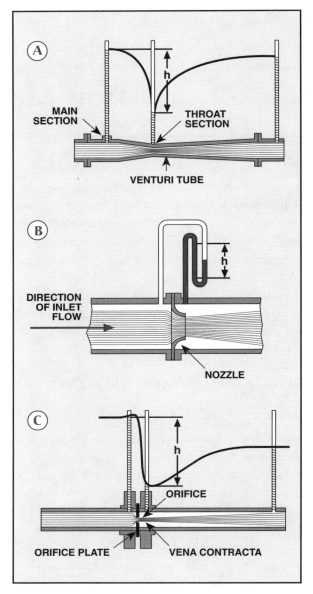

Figure 9.1 Flowmeter restrictive elements. A, venturi section; B, flow nozzle; C, flange-type orifice plate

5. their freedom from fouling by the settling of solids, and
6. the efficiency with which the element transmits fluid. An orifice plate, for example, has the greatest energy loss and a venturi section the least.

In head meters, differential pressure is measured at points called *pressure taps*. The point upstream is the high-pressure tap, and the point downstream is the low-pressure tap. Usually, operators install a mercury manometer at the pressure taps to measure

Flow Measurement

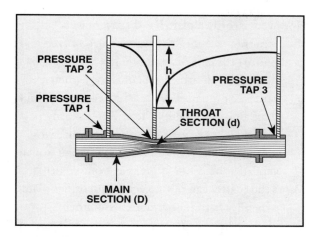

Figure 9.2   Typical venturi tube

differential pressure, but they also use other pressure-measuring devices.

Venturi Sections

A venturi section, which consists either of a venturi tube or a Dall tube (see figs. 9.2 and 9.3), is one of the most efficient restrictive elements. It is also difficult and expensive to manufacture because its specifications are critical. Further, venturi sections take a fair amount of time to replace.

A venturi tube has a main section, two pressure taps, and a throat section (fig. 9.2). The throat section has a given diameter, d. The input side of the tube begins with a main section of uniform diameter, D. A sharply tapering cone leads to the throat. After the throat is the discharge side. The discharge side is a cone section that is longer and has less slope than the throat. Pressure taps for venturi tubes are typically installed in the main section upstream and at the throat. (Fig. 9.2 shows a third tap in the straight section after the cone section.) The location of the taps is important for accurate measurement. Tap location depends on the type of fluid being measured and whether the venturi is installed in a vertical or a horizontal flow line.

The gradual velocity changes that the flowing fluid undergoes in a venturi tube accounts for the tube's efficiency. A gradual transition permits smooth flow, so little energy is wasted. In other words, a high percentage of the original pressure that the fluid entered the venturi with is recovered downstream.

A Dall tube (fig. 9.3) is similar to a venturi tube, but it has greater efficiency. A Dall tube's design differs significantly from the venturi in two respects.

1. An abrupt shoulder at the inlet cone comes after a short uniform section.
2. The low-pressure tap at the throat leads to an annular groove that encircles the throat.

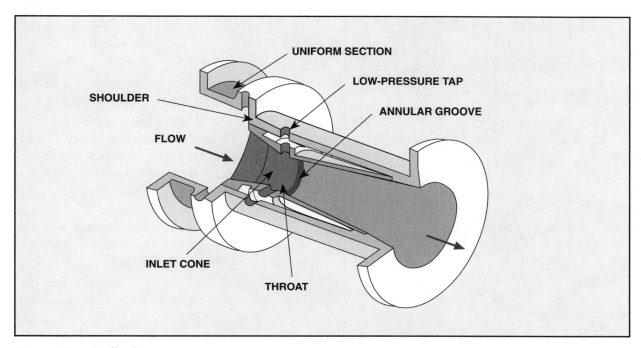

Figure 9.3   Dall tube

A Dall tube produces about twice the differential pressure of a venturi, and downstream pressure recovery is superior to other head meters.

The annular groove in the Dall tube rules out using it to measure fluids that contain dirt and debris. Dirt can clog the groove and ruin the tube's calibration. Faulty calibration leads to inaccurate measurement. What is more, the shoulder is a restriction that can accumulate dirt, which also leads to inaccuracies. Also, a Dall tube is very sensitive to pressure changes. Consequently, the piping leading to the pressure tap must be completely leakproof; even the smallest flow from a leak can lead to inaccurate differential-pressure indications.

Flow Nozzles

Because of their smooth inlet, flow nozzles (fig. 9.4) are almost as efficient as venturi tubes. However, manufacturing them is expensive because they must be made to very accurate specifications. The design requires the careful shaping of curved surfaces, so expensive machine tools are necessary.

Orifice Plates

Because orifice plates are simple and easy to manufacture, install, and replace, they are the most popular element in head meters that measure gas flow.

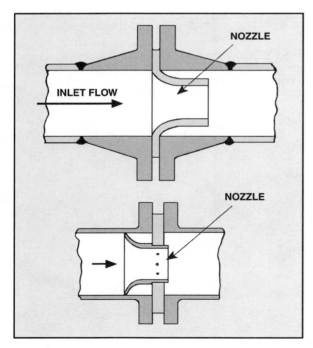

Figure 9.4   Flow nozzles

However, an orifice plate has disadvantages.
1. It cannot pass fluids as easily as flow nozzles and venturi tubes. For example, if a venturi tube's throat and an orifice are equal in diameter, the venturi tube passes about 65 percent more fluid than an orifice plate. The ability to easily pass fluids makes a venturi tube efficient, which means that the pressure loss downstream from a venturi tube is not as great as from an orifice.
2. Solid matter can pile up against an orifice plate, which leads to inaccurate measurement.
3. To measure accurately, the orifice must have a sharp edge. Thus, fluids with abrasive materials in them can rapidly erode the sharp edge and cause inaccuracies.

Despite disadvantages, the orifice plate has advantages that make it the primary choice for many applications. In metering natural gas on a large scale, for example, the permanent pressure loss it causes can be tolerated without serious consequences. Where large quantities of gas are metered, the pressure loss is insignificant. Although erosion and solid matter buildup create significant inaccuracies, the ease with which a damaged orifice can be replaced makes up for the disadvantage.

An orifice plate is installed between two flanges in a line (fig. 9.5A–C). Figure 9.5A shows the plate positioned within a circle formed by the flange bolts. Figure 9.5B is a cross section through the flange. It shows the orifice plate, the taps for the meter connection, and the drain connections. Drains are needed to remove condensation that can collect in the fitting. D stands for the inside diameter (ID) of the pipeline in which the flanges are installed. Orifice diameter is denoted by d. Figure 9.5C is a side view of the orifice plate between the flanges and one of the two jack studs that a technician turns to force the flange apart far enough to slip the orifice plate in and out of position.

Simple machine shops can manufacture and duplicate orifice plates. The plates are made of metal that resists erosion, corrosion, and distortion caused by differential pressure. Three principal factors important to plate manufacture and design are: (1) diameter of the orifice, (2) thickness of the plate, and (3) whether the edge of the orifice is rounded, beveled, or square.

Flow Measurement

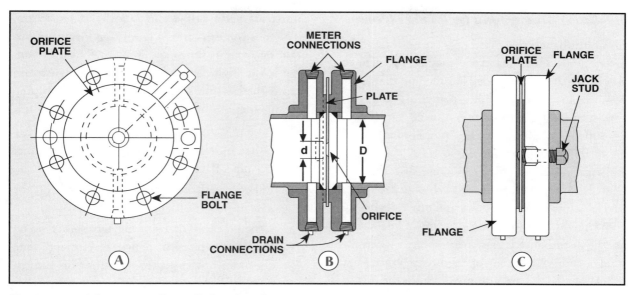

Figure 9.5 Three views of installed orifice plate

The diameter of the orifice, d, is closely related to the diameter of the pipe, D, in which the orifice is installed. The ratio of orifice-to-pipe diameter, d/D, is the beta factor and is designated by the Greek letter beta (β). Beta factors are less than 1.0 and commonly range from 0.125 to 0.75. The beta factor is one of several factors that enter into determining the fluid volume measured by an orifice meter.

Orifice plates are made only thick enough to resist buckling forces that cause distortion. A thin plate produces the best flow pattern, but it must meet certain standards. It must not exceed $\frac{1}{16} \times D$, $\frac{1}{8} \times d$, or $\frac{1}{8} \times (D-d)$.

Most orifices are made with square edges—that is, the orifice (the hole) is cut out of the plate so that the edges of the hole it makes in the plate are squared off. Square-edged holes are the easiest to make. If a plate is thin enough, the square edge allows the gas to exit the orifice and form a jet whose shape ensures accurate measurement. The best orifice is sharp edged—that is, it has virtually no edge to interfere with jet formation. Consequently, if an orifice plate has to be especially thick in relation to the diameter of its orifice, the downstream edge of the orifice is beveled. When the fluid from upstream encounters the sharp edge, it forms a suitably-shaped jet as it passes through the orifice. Unfortunately, erosion or corrosion sometimes rounds the edges of the orifice. Rounding of the upstream edge gives inaccurate measurement. An orifice with a rounded upstream edge makes the orifice behave as though its diameter is larger than it actually is.

The most widely used orifice is concentric—that is, the orifice is a circular hole in the middle of the plate. Other types of orifices are eccentric and segmental (fig. 9.6). An eccentric orifice is a circular hole that is off-center in the plate; a segmental orifice has a squared-off top. Another type of orifice plate is an annular orifice plate. It allows fluid to flow between the plate and the pipe wall. Each of these orifice plates possesses an advantage for a particular application.

Figure 9.6 Three types of orifice plate

Installation Arrangements for Primary Elements

Flow disturbances that occur near the upstream inlet adversely affect head meter accuracy. Near is defined as a distance that extends to 100 pipe diameters from the meter. For example, in an 8-in. (20.3-cm) pipe, near extends to about 66 ft or 20.3 m (100 × 8 = 800 in. and 800 ÷ 12 = 66.66 ft. Similarly, 20.3 cm × 100 = 2,030 cm and 2,030 ÷ 100 = 20.3 m). Obstructions in the pipe can cause disturbances. Obstructions include throttling valves and rough or uneven joints of pipe. Elbows and other deviations from straight-line flow also cause disturbances. Unless the disturbance is extremely severe, such as the rapid opening and closing of a throttling valve, a few yd (m) of straight, smooth pipe remove the disturbances before they reach the head meter's primary element. (The primary element in a head meter is the restriction used to create the pressure differential, such as an orifice, venturi tube, or flow nozzle.) Placing straight-run lengths of pipe upstream from the orifice plate reduces flow-line disturbances.

In addition, fluid-straightening vanes inside the pipe may be needed to break up swirls and vortices in the flow (fig. 9.7). As the fluid flows through the numerous small-area passages, the passages straighten and smooth the flow. The flowmeter manufacturer recommends the length, type, and size to use.

Location of Pressure Taps

In measuring differential pressure across an orifice plate or a similar device, a pressure tap is located upstream of the orifice and the other downstream. The location of two taps is very important in obtaining accurate flow measurement. The upstream tap is the high-pressure tap and the downstream tap is the low-pressure tap. For horizontal piping arrangements, the taps should be located as follows.

1. When measuring gas, locate the taps on top of the pipe, flange, or element.
2. When measuring liquid, put the taps on the side of the pipe, flange, or element.
3. When measuring steam, the taps should go on the top of the pipe, flange, or element when the differential-pressure-measuring instrument is located above the line. When the instrument is below the line, locate the taps on the side of the pipe, flange, or element.
4. When using flow nozzles, locate the upstream (high-pressure) tap one pipe diameter from the face of the nozzle and the downstream (low-pressure) tap one-half pipe diameter from the nozzle face.

Additional factors to consider about pressure taps include—

1. When measuring fluid flow in a vertical piping run, peripheral orientation of the pressure taps is not important.
2. When ordering flanges, consider ordering flange taps, too. Flange taps not only afford good pressure differential, but also are precisely located, if they are included as a part of the manufacture of the flanges.
3. Other tap locations include vena contracta taps used for maximum differential pressure; corner taps, which are primarily used in Europe; and pipe taps located on pipes adjacent to the orifice plate assembly. (A *vena contracta* is the part of the gas flow where its cross section is at a minimum as it emerges from the orifice. See fig. 9.1.)

Calculating Flow Velocity

The basis for flow calculations for head meters is the velocity of flow equation. The equation is—

$$v = \sqrt{2gh} \qquad \text{(Eq. 9.1)}$$

Figure 9.7   Straightening vanes

where

- v = fluid velocity, feet per second (ft/sec)
- g = acceleration because of gravity, ft/sec² (32.17 ft/sec²)
- h = differential pressure, ft.

To measure rate of flow with orifice plates, technicians must consider several corrective factors to the basic equation. Among these factors are velocity coefficient, orifice plate shape, and fluid properties. Fluid flow textbooks cover several other correction factors. But this basic manual presents only some of the factors.

The equation for liquid flow is—

$$Q = 45.47 E d^2 \sqrt{\frac{h_w}{G}} \qquad \text{(Eq. 9.2)}$$

where

- Q = rate of flow, ft³/hr
- E = efficiency factor of orifice plate
- d = diameter of orifice plate hole, in.
- h_w = differential pressure, in. of water
- G = specific gravity of flowing fluid.

The equation for gas flow is—

$$Q = 218.44 E d^2 \times \frac{p_f}{p_b} \times \frac{T_b}{T_f} \times \sqrt{\frac{h_w T_f}{p_f G}} \qquad \text{(Eq. 9.3)}$$

where

- Q = rate of flow in cubic feet per hour
- E = efficiency factor of orifice plate
- d = orifice plate hole diameter
- p_f = actual absolute pressure
- T_f = actual absolute temperature, °R
- T_b = base absolute temperature, °R
- p_b = base value of absolute pressure
- h_w = differential pressure, in. of water
- G = specific gravity of flowing fluid.

Tables, graphs, and other data are available to assist in correcting for the base values in the equations. Each application uses its own set of constants, but they all boil down to the fact that flow rate is proportional to the square root of differential pressure. Put another way, flow rate is not directly proportional to the differential pressure across a restriction such as an orifice plate. Instead, flow is proportional to the square root of diffential pressure across the restriction. This fact is important when using electronic transmitters or controllers because they may incorporate the square root function to display flow directly, or to transmit a linear signal to a remote PLC.

Mercury Manometer Orifice Meters

A popular mechanical means of measuring differential pressure is based on a mercury-filled U-tube manometer (fig. 9.8). For accurate measurement, three connections receive upstream, downstream, and static pressures. (Remember: static pressure is the pressure of the fluid as it flows through the line.) The meter uses these variables to measure the rate of flow. Upstream pressure goes to a range tube, which is a high-pressure chamber. Downstream pressure goes to a low-pressure chamber with a float in it. This float chamber's cross-sectional area is much larger than the range tube's. The float rises and falls with the mercury level in the range tube and actuates a recording pen shaft through special differential linkage. The static pressure line is attached to a Bourdon tube or other pressure-actuated device that drives a separate recording pen. Manometer-type orifice meters are rugged and capable of handling large static pressures. Manufacturers also build in protective features to prevent unusually high pressures from driving mercury out of the chambers.

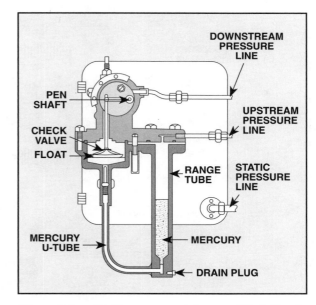

Figure 9.8 Orifice meter with a modified U-tube manometer

Bellows Orifice Meters

The bellows meter is a durable and accurate instrument. Figure 9.9 is a cross section of a bellows assembly that drives a recorder pen. The meter body consists of two opposed bellows that are screwed to a center plate and sealed against leakage by O-ring gaskets. A vacuum pump evacuates the interior spaces and the interior space of the center plate, which are then filled with an incompressible liquid that has a low freezing point.

During operation, upstream pressure is applied to the high-pressure housing and downstream pressure is applied to the low-pressure housing. Differ-ential pressure across the meter assembly compresses the high-pressure bellows, which drives incompressible liquid into the low-pressure bellows. An assembly of overrange valve plugs and linkages transmits motion of the free end of the low-pressure bellows to the drive arm. Bellows meters can measure the flow of liquids or gases, but care must be taken to prevent corrosive or extremely viscous fluids from entering the pressure housings and causing damage.

Variable-Area Meters

A variable-area flowmeter uses mechanisms that act against the force of fluid flow to drive a pen. Two types of variable-area meters are those that use pistons and those that use a rotameter. A rotameter consists of a float in a tapered tube, or a tapered float that moves in a fixed orifice. Flow passing around an annular space between the float and its container causes the float to rise until its weight counterbalances the pressure drop across it.

ELECTRONIC FLOW SENSORS AND METERS

A number of phenomena and forces are employed to detect fluid flow and convert flow information into a proportional electrical signal. Electronic flowmeters use such forces as magnetics, induced voltages, vortexes, and resistance changes.

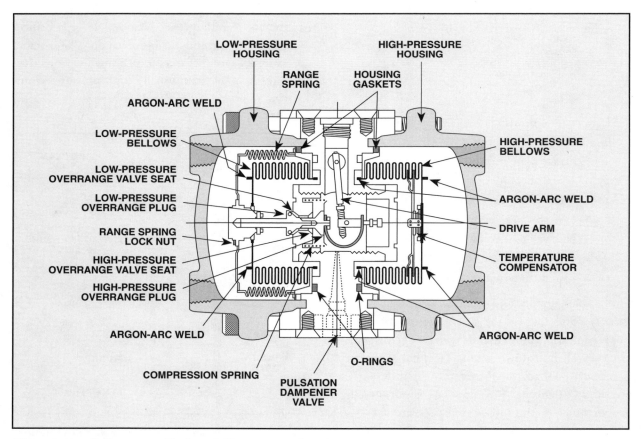

Figure 9.9 Bellows assembly for a bellows-type orifice meter

Electronic Differential-Pressure Flowmeters

Orifice meters and the differential pressure produced by them in response to fluid flow are widely utilized. In the past, operators used the output from these instruments to make charts or record local readings. Today, electronic transmitters measure the differential pressure and produce an electrical signal proportional to the pressure. Controllers and PLCs then process the information. Sometimes, the square root calculation required to provide a linear signal that is proportional to flow is incorporated within the transmitter instead of having it processed in a controller.

Electronic computers use electrical flow signals to perform complex calculations of the flow rate formulas for both liquids and gases. Further, they provide this information for use by an operator or for control purposes in a process. Since a PLC is a specialized computer, flow measurement and analysis is a relatively simple process when PLCs are available. Figure 9.10 shows a typical flow measurement system incorporating a square root extractor in the transmitter for linearizing flow from differential pressure.

Magnetic Flowmeters

A magnetic flowmeter operates on the principle that voltage is induced in an electrical conductor moving across a magnetic field. The conductor is the fluid (liquid) flowing in the line, and field coils mounted on diametrically opposite sides of a special section of line provide the magnetic field. Figure 9.11 is an exploded view of a magnetic flowmeter. It consists of a short section of nonmagnetic stainless steel tubing, which is fitted with flanges and lined with a durable insulating compound such as neoprene. Two electrodes, insulated from the tubing, fit through the lining and make electrical contact with the flowing liquid. Two coils produce a strong magnetic field. They are positioned to be perpendicular to the diameter represented by the line between the electrodes. A weatherproof cover completes the assembly.

The fluid to be measured must be a conductor, even a poor one. Alternating current applied to the coils provides the magnetic field. The result is an induced, alternating current voltage in the moving conductor, which is the liquid. Two electrodes detect this induced voltage, whose magnitude is proportional to fluid flow. Further processing of the electrode signals amplifies and converts the signals to a 4-to-20 mA control loop signal; or, the converted signals can drive a chart recorder.

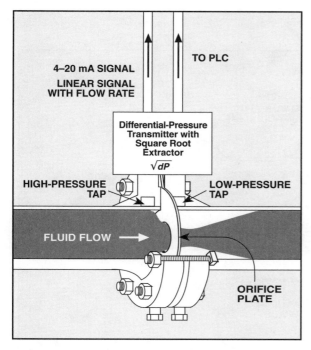

Figure 9.10 Electronic differential-pressure transmitter for measuring flow

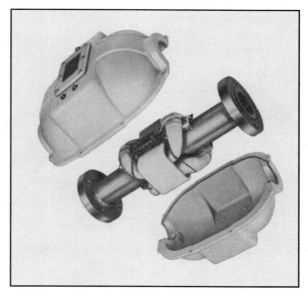

Figure 9.11 Exploded view of magnetic flowmeter (Courtesy Foxboro)

A magnetic flowmeter does not restrict liquid flow because only the very small contact surface of the electrodes is exposed to the flowing fluid. Also, it can handle rough-flowing liquids, such as slurries and sewage, and is immune to clogging. The signal output is linear with flow rate, is highly accurate, and can be transmitted to a remote location. A major disadvantage of magnetic flowmeters in the petrochemical industry is that the liquids this industry handles have conductivities that are so low that magnetic flowmeters cannot be used.

Mass Flowmeters

Manufacturers have developed several meters that respond to the mass of the flowing fluid rather than to the volume, area, or flow velocity to indicate flow rate. One type imparts angular momentum to the flowing fluid and then measures the torque required to remove this momentum (fig. 9.12). Basically, fluid flows through an impeller, which creates angular momentum in the fluid, and then past a decoupling disk to a turbine. The turbine runs a geosynchronous motor, which is in line with a synchronous motor attached to the impeller. The turbine spins and removes the angular momentum in the fluid. Additional instruments in the flowmeter measure the amount of torque required by the turbine to remove it. A close relationship between mass flow rate and torque exists in this flowmeter.

Another type uses liquid cooling and the amount of heat transferred by a resistor to measure mass flow rate (fig. 9.13). Two identical resistors

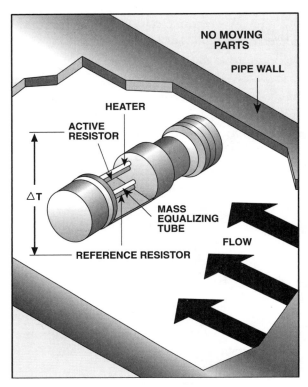

Figure 9.13   *Mass flowmeter*

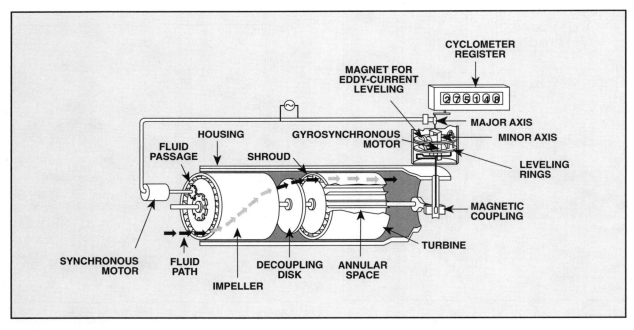

Figure 9.12   *Mass flowmeter that measures torque*

are placed in the flow stream. One resistor is the reference resistor and no external source heats it. An external source heats the other resistor, which is the active resistor. With no flow present, the difference between the resistor values is high. When fluid flows past these two resistors, the active resistor begins to cool down. Its cool-down rate is governed by heat transfer, the amount of which is based on the mass rate of flow of the fluid. As the active resistor cools, its resistance changes. The difference in resistor values is then calibrated in terms of flow rate.

Manufacturers usually custom-make mass flowmeters because such variables as fluid density, fluid type, pipe size, and velocity ranges in a particular application have to be taken into consideration. The output of a mass flowmeter is usually a 4-to-20 mA signal over the desired flow range of measurement. Flow accuracy near zero flow is very poor, which is the case with all flowmeters. However, accuracy is good in the middle to upper ranges.

Turbine Flowmeters

A turbine wheel placed in a flowline and allowed to freely rotate in proportion to the fluid's velocity can monitor fluid flow. When a magnetic pickup is placed near the rotating metal turbine blades, the blades induce a voltage in the pickup whose frequency in hertz (Hz) is directly related to fluid flow. A turbine flowmeter (fig. 9.14) is simple, easy to install and use, and accurate to within 99 percent over 10 to 100 percent of the flow range.

The key to a turbine meter's operation is a permanent magnet placed in a coil. When the metal turbine blades pass the tip of the magnet, the magnetic field changes, induces voltage in the coil, and produces an electrical frequency proportional to the flow rate. By selecting the proper constants from the application and construction of the turbine meter, this frequency can be directly related to flow rate in gal/min, bbl/day, m^3/hr, or whatever quantity and time rate is desired.

In some applications, the frequency is converted into a 4-to-20 mA control signal from a transmitter or is sent directly to a special PLC card that does the conversion.

Vortex Flowmeters

A blunt object placed in a fluid's flow stream creates vortices, or swirls, of fluid that fluctuate around the object (fig. 9.15). The frequency of these flucuations—how often they occur—is directly proportional to the flow rate of the fluid and can be used to create a flowmeter.

Since the vortices produce such a small force, very sensitive devices must be used to detect them and produce an electrical signal. Piezoelectric crystal material is suitable for this application because external pressures on it produce an electrical voltage. When incorporated into a vortex flowmeter

Figure 9.14   Turbine flowmeter (Courtesy Halliburton)

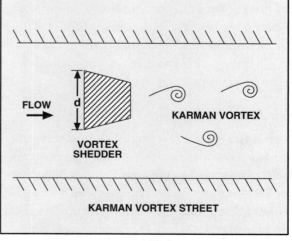

Figure 9.15   Demonstrating the vortex principle of flow

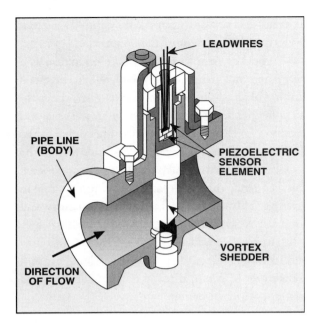

Figure 9.16   A typical vortex flowmeter (Courtesy Yokogawa)

(fig. 9.16), the piezoelectric material produces an alternating voltage whose frequency is directly proportional to flow rate. Using conversion constants for the proper flow units, this type of unit can be used to measure the flow of many types of fluids.

Positive-Displacement Meters

Some flowmeters cannot be used in applications where wide flow ranges occur or in applications where power for operation is required. Positive-displacement flowmeters utilize the energy from the flowing fluid to operate indicating or totalizing meters. They typically rotate a fixed quantity of fluid for each segment of rotation, operate an electrical switch during each rotation, and are calibrated in terms of quantity, not rate, of fluid.

Gas Meters

A gas meter (fig. 9.17) is familiar to most consumers of natural gas who use it to monitor gas usage and which forms the basis for monthly gas bills. It accurately measures volumetric quantities over a wide range of flow values and does not require electrical power to operate. Where the gas pressure is too low, this meter is quite adequate. Figure 9.18 shows a cutaway section of the main case.

Figure 9.17   Gas meter (Courtesy American Meter)

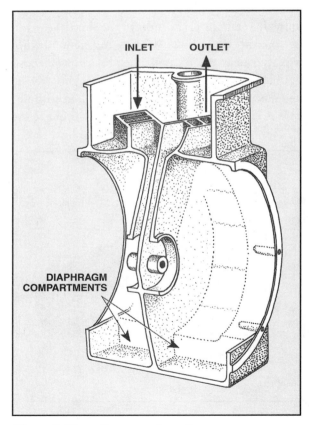

Figure 9.18   Cutaway view of gas meter case

Figure 9.19A–D explains the meter's operation. Because the meter has four separate compartments and two diaphragms, which operate in the two lower compartments, it is termed a two-diaphragm, four-compartment meter.

Each compartment can be connected to inlet or outlet piping, or sliding valves can close each compartment, depending on their position. The two sliding valves work together in a definite relation through a system of levers that are attached to free ends of the diaphragms.

In figure 9.19A, gas is flowing into the interior compartment of the left diaphragm assembly, and expansion of this assembly is forcing gas out of the exterior compartment into the outlet piping. When the free end of the left diaphragm approaches its maximum travel on the expansion phase, an attached lever operates the two interconnected sliding valves, driving them to the position shown in figure 9.19B. This action shuts off the entire left side of the meter and simultaneously opens the appropriate right-hand compartments to the inlet and outlet piping.

Full expansion of the right-hand diaphragm assembly drives both slide valves toward the center. The entire right side of the meter is closed off, while the left side receives gas into its exterior compartment, causing a compression of the diaphragm assembly and forcing its gas into the outlet piping (fig. 9.19C).

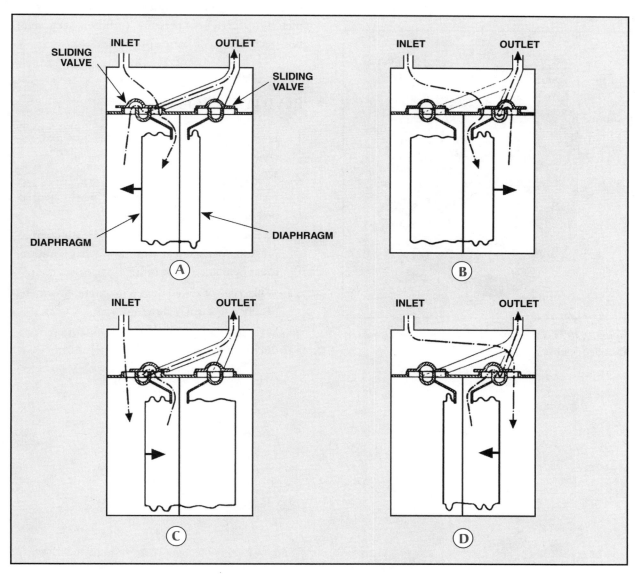

Figure 9.19   Operation of the gas meter

Full compression of the left diaphragm again actuates the valves, driving them to the positions shown in figure 9.19D. Gas then flows into the right-hand exterior compartment, compressing the diaphragm and operating the valves to the positions shown in figure 9.19A. The cycle of operation is then complete.

The use of two diaphragms and four compartments provides satisfactory gas delivery. Output pressure does not pulsate significantly because valve operation is timed so that the rapid closing off of a compartment near the end of its delivery phase is accompanied by an equally rapid opening of the compartment next in line for delivery.

A phantom view of a gas meter (fig. 9.20) shows the Bakelite sliding valves, which are called D valves, and the levers that operate them. Diaphragm motion moves a crank, which, in turn, rotates a shaft. The rotating shaft operates the valves and drives a counter. The counter indicates the amount of gas the meter has measured. Incidentally, when used on flowmeters, a counter is called an index.

SUMMARY

Flowmeters measure the rate of flow of fluids or monitor the total quantity of fluid delivered over a specific time. Mechanical flowmeters use differential pressure or fluid forces to actuate them. Electrical meters use several methods to display flow rate, including differential pressure, turbine wheels, mass flow techniques, or vortex principles.

REVIEW EXERCISE

1. Describe the operation of a mercury manometer.
2. What is the function of an orifice plate?
3. Name the two pressure taps used in a differential-pressure flowmeter.
4. To make the output of an electronic differential-pressure meter linear with flow, what additions must be made to the unit?
5. What type of signal does a magnetic flowmeter produce to display flow?
6. What type of signal does a turbine meter produce to display flow?

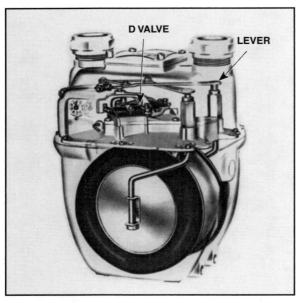

Figure 9.20   Phantom view of a gas meter (Courtesy Rockwell Manufacturing)

10
Flow Control

Controlling the flow of fluids is important when controlling such process variables as pressure, temperature, and liquid level. When fluid flow controls process variables, it is a *manipulated variable*. When fluid flow produces a change in the rate of flow from a set point to bring about a corrective action in a control system, it is a *controlled variable*. Flow control's use as a controlled variable is limited.

This chapter discusses types of flow-control devices, considerations involved in flow control, and applications of flow control.

MECHANICAL FLOW CONTROL ELEMENTS

Many mechanical devices control fluid flow. One such device is a manual valve that an operator adjusts (opens or closes) to control the flow rate and quantity of fluid. A simple water faucet, or tap, is an example of a mechanical flow-control device. It not only controls the quantity of water applied to a lawn or garden, but also the rate at which water is applied. The position of the water tap's adjustment valve is important. For example, if you open the valve too wide, water runs off and is wasted.

Fixed Flow Beans

A *flow bean*, or *choke*, provides fixed flow control—that is, the bean, or choke's opening is not adjustable; it is a fixed size. Of course, flow beans are available in several fixed sizes, so that operators can select a size that is appropriate for a particular application. Flow beans often control the flow of natural gas from a well (fig. 10.1). The flow bean is a constriction that is placed in a special nipple. The nipple is part of the piping. The flow bean is a metal plug with a hole drilled through it. It has external threads and a socket for an Allen, or hex, wrench. The threads and wrench socket allow an operator to easily change the flow bean.

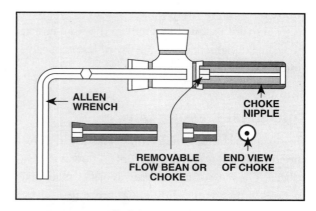

Figure 10.1   Fixed flow bean

Variable Flow Beans

Another flow bean is adjustable (fig. 10.2). It is a needle valve in a right-angle body. Adjustable flow

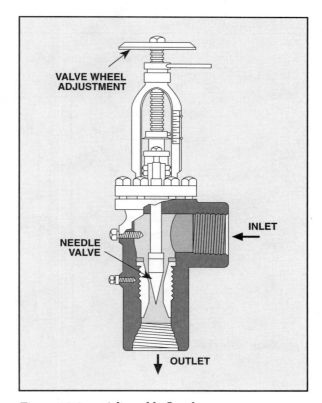

Figure 10.2   Adjustable flow bean

beans are often placed in the piping between a producing well and a low-pressure vessel (a separator). Well fluids are sent to such vessels to remove water and separate oil and gas. Well pressure and handling capacity of the processing equipment determine a fixed bean's hole size or an adjustable bean's adjusted size. Because well pressures are typically high—they can be in excess of 10,000 psi (70,000 kPa)—flow beans are very useful. They reduce the flow rate at the wellhead, which allows the use of less expensive piping leading to process vessels.

Differential-Pressure Devices

Measuring the head, or differential, pressure in a flow line and using this pressure to control flow is widely utilized. The measuring means is an orifice plate that operates with suitable differential-pressure devices. Differential-pressure devices are readily adapted to control and measuring functions.

Mechanical measuring devices installed with orifice plates use feedback mechanisms to produce a force balance. The force balance, in turn, produces a control output pressure of 3 to 15 psi (20 to 100 kPa). This 3- to 15-psi (20- to 100-kPa) control signal is used to further control an adjustable flow valve for regulating flow from the desired set point.

A force-balance sensing device (fig. 10.3) incorporates a set-point adjustment; a range, or span, adjustment; and mechanical feedback to regulate the output control pressure of 3 to 15 psi (20 to 100 kPa). Two upstream (hi) and downstream (lo) pressures from the orifice plate's pipe taps are admitted to two cavities, which a sensing diaphragm separates. The diaphragm is clamped between a pair of baffle plates equipped with drive rods that extend out of the cavities through sealing bellows. The rod on the right side has a flanged fitting on which a zero-adjusting spring rests. The flange on the fitting serves as an overrange and underrange stop.

The rod from the left cavity is attached to a baffle and a feedback diaphragm. The baffle's outer surface acts as a flapper on the nozzle and the two form a nozzle-flapper assembly. Changes in differential pressure applied to the instrument move the two diaphragms and vary nozzle-flapper clearance. The supply line's air pressure goes through a small orifice to the left side of the feedback diaphragm. This air leaks through the nozzle at a rate determined by flapper-nozzle clearance and the

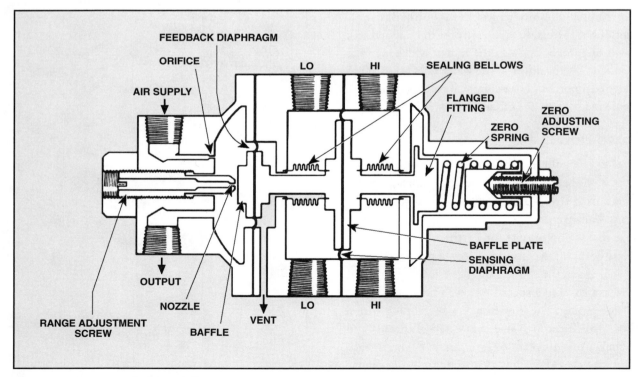

Figure 10.3 Force-balance differential-pressure sensing instrument

position of the range-adjustment screw.

During a condition of equilibrium, forces acting on the sensing diaphragm and on the feedback diaphragm have a net value of zero. Should differential pressure rise or fall, the change in force acting on the sensing diaphragm is opposed by a change in force acting on the feedback diaphragm.

The zero-adjusting screw establishes the set point of the system, while the range adjustment screw varies the sensitivity of the instrument. Air output from the instrument can be used to actuate a remote recorder or an air relay for control purposes.

ELECTRONIC FLOW CONTROLLERS

Electronic differential-pressure transmitters are the most popular means of measuring flow. Their signal is used to control other elements in a flow control process. Figure 10.4 shows the electronic flow control elements that control the rate of flow in a pipeline. A pressure transmitter, which has a square-root extractor to linearize flow, senses the differential pressure across an orifice plate. It then puts out a 4-to-20-mA signal that is proportional to the range of flow in the pipe. This signal goes to the analog input of a PLC processor. The PLC compares the signals to an electronic set-point reference it receives as a 1-to-5 VDC signal from the analog input. It then delivers the corrected signal to the PLC's analog output. The electrical analog output of the PLC is, like the transmitter's output, a 4-to-20 mA signal. This signal represents the error, or deviation, from the desired set point and sets a proportional control valve.

To adjust the control valve, a device called an I/P transducer receives the analog output signal and converts the current to a 3-to-15 psi signal that goes to the proportional valve positioner. The positioner moves the valve as required to control flow.

In equation form, flow in a differential-pressure instrument is—

$$Q = K \sqrt{P_d} \qquad (Eq.\ 10.1)$$

where

Q = volume flow rate
K = flow coefficient
P_d = differential pressure.

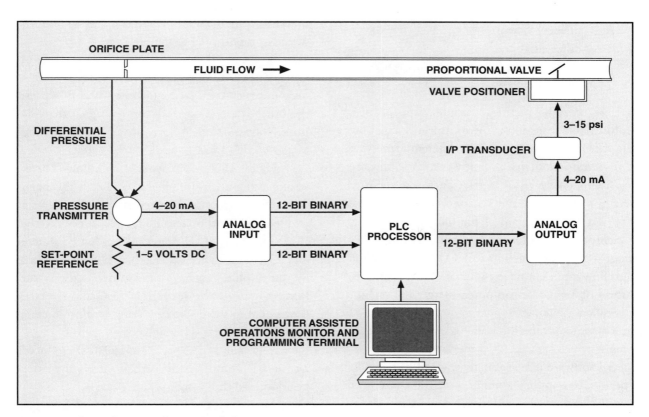

Figure 10.4   Electronic flow control elements

Measurement of incompressible water flow using equation 10.1 and the electronic flow control elements shown in figure 10.4 is quite accurate. However, some fluids, such as gas and steam, change their flow characteristics when subjected to differing temperatures and pressures. Thus, to achieve greater flow measurement accuracy using a differential-pressure flowmeter, flow-line temperature and pressure need to be compensated for. In a PLC system, a differential-pressure signal, temperature signal, and a static pressure signal can be sent to the PLC, which combines the signals to achieve an accurate measurement of flow.

Multivariable flowmetering solves the complexity of combining differential pressure, temperature, and static pressure for a more accurate flow measurement. Combining these three variables into a single transmitter produces the final signal output. Multivariable metering solves the mass flow rate equation, which is—

$$Q = K \sqrt{P_d \times \rho} \qquad \text{(Eq. 10.2)}$$

where

Q = volume flow rate
K = flow coefficient
P_d = differential pressure
ρ = flowing density based on temperature and static pressure.

Figure 10.5 pictures a popular version of a multivariable transmitter for measuring mass flow rate. The unit not only measures differential pressure across a restriction, but also measures static pressure and temperature using a resistance temperature detector (RTD) in the sensor assembly. Figure 10.6 is a block diagram of the multivariable unit. It incorporates a differential-pressure sensing cell, a gauge pressure sensor, and an RTD to calculate the mass flow rate. A built-in multivariable transmitter, instead of a separate computer, merges the signals. It is called a *smart mass-flow transmitter* because its built-in microprocessor uses highway addressable remote transducer (HART) programming language. Special software is available that, when installed in a personal computer, configures the transmitter for flow applications. The output is a 4-to-20 mA signal that controls flow.

Figure 10.5 Multivariable transmitter for measuring mass flow rate (Courtesy Fisher-Rosemount)

Turbine Flowmetering

Turbine flowmetering is another electronic method for measuring and controlling flow. A turbine flowmeter is inserted directly in the flowstream. The flowstream rotates the turbine and this rotation is directly proportional to the flow rate.

A turbine flowmeter puts out two forms of signal: one is AC voltage at a given frequency that a magnetic pickup from the turbine wheel produces; the other is a 4-to-20 mA control signal that the meter converts from AC frequency. Either method can be used for flow control.

Figure 10.7 is a schematic of a flow-control system using a turbine meter. A PLC is the main controller. A turbine flowmeter, which produces a frequency proportional to flow, feeds directly to the PLC's input card. This input card is called a quadrate rate (QR) card. It counts electrical pulses produced by the turbine flowmeter, and converts the count into a calibrated flow signal. The PLC compares the flow signal to the desired set point and produces a 4-to-20 mA signal. This signal then goes to an I/P transducer. The 3 to 15 psi (20 to 100 kPa) produced by the I/P transducer, in turn, controls the valve position, and thus flow.

Another type of flowmetering, which is similar to the systems illustrated in figures 10.4 and 10.7,

Flow Control

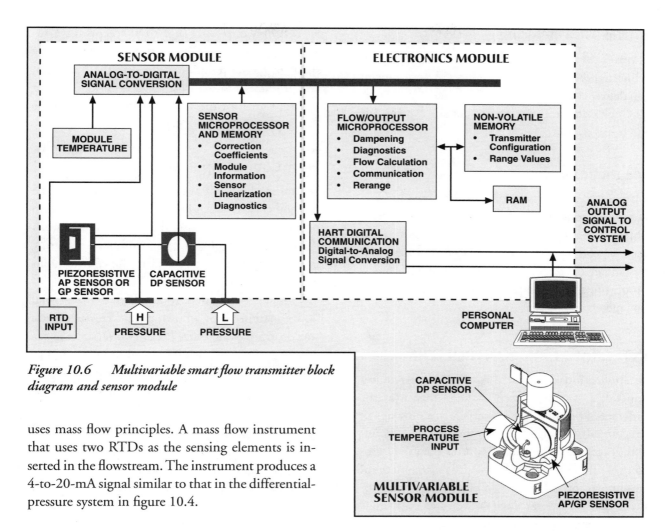

Figure 10.6 Multivariable smart flow transmitter block diagram and sensor module

uses mass flow principles. A mass flow instrument that uses two RTDs as the sensing elements is inserted in the flowstream. The instrument produces a 4-to-20-mA signal similar to that in the differential-pressure system in figure 10.4.

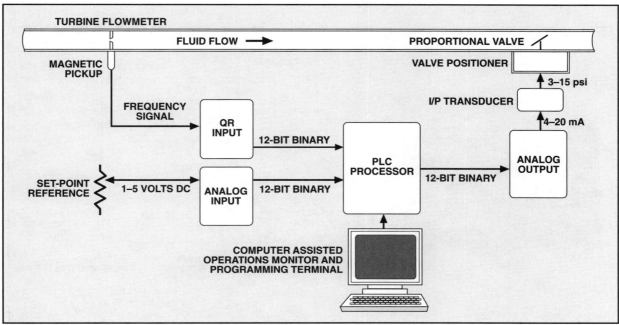

Figure 10.7 Flow control system using turbine flowmeter feedback and PLC

Vortex Flowmetering

A vortex flowmeter produces a signal similar to that which a turbine flowmeter produces. A vortex meter can deliver either an amplified frequency signal to the PLC, or a device at the instrument site can convert the signal to 4 to 20 mA.

Magnetic Flowmetering

Magnetic flowmeters can be used to measure and control conductive fluids. A magnetic flowmeter's output controls flow in the same manner as the meters previously discussed. Indeed, if a magnetic flowmeter produces a 4-to-20-mA signal that is proportional to flow, figure 10.4 applies to magnetic flowmeters, too.

INTEGRAL FLOW CONTROLLERS

In applications where a throttling valve requires control, a system is used that (1) measures the differential pressure, (2) converts it to a flow signal, (3) compares the signal against the desired set point, and (4) produces a signal to control the valve. This control can be accomplished with separate pieces of equipment, or the pieces can be combined. Equipment that incorporates all the features in a single enclosure is a *flow controller*.

A flow controller (fig. 10.8) accepts the 4-to-20-mA signal from a flow transmitter, compares the signal to the set point established in the controller, and produces a 3-to-15-psi (20-to-100-kPa) output control signal for the valve actuator. Adjusting the valve's position achieves flow control.

Flow Control Applications

Several methods are available to control flow. The method chosen depends largely on what the system needs to accomplish. Also, the relationship between the flow rate and the variable conditions occurring upstream and downstream from the control point must be considered.

Refer to equation 10.2, which is the general flow equation when a flowing fluid creates a pressure drop across an orifice plate. Two variables in the equation, P_d, which is differential pressure, and ρ, which is flowing density based on temperature and static pressure, affect control. Flowing density, ρ, is obtained at the downstream tap.

Gas and Steam Flow Control

A gas or steam flow installation may require that a constant flow rate be maintained into a downstream line having a constant pressure. In other words, if flow into the line is constant, downstream pressure

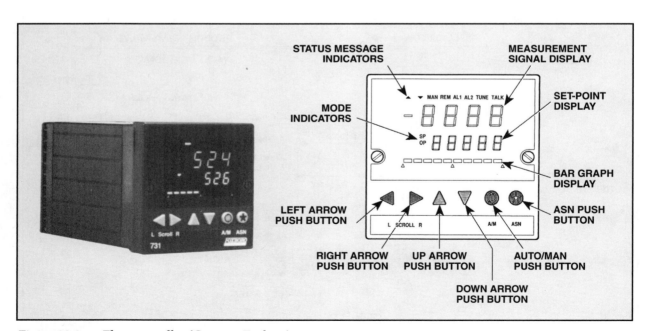

Figure 10.8 *Flow controller (Courtesy Foxboro)*

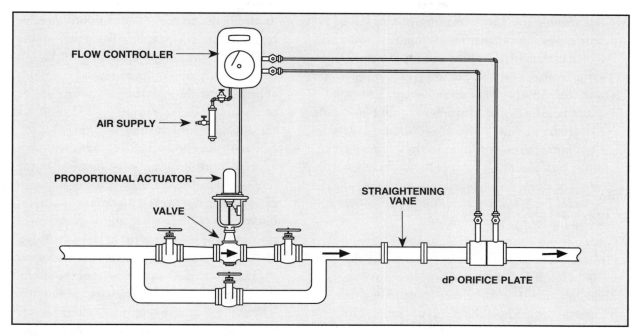

Figure 10.9   Installation for maintaining constant downstream pressure

must also remain constant. Such an installation functions best if a throttling valve is placed upstream from the orifice plate (fig. 10.9). The throttling valve keeps the pressure ahead of the orifice plate at a constant value, regardless of pressure fluctuations upstream from the valve. The same arrangement can be used to deliver gas into fluctuating downstream pressure where it is preferable that the flow rate be related to downstream pressure.

If an installation has constant upstream pressure and fluctuating downstream pressure, the throttling valve should be downstream from the orifice plate to maintain a constant flow rate (fig. 10.10). In this way, constant pressure is maintained on either side of the orifice plate, so that flowing density, ρ, remains constant.

If variable pressures occur both upstream and downstream, the throttling valve should be installed upstream and a pressure-compensated flow controller used to control it. A pressure-actuated device that automatically corrects for variations in downstream pressure accomplishes pressure compensation in a controller.

Liquid Flow Control

In controlling gas flow, both the control valve and the primary element are located in the line in which flow is to be controlled. Some installations that control liquids cannot use this type of system. For example, in a liquid flow line where a reciprocating or other positive-displacement pump drives the liquid, a throttling valve should not be installed in the pump's discharge line. Excessive line pressure could build up against a closed valve. The amount of pressure buildup is limited only by the pump's power source and the rupture strength of the line or valve.

Controlling the flow rate of liquid from the discharge of positive-displacement pumps can be achieved in several ways.

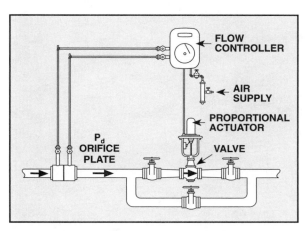

Figure 10.10   Installation for maintaining constant upstream pressure

1. For steam-driven reciprocating pumps, the orifice plate or other primary element is installed in the discharge line, and the throttling valve is placed in the pump's steam-supply line (fig. 10.11). This arrangement assures that a failed valve or controller cannot block the discharge line and damage the system. Regulating the energy input to the steam end of the pump controls its speed and discharge rate. The downstream pressure-sensing device provides feedback to a throttling valve in the steam line of the pump engine.
2. For steam-, gas-, or diesel-engine-driven pumps, the orifice plate is installed in the discharge line, and the controller drives a diaphragm motor that either resets an engine governor or controls the steam or fuel feed to the engine.
3. For positive-displacement pumps, regardless of their drive, and particularly for a constant-speed electric-motor drive, the orifice plate is located in the discharge line. A throttling valve is located in a bypass line from the pump's discharge to the pump's intake (fig. 10.12). In this system, a constant pump speed maintains constant flow through the pump. The degree of closure of the throttling valve determines the discharge into the line.
4. If a centrifugal pump drives liquid through a line, a control valve can be installed in the discharge line (fig. 10.13). Centrifugal pumps can be set up not to exceed a certain maximum discharge pressure, even against a tightly closed discharge line. As long as the line, control valve, and other system components can withstand the maximum discharge pressure of the pump, direct throttling of the discharge is suitable. Throttling the steam supply to a turbine sometimes controls discharge from steam-driven centrifugal pumps. Or, a diaphragm-motor device resets a turbine-speed governor. Such arrangements save input energy.

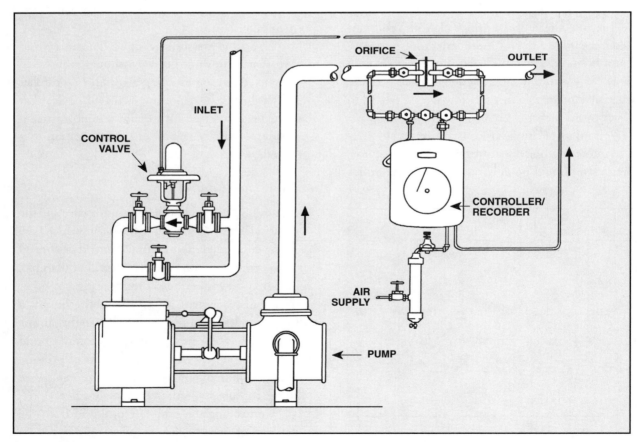

Figure 10.11   *Downstream pressure sensing device for a steam-driven pump*

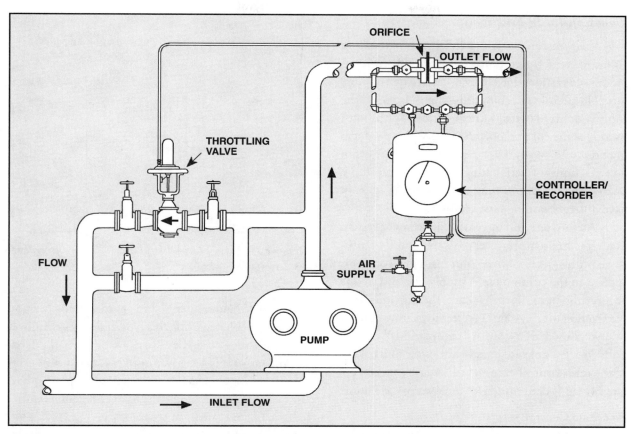

Figure 10.12   *Throttling valve in bypass line for a postive-displacement pump*

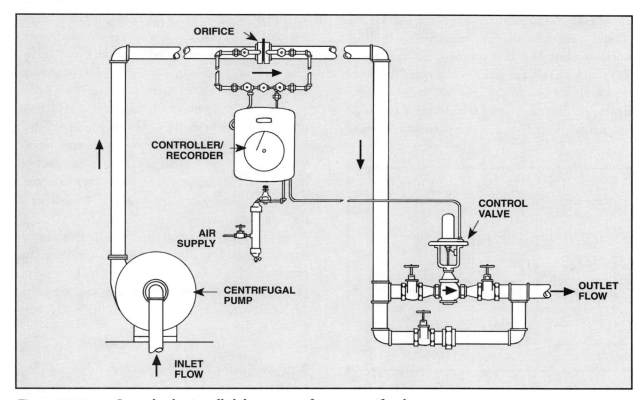

Figure 10.13   *Control valve installed downstream from a centrifugal pump*

Flow Control In Fractionating Columns

A special pressurized vessel called a fractionating column, or a fractionator, processes petroleum products. Fractionating columns receive two or more miscible petroleum compounds and separate the compounds into two or more products, or fractions. As long as the column operates properly, it produces fractions at a steady rate. Separation of petroleum into fractions—fractionating—occurs because the hydrocarbon fractions that make up the petroleum have different boiling temperatures.

An intricate and integrated instrument system that regulates temperature, pressure, and flow usually controls fractionating columns. Flow control is applied to the column's feed supply line and to one or more discharge lines that carry the fractions from the fractionating system. Frequently, a flow-control system is used that consists of individual control units that are not integrated with other units. That is, the units control entirely on the basis of the flow rate, which is determined by the measuring means.

Feed-Rate Control

Figure 10.14 diagrams a low-volume feed-rate controller. That is, the feed rate is not high enough to tax the system. The fractionator could run indefinitely without exhausting the feed supply. This form of control is usually installed on the initial column of a fractionating train, or series of columns. The flow-control system has an orifice plate as the primary element, a differential-pressure transmitter that sends its signal to a proportional or proportional-plus-reset

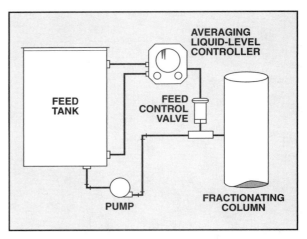

Figure 10.15   Liquid-level controller for succeeding fractionating columns

controller, and a diaphragm-motor valve. A fractionating column using this type of feed-rate control requires time to settle down, even when the feed and heat supply are uniform. Once a condition of equilibrium is achieved, however, the column has little trouble maintaining the balance of temperature, pressure, and rate of output.

When the feed supply to a column is relatively limited or sporadic in nature, which may occur in the intermediate column of a fractionating train, a low-volume feed-rate controller is usually not satisfactory. A temporarily exhausted feed supply, or a feed supply that is not at an optimum rate, can upset the process occurring in the intermediate column.

Figure 10.15 shows a system that controls flow in accordance with the liquid level in a feed tank that is supplied from the first fractionating column. Regulated discharge from the feed tank does not control liquid level. Flow control is still the primary concern, and its rate should be uniform and consistent with the feed supply available.

The liquid level in the feed tank determines the set point of the flow controller. Thus, an electronic or pneumatic flow controller is equipped with set-point adjustments that allow the operator to choose the desired value of the controlled variable (fig. 10.16).

Control of Fraction Withdrawal Rate

A simple fractionating column splits the feed into two fractions; one is withdrawn from the bottom of the column and the other from the top. Withdrawal

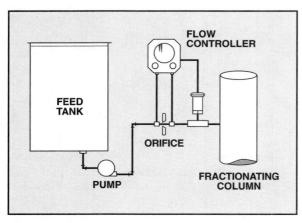

Figure 10.14   Feed-rate control for initial column of fractionating columns

Flow Control

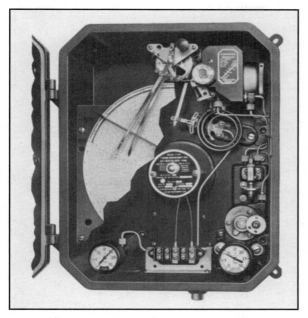

Figure 10.16   Electronic/pneumatic controller

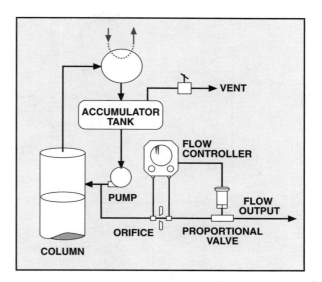

Figure 10.17   A system with simple flow-rate control

rate from the top of the column can be controlled on the basis of temperature and pressure or by using flow rate as the controlled variable. Withdrawal rate from the bottom of a column is usually controlled as a function of liquid level in the column bottom, but a cascade system between feed rate and bottom product withdrawal rate is feasible.

Control of Top Product Discharge Rate

If the feed rate to a column is constant, then a proportional controller or a proportional-plus-reset controller can be used to maintain steady output of product from the top of the column (fig. 10.17). The primary element is an orifice plate. Some of the product is allowed to flow back into the top of the tower. This flowback is called reflux and a certain amount of the condensed product is returned to the column to enrich the output.

To control the output of columns that do not have a constant rate of feed, it is usually necessary to provide a cascade control system. In one method of control (fig. 10.18), the signal from the liquid-level transmitter of the feed tank is applied to the flow controller.

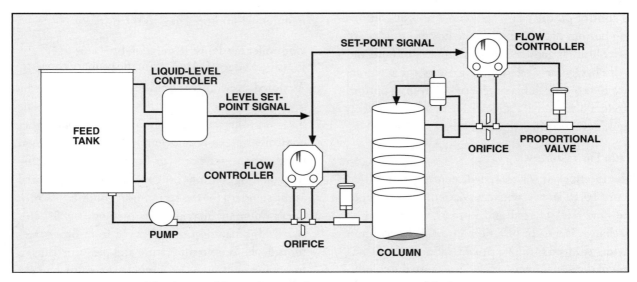

Figure 10.18   Liquid-level control for a system without a constant rate of feed

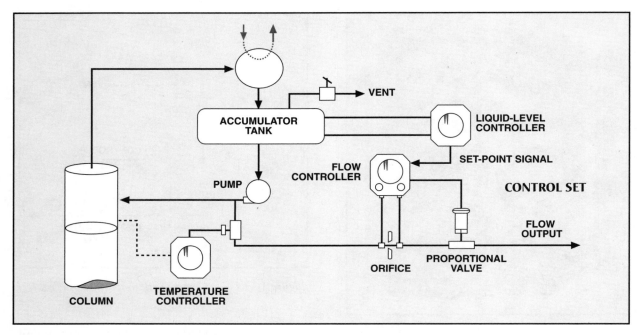

Figure 10.19   Liquid-level control system with accumulator tank

In another form of flow control of output, the liquid level in an accumulator tank establishes the set point of the flow controller (fig. 10.19). The temperature near the upper end of the column's interior, or the temperature at another point, controls the amount of product returned to the tower as reflux. Thus, the reflux rate has an indirect controlling effect on the flow rate of product output.

Control of Bottom Product Discharge Rate

Using an interlocked feed rate and bottom discharge to control product flow is one method of controlling bottom discharge rate. The bottom product of the column is gathered in an accumulator tank and is discharged at a rate determined by the feed rate into the column. This system of controlling bottom product discharge is less desirable than one based on liquid level existing in the reboiler.

Ratio Flow Control

The interlocked, or cascaded, control system that figure 10.18 shows, where a relation exists between the flow rates of feed and upper column product discharge, is a form of ratio control. If the column divides the feed into two equal fractions and 50 percent of the upper fraction is fed back into the column reflux, then 25 percent of the original feed material is left for discharge as upper column product. So, for every quantity of liquid passed by the feed-control valve, one-fourth of that quantity must pass through the discharge as the upper product. Thus, this system is a true example of ratio flow control.

Ratio flow control is used in many processes. Usually, ratio flow control is simply a matter of allowing a primary flow rate to establish the set point of a secondary flow controller. The primary rate varies according to the needs of the process, and the secondary, or ratio-controlled, rate is automatically maintained in proportion to the primary.

Controller Set-Point Regulation by Vapor Pressure Differential

As noted previously, liquid level can establish the set point of one or more flow controllers, or one flow controller can establish the set point of another controller. Pressure and temperature are also used to adjust the set point of flow controllers. Some fractionating columns use vapor pressure differential as the standard for the set point. A column that produces isobutane may use this method (fig. 10.20). As usual, separation of isobutane from other fractions depends on temperature and pressure. But, in isobutane columns, the temperature must be very accurately controlled for high purity. In other words,

Flow Control

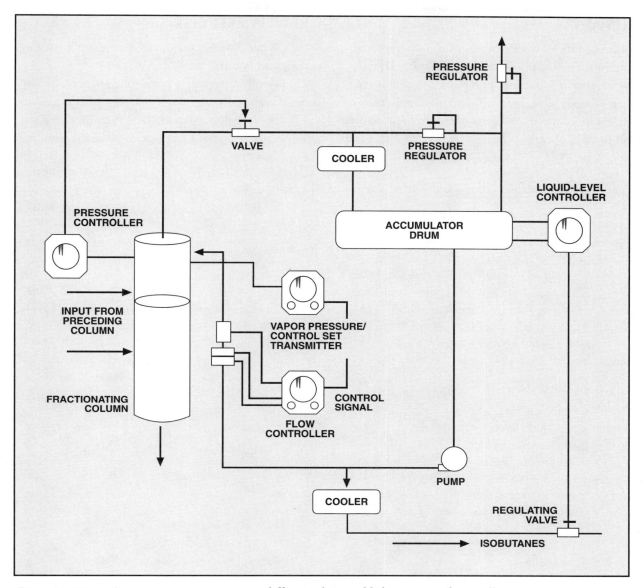

Figure 10.20   System using vapor pressure differential to establish set point of controller

deviation of a few fractions of a degree Fahrenheit can make a big difference in the isobutane's purity. Using vapor pressure differential is one way to accurately control temperature.

In the system, a vapor pressure bulb on one of the upper trays of the column is filled with the liquid that normally exists on the tray. If the temperature on this tray varies enough to change the composition of the liquid on the tray, it also changes the vapor pressure in the bulb. Consequently, a differential vapor pressure exists, and the differential-pressure-measuring means readily detects it. A signal then goes to the flow controller's control set in the reflux line and more or less reflux is returned to the column to bring about the needed correction.

Vapor pressure differential, instead of temperature variation, is used to make the set-point change in systems where small temperature changes make a big difference in product purity. Temperature controllers cannot easily maintain such close control. Even small temperature variations produce a pressure change of about 1.5 psi per °F (18.6 kPa per °C). Under such conditions, using vapor pressure differential to control the change is much easier than using temperature variations.

SUMMARY

The applications discussed in this chapter represent common methods of flow control. Piping arrangements and installation practices covered in the chapter are typical, but, for particular applications, readers should refer to the manufacturer's literature and to an application engineer.

REVIEW EXERCISE

1. What device represents the simplest means of flow control?
2. While differential pressure produced by the flow of a fluid through an orifice plate is used to measure flow, what mathematical modification to pressure must be made to make the relationship linear?

11

Gravity, Viscosity, Humidity, and pH

Variable factors such as specific gravity, density, viscosity, humidity, and pH often modify automatic control of pressure, temperature, liquid level, and flow rate. Consequently, these variable factors must also be accurately measured and controlled.

MEASURING SPECIFIC GRAVITY AND DENSITY

Specific gravity expresses a comparison between the densities of a particular substance and a reference substance, which is usually water or air. If water is the reference substance, its specific gravity is 1. In gas-flow measurement, air is the reference substance. Water and air are used almost exclusively for specific gravity measurements, although oxygen is sometimes used for critical scientific measurement of gases.

Temperature and pressure affect density, and therefore they must be taken into account when making specific gravity measurements. However, ordinary pressures can be ignored when dealing with incompressible liquids. For accurate measurement of liquid density, scientists usually specify double-distilled water at 4°C (39.2°F) as the standard. (Water is densest at 4°C.) For accurate measurements of gas density, they usually specify air at a standard temperature of 0°C and a pressure of 760 mm of mercury. On the other hand, U.S. engineering standards often specify 60°F and 14.73 psia for temperature and pressure, although deviations from these values are common.

Measuring Scales

Ordinarily, people do not measure specific gravity as often as they do temperature, humidity, or atmospheric pressure, which are pertinent to weather forecasting. However, automobile enthusiasts may be aware of making specific gravity measurements to determine the charge in a lead-acid battery or to establish the strength of an antifreeze solution in a cooling system. Although specific gravity measurements may not be important in everyday life, such measurements are very important in science and technology.

In industrial processes, measuring a solution's specific gravity is often the simplest and most accurate way to determine the solution's composition. The petroleum industry and the Bureau of Mines measure the specific gravity of petroleum using *API gravity*, which is based on the specific gravity of water. The strength of acid solutions is readily determined by specific gravity. The higher the specific gravity, the higher is the acid concentration. The charge of a lead-acid storage battery is inferred by measuring the specific gravity of its acid.

API Scale

During the 1920s, the American Petroleum Institute (API) devised and adopted a scale of specific gravity measurement units called degrees (°) API. Although the scale is different from the ordinary specific gravity scale, it bears a definite relation to it. The equation for determining API gravity is—

$$°API = \frac{140}{1} - 130 \qquad (Eq.\ 11.1)$$

where

G = specific gravity of petroleum with reference to water, both at 60°F (15.55°C).

As an example, determine the API gravity of water that has a specific gravity of 1.

$$°API = \frac{140}{1} - 130$$
$$= 140 - 130$$
$$°API = 10.$$

As another example, determine the API gravity of oil whose specific gravity is 0.9462.

$$°API = \frac{140}{0.9462} - 130$$
$$= 146.1073 - 130$$
$$°API = 16.1.$$

Finally, determine the API gravity of oil whose specific gravity is 0.9288.

$$°API = \frac{140}{0.9462} - 130$$
$$= 150.7321 - 130$$
$$°API = 20.7.$$

Note that as the specific gravity of a petroleum liquid *decreases*, its API gravity *increases*. (In the examples, oil with a specific gravity of 0.9462 has an API gravity of 16.1°, while the oil with a specific gravity of 0.9288 has an API gravity of 20.7°.)

The API scale has advantages for particular applications. For one thing, it provides finer gradations between whole number units. Consider a gravity change of 1° API, say from 20° to 21°, for example. The equivalent change in specific gravity is from 0.9333 to 0.9272. Clearly, the API scale is easier to deal with. Besides this advantage, the scale also makes it easier to correct an oil's density to a temperature standard of 60°F.

Baumé Scale

Another specific gravity scale that has been in use for almost two centuries is the Baumé scale. The French chemist Antoine Baumé invented it in the 1800s. Technicians frequently use this scale to measure the specific gravity of acids and other liquids such as syrup. The Baumé scale uses one equation for light liquids and another for heavy liquids. For light liquids, the equation is—

$$°Baumé = \frac{140}{G} - 130 \qquad \text{(Eq. 11.2)}$$

where

G = specific gravity.

For heavy liquids, the equation is—

$$°Baumé = 145 - \frac{145}{G}. \qquad \text{(Eq. 11.2)}$$

In both Baumé equations, specific gravity (G) is usually measured at 60°F.

Besides the API and Baumé scales, industry employs additional specific gravity scales to suit specific needs. Handbooks devoted to the needs contain the scales. Usually, the equation is simple.

Measuring Devices

A *hydrometer* is the most common instrument used to measure the specific gravity of liquids. Also, some of the devices that measure liquid levels can be adapted to measure specific gravity. Examples include air-bubbler systems, displacer-float devices, and methods using hydrostatic head. Each can operate an electronic transmitter, a recorder, or a computer memory device to record specific gravity.

Hydrometers

Hydrometers have a large bulblike section on one end and a thin, straight stem on the other (fig. 11.1). The stem is graduated with a scale that is suitable for its particular application. For example, a hydrometer measuring oil's specific gravity is usually graduated in °API, which is the oil industry's standard. A hydrometer balances the total weight of the hydrometer against the weight of the liquid it displaces when upright and floating in the liquid.

The bulb of the hydrometer displaces a large portion of the liquid, while the thin stem spreads out the scale for accurate determination of specific gravity. The weight and other design features of a hydrometer depend on the range of values to be measured. An instrument that measures storage batteries usually has a scale range of 1.050 to 1.310, since this range encompasses the significant range between a dead and a fully-charged battery. A hydrometer scale for degrees API might contain any portion of the range from 0° to 100° API. The two hydrometers, shown in figure 11.1, range from 29° to 41° API.

Because temperature affects specific gravity, some hydrometers contain a thermometer. The range of the thermometer is appropriate for the hydrometer's use. For example, the hydrometer shown in figure 11.1 has a temperature range of 20° to 140°F. Instructions and tables for making temperature corrections to specific gravity are widely available.

Air-Bubbler Systems

An air-bubbler system forces air into a tank containing the liquid whose specific gravity needs to be determined. One air-bubbler system that measures

Gravity, Viscosity, Humidity, and pH

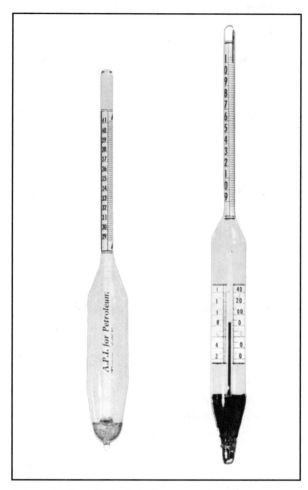

Figure 11.1 Typical hydrometers used to measure specific gravity (Courtesy Taylor Instrument)

specific gravity has two tanks, each of which contains liquids of equal heights above a common baseline (fig. 11.2). One tank is a sampling tank, which contains the liquid to be measured. The other tank is a reference tank, which contains a reference liquid. The reference liquid may be water or another liquid suited to the needs. Air is supplied to the bubble tubes at a rate just great enough to produce a steady stream of bubbles. Two inputs connect the differential-pressure instrument to the bubble-tube lines. In conventional units, the equation to determine the pressure needed to overcome the hydrostatic head in either tank is—

$$p = 0.433 \times G \times H \quad \text{(Eq. 11.4)}$$

where

p = pressure, psi
G = specific gravity, sampling tank liquid
H = height, ft of liquid surface above lower end of bubble tube.

In SI units, the equation is—

$$p = 9.795 \times G \times H \quad \text{(Eq. 11.5)}$$

where

p = pressure, kPa
G = specific gravity, sampling tank liquid
H = height, m of liquid above lower end of bubble tube.

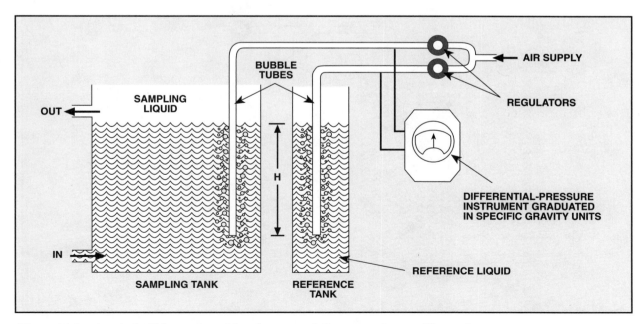

Figure 11.2 An air-bubbler system with reference tank for measuring specific gravity

To understand how equations 11.4 and 11.5 determine specific gravity of the liquid in the sampling tank, use conventional units (eq. 11.4) and assume that—

1. a liquid of unknown specific gravity is in the sampling tank with a liquid height of 10 ft.
2. water is in the reference tank and its specific gravity is 1.
3. a differential pressure of 0.5 psi exists, and the liquid has a specific gravity greater than 1.
4. p_1 is the pressure in the reference tank and p_2 is the pressure in the sampling tank.

From this information, it holds that—

$$p_1 = 0.433 \times 1 \times H.$$

As stated earlier, the differential-pressure indicator shows that pressure in the sampling tank is 0.5 psi greater than the pressure in the reference tank at a liquid height of 10 ft. With this knowledge, the following equation can be set up—

$$p_1 = p_2$$

which calculates as—

$$p_1 = 0.433 \times G \times H$$
$$p_1 = (0.433 \times 1 \times H) + 0.5$$
$$p_2 = 0.433 \times G \times H$$
$$p_2 = (0.433 \times H) + 0.5.$$

To solve the equations for G, divide both sides by $0.433 \times H$, which yields—

$$G \times \frac{0.433 \times H}{0.433 \times H} = \frac{0.433 \times H}{0.433 \times H} + \frac{0.5}{0.433 \times H}$$
$$= 1 + \frac{0.5}{0.433 \times 10}$$
$$G \times 1 = 1 + 0.115$$
$$G = 1.115,\text{ which is the specific gravity of the liquid in the sampling tank.}$$

Another type of bubbler system for measuring specific gravity does not require a reference tank. It simply has two bubble tubes in a single sampling tank (fig. 11.3). In this system, as in the two-tank system, it is important to maintain a fixed level in the sampling tank. The differential pressure that exists between the two bubble tubes depends not only on the vertical separation of their open ends, but also on the specific gravity of the liquid.

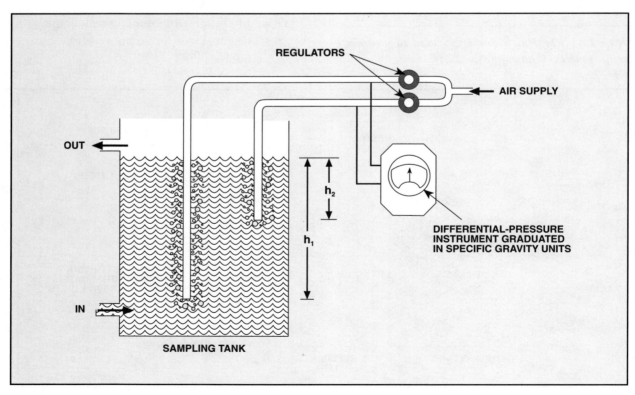

Figure 11.3 Air-bubbler system without reference tank

An example, using conventional units of measure, shows how the system works. (SI units work the same way, so to keep it simple, only conventional units are used.) Suppose the bubbler tube ends are separated by 2 ft of water, one at 2 ft from the surface (p_1), the other at 4 ft (p_2). The pressures required to overcome the heads in each tube are—

$p_1 = 0.433 \times 1 \times 2 = 0.866$ psig
$p_2 = 0.433 \times 1 \times 4 = 1.732$ psig.

These numbers can be confirmed by checking the differential-pressure instrument. It should indicate $1.732 - 0.866 = 0.866$ psig.

Suppose salt is dissolved in the water until the specific gravity becomes 1.25. The pressures now needed to overcome the hydrostatic head are—

$p_1 = 0.433 \times 1.25 \times 2 = 1.082$ psig
$p_2 = 0.433 \times 1.25 \times 4 = 2.165$ psig.

This situation produces a differential pressure of 1.083 psig because $2.165 - 1.082 = 1.083$ psig.

So, a change in specific gravity from 1 to 1.25 produces a change in differential pressure of 0.217 psig because $1.083 - 0.866 = 0.217$ psig. Note that the greater the separation of the bubble tube ends in the liquid, the greater is the differential-pressure indication for a given change in specific gravity of the liquid.

Displacer Floats

Displacer floats are buoyant elements that are heavier than the maximum amount of liquid they can displace (fig. 11.4A and B). They are often used to measure liquid levels in tanks. In the cutaway of the displacer-float element and cage (fig. 11.4B), note that liquid flows into the middle of the cage and out each end. When liquid enters at the middle of this type of cage, it is called a piezometer ring. A piezometer ring reduces or eliminates the adverse effect on the response of instruments of this type caused by rapidly flowing liquids.

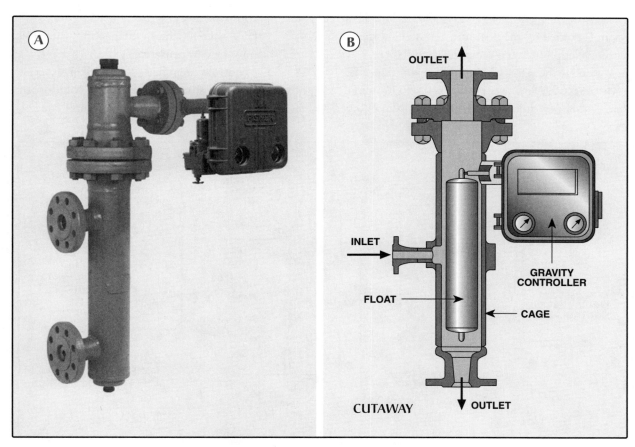

Figure 11.4 A displacer-float specific gravity meter (Courtesy Fisher Controls)

The specific gravity measuring and control system in figure 11.5 uses a displacer float in a piezometer ring cage. The displacer float exerts a downward force on the float rod. The rod twists the torque tube in proportion to the apparent weight of the float, and this action causes the free end of the rotary shaft to position a flapper in the controller or transmitter. The apparent weight of the float is its true weight minus the buoyant force exerted on it by the liquid in the cage.

When used for measuring or controlling specific gravity, displacer floats are entirely submerged in the liquid. The downward force exerted by the element then becomes a function of the specific gravity of the liquid. As an example, consider a displacer float of 100 in.^{3} (1,638.7 cm^3) and 4.75 lb (2.15 kg), true weight. When submerged in water, the float displaces 100 in.^{3} (1,638.7 cm^3) of water, and, since water weighs 0.036 lb/in.^{3} (1 g/cm^3), the total amount displaced weighs 3.6 lb (1,639 g or 1.64 kg).

The weight of the water exerts a buoyant, or upward, force on the float. Thus, the apparent weight of the float when submerged is 1.15 lb (0.52 kg) because 4.75 − 3.6 = 1.15 lb (2.15 − 1.63 = 0.52 kg). This weight is exerted on the float rod, and it establishes a definite relation between the nozzle and flapper.

Replace the water with a liquid having a specific gravity of 0.600. Since water weighs 0.036 lb/in.^{3} (1 g/cm^3), this new liquid weighs 0.0216 lb/in.^{3} (0.6 g/cm^3) because 0.6 × 0.036 = 0.0216 lb/in.^{3} (1 × 0.6 = 0.6 g/cm^3). So, the new value of downward force, or apparent weight, of the float element is 2.59 lb (1.17 kg) because 4.75 − (100 × 0.0216) = 2.59 lb (1.17 kg). A change in weight from 1.15 lb (0.57 kg) to 2.59 lb (1.17 kg) is considerable.

As the specific gravity of a liquid exceeds 1.00, an upper limit for the displacer float is reached when the buoyant force of the liquid equals the weight of the float. That is, when 0.036 × G × 100 = 4.75, where G is the specific gravity of the liquid and is equal to about 1.32. The practical limit of the liquid's specific gravity when using a displacer float is about 1.2. However, using floats with a greater effective density provides a practical way of measuring any reasonable specific gravity.

In practice, the choice of displacer float depends on several considerations, which include the—

- desired proportional band setting,
- range of specific gravity values to be measured,
- nature of the liquid—that is, whether it is corrosive, and
- effects of temperature change and whether it must be compensated.

The control system in figure 11.5 is an example of regulating the strength of a caustic solution by means of specific gravity measurement and control.

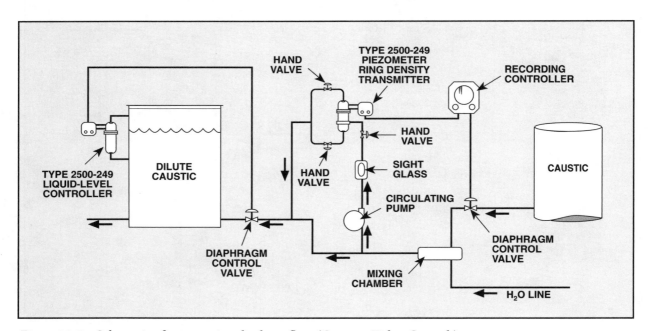

Figure 11.5 Schematic of system using displacer float (Courtesy Fisher Controls)

Note the manual-control (hand) flow valves in lines leading from the displacer-float cage. They are adjusted so that a balanced flow comes from top and bottom of the cage, thus assuring a completely filled cage at all times. The sight glass (sometimes a rotameter is used instead of a sight glass) and the hand valve above it, regulate the rate of flow through the float cage. The recording controller supplements the density transmitter, providing continuous indication and a permanent record of specific gravity.

MEASURING VISCOSITY

Viscosity is a property that everyone recognizes as belonging to syrup, heavy oils, mucilage, ketchup, and other liquids that seem reluctant to flow or to pour as easily as water. Ordinarily, water, gasoline, and air are not thought of as viscous fluids, but these fluids also possess viscosity; indeed, all fluids are viscous, but some are more viscous than others.

Dimensions of Viscosity

Simply put, viscosity is a fluid's resistance to flow. Fluids with high viscosity, like syrup, have high resistance to flow. Fluids with low viscosity, like water, have low resistance to flow. More technically, however, viscosity is related to force, time, and area.

In the nineteenth century, English engineer Sir Frederick Stokes carried out experiments to determine the viscosity of various liquids, using simple laboratory equipment (fig. 11.6). Stokes' equipment provides a way to explain viscosity and how the factors of force, time, and area are derived. The long glass cylinder, whose length is graduated in centimetres, is filled with a liquid whose viscosity is to be determined. A polished ball, made of material having a density somewhat greater than the liquid, is allowed to fall through the column of liquid. Note that if the ball is permitted to fall freely in a vacuum and not in a fluid, its velocity increases indefinitely according to the equation—

$$v = g \times t \qquad \text{(Eq. 11.6)}$$

where
- v = velocity
- g = acceleration due to gravity (about 32 ft/sec² or 9.75 m/sec²)

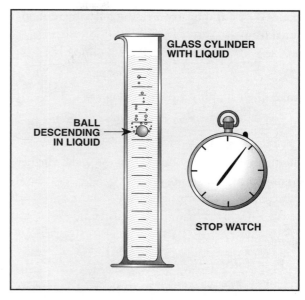

Figure 11.6 Lab equipment used over a century ago to determine viscosity of liquids

- t = elapsed time after the free fall began from a position of rest, or zero velocity.

On the other hand, an object falling through air (or other gas) or a liquid eventually attains a terminal velocity where velocity becomes a constant. This condition occurs when the weight of the object is exactly balanced by the frictional forces of the air or other fluid acting against the descent of the object. Because weight is the product of mass and acceleration, which is caused by gravity, weight is a force. Thus, at terminal velocity, the net force acting on the object is zero.

Stokes chose a liquid column long enough and a ball density of just the right weight so that he could accurately observe the time it took the ball to fall a given distance after it had attained terminal velocity. He found the velocity to be inversely proportional to the absolute viscosity of the liquid and arrived at the following equation for determining absolute viscosity—

$$\mu = \frac{2r^2\,(p - p')\,g}{9v} \qquad \text{(Eq. 11.7)}$$

where
- μ = viscosity, poise
- r = radius of ball, cm
- p = density of ball, g/cm³
- p' = density of liquid, g/cm³
- g = acceleration due to gravity, cm/sec²
- v = velocity, cm/sec.

Stokes then put the factors in the following dimensional form—

$$\text{poises} = \frac{2 \times cm^2 \times g/cm^3 \times cm/sec^2}{9 \times cm/sec}.$$

Stokes then simplified the equation as—

$$\text{poises} = \frac{2 \times (cm^2/cm^3) \times (g \times cm/sec^2)}{9 \times cm/sec}.$$

He then noted that $g \times cm/sec^2$ = force and further simplified the equation as—

$$\text{poises} = \frac{2 \times F \times sec}{9 \times cm^2}.$$

where

F = force.

The factors of force, time, and area that Stokes developed in the equation came from (1) the gravitational and frictional forces acting on the ball, (2) the time-related factors of acceleration and velocity, and (3) the linear distances represented by the radius of the ball and the terminal velocity. Bear in mind, too, that forces and velocities caused by factors other than gravity are equally applicable. For example, an additional factor affecting viscosity is the force or pressure (which is force per unit area) needed to push fluids through a horizontal pipe.

Units and Scales

So far, only absolute viscosity, or dynamic viscosity, as it is sometimes called, has been discussed. Just as temperature and pressure have scales besides those based on absolute values, viscosity also has additional scales. Indeed, no other variable besides viscosity likely has as many variations in scales or means of expression. To better understand viscosity, consider a few definitions.

Kinematic viscosity. Kinematics is a physics term. It is the study of motion without regard to forces or mass. Kinematic viscosity dimensions are quite different from absolute viscosity dimensions. Kinematic viscosity is the ratio of absolute viscosity to the density of the fluid, so its dimensions are length (L) squared divided by time (t) ($L^2 \div t$).

In the old metric system (the centimetre-gram-second system and not the SI system), the unit of kinematic viscosity is the stoke, named after Stokes. A centistoke, which is 0.01 stoke, is also commonly used.

Specific viscosity. Specific viscosity is the ratio of the absolute viscosity of a substance to that of a standard fluid, like water, with the viscosity of both fluids being measured at the same temperature.

Relative viscosity. Relative viscosity is similar to specific viscosity, but water at 68°F (20°C) is used as the standard. Since the absolute viscosity of water at this temperature is approximately 1 centipoise, the relative viscosity of a fluid is practically equal to its absolute viscosity.

Viscosity index. Viscosity index is frequently applied to petroleum-based lubricants. It is a number that indicates the effect temperature changes have on the oil's viscosity. Oils that undergo a minimum change in viscosity for a given change in temperature have a high viscosity index.

Fluidity. The reciprocal of absolute viscosity, fluidity is a measure of the ease with which fluids flow. The unit of fluidity in the centimetre-gram-second system is the *rhe*; it equals—

$$\frac{1}{\text{poise}}.$$

Saybolt Seconds Universal. The American chemist George M. Saybolt devised this scale in the 1920s. The scale uses Saybolt seconds for units of measurement. Liquid at a constant temperature and pressure is permitted to flow through a small orifice that forms part of an instrument called a viscosimeter. The number of seconds required for a given volume of the liquid to pass through the orifice is used as the measure of viscosity. Another Saybolt system, called *Saybolt Furol*, is similar to the Saybolt Universal system, but a larger orifice is used. It is designed to measure the viscosity of heavy oils; indeed, the word furol is coined from the words fuel and road oil.

Redwood, Redwood Admiralty, and Engler seconds. Similar to the Saybolt system are the Redwood, Redwood Admiralty, and Engler seconds systems. Sir Bovington Redwood developed special viscosity-measuring scales that bear his name for the British Navy in the 1920s. Redwood and Engler seconds are the British or European counterparts of Saybolt seconds.

Engler degrees. The Engler degree system expresses viscosity for a given fluid under specific conditions as a ratio of the time taken for a given volume of the fluid to pass through an Engler viscosimeter to the time taken for the same volume of water at 68°F to pass through the same viscosimeter.

Viscosity-Measuring Instruments

Any instrument that measures viscosity is called a *viscometer* or a *viscosimeter*. Stokes' apparatus is one kind of viscosimeter, and similar ones are available. One substitutes a piston for the ball, while another requires the timing of the rise of a bubble of air through the liquid. Measuring viscosity by noting the time required for a specified quantity of liquid to pass through a thin channel or an orifice is popular. Such a method is simple, inexpensive, and accurate.

The most popular viscosity meters are the Saybolt viscosimeter and direct-reading instruments. A Saybolt viscosimeter uses a temperature-controlled bath to regulate the liquid's temperature. In operation, a stopwatch measures the time for 60 mL of liquid to flow into a receiving flask. The results of this test are expressed in Saybolt Seconds Universal (SSU) or in Saybolt Seconds Furol (SSF) and then converted into kinematic viscosity and absolute viscosity with simple equations.

A direct-reading instrument (fig. 11.7A and B) measures viscosity without a stopwatch. The instrument uses the principle of relative motion between concentric cylinders separated by annular volumes of the liquid being measured (fig. 11.7B). Liquid fills a large part of the volume between two cylinders. As the lower cylinder rotates, the liquid it contains is also set in motion and exerts torque on the upper

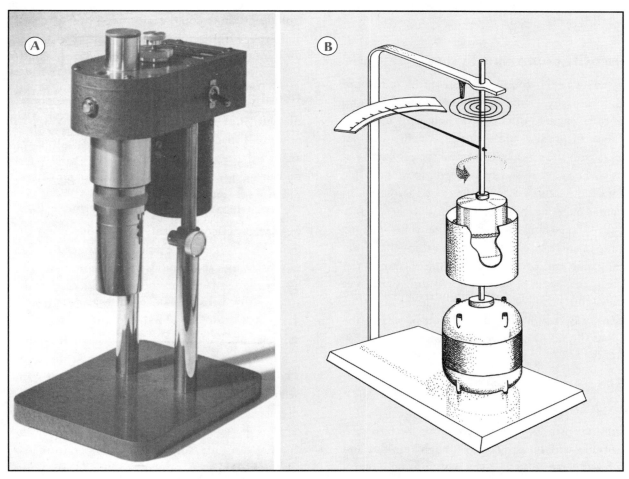

Figure 11.7 Direct-reading instruments for measuring viscosity. A, one type of direct-reading instrument; B, a simplified schematic of the principle involved.

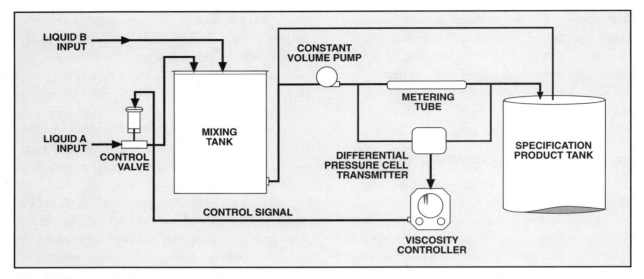

Figure 11.8   Control system for maintaining viscosity in a liquid

cylinder, which is restrained by a calibrated spring. The torque is proportional to the speed of rotation and the viscosity of the liquid.

Viscosity Controlling Systems

In some cases, it is desirable to control and record liquid viscosity. Figure 11.8 shows a typical viscosity control system. In this case, two liquids are involved. One is the product and the other is a chemical that mixes with the product to control the product's viscosity. The control system is based on the principle that the differential pressure developed across a thin-channel metering tube is proportional to the viscosity of the fluid passing through it. A controlled-volume pump maintains a constant flow rate through the metering tube, and a differential-pressure cell transmitter detects changes in pressure. The resulting signal goes to a viscosity controller. The controller regulates the viscosity by controlling the amount of one of the liquids (liquid A) entering the mixing tank.

MEASURING HUMIDITY AND DEW POINT

Humidity is a well-known condition because it is associated with human comfort. High humidity and high temperature are charged with causing physical discomfort, as in, "It isn't the heat, it's the humidity." Humidity is, however, also important to many industrial processes. The preparation or manufacture of some industrial gases requires measuring and controlling moisture content, or humidity. The textile and paper manufacturing industries also have need for humidity control.

Absolute and Relative Humidity

Absolute humidity is an expression for the weight of water dispersed as vapor in a unit weight of dry air or other gas. A common system of units for measuring it is grains of water per lb of air (mL of water per g of air).

Relative humidity is a familiar term because weather reports often state it. To understand the term, be aware that the amount of water that can exist as a vapor in air varies with temperature. The higher the temperature, the more water vapor the air supports. Relative humidity is the percentage of the total moisture that the air can support at a given temperature. For example, if the relative humidity is 50 percent at 90°F (32°C), then half the air at this temperature has water vapor in it. As the temperature rises, the relative humidity decreases, and vice versa, assuming the absolute moisture content remains constant. Figure 11.9 charts the relationship between dry- and wet-bulb temperatures and relative humidity. (Comparing dry and wet bulbs at a given temperature is a method of determining relative humidity; this method is discussed in detail

Gravity, Viscosity, Humidity, and pH

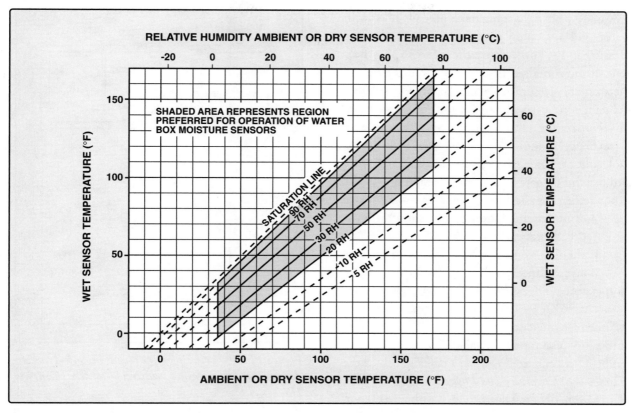

Figure 11.9 Chart of relative humidity versus dry- and wet-bulb temperatures

shortly.) Saturation is the maximum relative humidity (100 percent RH) that the air can support at a given temperature.

To better understand relative humidity, an understanding of water-vapor pressure is necessary. Figure 11.10 shows equipment that determines vapor

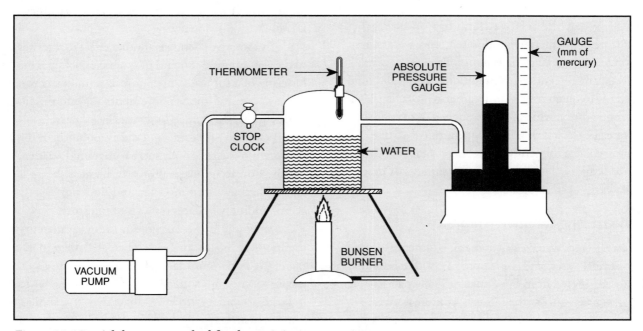

Figure 11.10 A laboratory method for determining vapor pressure

pressure. An air-tight container is partially filled with water, and a vacuum pump reduces the absolute pressure in the space above the water to very near zero. If the container is sealed off from the pump, the absolute pressure gauge soon stabilizes. At 10°C, the gauge indicates about 9.2 mm of mercury. At this temperature and pressure, a *state of equilibrium* exists between liquid and vapor—that is, for every molecule of water that escapes the liquid surface and becomes a part of the vapor, a molecule of the vapor returns to the liquid state. The space above the water under the conditions of equilibrium is saturated with water vapor.

If the temperature is changed, vapor pressure also changes. Vapor pressure increases with water temperature, and vice versa, but this change is not linear. At 20°C, for example, vapor pressure is 17.5 millimetres of mercury; at 100°C, it is 760 millimetres of mercury. Vapor pressure tables list a wide range of temperatures and are found in manuals such as the *Handbook of Chemistry and Physics*.

Ordinarily, air is not saturated with water vapor, although saturation occurs frequently and accounts for the dew sometimes found on cool mornings. In such cases, the temperature has fallen below the dew point, so the air is no longer capable of maintaining its moisture content. Later in the day, with rising temperatures, the dew evaporates and again becomes vapor in the atmosphere.

When the relative humidity of air reaches 100 percent, the air is saturated, and water vapor condenses and falls out as liquid. Fog and dew result from 100-percent relative humidity. The dew point, a term frequently used in weather reports, is the air temperature at which water vapor begins to condense. For example, a dew point temperature of 43°F (6°C) means that at this temperature, water vapor condenses into liquid water and collects on surfaces as dew or shows up in the air as fog.

Measuring Relative Humidity

The effects of water vapor in the air or other gas appear in several ways, and a measure of relative humidity can be inferred from these effects. Some materials are *hygroscopic*—that is, they absorb moisture from the air. Some of these materials undergo significant changes in linear dimensions as their moisture content varies.

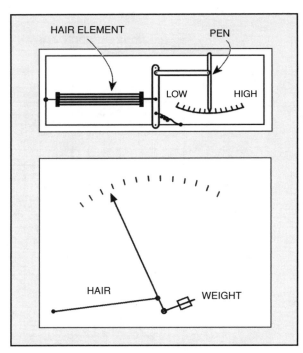

Figure 11.11 Principles involved in hair-type indicator used to measure relative humidity (Courtesy Honeywell)

Human hair is one substance that changes dimensions as the temperature and moisture content change. Hair increases in length with increasing relative humidity and consequently is used as the sensitive element in humidity indicators (fig. 11.11). To attain reasonable accuracy with such an arrangement, each instrument is calibrated individually.

Evaporative cooling is another effect that relates to the air's moisture content. Evaporative cooling is the lowering of a surface temperature as liquid water on the surface turns into vapor (evaporates) into the air. For example, the human body uses evaporative cooling. It emits perspiration, and the evaporation of the perspiration cools the body surface. In humid weather, perspiration accumulates profusely, because the air is already heavily moisture laden and does not eagerly accept additional concentrations of moisture.

Similarly, if the bulb of an ordinary mercury thermometer is surrounded with an absorbent material, such as cotton, that has been soaked in water, evaporation of this water causes the thermometer to indicate a temperature lower than that of a similar, but dry, thermometer placed beside it. Thermometers that have a wetting device attached to their bulbs are called wet-bulb thermometers. Dry-bulb thermometers do not have a wetting device.

Evaporative cooling occurs because heat is required to evaporate (or vaporize) water and other substances. Thus, the vaporizing of moisture in the wet material around the thermometer bulb causes a heat loss at that point and the temperature decreases. A *psychrometer* uses the evaporation principle to measure relative humidity. For example, a sling psychrometer has two thermometers fitted into a frame (fig. 11.12). One thermometer, which has a piece of cloth over its bulb, is the wet-bulb thermometer. The other is the dry-bulb thermometer. To operate the sling psychrometer, one thoroughly wets the cloth with clean water, then whirls the assembly for 20 to 30 sec so that a steady flow of air vaporizes some of the moisture. One quickly reads the wet-bulb thermometer and then the dry-bulb unit. The temperature of the wet-bulb thermometer is lower than the dry-bulb's temperature. Finally, one refers to a table or a calculator that accompanies the sling psychrometer to determine the relative humidity. The greater the temperature difference, the lower the relative humidity.

An installation for a recording psychrometer includes wet and dry bulbs that actuate pens in an indicator-recorder, which displays and records the two temperatures (fig. 11.13). The porous sleeve that fits over the wet bulb serves the same purpose as the cloth sleeve in the sling psychrometer. Water goes through a filter and onto the porous sleeve at a constant low pressure. The low-pressure source is a 12- to 18-in. (300- to 450-mm) long standpipe, which provides

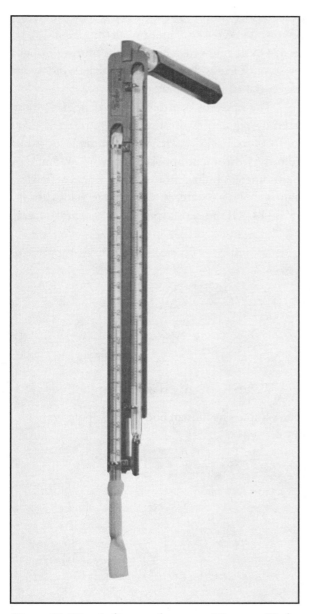

Figure 11.12   A sling psychrometer (Courtesy Taylor Instrument)

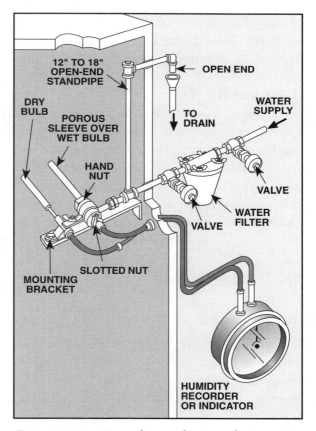

Figure 11.13   A recording psychrometer for measuring and recording humidity (Courtesy Foxboro)

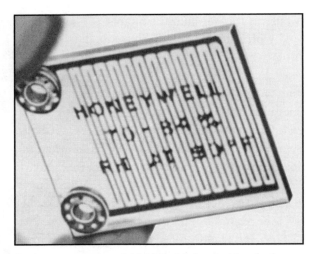

Figure 11.14   A gold-leaf grid, lithium chloride element used for measuring humidity (Courtesy Honeywell)

head pressure. The small amount of pressure keeps the sleeve moist without flooding it with excess water. If air movement is insufficient in the space around the thermometers, fans or blowers are installed to stir the air near the wet bulb.

Electric elements can also measure relative humidity. One consists of a pair of gold-leaf grids electrically separated from one another by a coating of lithium chloride (fig. 11.14). When perfectly dry (zero relative humidity), the lithium chloride coating is a nonconductor of electricity. When moist, however, it conducts current. Because lithium chloride is hygroscopic, it absorbs moisture from the surrounding atmosphere, and the amount it absorbs is directly related to the humidity of the atmosphere. The electrical resistance between the two gold-leaf grids varies with the moisture content of the lithium chloride and therefore also with the humidity. Compensators for temperature variations are incorporated into electrical elements of this type to give accurate relative humidity readings.

Measuring Dew Point

Dew point is the temperature at which moisture completely saturates the air or other gas. Dew-point measurement is often more important than humidity measurement, and a number of methods are available for measuring dew point. One method heats and cools a polished surface and then photoelectric devices detect condensation as it occurs on the surface.

Dew point is of great importance in the natural gas industry because liquid condensing from natural gas can cause problems. For example, sometimes gas liquids (condensates) freeze to form solids called *hydrates*, which can block flow in a line. By monitoring dew-point temperature, operators can note the formation of condensation and take steps to prevent it, perhaps by installing a heater.

A Bureau of Mines dew-point tester (fig. 11.15) can be used at gas pressures as high as 3,000 psi (21,000 kPa) and at any temperature below 140°F (60°C). Gas bleeds into the instrument and impinges on a polished mirror surface (the target). Mirror temperature is lowered until moisture in the gas condenses on it. The operator then observes the mirror through a system of lenses that forms a gas-tight window in the central cavity containing the mirror.

The operator controls mirror temperature by carefully regulating the flow of a refrigerant that contacts its backside. A thermometer protrudes from the instrument opposite the window, and the thermometer's bulb contacts the mirror. The refrigerant may be ice water if temperature values above 35°F (1.6°C) are satisfactory. Lower temperatures

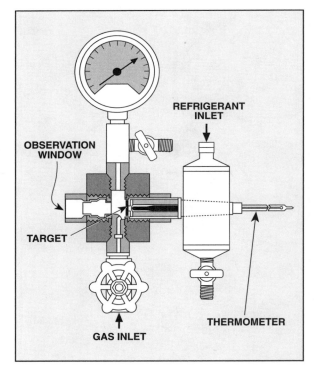

Figure 11.15   A Bureau of Mines type of dew-point tester

may be obtained by allowing compressed propane gas to expand in the cooling chamber or by using a solution of dry ice and acetone.

Operators must exercise care to obtain accurate measurement. They must be alert to the first sign of condensation and be capable of distinguishing between a hydrocarbon condensate, such as butane, and water condensate.

Manufacturers have incorporated the principles of the Bureau of Mines dew-point tester in automatic systems that measure and record dew-point temperature. Some systems use automatic refrigeration and heating units that adjust the temperature of the mirror to the dew-point value. A photoelectric device controls the heating and cooling effects by observing the formation of dew on the mirror, while a thermocouple circuit monitors the mirror temperature and transmits its value to a recorder-indicator.

One dew-point measuring device (fig. 11.16) keeps a saturated solution of hygroscopic lithium chloride at a temperature that is in equilibrium with the vapor pressure of the measured atmosphere. A temperature bulb, which contains an RTD, thermocouple, or a filled-thermal system, senses this equilibrium temperature. The sensor sends the equilibrium temperature to a recorder-indicator, such as a computer, display, or recording device, which records the dew-point temperature.

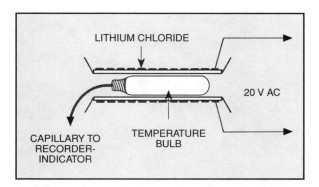

Figure 11.16 Dew-point electrical measuring device

MEASURING pH

A variable that some processes measure is whether a substance is acid or base (alkaline) and to what degree. The degree of acidity or alkalinity of a substance is its pH factor.

An acid neutralizes a base, and vice versa. This neutralizing effect explains why a strong solution of baking soda and water is a first-aid measure in treating a person who has spilled acid on his or her skin. The baking soda is a base and neutralizes the acid. Neutralizing forms a salt. For example, hydrochloric acid (HCl) added to sodium hydroxide (NaOH) results in the formation of common table salt (NaCl).

If acid is poured into water, the acid breaks down chemically, or *dissociates*, to produce hydrogen (H) ions. An acid is characterized chemically by having a wealth of hydrogen ions (H^+). If a base is poured into water, it also dissociates and produces hydroxyl ions (OH^-). The symbol pH is derived from the common chemical symbol for hydrogen (H) in both ions. In short, pH is a chemical abbreviation for hydrogen ion concentration.

The pH Scale

Pure water ionizes only slightly because it has only a small percentage of hydrogen and hydroxyl ions. In mathematical terms, 10^{-7} gram-ions of H^+ ions exist in a litre of pure water. (10^{-7} is the number 10 followed by seven zeros, which is 100,000,000.) The same quantity of gram-ions exists for hydroxyl ions in the same quantity of water.

The pH factor is derived from an expression (a *logarithm*) that eliminates negative powers of 10 when dealing with acids and bases. A logarithm is an exponent that indicates the power to which a number is raised to produce a given number. For example, the logarithm, or *log*, of 100 to the base 10 is 2. Put another way, $10^2 = 100$, so the term, "log 2 to the base 10," or, simply, "log 2," is another way to express 100. Logarithms are useful in making scientific and mathematical calculations. The equation for deriving pH is—

$$pH = \log[H^+] \qquad \text{(Eq. 11.8)}$$

where

pH = acidity or basicity of a solution
log = logarithm to base 10
$[H^+]$ = concentration of hydrogen ions per L of water or other solution.

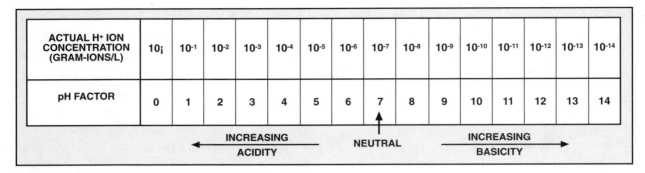

Figure 11.17 pH factor and ion concentration

Since the hydrogen ion concentration of water is 10^{-7}, the pH factor can be calculated to be—

$$pH = \log [H^+]$$
$$= -\log 10^7$$
$$= (7)$$
$$pH = 7.$$

Therefore, water is neutral. It is neither acidic nor basic and has a pH of 7. A substance that is not neutral has a pH different from 7. Substances with pH numbers less than 7 are acidic and substances with pH numbers greater than 7 are basic. For example, a liquid with a pH of 7.1 is basic and a substance with a pH of 6.9 is acidic.

Common substances and their pH factors include—

- Lemon juice: 2.0 to 2.2
- Other citrus juices: 3.0 to 4.5
- Vegetable and melon juices: 5.0 to 7.0
- Fresh milk: 6.50 to 6.65

Figure 11.17 shows pH factors from 0 to 14 and the ion concentration at the various pH factors.

Indicators of pH

Litmus paper is one material used in laboratory work that gives a rough estimate of pH. It is a strip of paper that is saturated with weak acids, bases, and salts. The acids, bases, and salts are allowed to dry and tint the litmus strip a certain color—often, but not always, pink or pale blue. (Laboratory workers simply say red or blue.) The paper changes its base color when exposed to acids and bases. For example, a strip of red litmus changes to blue when dipped in a base solution. Similarly, a strip of blue litmus paper changes to red if it is dipped in an acid solution.

Electrical devices can also measure pH. A solution called an *electrolyte* can be placed in a voltage cell. (An electrolyte is a nonmetallic electric conductor—usually a liquid—in which current is carried by the ions in the liquid.) The voltage the cell produces depends on the hydrogen ion concentration of the solution into which it is placed. One electric device for measuring pH employs a glass electrode as the sensing element (fig. 11.18.) When the glass electrode is placed

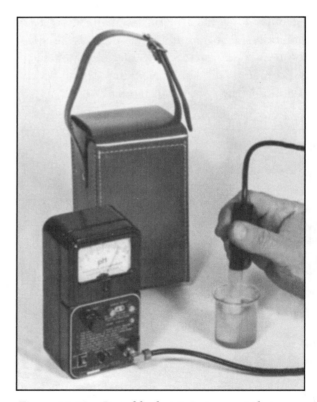

Figure 11.18 Portable electric instrument that uses a glass electrode to measure pH factor (Courtesy Analytical Measurements)

in the solution whose pH is to be determined, a needle indicator on a meter swings in one direction if the solution is basic and another direction if the solution is acid. The pH is read directly from the scale on the meter.

SUMMARY

Although temperature, pressure, flow, and liquid level are the main factors in measuring and controlling a process, other variables also affect processes. These include humidity, specific gravity, viscosity, and pH factor.

REVIEW EXERCISE

1. Define dew point.
2. Define relative humidity.
3. If a substance's pH is 7.2, is it base or acid?
4. Is molasses more or less viscous than water?
5. To determine the specific gravity of a liquid, what fluid is used as a standard for comparison?

12
Programmable Logic Controllers

The programmable logic controller (PLC) represents a significant advance in instrumentation. Since the PLC's introduction into automobiles in 1969, it has virtually replaced electromechanical relays in control circuits. Using solid-state electronic components, a PLC's reliability and flexibility are ideally suited for harsh industrial environments. Further, with only minimal hardware changes, technicians can easily reprogram the control circuit's ladder logic to suit a particular application.

A computer is the heart of a PLC, and those who first marketed it knew that people were initially skeptical of computer devices. So, they named it a controller to make it sound familiar to field operators and engineers. In addition, they added the terms programmable logic to indicate that operators could change the device's operation with software.

Early PLCs replaced relay logic circuits and hard-wired, solid-state controllers and were known as discrete, or on-off, controllers. Today's PLCs are more complex and powerful, and can handle analog signals from instruments in the form of current, frequency, and resistance. They can also perform mathematical comparisons; multiply and divide; extract square roots; and perform proportional, integral, and derivative (PID) functions.

PLC OPERATING CONCEPTS

Most PLCs have five common building blocks that originated from relay ladder logic in control circuits. Figure 12.1 is a ladder logic diagram that shows several functions. The two vertical lines on either end of the diagram are bus voltage, or power supply, lines. The left line is the hot bus and the right line is the common, or neutral, bus. These lines are also called rails in ladder logic terminology.

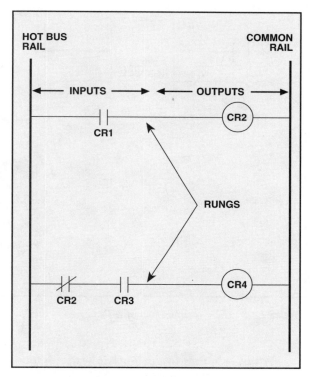

Figure 12.1 *Typical relay ladder logic diagram*

The two horizontal lines (the rungs) contain the logic control circuit. Figure 12.1 shows, in symbol form, five relays, contacts, or coils on the rungs. They are labeled CR1, CR2, CR3, and CR4. Devices and contacts (such as CR1) on the left side of the rung are inputs. The devices on the right side, such as the coil labeled CR2, are outputs. The lines that connect the input devices to the output devices on a particular rung are hard wired in relay circuits. However, software in the PLC's programming terminal also logically connects them. (Logic, in this sense, means the computer and its software not only recognize the electrical connection, but also recognize the function each component is designed to perform, and ensures that the components perform them properly.)

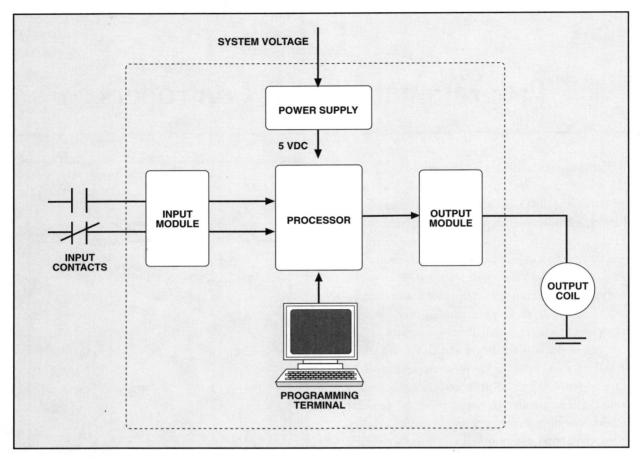

Figure 12.2 Five components of a PLC system

A PLC requires five major hardware components (fig. 12.2). They are—

1. A power supply that converts system voltage to 5 VDC for the integrated computer components.
2. Input modules that accept input field signals from such devices as switches, analog transmitters, frequency transmitters, and RTD sensors, and deliver digital signals to the processor.
3. A processor that uses ladder logic programming to sense digital input signals and issue digital output commands based on the ladder logic arrangement. The processor consists of a microprocessor, memory components, and EEPROM. (EEPROM is a special type of nonvolatile computer memory. Storing data in EEPROM simply means that a power loss or other interruption does not destroy or disrupt the information.)
4. Output modules that convert digital commands from the processor to operate field devices in the form of on-off control or analog control signals.
5. A programming terminal that allows entry of the desired ladder logic, or program, from the user to the processor. Usually, the user plugs in a personal computer or a hand-held programming device into the terminal.

Ladder Logic Programming

Technicians mainly use two methods to program PLCs. One programming method is based on Boolean symbols, or logic. Boolean symbols, some of which are written as OR, AND, NOR, NAND, and INVERTER, represent input-contact-logic arrangements in the program. The other programming method is based on ladder logic. Ladder logic uses contacts and coils such as those found in relay control diagrams.

Technicians use the Boolean method when a dedicated hand-held programming terminal programs the PLC. They employ the ladder logic method when a personal computer (PC) is used as the programmer. Because PCs are widely available, ladder logic programming is the most popular method.

Ladder logic is simply a diagram. By using information from the field, technicians develop a ladder logic diagram to solve a particular control problem. They then use the ladder logic diagram to program the PLC. Technicians must adhere to several rules when creating and reading a ladder logic diagram.

1. Show all device symbols in their shelf, or deenergized, condition.
2. Show input devices on the left part of the rung.
3. Show output devices on the right part of the rung.
4. Connect input devices so they create logic to accomplish a desired function and produce an output.
5. Read the ladder logic diagram from top to bottom and left to right.

As an example of developing a ladder logic diagram, let's say that we want to control air pressure in an air compressor. The air compressor uses an AC motor that must be turned on and off. When on, the motor runs the compressor to produce pressure. An air volume tank stores the pressure. The controls consist of a manual-off-auto switch, regulating pressure switch, high air temperature switch, low oil pressure switch, high-pressure shutdown switch, reset pushbutton, and related hardware. When the air compressor is first turned on in the auto mode, and the air pressure is below the regulating pressure switch set point, a timer is required to bypass the low oil pressure switch until pressure is built up. Figure 12.3 shows the ladder logic diagram developed to regulate and protect the air compressor.

To learn about the diagram, first note the reset switch at upper left and coil CR1 at upper right. Momentarily closing the reset switch energizes output coil CR1. Now note that TD1 is a time delay switch, HAT is a high air temperature switch, and H-H pressure is a high-high pressure switch. If the TD1, HAT, and H-H pressure switches are closed

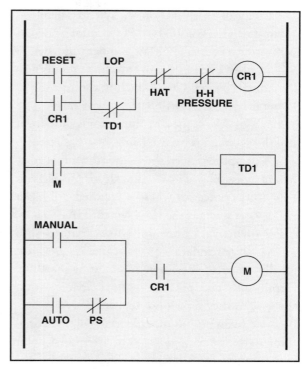

Figure 12.3 Ladder logic diagram of air compressor control

and conducting, coil CR1 is energized. Contact CR1, which is below the reset switch at upper left, is in the same rung and is in parallel with the reset switch. Contact CR1 also closes to provide an interlock for this rung.

Note the third rung at the bottom of the diagram. This rung also contains a contact, which is labeled CR1. Contact CR1 in this rung closes when its coil becomes energized. If regulating pressure switch PS is normally closed when the pressure is low, and if the selector switch is in auto and its contact is closed, motor control relay coil M is energized. Relay M operates the motor starter to start the air compressor motor. Note the second rung in the middle of the diagram. Contact M in this rung also closes and begins the timing of the second rung's time delay relay TD1.

After sufficient time, the low oil pressure switch LOP in the first (top) rung closes. Shortly thereafter, time delay relay contact TD1 in rung 1 (top rung) opens. The system continues to build pressure until regulating pressure switch PS in rung 3 (bottom rung) opens. PS's opening turns off motor relay M and resets timer TD1 in rung 1 for the next operating period.

In case of a malfunction, the system is designed to shut down the compressor. For example, say that overheating occurs. High air temperature switch HAT in rung 1 senses the overheating and deenergizes contact switch CR1, which is also in rung 1. Deenergized CR1 turns off the motor relay M in rung 2. With M off, the air compressor is shut down until the problem is resolved and rung 1 is reset.

A computer screen shows the ladder logic diagram in figure 12.3. The PC downloads the diagram to the PLC's processor where it is stored. All input field devices, such as the reset switch, HAT switch, LOP switch, H-H pressure switch, manual-auto switches, and regulating pressure switch are individually wired to the input modules of the PLC. Outputs CR1 and TD1 are internal devices within the software; they are not external devices. Also, they do not go to an output module terminal. However, motor starter relay M does go to a terminal point on the output module and is wired to the motor starter panel.

Once the run command is given to the PLC, it begins operating the air compressor. The internal clock in the processor reads the diagram inputs. It scans the program from top to bottom and left to right in sequence, issuing output commands required by the input devices. Scan time is rapid—typically a few milliseconds.

Processor Characteristics

The processor's memory section generally has two areas: program files and data files. Program files store the ladder logic program and subroutine and error files. Data files store data associated with the control program, such as the status of input and output bits, timer values, and other constants and variables.

Devices in the previously discussed compressor system are either on or off. Thus, they are discrete devices because they have only two states. Binary and digital are other terms for discrete. The numbers 1 and 0, which in the computer world are symbols, identify whether a device is energized or deenergized. The binary symbols 1 and 0 for each switch or contact is one bit, or one binary digit.

The amount of PLC processor memory required depends on the size of the program. A simple program may require only a few words of memory while a large system may require 64,000 words or more of memory. The complexity of the control program or plan determines the amount of memory required and the size of the processor.

One of the biggest advantages of a PLC is that a technician can change the ladder logic without reconnecting input and output devices. Since each component in a rung is logically connected and not hard wired, technicians can use the programming terminal to make reconnections. And they can easily change time delay values and other parameters from the PC keyboard.

Every PLC manufacturer has a programming language that allows the programming terminal to communicate with the processor. Each language utilizes ladder logic and structured texts such as Boolean logic, ladder logic with advanced function blocks, or sequential function charts.

Logic Numbering Systems

To identify the location of terminal points, modules, and modular assemblies, a method of addressing or locating these devices is needed. In a digital identification system, binary numbers locate devices.

The decimal numbering system normally used allows us to count from 0 to 9. When the count exceeds 9, add a 1 in front of the numbers from 0 to 19 to achieve a 10, 11, 12, and so on to 19. When the count exceeds 19, change the 1 to a 2 and count 20, 21, 22, and so on. When 99 is reached, add a third column of numbers beginning with 1 and start with 100, 101, and so forth.

The value of each column of numbers in the decimal system is based on the number 10. For example, the rightmost column is referred to as 1s and represents 10^0. (Any number raised to the power of zero is 1.) The next column is 10s and represents 10^1. The next column is 100s and represents 10^2, the next column is 1,000s and represents 10^3, and so forth. For example, the number 747 consists of two 7 symbols whose value is not the same because of their position. For convenience, we call this number "seven hundred forty seven," but 747 really means seven hundreds, four tens, and seven ones. Incidentally, the number in the rightmost column is called the least significant number, while the number in the leftmost column is called the most significant number. At any rate, 747 can be shown in tabular form as—

Base ten value:	10^2	10^1	10^0
Column value:	100	10	1
Decimal number:	7	4	7

Unfortunately, computers and PLCs cannot work with decimal number language. Just as a person from Japan needs English translated into Japanese to understand someone from the U.S., computers must have decimal numbers translated into binary numbers to understand them.

A device that has only two states, or conditions, is a binary, or digital, device. These two states may be on-off, energized-deenergized, true-false, yes-no, voltage-no voltage, and so forth. Notice that the two conditions are exactly opposite in meaning. Instead of using words to identify the two states, or conditions, computers use the symbols 1 or 0. One and 0 do not have any numerical value; instead, they are logic states, or conditions.

To further develop a method of communicating with a computer and translation language, let's establish an example. Figure 12.4 symbolizes four pipeline pumps operating into a common flow line under various on-off conditions. These pumps can be turned on or off, and the symbols 1 and 0 identify them as on or off. When on, the pumps deliver a liquid that is measured in gallons per minute. This flow is identified as total gpm. With all pumps off, 0 is in each pump's column and total gpm is zero. When the 1-gpm pump is turned on, the symbol 1 is in its column, while the other columns retain the 0 symbols. Thus, total gpm is 1. When the 2-gpm pump is turned on by itself, a 1 is in its column, and the total gpm is 2. When both the 1-gpm pump and the 2-gpm pump are on, the total flow is 3 gpm. In figure 12.4, note that many on-off pump combinations exist and that the total gpm varies accordingly. Also note that the 0s and 1s indicate which of the pumps is on and which is off. Further, these binary 0s and 1s correspond to the flow rate in decimal numbers. For example, 1100 corresponds to 12 gpm.

The column labeled with a decimal value of 1 represents the first digit in a binary number. In this case, 1^0 is 1. The column with the decimal number 2 is the second digit in a binary number because 2^1 is 2. The third number is 4, or 2^2, and the fourth number is 2^3, or 8. This process can continue to produce decimal column values of 16, 32, 64, 128, 256, and so forth, doubling the value of the preceding number. As previously mentioned, the lowest significant digit is on the right and the most significant digit is on the left.

As an exercise, try creating digital and decimal numbers from the following.

Binary Number	Decimal Number
1101	?
101101	?
?	27
?	132

The answers—

Binary Number	Decimal Number
1101	13
101101	45
11011	27
10000100	132

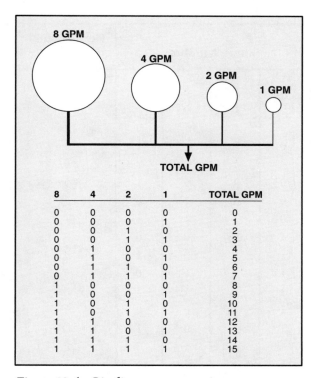

Figure 12.4 Pipeline pump operation

Binary numbers communicate with the computer in a PLC because the binary digits not only represent decimal numbers, but also they can represent alphabetic characters, symbols, graphics, and the like.

Octal and Hexadecimal Numbers

If a binary number becomes too long, it can be cumbersome. Such long numbers can be shortened into a new number using three binary digits. With three binary digits, counting is done in decimal form from 0 to 7, or in eight counts. Use of eight counts leads to the name, octal number. An octal number is a shortened version of a binary number. The base number for octal is 8, or 2^3. As an example of shortening a binary number to octal, take the binary number 110101010111101. First, rearrange the numbers into groups of three: 110 101 010 111 101. Finally, give the binary numbers their equivalent octal numbers. Thus, 110 is 6, 101 is 5, 010 is 2, 111 is 7, and 101 is 5. The new octal number is written as 65275_8. It has no numerical value; it only represents the original binary number, which consists of 1s and 0s. Be aware that octal digits are counted from 0 to 7; 8 and 9 do not exist in this system. Table 12.1 shows binary numbers and their equivalent octal numbers.

Hexadecimal numbers are another way to represent binary digits and reduce the number of binary symbols. Hexadecimal (hex, for short) numbers are created by combining and grouping four binary digits, instead of three, as with octal numbers. The base for hex numbers is 16 (2^4). Table 12.2 shows the binary numbers and the hex symbols.

Beyond 9 in the hex system, numbers become alphabetical characters because single digit symbols are required for programming computers. The subscripts H or 16 identify hex numbers.

Table 12.3 summarizes the four numbering systems.

The four numbering systems discussed are the binary whose base is 2, the octal whose base is 8, the decimal whose base is 10, and the hexadecimal whose base is 16. Many PLC manufacturers identify field terminal points on their input and output modules with octal numbers. Octal and hexadecimal numbers are also used in data table identifications, digital switch inputs, and other applications associated with PLCs.

Table 12.1
Binary and Octal Numbers

Binary Numbers			Octal Numbers
0	0	0	0
0	0	1	1
0	1	0	2
0	1	1	3
1	0	0	4
1	0	1	5
1	1	0	6
1	1	1	7

Once 111 is reached in the binary rows to establish a count of 7, no further digits are allowed. After 0 to 7, subsequent octal number sequences are 10 to 17, 20 to 27, 30 to 37, 40 to 47, etc.

Table 12.2
Binary and Hexadecimal Numbers

Binary Numbers				Hexadecimal Numbers
0	0	0	0	0
0	0	0	1	1
0	0	1	0	2
0	0	1	1	3
0	1	0	0	4
0	1	0	1	5
0	1	1	0	6
0	1	1	1	7
1	0	0	0	8
1	0	0	1	9
1	0	1	0	A
1	0	1	1	B
1	1	0	0	C
1	1	0	1	D
1	1	1	0	E
1	1	1	1	F

Table 12.3
Binary, Octal, Hexadecimal, and Decimal Equivalents

Binary Number	Octal Number	Hexadecimal Number	Decimal Number
1100011101	1435_8	$31D_H$	797
1110011100110	16346_8	$1CE6_H$	7398

PLCs can also communicate with external devices to provide or receive information from human-machine-interface (HMI) devices. Machine-to-machine communication can be accomplished digitally, but HMI requires a more interpretable form. When HMI communication is required, the binary numbers of 1 and 0 can be arranged in combinations of 1s and 0s that are unique. This arrangement is referred to as binary coding. Binary coding can be grouped in two categories of numbers only or in decimal numbers, symbols, and alphabetical letters. The most common binary codes are American Standard Code for Information Interchange (ASCII), binary coded decimal (BCD), and Gray codes.

The ASCII binary code, uses 6, 7, or 8 bits to represent alphanumeric symbols, other symbols, and control functions. The most common ASCII character set is the seven-bit code, which provides 2^7, or 128, possible characters. This seven-bit code provides for upper and lower case alphabetical characters, decimal numbers, special symbols, and control characters. This scheme provides all combinations of characters used when communicating with peripherals and interfaces. If error checking during transmission is needed, an eighth bit can be used for parity checks. Transmission of data using ASCII code is usually in serial format between the PLC and peripheral equipment.

Another common code is BCD. The BCD code system allows easy interface between machines and humans by using a four-bit binary number to create equivalent decimal numbers from 0 to 9. When data is entered with thumbwheel switches, encoders, or analog input-output instructions, this form of conversion is easy and convenient. Table 12.4 converts numbers from decimal to binary to BCD.

Table 12.4
Decimal to Binary to BCD Conversions

Decimal	Binary	BCD
0	0	0000
1	1	0001
2	10	0010
3	11	0011
4	100	0100
5	101	0101
6	110	0110
7	111	0111
8	1000	1000
9	1001	1001

The primary application for the Gray binary code is in angular shaft positioning systems. When digital encoders are used in positioning transducers, the output 1s and 0s are used in a four-bit format to identify angular position. If standard binary code is used, a dramatic change in the position of the 1 would occur between binary 7 and 8 because four bits can change at once, resulting in possible errors in transmission. By using the four-bit Gray code, which differs from standard binary code, only one bit changes as the counting number increases.

PLC BRANDS

Many brands of PLCs are manufactured. They have features that make them suitable for various applications. Also, they generally have unique programming schemes.

Allen-Bradley PLC-5

One of the largest manufacturers of PLCs in the United States is Allen-Bradley whose product line is broad. They manufacture such products as large-scale PLCs, small machine tool versions, and annunciators. Within the PLC products, input-output (I/O) modules include discrete types, analog types, totalizing or frequency modules, and RTD modules. The programming capability includes on-off functions, timers, counters, 12-bit input and output analog functions, arithmetic functions, comparators, and so forth. Figure 12.5A and B shows two Allen-Bradley processors in its PLC-5 line. Figure 12.6 shows two Allen-Bradley I/O modules.

Figure 12.7 shows how a PLC-5 and an I/O module can be arranged in a chassis. Modules can be arranged in the chassis as desired. However, the processor is always placed in the first slot on the left side. The I/O modules can be arranged in any order, with input modules intermixed with the output modules. Using Allen-Bradley's single-slot arrangement, each module is referred to as a group. Further, eight groups are referred to as a rack and each group can have eight or 16 logic terminal points for field wiring.

Input field contacts or switches are identified or addressed as—

$$I:xxy/zz$$

where

I is the designation for input devices
xx is the rack number of the assembly
y is the group number
zz is the terminal point number.

Output field devices are addressed in a similar fashion as—

$$O:xxy/zz$$

where

O is the designation for output devices
xx is the rack number of the assembly
y is the group number
zz is the terminal point number.

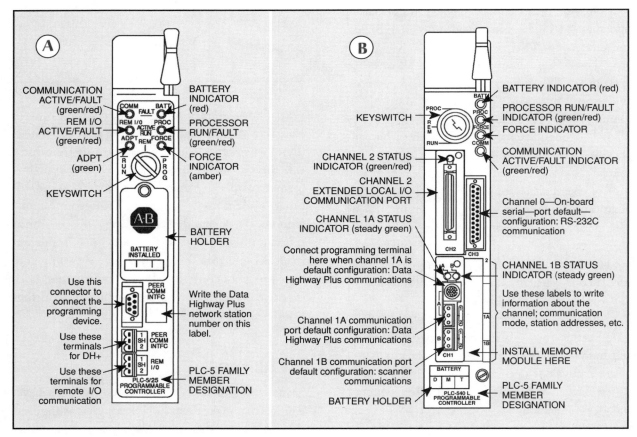

Figure 12.5 Allen-Bradley processor modules

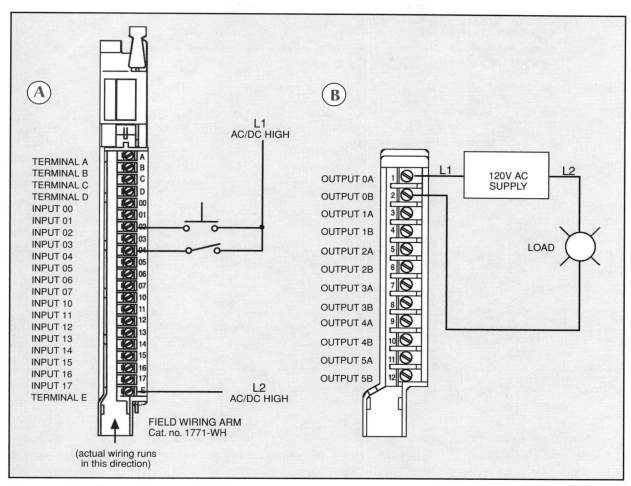

Figure 12.6 Allen-Bradley I/O modules

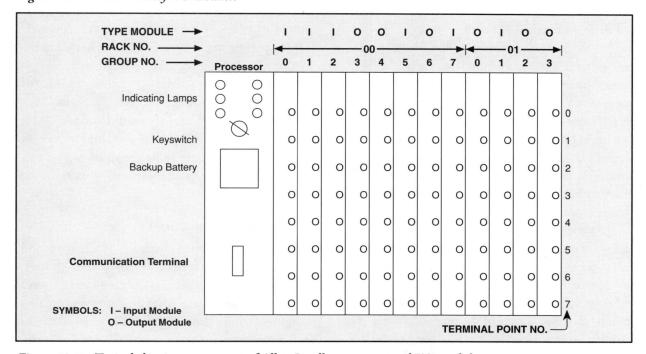

Figure 12.7 Typical chassis arrangement of Allen-Bradley processor and I/O modules

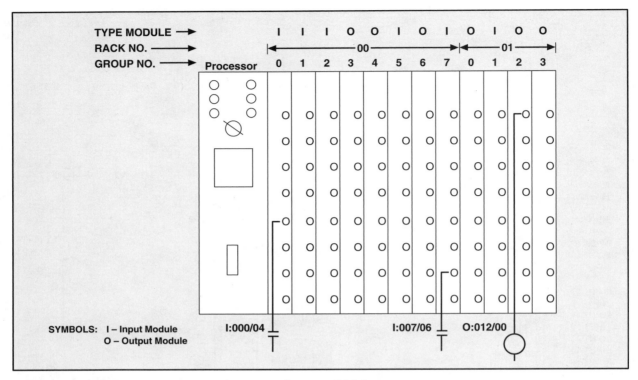

Figure 12.8 Chassis arrangement with input and output field devices

By referring to figures 12.7 and 12.8, we can address the racks, groups, and terminal points along with the field devices that terminate to the terminal points. The switches shown terminated to the input modules have been assigned their address numbers as well as the output devices.

After the control designer has created ladder logic to solve a particular problem, the ladder logic is graphically created on the programming terminal—usually a PC—that contains the base software for the Allen-Bradley system. With the processor selector switch placed in the programming position, the program is downloaded into the processor memory. To operate the program, the processor selector switch is turned to the run position and the unit is placed in operation.

As the processor internal clock produces the timing for scanning the program, the processor reads the ladder logic from top to bottom and left to right in sequence. As the processor reads input device states and compares them to the program, it issues output commands for each rung of the program. When the scanning process completes the final rung of the ladder logic program, it starts all over at the first rung.

When operators view the ladder logic schematic during the run operation on the computer monitor, they will note that the ladder logic symbols do not show opening or closing contacts. Instead, the monitor shows a change in the contact's surrounding color. Generally, if a contact is conducting, or permitting, a logic 1 to be transferred to the next device in the rung, it becomes green in color. If all input devices in a series are colored (or intensified) in green, the rung becomes true, establishes itself as a logic 1, and activates the output device to its true or logic 1 state. If any input device in the series string ceases to be green, or reverts to a nonconducting state, the output will turn off or become false.

Analog Inputs and Outputs

Previously, this chapter discussed PLC inputs and outputs and emphasized on-off, or discrete, devices. Recall that these input and output devices have only two states. These states are, however, definite conditions and are defined digitally as 1 and 0. On the other hand, process transmitters are analog devices. Their signals are continuous with an infinite number of points between any two limits. When calibrated, for example, a pressure transmitter operating between 0 and 200 psi (0 and 1,500 kPa) produces a current output signal of 4 to 20 mA. The input pressure and the output current are analog (or analogous) signals of an event that is taking place. Typical devices that produce analog signals are pressure, temperature, level, and flow transmitters, RTD devices, vibration transducers, and load cells.

Since the PLC processor is a digital computer that can only recognize 1s and 0s, some device must be inserted that converts the analog mA signal into a group of 1s and 0s that correspond to the pressure. This interface between analog current and the digital processor must be capable of translating the analog signal into discrete, or digital, values that can be interpreted by the processor.

Analog input modules are used in PLC systems to perform the conversion from analog signals to digital values (fig. 12.9). Circuitry in the modules senses the range of analog signals (typically, 4 to 20 mA), converts it to an analog voltage of 1 to 5 VDC, and produces a multiple-bit digital signal whose digital value corresponds to the analog input level. The conversion from analog to digital is called an analog-to-digital, or A/D, converter. Figure 12.10 is a functional diagram of an analog module, which converts analog signals to digital. Note that the 4-

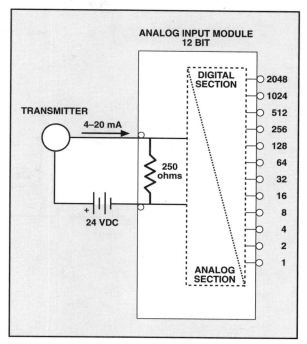

Figure 12.10 Functional diagram of analog input modules

to-20 mA current signal is converted into voltage through a 250-Ω resistor. This analog voltage has a normal range of 1 to 5 VDC and 0 to 5 VDC if the signal wire becomes disconnected. The conversion of analog to digital is usually made with a four-bit, eight-bit, or 12-bit converter, depending on the resolution and accuracy desired.

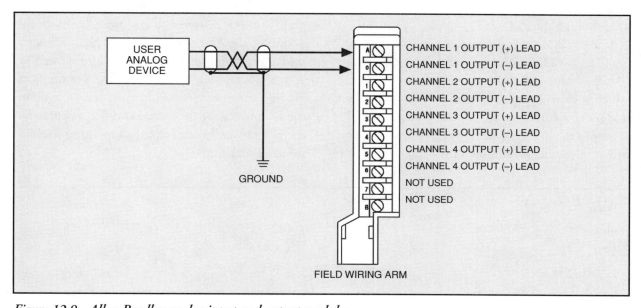

Figure 12.9 Allen-Bradley analog input and output modules

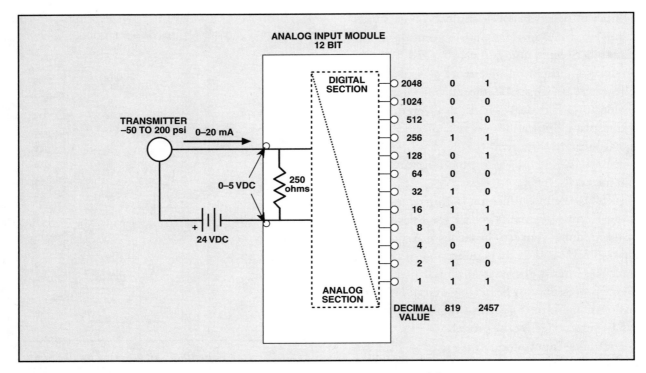

Figure 12.11 Process transmitter connection to a 12-bit analog input module

For example, a 12-bit A/D converter means that the analog signal range is converted into 2^{12}, or 4,096, increments. Thus, its resolution is 1 part in 4,096 parts. Put another way, the analog signal is reproduced in digital form within 0.025 percent of its value. When the overall accuracy of a measuring system is considered, a 1 percent accurate transmitter has a very small 0.025-percent error added to its value through the PLC.

If the input signal covers a range of 0 to 20 mA (0 to 5 VDC), the A/D converter begins with a decimal value of 0 and ends with a value of 4,095 (which is close enough to 4,096 to be suitable) by producing a combination of twelve 1s and 0s. A register in the processor stores the digital values for further operations within the PLC's circuitry.

Figure 12.11 shows a process pressure transmitter that has been calibrated over the range of 0 to 200 psi to produce an output signal of 4 to 20 mA. Its output signal is connected to an analog input module of the PLC where the signal is converted to a voltage of 1 to 5 VDC (0 to 5 VDC when a broken wire is taken into account). Using a 12-bit analog input module, the digital output signal is produced prior to being delivered to the processor. Each bit output of the A/D converter is weighted in binary format beginning with 1 at the first bit followed by 2, 4, 8, 16, etc., and ending with 2,048 at the 12th bit. Since a linear relationship exists at each point in the system, scaling the system at different points in the process can be accomplished by using selected values of pressure.

Pressure	-50	0	100	150	200
mA	0	4	12	16	20
VDC	0	1	3	4	5
Decimal Value	0	819	2,457	3,276	4,095
Binary Value	0	1100110011	100110011001	110011001100	111111111111

Note that the pressure of -50 in the chart does not physically exist; instead, it represents 0 mA, which occurs if a signal wire breaks. The system sends binary values to the processor in a block format, which the software in the processors uses as a value, and inserts them in the ladder logic program. The numbers to the right in Figure 12.11 show typical binary and decimal values for pressures of 0 psi and 100 psi.

The processor's logic uses the binary value at each pressure level in several ways, depending on what is being accomplished. For example, the logic can—

1. compare values with a preset amount to indicate level or sound an alarm.
2. indicate pressure values on a monitor chart or flow diagram.
3. compare values with a set point in a control mode and forward the result to an analog output module.
4. determine flow rate by performing mathematical calculations, such as square root functions, when an orifice plate is used.

Another analog module is the analog output module that delivers a 4-to-20-mA or 15-VDC signal to an external instrument or a transducer (fig. 12.12). This module is particularly useful in flow control applications where the analog output signal is delivered to an I/P transducer that converts the input signal to a 315-psig (2,000-kPa, gauge) valve positioner.

Other analog modules available in PLCs include frequency-, or pulse-input, modules. These modules accept turbine or vortex flowmeter frequency signals that are proportional to the flow rate. A frequency-counter-to-binary converter changes the module's proportional frequency signals to a binary signal. Also, RTD and thermocouple input

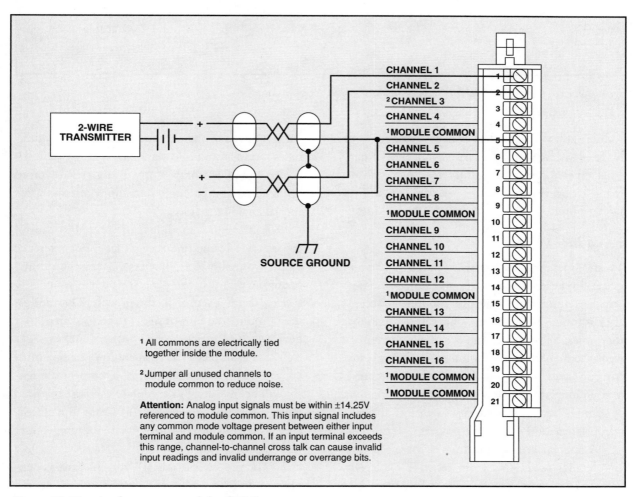

Figure 12.12 Analog output module of PLC

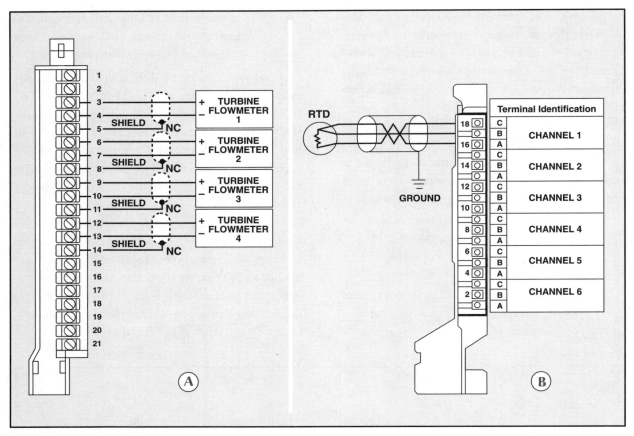

Figure 12.13 Frequency and RTD/thermocouple input PLC modules

modules are available that convert resistance, or low-level voltage, to a binary signal. The binary signal gives direct temperature readings without having to go through a temperature transmitter (fig. 12.13).

Special Interface Modules

PLC modules that interface with special sensors and devices are available for particular functions. One such discrete module is a pulse module. It accepts very fast input pulses and stretches them into longer pulse times. Because a processor's scanner requires a given amount of time to view input conditions before acting on them, it may not observe a fast input pulse. By stretching the pulse within the input module, the scanner has an opportunity to view it and take appropriate action within the program.

A special analog module is a weigh-scale module that reads data from resistive load cells. Weight-scale modules support the industry standard of 2- or 3-millivolt (mV) per V (mV/V) load cells.

An advantage of this input module is that it eliminates a transmitter and the need for a 4-to-20 mA analog module. Storage tanks and vessels involved in blending and batching operations often use this special analog module.

A PID input module is often used in control processes. The module accepts an input reference signal from the PLC that establishes the desired operating point. It compares the reference signal to a signal from a sensor in the process. The module then produces an output signal to control an output device such as an actuator or proportional valve.

For example, if the temperature of a tank filled with a fluid is to be maintained, a temperature sensor will be installed in the tank to measure the fluid temperature and the amount of steam in a closed-pipe system is controlled to maintain the temperature (fig. 12.14).

The process variable (PV) in this case is temperature and the module compares it with the desired temperature set point (SP). If the process variable

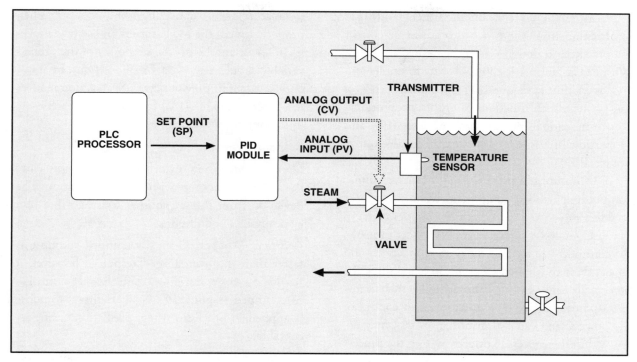

Figure 12.14   Temperature control using PID module

deviates from the set point, the system produces an error signal. Subsequently, the control loop modifies the controlled variable's (CV) output to bring the error to zero. During the course of control in the PID module, a control algorithm, or equation, is solved by the control loop using proportional, integral, and derivative action.

The PID action is simultaneous in the module. First, the PID module reads the control variable error and produces a proportional control action. An integral control, or reset, action follows, which provides additional correction to the control variable. The additional correction changes the process variable over a period of time. Then, the PID module introduces a derivative, or rate, action, which further corrects the control output. This correction is proportional to the rate of change of error.

PLC APPLICATIONS AND LOOP TUNING

PLCs have changed the way process data are obtained, controlled, used, and recorded. In the past, mechanical devices measured such variables as temperature, level, flow, and pressure. Moreover, the mechanical devices controlled the variables and recorded them on chart recorders that mechanical linkages operated. While many mechanical devices are still in use, PLCs and computer-controlled systems are rapidly replacing them.

Because a PLC's foundation is an industrial computer located in its processor, technicians can easily modify a PLC's instructions with software commands. Further, a PLC can control processes with specialized functions, and record process data in its electronic memory or on CD-ROMs, disks, or magnetic tape. What is more, operators can modify processes if necessary by observing historical data from a PLC's trending program. Indeed, computer assisted operations (CAOs) of processes are becoming quite common.

Modes of Operation

Process control is characterized by the way in which a control responds to process variables. In open-loop control, for example, a set point is established and the process selects a value or position until an operator intervenes. Thus, open-loop control is a manual process: a human must intervene to make a correction or adjustment.

Automatic control, on the other hand, uses information from the process to adjust or correct the process as it deviates from the desired set point. With PLCs, automatic control is easy to implement and to modify if required. PLC automatic control can perform such functions as on/off, proportional (P), proportional-plus integral, or reset (PI), and proportional-plus-integral (or reset) plus derivative, or rate (PID).

To optimize a process using PLCs or similar computer-controlled equipment, technicians must establish and enter process constants as input data to the PLC processor or controller. When set properly, the controller can regulate the process effectively and efficiently to the desired set point. In establishing the process constants, operators must consider such factors as set-point changes, desired closeness of control of the process, and external disturbances that may occur within the process. Correctly setting the process constants ensures the desired results. The process will respond as expected with minimum overshoot and stable operation.

Figure 12.15 shows overdamped, critically damped, or underdamped responses to changes above and below the set point (SP). When a process variable strays from the set point, the system responds to the change. However, the process does not immediately change; a certain lag occurs. Moreover, once the change begins, it does not immediately reach the set point and stabilize. In short, the process variable responds by going above and below the set point (oscillating) before stabilizing at the set point. When properly tuned, the oscillatory response (the movement above and below the set point) of the process variable should attenuate, or smooth out, to about one-fourth of its previous cycle. Such action reduces the process variable (PV) by one-fourth each cycle until constant PV is achieved.

Response types include—

Overdamped. When disturbed, the PV approaches the set point smoothly without any oscillation. The deviation from the set point is less than that of a critically-damped response.

Critically Damped. Critically-damped response is faster than the overdamped response; however, it results in a larger deviation from the set point than overdamped response. Critically-damped response is applied in processes where oscillation is not acceptable.

Underdamped. Underdamped response oscillates several times as the system recovers from a disturbance. The PV produces several oscillations before settling to a constant value. The advantage is a fast response to disturbances.

One-Fourth Amplitude Response. (Not shown in fig. 12.15.) This response represents the ideal response to PV disturbances because the closed-loop response attenuates oscillation by one-fourth for each succeeding oscillation.

Technicians use several testing or analytical methods to determine the gain, integral time constant, and derivative time constant involved in producing the desired response. Tests include the (1) Ziegler-Nichols open-loop tuning method, (2) the integral of time and absolute error open-loop tuning method, and (3) the Ziegler-Nichols closed-loop tuning method. Descriptions of these methods are beyond the scope of this book. However, manufacturers publish references and advanced textbooks cover them. Also, several vendors provide software for determining good performance constants that can be run on personal computers and coordinated with PLC control systems. Figure 12.16 is a schematic of such a system. This type of testing saves time and provides optimum tuning performance. Technicians can enter the constants obtained from response tests into the PLC processor program, PID input module

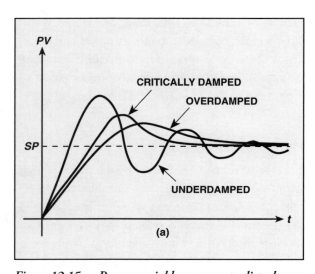

Figure 12.15   Process variable responses to disturbances

Programmable Logic Controllers

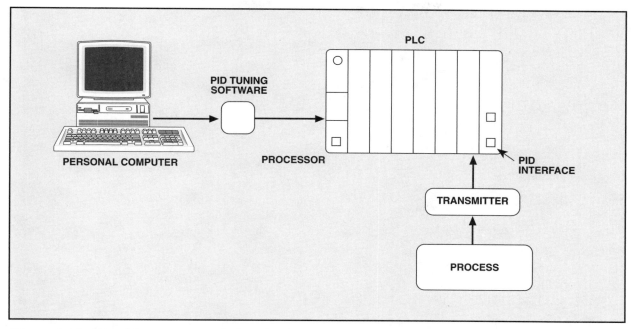

Figure 12.16 Closed-loop tuning methods with software

or process controller through a digital thumbwheel switch or rotary dial. Then, they can fine tune the values by observing the actual process.

SUMMARY

The PLC has significantly advanced automation and control in instrumentation. When used with discrete devices, such as level, temperature, and pressure switches, the PLC can turn on and off alarms, motors, and other process equipment. When used with analog devices, such as process transmitters, RTDs, turbine flow meters, and I/P transducers, it extends the use of PID control.

REVIEW EXERCISE

1. What numerical language does a PLC processor use?
2. Field devices are connected to certain PLC modules. Name them.
3. Pressure, level, and temperature transmitters produce a 4-to-20 mA signal and are connected to this type of PLC module. Name it.
4. Addressing field switches and inputs and outputs of PLC modules is generally achieved with this numbering system. What is it?
5. Name four modes of controller operation.
6. Identify two methods of obtaining gain, integral, and derivative time constants.

Appendix A

NUMBERING SYSTEMS AND CODES

The tables and other reference information contained in this appendix are compiled for the purpose of supporting operation, programming, and maintenance of PLCs, industrial computer systems, distributed control systems, and digital field bus systems.

Binary Number System

A binary number system has these characteristics:
1. It is a base 2 numbering system.
2. It uses 1 and 0 as symbols to make discrete decisions or show states of devices.
3. The largest valued symbol is 1. The lowest valued symbol is 0.
4. A binary number can be converted to decimal, and decimal to binary, by using the following table that counts through four binary digits (4 bits). Larger binary numbers can be achieved by adding more columns whose values are 2^4, 2^5, 2^6, 2^7, 2^8, etc.

2^3	2^2	2^1	2^0	
8	4	2	1	Decimal Value
0	0	0	0	0
0	0	0	1	1
0	0	1	0	2
0	0	1	1	3
0	1	0	0	4
0	1	0	1	5
0	1	1	0	6
0	1	1	1	7
1	0	0	0	8
1	0	0	1	9
1	0	1	0	10
1	0	1	1	11
1	1	0	0	12
1	1	0	1	13
1	1	1	0	14
1	1	1	1	15

Octal Number System

An octal number system has these characteristics:
1. It is a base 8 numbering system.
2. It consists of eight symbols: 0, 1, 2, 3, 4, 5, 6, 7

3. Used in many PLCs for addressing word and bit locations in data tables and for I/O points.
4. Octal number uses decimal symbols with a subscript of 8.
5. Conversion from binary to octal is performed by combining the binary number into groups of three bits.

An example of this is:

Binary Number		Regrouping Bits in Groups of Three					
110111101001001111	or	110	111	101	001	001	111
Equivalent decimal number:		6	7	5	1	1	7

The octal number for the above binary number is then: 675117_8

Hexadecimal Number System

A hex number system has these characteristics:
1. It is a base 16 numbering system.
2. It consists of sixteen symbols: 0, 1, 2, 3, 4, 5, 6, 7, 8, 9, A, B, C, D, E, F
3. Used in many display systems.
4. Hexadecimal number uses alphanumeric symbols with a subscript of 16.
5. Conversion from binary to hex is performed by combining the binary number in groups of four bits with the following values:

2^3	2^2	2^1	2^0	
8	4	2	1	Hex Value
0	0	0	0	0
0	0	0	1	1
0	0	1	0	2
0	0	1	1	3
0	1	0	0	4
0	1	0	1	5
0	1	1	0	6
0	1	1	1	7
1	0	0	0	8
1	0	0	1	9
1	0	1	0	A
1	0	1	1	B
1	1	0	0	C
1	1	0	1	D
1	1	1	0	E
1	1	1	1	F

An example of this is:

Binary Number Regrouping Bits in Groups of Four

110111101001001111	or	0011	0111	1010	0100	1111
Equivalent alpha numeric number:		3	7	A	4	F

The hex number for the above binary number is then: **$37A4F_{16}$**

Binary Coded Decimal (BCD) Code

A BCD Code has these characteristics:
1. Used to store information or used with input devices such as thumbwheel switches and 7-segment displays.
2. It consists of the 4-digit binary equivalent of the ten decimal numbers, 0-9.
3. Showing a decimal number in BCD is to write the equivalent BCD code for each digit. For example, the decimal number 72 becomes 0111 0010. The space between digits signifies a BCD coded number.

BCD	Binary	Decimal
0000	0	0
0001	1	1
0010	10	2
0011	11	3
0100	100	4
0101	101	5
0110	110	6
0111	111	7
1000	1000	8
1001	1001	9

American Standard Code for Information Interchange (ASCII)

An ASCII Code has these characteristics:
1. Used in PLCs to represent numbers, letters, and symbols. It is also used to send and receive alphanumeric data, such as the letters and symbols found on a computer keyboard.

LIST OF ASCII CODES

Symbol, Name of Character	Binary Number	Octal Number	Hex Number	Decimal Number
NUL	000 0000	000	0x00	0
SOH, Control-A	000 0001	001	0x01	1
STX, Control-B	000 0010	002	0x02	2
ETX, Control-C	000 0011	003	0x03	3
EOT, Control-D	000 0100	004	0x04	4
ENQ, Control-E	000 0101	005	0x05	5
ACK, Control-F	000 0110	006	0x06	6
BEL, Control-G	000 0111	007	0x07	7
BS, Backspace, Ctrl-H	000 1000	010	0x08	8
HT, Tab, Ctrl-I	000 1001	011	0x09	9
LF, Line Feed, Ctrl-J	000 1010	012	0x0a	10
VT, Ctrl-K	000 1011	013	0x0b	11
FF, Form feed, NP, Ctrl-L	000 1100	014	0x0c	12
CR, Carriage rtrn, Ctrl-M	000 1101	015	0x0d	13
SO, Ctrl-N	000 1110	016	0x0e	14
SI, Ctrl-O	000 1111	017	0x0f	15
DLE, Ctrl-P	001 0000	020	0x10	16
DC1, XON, Ctrl-Q	001 0001	021	0x11	17
DC2, Ctrl-R	001 0010	022	0x12	18
DC3, XOFF, Ctrl-S	001 0011	023	0x13	19
DC4, Ctrl-T	001 0100	024	0x14	20
NAK, Ctrl-U	001 0101	025	0x15	21
SYN, Ctrl-V	001 0110	026	0x16	22
ETB, Ctrl-W	001 0111	027	0x17	23
CAN, Ctrl-X	001 1000	030	0x18	24
EM, Ctrl-Y	001 1001	031	0x19	25
SUB, Ctrl-Z	001 1010	032	0x1a	26
ESC, Escape	001 1011	033	0x1b	27
FS	001 1100	034	0x1c	28
GS	001 1101	035	0x1d	29
RS	001 1110	036	0x1e	30
US	001 1111	037	0x1f	31
Space	010 0000	040	0x20	32
! Exclamation mark	010 0001	041	0x21	33
" Double quote	010 0010	042	0x22	34
# Hash	010 0011	043	0x23	35
$ Dollar	010 0100	044	0x24	36
% Percent	010 0101	045	0x25	37
& Ampersand	010 0110	046	0x26	38
' Quote	010 0111	047	0x27	39
(Open parenthesis	010 1000	050	0x28	40
) Close parenthesis	010 1001	051	0x29	41
* Asterisk	010 1010	052	0x2a	42
+ Plus	010 1011	053	0x2b	43
, Comma	010 1100	054	0x2c	44
- Minus	010 1101	055	0x2d	45
. Full Stop	010 1110	056	0x2e	46
/ Oblique stroke	010 1111	057	0x2f	47
0 Zero	011 0000	060	0x30	48
1	011 0001	061	0x31	49
2	011 0010	062	0x32	50
3	011 0011	063	0x33	51
4	011 0100	064	0x34	52
5	011 0101	065	0x35	53
6	011 0110	066	0x36	54
7	011 0111	067	0x37	55

LIST OF ASCII CODES, cont.

Symbol, Name of Character	Binary Number	Octal Number	Hex Number	Decimal Number
8	011 1000	070	0x38	56
9	011 1001	071	0x39	57
: Colon	011 1010	072	0x3a	58
; Semicolon	011 1011	073	0x3b	59
< Less than	011 1100	074	0x3c	60
= Equals	011 1101	075	0x3d	61
> Greater than	011 1110	076	0x3e	62
? Question mark	011 1111	077	0x3f	63
@ Commercial at	100 0000	0100	0x40	64
A	100 0001	0101	0x41	65
B	100 0010	0102	0x42	66
C	100 0011	0103	0x43	67
D	100 0100	0104	0x44	68
E	100 0101	0105	0x45	69
F	100 0110	0106	0x46	70
G	100 0111	0107	0x47	71
H	100 1000	0110	0x48	72
I	100 1001	0111	0x49	73
J	100 1010	0112	0x4a	74
K	100 1011	0113	0x4b	75
L	100 1100	0114	0x4c	76
M	100 1101	0115	0x4d	77
N	100 1110	0116	0x4e	78
O	100 1111	0117	0x4f	79
P	101 0000	0120	0x50	80
Q	101 0001	0121	0x51	81
R	101 0010	0122	0x52	82
S	101 0011	0123	0x53	83
T	101 0100	0124	0x54	84
U	101 0101	0125	0x55	85
V	101 0110	0126	0x56	86
W	101 0111	0127	0x57	87
X	101 1000	0130	0x58	88
Y	101 1001	0131	0x59	89
Z	101 1010	0132	0x5a	90
[Open square bracket	101 1011	0133	0x5b	91
\ Backslash	101 1100	0134	0x5c	92
] Close square bracket	101 1101	0135	0x5d	93
^ Caret	101 1110	0136	0x5e	94
_ Underscore	101 1111	0137	0x5f	95
` Back quote	110 0000	0140	0x60	96
a	110 0001	0141	0x61	97
b	110 0010	0142	0x62	98
c	110 0011	0143	0x63	99
d	110 0100	0144	0x64	100
e	110 0101	0145	0x65	101
f	110 0110	0146	0x66	102
g	110 0111	0147	0x67	103
h	110 1000	0150	0x68	104
i	110 1001	0151	0x69	105
j	110 1010	0152	0x6a	106
k	110 1011	0153	0x6b	107
l	110 1100	0154	0x6c	108
m	110 1101	0155	0x6d	109
n	110 1110	0156	0x6e	110
o	110 1111	0157	0x6f	111

LIST OF ASCII CODES, cont.

Symbol, Name of Character	Binary Number	Octal Number	Hex Number	Decimal Number
p	111 0000	0160	0x70	112
q	111 0001	0161	0x71	113
r	111 0010	0162	0x72	114
s	111 0011	0163	0x73	115
t	111 0100	0164	0x74	116
u	111 0101	0165	0x75	117
v	111 0110	0166	0x76	118
w	111 0111	0167	0x77	119
x	111 1000	0170	0x78	120
y	111 1001	0171	0x79	121
z	111 1010	0172	0x7a	122
{ Open curly bracket	111 1011	0173	0x7b	123
| Vertical bar	111 1100	0174	0x7c	124
} Close curly bracket	111 1101	0175	0x7d	125
~ Tilde	111 1110	0176	0x7e	126
Delete	111 1111	0177	0x7f	127

Gray Code

A gray code has these characteristics:

> Used to transmit binary code from rotating equipment when tracking position of the shaft. As the shaft increments from one position to another, only one bit changes in the code to permit a more stable and precise monitoring of equipment position.

Gray Code	Binary
0000	0000
0001	0001
0011	0010
0010	0011
0110	0100
0111	0101
0101	0110
0100	0111
1100	1000
1101	1001
1111	1010
1110	1011
1010	1100
1011	1101
1001	1110
1000	1111

Appendix B

Temperature Sensor Reference Tables

The following information on thermocouples and RTDs are conversion tables for standard devices used in the process industries.

Thermocouple Wire Identification

ANSI Code	Alloy Combination +Lead	Alloy Combination −Lead	Color Coding	DIN 4
B	Platinum-30% Rhodium Pt-30% Rh	Platinum-6% Rhodium Pt-6% Rh	None Established	
E	Chromel Nickel-Chromium Ni-Cr	Constantan Copper-Nickel Cu-Ni	+Purple −Red	
J	Iron Fe	Constantan Copper-Nickel Cu-Ni	+White −Red	+Red −Blue
K	Chromel Nickel-Chromium Ni-Cr	Alumel Nickel-Alumel Ni-Al	+Yellow −Red	+Red −Green
N*	Omega-P™ Nicrosil Ni-Cr-Si	Omega-N™ NISIL Ni-Si-Mg	+Orange −Red	
R	Platinum-13% Rhodium Pt-13% Rh	Platinum Pt	None Established	
S	Platinum-10% Rhodium Pt-10% Rh	Platinum Pt	None Established	†+Red †−White
T	Copper Cu	Constantan Copper-Nickel Cu-Ni	+Blue −Red	+Red −Brown

Resistance vs. Wire Diameter

Resistance in Ohms per Double Foot at 68°F

AWG. No.	Diameter (In.)	Type K Chromel/ Alumel	Type J Iron/ Constantan	Type T Copper/ Constantan	Type E Chromel/ Constantan	Type S Pt/ Pt 10% Rh	Type R Pt/ Pt 13% Rh
6	0.162	0.023	0.014	0.012	0.027	0.007	0.007
8	0.128	0.037	0.022	0.019	0.044	0.011	0.011
10	0.102	0.058	0.034	0.029	0.069	0.018	0.018
12	0.081	0.091	0.054	0.046	0.109	0.028	0.029
14	0.064	0.146	0.087	0.074	0.175	0.045	0.047
16	0.051	0.230	0.137	0.117	0.276	0.071	0.073
18	0.040	0.374	0.222	0.190	0.448	0.116	0.119
20	0.032	0.586	0.357	0.298	0.707	0.185	0.190
24	0.0201	1.490	0.878	0.7526	1.78	0.464	0.478
26	0.0159	2.381	1.405	1.204	2.836	0.740	0.760
30	0.0100	5.984	3.551	3.043	7.169	1.85	1.91
32	0.0080	9.524	5.599	4.758	11.31	1.96	3.04
34	0.0063	15.17	8.946	7.66	18.09	4.66	4.82
36	0.0050	24.08	14.20	12.17	28.76	7.40	7.64
38	0.0039	38.20	23.35	19.99	45.41	11.6	11.95
40	0.00315	60.88	37.01	31.64	73.57	18.6	19.3
44	0.0020	149.6	88.78	76.09	179.2	74.0	76.5
50	0.0010	598.4	355.1	304.3	716.9	185	191
56	0.00049	2408	1420	1217	2816	740	764

Wire Table for Standard Annealed Copper
American Wire Gauge English Units

Gauge No. A.W.G.	Diameter in Mils at 20°C	Ohms per 1,000 ft* at 20°C (68°F)	Feet per Ohm† at 20°C (68°F)
0000	460.0	0.049 01	20,400.0
000	409.6	0.061 80	16,180.0
00	364.8	0.077 93	12,830.0
0	324.9	0.098 27	10,180.0
1	289.3	0.1239	8,070.0
2	257.6	0.1563	6,400.0
3	229.4	0.1970	5,075.0
4	204.3	0.2485	4,025.0
5	181.9	0.3133	3,192.0
6	162.0	0.3951	2,531.0
7	144.3	0.4982	2,007.0
8	128.5	0.6282	1,592.0
9	114.4	0.7921	1,262.0
10	101.9	0.9989	1,001.0
11	90.74	1.260	794.0
12	80.81	1.588	629.6
13	71.96	2.003	499.3
14	64.08	2.525	396.0
15	57.07	3.184	314.0
16	50.82	4.016	249.0
17	45.26	5.064	197.5
18	40.30	6.385	156.6
19	35.89	8.051	124.2
20	31.96	10.15	98.50
21	28.46	12.80	78.11
22	25.35	16.14	61.95
23	22.57	20.36	49.13
24	20.10	25.67	38.96
25	17.90	32.37	30.90
26	15.94	40.81	24.50
27	14.20	51.47	19.43
28	12.64	64.90	15.41
29	11.26	81.83	12.22
30	10.03	103.2	9.691
31	8.928	130.1	7.685
32	7.950	164.1	6.095
33	7.080	206.9	4.833
34	6.305	260.9	3.833
35	5.615	329.0	3.040
36	5.000	414.8	2.411
37	4.453	523.1	1.912
38	3.965	659.6	1.516
39	3.531	831.8	1.202
40	3.145	049.0	0.9534

* Resistance at the stated temperatures of a wire whose length is 1,000 ft at 20°C.

† Length at 20°C of a wire whose resistance is 1 ohm at the stated temperatures.

Type B Thermocouple
Thermoelectric Voltage as a Function of Temperature (°C)
Reference Junctions at 0°C
THERMOELECTRIC VOLTAGE IN ABSOLUTE MILLIVOLTS

°C	0	1	2	3	4	5	6	7	8	9	10	°C
0	0.000	−0.000	−0.000	−0.001	−0.001	−0.001	−0.001	−0.001	−0.002	−0.002	−0.002	0
10	−0.002	−0.002	−0.002	−0.002	−0.002	−0.002	−0.002	−0.002	−0.003	−0.003	−0.003	10
20	−0.003	−0.003	−0.003	−0.003	−0.003	−0.002	−0.002	−0.002	−0.002	−0.002	−0.002	20
30	−0.002	−0.002	−0.002	−0.002	−0.002	−0.001	−0.001	−0.001	−0.001	−0.001	−0.000	30
40	−0.000	−0.000	−0.000	0.000	0.000	0.001	0.001	0.001	0.002	0.002	0.002	40
50	0.002	0.003	0.003	0.003	0.004	0.004	0.004	0.005	0.005	0.006	0.006	50
60	0.006	0.007	0.007	0.008	0.008	0.009	0.009	0.010	0.010	0.011	0.011	60
70	0.011	0.012	0.012	0.013	0.014	0.014	0.015	0.015	0.016	0.017	0.017	70
80	0.017	0.018	0.019	0.020	0.020	0.021	0.022	0.022	0.023	0.024	0.025	80
90	0.025	0.026	0.026	0.027	0.028	0.029	0.030	0.031	0.031	0.032	0.033	90
100	0.033	0.034	0.035	0.036	0.037	0.038	0.039	0.040	0.041	0.042	0.043	100
110	0.043	0.044	0.045	0.046	0.047	0.048	0.049	0.050	0.051	0.052	0.053	110
120	0.053	0.055	0.056	0.057	0.058	0.059	0.060	0.062	0.063	0.064	0.065	120
130	0.065	0.066	0.068	0.069	0.070	0.071	0.073	0.074	0.075	0.077	0.078	130
140	0.078	0.079	0.081	0.082	0.083	0.085	0.086	0.088	0.089	0.091	0.092	140
150	0.092	0.093	0.095	0.096	0.098	0.099	0.101	0.102	0.104	0.106	0.107	150
160	0.107	0.109	0.110	0.112	0.113	0.115	0.117	0.118	0.120	0.122	0.123	160
170	0.123	0.125	0.127	0.128	0.130	0.132	0.133	0.135	0.137	0.139	0.140	170
180	0.140	0.142	0.144	0.146	0.148	0.149	0.151	0.153	0.155	0.157	0.159	180
190	0.159	0.161	0.163	0.164	0.166	0.168	0.170	0.172	0.174	0.176	0.178	190
200	0.178	0.180	0.182	0.184	0.186	0.188	0.190	0.192	0.194	0.197	0.199	200
210	0.199	0.201	0.203	0.205	0.207	0.209	0.211	0.214	0.216	0.218	0.220	210
220	0.220	0.222	0.225	0.227	0.229	0.231	0.234	0.236	0.238	0.240	0.243	220
230	0.243	0.245	0.247	0.250	0.252	0.254	0.257	0.259	0.262	0.264	0.266	230
240	0.266	0.269	0.271	0.274	0.276	0.279	0.281	0.284	0.286	0.289	0.291	240
250	0.291	0.294	0.296	0.299	0.301	0.304	0.307	0.309	0.312	0.314	0.317	250
260	0.317	0.320	0.322	0.325	0.328	0.330	0.333	0.336	0.338	0.341	0.344	260
270	0.344	0.347	0.349	0.352	0.355	0.358	0.360	0.363	0.366	0.369	0.372	270
280	0.372	0.375	0.377	0.380	0.383	0.386	0.389	0.392	0.395	0.398	0.401	280
290	0.401	0.404	0.406	0.409	0.412	0.415	0.418	0.421	0.424	0.427	0.431	290
300	0.431	0.434	0.437	0.440	0.443	0.446	0.449	0.452	0.455	0.458	0.462	300
310	0.462	0.465	0.468	0.471	0.474	0.477	0.481	0.484	0.487	0.490	0.494	310
320	0.494	0.497	0.500	0.503	0.507	0.510	0.513	0.517	0.520	0.523	0.527	320
330	0.527	0.530	0.533	0.537	0.540	0.544	0.547	0.550	0.554	0.557	0.561	330
340	0.561	0.564	0.568	0.571	0.575	0.578	0.582	0.585	0.589	0.592	0.596	340
350	0.596	0.599	0.603	0.606	0.610	0.614	0.617	0.621	0.625	0.628	0.632	350
360	0.632	0.636	0.639	0.643	0.647	0.650	0.654	0.658	0.661	0.665	0.669	360
370	0.669	0.673	0.677	0.680	0.684	0.688	0.692	0.696	0.699	0.703	0.707	370
380	0.707	0.711	0.715	0.719	0.723	0.727	0.730	0.734	0.738	0.742	0.746	380
390	0.746	0.750	0.754	0.758	0.762	0.766	0.770	0.774	0.778	0.782	0.786	390
400	0.786	0.790	0.794	0.799	0.803	0.807	0.811	0.815	0.819	0.823	0.827	400
410	0.827	0.832	0.836	0.840	0.844	0.848	0.853	0.857	0.861	0.865	0.870	410
420	0.870	0.874	0.878	0.882	0.887	0.891	0.895	0.900	0.904	0.908	0.913	420
430	0.913	0.917	0.921	0.926	0.930	0.935	0.939	0.943	0.948	0.952	0.957	430
440	0.957	0.961	0.966	0.970	0.975	0.979	0.984	0.988	0.993	0.997	1.002	440
450	1.002	1.006	1.011	1.015	1.020	1.025	1.029	1.034	1.039	1.043	1.048	450
460	1.048	1.052	1.057	1.062	1.066	1.071	1.076	1.081	1.085	1.090	1.095	460
470	1.095	1.100	1.104	1.109	1.114	1.119	1.123	1.128	1.133	1.138	1.143	470
480	1.143	1.148	1.152	1.157	1.162	1.167	1.172	1.177	1.182	1.187	1.192	480
490	1.192	1.197	1.202	1.206	1.211	1.216	1.221	1.226	1.231	1.236	1.241	490
500	1.241	1.246	1.252	1.257	1.262	1.267	1.272	1.277	1.282	1.287	1.292	500
510	1.292	1.297	1.303	1.308	1.313	1.318	1.323	1.328	1.334	1.339	1.344	510
520	1.344	1.349	1.354	1.360	1.365	1.370	1.375	1.381	1.386	1.391	1.397	520
530	1.397	1.402	1.407	1.413	1.418	1.423	1.429	1.434	1.439	1.445	1.450	530
540	1.450	1.456	1.461	1.467	1.472	1.477	1.483	1.488	1.494	1.499	1.505	540
550	1.505	1.510	1.516	1.521	1.527	1.532	1.538	1.544	1.549	1.555	1.560	550
560	1.560	1.566	1.571	1.577	1.583	1.588	1.594	1.600	1.605	1.611	1.617	560
570	1.617	1.622	1.628	1.634	1.639	1.645	1.651	1.657	1.662	1.668	1.674	570
580	1.674	1.680	1.685	1.691	1.697	1.703	1.709	1.715	1.720	1.726	1.732	580
590	1.732	1.738	1.744	1.750	1.756	1.762	1.767	1.773	1.779	1.785	1.791	590

Type B Thermocouple
Thermoelectric Voltage as a Function of Temperature (°C)
Reference Junctions at 0°C
THERMOELECTRIC VOLTAGE IN ABSOLUTE MILLIVOLTS

°C	0	1	2	3	4	5	6	7	8	9	10	°C
600	1.791	1.797	1.803	1.809	1.815	1.821	1.827	1.833	1.839	1.845	1.851	600
610	1.851	1.857	1.863	1.869	1.875	1.882	1.888	1.894	1.900	1.906	1.912	610
620	1.912	1.918	1.924	1.931	1.937	1.943	1.949	1.955	1.961	1.968	1.974	620
630	1.974	1.980	1.986	1.993	1.999	2.005	2.011	2.018	2.024	2.030	2.036	630
640	2.036	2.043	2.049	2.055	2.062	2.068	2.074	2.081	2.087	2.094	2.100	640
650	2.100	2.106	2.113	2.119	2.126	2.132	2.139	2.145	2.151	2.158	2.164	650
660	2.164	2.171	2.177	2.184	2.190	2.197	2.203	2.210	2.216	2.223	2.230	660
670	2.230	2.236	2.243	2.249	2.256	2.263	2.269	2.276	2.282	2.289	2.296	670
680	2.296	2.302	2.309	2.316	2.322	2.329	2.336	2.343	2.349	2.356	2.363	680
690	2.363	2.369	2.376	2.383	2.390	2.396	2.403	2.410	2.417	2.424	2.430	690
700	2.430	2.437	2.444	2.451	2.458	2.465	2.472	2.478	2.485	2.492	2.499	700
710	2.499	2.506	2.513	2.520	2.527	2.534	2.541	2.548	2.555	2.562	2.569	710
720	2.569	2.576	2.583	2.590	2.597	2.604	2.611	2.618	2.625	2.632	2.639	720
730	2.639	2.646	2.653	2.660	2.667	2.674	2.682	2.689	2.696	2.703	2.710	730
740	2.710	2.717	2.724	2.732	2.739	2.746	2.753	2.760	2.768	2.775	2.782	740
750	2.782	2.789	2.797	2.804	2.811	2.818	2.826	2.833	2.840	2.848	2.855	750
760	2.855	2.862	2.869	2.877	2.884	2.892	2.899	2.906	2.914	2.921	2.928	760
770	2.928	2.936	2.943	2.951	2.958	2.966	2.973	2.980	2.988	2.995	3.003	770
780	3.003	3.010	3.018	3.025	3.033	3.040	3.048	3.055	3.063	3.070	3.078	780
790	3.078	3.086	3.093	3.101	3.108	3.116	3.124	3.131	3.139	3.146	3.154	790
800	3.154	3.162	3.169	3.177	3.185	3.192	3.200	3.208	3.215	3.223	3.231	800
810	3.231	3.239	3.246	3.254	3.262	3.269	3.277	3.285	3.293	3.301	3.308	810
820	3.308	3.316	3.324	3.332	3.340	3.347	3.355	3.363	3.371	3.379	3.387	820
830	3.387	3.395	3.402	3.410	3.418	3.426	3.434	3.442	3.450	3.458	3.466	830
840	3.466	3.474	3.482	3.490	3.498	3.506	3.514	3.522	3.530	3.538	3.546	840
850	3.546	3.554	3.562	3.570	3.578	3.586	3.594	3.602	3.610	3.618	3.626	850
860	3.626	3.634	3.643	3.651	3.659	3.667	3.675	3.683	3.691	3.700	3.708	860
870	3.708	3.716	3.724	3.732	3.741	3.749	3.757	3.765	3.773	3.782	3.790	870
880	3.790	3.798	3.806	3.815	3.823	3.831	3.840	3.848	3.856	3.865	3.873	880
890	3.873	3.881	3.890	3.898	3.906	3.915	3.923	3.931	3.940	3.948	3.957	890
900	3.957	3.965	3.973	3.982	3.990	3.999	4.007	4.016	4.024	4.032	4.041	900
910	4.041	4.049	4.058	4.066	4.075	4.083	4.092	4.100	4.109	4.117	4.126	910
920	4.126	4.135	4.143	4.152	4.160	4.169	4.177	4.186	4.195	4.203	4.212	920
930	4.212	4.220	4.229	4.238	4.246	4.255	4.264	4.272	4.281	4.290	4.298	930
940	4.298	4.307	4.316	4.325	4.333	4.342	4.351	4.359	4.368	4.377	4.386	940
950	4.386	4.394	4.403	4.412	4.421	4.430	4.438	4.447	4.456	4.465	4.474	950
960	4.474	4.483	4.491	4.500	4.509	4.518	4.527	4.536	4.545	4.553	4.562	960
970	4.562	4.571	4.580	4.589	4.598	4.607	4.616	4.625	4.634	4.643	4.652	970
980	4.652	4.661	4.670	4.679	4.688	4.697	4.706	4.715	4.724	4.733	4.742	980
990	4.742	4.751	4.760	4.769	4.778	4.787	4.796	4.805	4.814	4.824	4.833	990
1000	4.833	4.842	4.851	4.860	4.869	4.878	4.887	4.897	4.906	4.915	4.924	1000
1010	4.924	4.933	4.942	4.952	4.961	4.970	4.979	4.989	4.998	5.007	5.016	1010
1020	5.016	5.025	5.035	5.044	5.053	5.063	5.072	5.081	5.090	5.100	5.109	1020
1030	5.109	5.118	5.128	5.137	5.146	5.156	5.165	5.174	5.184	5.193	5.202	1030
1040	5.202	5.212	5.221	5.231	5.240	5.249	5.259	5.268	5.278	5.287	5.297	1040
1050	5.297	5.306	5.316	5.325	5.334	5.344	5.353	5.363	5.372	5.382	5.391	1050
1060	5.391	5.401	5.410	5.420	5.429	5.439	5.449	5.458	5.468	5.477	5.487	1060
1070	5.487	5.496	5.506	5.516	5.525	5.535	5.544	5.554	5.564	5.573	5.583	1070
1080	5.583	5.593	5.602	5.612	5.621	5.631	5.641	5.651	5.660	5.670	5.680	1080
1090	5.680	5.689	5.699	5.709	5.718	5.728	5.738	5.748	5.757	5.767	5.777	1090
1100	5.777	5.787	5.796	5.806	5.816	5.826	5.836	5.845	5.855	5.865	5.875	1100
1110	5.875	5.885	5.895	5.904	5.914	5.924	5.934	5.944	5.954	5.964	5.973	1110
1120	5.973	5.983	5.993	6.003	6.013	6.023	6.033	6.043	6.053	6.063	6.073	1120
1130	6.073	6.083	6.093	6.102	6.112	6.122	6.132	6.142	6.152	6.162	6.172	1130
1140	6.172	6.182	6.192	6.202	6.212	6.223	6.233	6.243	6.253	6.263	6.273	1140
1150	6.273	6.283	6.293	6.303	6.313	6.323	6.333	6.343	6.353	6.364	6.374	1150
1160	6.374	6.384	6.394	6.404	6.414	6.424	6.435	6.445	6.455	6.465	6.475	1160
1170	6.475	6.485	6.496	6.506	6.516	6.526	6.536	6.547	6.557	6.567	6.577	1170
1180	6.577	6.588	6.598	6.608	6.618	6.629	6.639	6.649	6.659	6.670	6.680	1180
1190	6.680	6.690	6.701	6.711	6.721	6.732	6.742	6.752	6.763	6.773	6.783	1190

Type B Thermocouple
Thermoelectric Voltage as a Function of Temperature (°C)
Reference Junctions at 0°C
THERMOELECTRIC VOLTAGE IN ABSOLUTE MILLIVOLTS

°C	0	1	2	3	4	5	6	7	8	9	10	°C
1200	6.783	6.794	6.804	6.814	6.825	6.835	6.846	6.856	6.866	6.877	6.887	1200
1210	6.887	6.898	6.908	6.918	6.929	6.939	6.950	6.960	6.971	6.981	6.991	1210
1220	6.991	7.002	7.012	7.023	7.033	7.044	7.054	7.065	7.075	7.086	7.096	1220
1230	7.096	7.107	7.117	7.128	7.138	7.149	7.159	7.170	7.181	7.191	7.202	1230
1240	7.202	7.212	7.223	7.233	7.244	7.255	7.265	7.276	7.286	7.297	7.308	1240
1250	7.308	7.318	7.329	7.339	7.350	7.361	7.371	7.382	7.393	7.403	7.414	1250
1260	7.414	7.425	7.435	7.446	7.457	7.467	7.478	7.489	7.500	7.510	7.521	1260
1270	7.521	7.532	7.542	7.553	7.564	7.575	7.585	7.596	7.607	7.618	7.628	1270
1280	7.628	7.639	7.650	7.661	7.671	7.682	7.693	7.704	7.715	7.725	7.736	1280
1290	7.736	7.747	7.758	7.769	7.780	7.790	7.801	7.812	7.823	7.834	7.845	1290
1300	7.845	7.855	7.866	7.877	7.888	7.899	7.910	7.921	7.932	7.943	7.953	1300
1310	7.953	7.964	7.975	7.986	7.997	8.008	8.019	8.030	8.041	8.052	8.063	1310
1320	8.063	8.074	8.085	8.096	8.107	8.118	8.128	8.139	8.150	8.161	8.172	1320
1330	8.172	8.183	8.194	8.205	8.216	8.227	8.238	8.249	8.261	8.272	8.283	1330
1340	8.283	8.294	8.305	8.316	8.327	8.338	8.349	8.360	8.371	8.382	8.393	1340
1350	8.393	8.404	8.415	8.426	8.437	8.449	8.460	8.471	8.482	8.493	8.504	1350
1360	8.504	8.515	8.526	8.538	8.549	8.560	8.571	8.582	8.593	8.604	8.616	1360
1370	8.616	8.627	8.638	8.649	8.660	8.671	8.683	8.694	8.705	8.716	8.727	1370
1380	8.727	8.738	8.750	8.761	8.772	8.783	8.795	8.806	8.817	8.828	8.839	1380
1390	8.839	8.851	8.862	8.873	8.884	8.896	8.907	8.918	8.929	8.941	8.952	1390
1400	8.952	8.963	8.974	8.986	8.997	9.008	9.020	9.031	9.042	9.053	9.065	1400
1410	9.065	9.076	9.087	9.099	9.110	9.121	9.133	9.144	9.155	9.167	9.178	1410
1420	9.178	9.189	9.201	9.212	9.223	9.235	9.246	9.257	9.269	9.280	9.291	1420
1430	9.291	9.303	9.314	9.326	9.337	9.348	9.360	9.371	9.382	9.394	9.405	1430
1440	9.405	9.417	9.428	9.439	9.451	9.462	9.474	9.485	9.497	9.508	9.519	1440
1450	9.519	9.531	9.542	9.554	9.565	9.577	9.588	9.599	9.611	9.622	9.634	1450
1460	9.634	9.645	9.657	9.668	9.680	9.691	9.703	9.714	9.726	9.737	9.748	1460
1470	9.748	9.760	9.771	9.783	9.794	9.806	9.817	9.829	9.840	9.852	9.863	1470
1480	9.863	9.875	9.886	9.898	9.909	9.921	9.933	9.944	9.956	9.967	9.979	1480
1490	9.979	9.990	10.002	10.013	10.025	10.036	10.048	10.059	10.071	10.082	10.094	1490
1500	10.094	10.106	10.117	10.129	10.140	10.152	10.163	10.175	10.187	10.198	10.210	1500
1510	10.210	10.221	10.233	10.244	10.256	10.268	10.279	10.291	10.302	10.314	10.325	1510
1520	10.325	10.337	10.349	10.360	10.372	10.383	10.395	10.407	10.418	10.430	10.441	1520
1530	10.441	10.453	10.465	10.476	10.488	10.500	10.511	10.523	10.534	10.546	10.558	1530
1540	10.558	10.569	10.581	10.593	10.604	10.616	10.627	10.639	10.651	10.662	10.674	1540
1550	10.674	10.686	10.697	10.709	10.721	10.732	10.744	10.756	10.767	10.779	10.790	1550
1560	10.790	10.802	10.814	10.825	10.837	10.849	10.860	10.872	10.884	10.895	10.907	1560
1570	10.907	10.919	10.930	10.942	10.954	10.965	10.977	10.989	11.000	11.012	11.024	1570
1580	11.024	11.035	11.047	11.059	11.070	11.082	11.094	11.105	11.117	11.129	11.141	1580
1590	11.141	11.152	11.164	11.176	11.187	11.199	11.211	11.222	11.234	11.246	11.257	1590
1600	11.257	11.269	11.281	11.292	11.304	11.316	11.328	11.339	11.351	11.363	11.374	1600
1610	11.374	11.386	11.398	11.409	11.421	11.433	11.444	11.456	11.468	11.480	11.491	1610
1620	11.491	11.503	11.515	11.526	11.538	11.550	11.561	11.573	11.585	11.597	11.608	1620
1630	11.608	11.620	11.632	11.643	11.655	11.667	11.678	11.690	11.702	11.714	11.725	1630
1640	11.725	11.737	11.749	11.760	11.772	11.784	11.795	11.807	11.819	11.830	11.842	1640
1650	11.842	11.854	11.866	11.877	11.889	11.901	11.912	11.924	11.936	11.947	11.959	1650
1660	11.959	11.971	11.983	11.994	12.006	12.018	12.029	12.041	12.053	12.064	12.076	1660
1670	12.076	12.088	12.099	12.111	12.123	12.134	12.146	12.158	12.170	12.181	12.193	1670
1680	12.193	12.205	12.216	12.228	12.240	12.251	12.263	12.275	12.286	12.298	12.310	1680
1690	12.310	12.321	12.333	12.345	12.356	12.368	12.380	12.391	12.403	12.415	12.426	1690
1700	12.426	12.438	12.450	12.461	12.473	12.485	12.496	12.508	12.520	12.531	12.543	1700
1710	12.543	12.555	12.566	12.578	12.590	12.601	12.613	12.624	12.636	12.648	12.659	1710
1720	12.659	12.671	12.683	12.694	12.706	12.718	12.729	12.741	12.752	12.764	12.776	1720
1730	12.776	12.787	12.799	12.811	12.822	12.834	12.845	12.857	12.869	12.880	12.892	1730
1740	12.892	12.903	12.915	12.927	12.938	12.950	12.961	12.973	12.985	12.996	13.008	1740
1750	13.008	13.019	13.031	13.043	13.054	13.066	13.077	13.089	13.100	13.112	13.124	1750
1760	13.124	13.135	13.147	13.158	13.170	13.181	13.193	13.204	13.216	13.228	13.239	1760
1770	13.239	13.251	13.262	13.274	13.285	13.297	13.308	13.320	13.331	13.343	13.354	1770
1780	13.354	13.366	13.378	13.389	13.401	13.412	13.424	13.435	13.447	13.458	13.470	1780
1790	13.470	13.481	13.493	13.504	13.516	13.527	13.539	13.550	13.562	13.573	13.585	1790
1800	13.585	13.596	13.607	13.619	13.630	13.642	13.653	13.665	13.676	13.688	13.699	1800
1810	13.699	13.711	13.722	13.733	13.745	13.756	13.768	13.779	13.791	13.802	13.814	1810

Type B Thermocouple
Thermoelectric Voltage as a Function of Temperature (°F)
Reference Junctions at 32°F
THERMOELECTRIC VOLTAGE IN ABSOLUTE MILLIVOLTS

°F	0	1	2	3	4	5	6	7	8	9	10	°F
30			0.000	−0.000	−0.000	−0.000	−0.001	−0.001	−0.001	−0.001	−0.001	30
40	−0.001	−0.001	−0.001	−0.001	−0.001	−0.001	−0.002	−0.002	−0.002	−0.002	−0.002	40
50	−0.002	−0.002	−0.002	−0.002	−0.002	−0.002	−0.002	−0.002	−0.002	−0.002	−0.002	50
60	−0.002	−0.002	−0.002	−0.003	−0.003	−0.003	−0.003	−0.003	−0.003	−0.003	−0.003	60
70	−0.003	−0.003	−0.003	−0.003	−0.003	−0.003	−0.003	−0.002	−0.002	−0.002	−0.002	70
80	−0.002	−0.002	−0.002	−0.002	−0.002	−0.002	−0.002	−0.002	−0.002	−0.002	−0.002	80
90	−0.002	−0.002	−0.002	−0.002	−0.002	−0.001	−0.001	−0.001	−0.001	−0.001	−0.001	90
100	−0.001	−0.001	−0.001	−0.001	−0.000	−0.000	−0.000	−0.000	0.000	0.000	0.000	100
110	0.000	0.000	0.001	0.001	0.001	0.001	0.001	0.001	0.002	0.002	0.002	110
120	0.002	0.002	0.002	0.002	0.003	0.003	0.003	0.003	0.003	0.004	0.004	120
130	0.004	0.004	0.004	0.005	0.005	0.005	0.005	0.005	0.006	0.006	0.006	130
140	0.006	0.006	0.007	0.007	0.007	0.007	0.008	0.008	0.008	0.009	0.009	140
150	0.009	0.009	0.009	0.010	0.010	0.010	0.011	0.011	0.011	0.012	0.012	150
160	0.012	0.012	0.013	0.013	0.013	0.014	0.014	0.014	0.015	0.015	0.015	160
170	0.015	0.016	0.016	0.016	0.017	0.017	0.017	0.018	0.018	0.019	0.019	170
180	0.019	0.019	0.020	0.020	0.021	0.021	0.021	0.022	0.022	0.023	0.023	180
190	0.023	0.023	0.024	0.024	0.025	0.025	0.026	0.026	0.027	0.027	0.027	190
200	0.027	0.028	0.028	0.029	0.029	0.030	0.030	0.031	0.031	0.032	0.032	200
210	0.032	0.033	0.033	0.034	0.034	0.035	0.035	0.036	0.036	0.037	0.037	210
220	0.037	0.038	0.038	0.039	0.039	0.040	0.041	0.041	0.042	0.042	0.043	220
230	0.043	0.043	0.044	0.044	0.045	0.046	0.046	0.047	0.047	0.048	0.049	230
240	0.049	0.049	0.050	0.050	0.051	0.052	0.052	0.053	0.053	0.054	0.055	240
250	0.055	0.055	0.056	0.057	0.057	0.058	0.058	0.059	0.060	0.060	0.061	250
260	0.061	0.062	0.062	0.063	0.064	0.064	0.065	0.066	0.067	0.067	0.068	260
270	0.068	0.069	0.069	0.070	0.071	0.071	0.072	0.073	0.074	0.074	0.075	270
280	0.075	0.076	0.077	0.077	0.078	0.079	0.080	0.080	0.081	0.082	0.083	280
290	0.083	0.083	0.084	0.085	0.086	0.086	0.087	0.088	0.089	0.090	0.090	290
300	0.090	0.091	0.092	0.093	0.094	0.094	0.095	0.096	0.097	0.098	0.099	300
310	0.099	0.099	0.100	0.101	0.102	0.103	0.104	0.104	0.105	0.106	0.107	310
320	0.107	0.108	0.109	0.110	0.111	0.111	0.112	0.113	0.114	0.115	0.116	320
330	0.116	0.117	0.118	0.119	0.120	0.120	0.121	0.122	0.123	0.124	0.125	330
340	0.125	0.126	0.127	0.128	0.129	0.130	0.131	0.132	0.133	0.134	0.135	340
350	0.135	0.136	0.137	0.138	0.138	0.139	0.140	0.141	0.142	0.143	0.144	350
360	0.144	0.145	0.146	0.147	0.148	0.149	0.151	0.152	0.153	0.154	0.155	360
370	0.155	0.156	0.157	0.158	0.159	0.160	0.161	0.162	0.163	0.164	0.165	370
380	0.165	0.166	0.167	0.168	0.169	0.171	0.172	0.173	0.174	0.175	0.176	380
390	0.176	0.177	0.178	0.179	0.180	0.182	0.183	0.184	0.185	0.186	0.187	390
400	0.187	0.188	0.189	0.191	0.192	0.193	0.194	0.195	0.196	0.197	0.199	400
410	0.199	0.200	0.201	0.202	0.203	0.205	0.206	0.207	0.208	0.209	0.210	410
420	0.210	0.212	0.213	0.214	0.215	0.217	0.218	0.219	0.220	0.221	0.223	420
430	0.223	0.224	0.225	0.226	0.228	0.229	0.230	0.231	0.233	0.234	0.235	430
440	0.235	0.236	0.238	0.239	0.240	0.242	0.243	0.244	0.245	0.247	0.248	440
450	0.248	0.249	0.251	0.252	0.253	0.254	0.256	0.257	0.258	0.260	0.261	450
460	0.261	0.262	0.264	0.265	0.266	0.268	0.269	0.271	0.272	0.273	0.275	460
470	0.275	0.276	0.277	0.279	0.280	0.281	0.283	0.284	0.286	0.287	0.288	470
480	0.288	0.290	0.291	0.293	0.294	0.295	0.297	0.298	0.300	0.301	0.303	480
490	0.303	0.304	0.305	0.307	0.308	0.310	0.311	0.313	0.314	0.315	0.317	490
500	0.317	0.318	0.320	0.321	0.323	0.324	0.326	0.327	0.329	0.330	0.332	500
510	0.332	0.333	0.335	0.336	0.338	0.339	0.341	0.342	0.344	0.345	0.347	510
520	0.347	0.348	0.350	0.351	0.353	0.355	0.356	0.358	0.359	0.361	0.362	520
530	0.362	0.364	0.365	0.367	0.369	0.370	0.372	0.373	0.375	0.376	0.378	530
540	0.378	0.380	0.381	0.383	0.384	0.386	0.388	0.389	0.391	0.392	0.394	540
550	0.394	0.396	0.397	0.399	0.401	0.402	0.404	0.405	0.407	0.409	0.410	550
560	0.410	0.412	0.414	0.415	0.417	0.419	0.420	0.422	0.424	0.425	0.427	560
570	0.427	0.429	0.431	0.432	0.434	0.436	0.437	0.439	0.441	0.442	0.444	570
580	0.444	0.446	0.448	0.449	0.451	0.453	0.455	0.456	0.458	0.460	0.462	580
590	0.462	0.463	0.465	0.467	0.469	0.470	0.472	0.474	0.476	0.477	0.479	590
600	0.479	0.481	0.483	0.485	0.486	0.488	0.490	0.492	0.494	0.495	0.497	600
610	0.497	0.499	0.501	0.503	0.504	0.506	0.508	0.510	0.512	0.514	0.515	610
620	0.515	0.517	0.519	0.521	0.523	0.525	0.527	0.528	0.530	0.532	0.534	620
630	0.534	0.536	0.538	0.540	0.542	0.544	0.545	0.547	0.549	0.551	0.553	630
640	0.553	0.555	0.557	0.559	0.561	0.563	0.565	0.566	0.568	0.570	0.572	640

Type B Thermocouple
Thermoelectric Voltage as a Function of Temperature (°F)
Reference Junctions at 32°F
THERMOELECTRIC VOLTAGE IN ABSOLUTE MILLIVOLTS

°F	0	1	2	3	4	5	6	7	8	9	10	°F
650	0.572	0.574	0.576	0.578	0.580	0.582	0.584	0.586	0.588	0.590	0.592	650
660	0.592	0.594	0.596	0.598	0.600	0.602	0.604	0.606	0.608	0.610	0.612	660
670	0.612	0.614	0.616	0.618	0.620	0.622	0.624	0.626	0.628	0.630	0.632	670
680	0.632	0.634	0.636	0.638	0.640	0.642	0.644	0.646	0.648	0.650	0.652	680
690	0.652	0.654	0.656	0.659	0.661	0.663	0.665	0.667	0.669	0.671	0.673	690
700	0.673	0.675	0.677	0.679	0.682	0.684	0.686	0.688	0.690	0.692	0.694	700
710	0.694	0.696	0.699	0.701	0.703	0.705	0.707	0.709	0.711	0.714	0.716	710
720	0.716	0.718	0.720	0.722	0.724	0.727	0.729	0.731	0.733	0.735	0.737	720
730	0.737	0.740	0.742	0.744	0.746	0.748	0.751	0.753	0.755	0.757	0.759	730
740	0.759	0.762	0.764	0.766	0.768	0.771	0.773	0.775	0.777	0.780	0.782	740
750	0.782	0.784	0.786	0.789	0.791	0.793	0.795	0.798	0.800	0.802	0.804	750
760	0.804	0.807	0.809	0.811	0.814	0.816	0.818	0.821	0.823	0.825	0.827	760
770	0.827	0.830	0.832	0.834	0.837	0.839	0.841	0.844	0.846	0.848	0.851	770
780	0.851	0.853	0.855	0.858	0.860	0.862	0.865	0.867	0.870	0.872	0.874	780
790	0.874	0.877	0.879	0.881	0.884	0.886	0.889	0.891	0.893	0.896	0.898	790
800	0.898	0.901	0.903	0.905	0.908	0.910	0.913	0.915	0.918	0.920	0.922	800
810	0.922	0.925	0.927	0.930	0.932	0.935	0.937	0.939	0.942	0.944	0.947	810
820	0.947	0.949	0.952	0.954	0.957	0.959	0.962	0.964	0.967	0.969	0.972	820
830	0.972	0.974	0.977	0.979	0.982	0.984	0.987	0.989	0.992	0.994	0.997	830
840	0.997	0.999	1.002	1.004	1.007	1.009	1.012	1.014	1.017	1.020	1.022	840
850	1.022	1.025	1.027	1.030	1.032	1.035	1.037	1.040	1.043	1.045	1.048	850
860	1.048	1.050	1.053	1.056	1.058	1.061	1.063	1.066	1.069	1.071	1.074	860
870	1.074	1.076	1.079	1.082	1.084	1.087	1.090	1.092	1.095	1.097	1.100	870
880	1.100	1.103	1.105	1.108	1.111	1.113	1.116	1.119	1.121	1.124	1.127	880
890	1.127	1.129	1.132	1.135	1.137	1.140	1.143	1.145	1.148	1.151	1.153	890
900	1.153	1.156	1.159	1.162	1.164	1.167	1.170	1.172	1.175	1.178	1.181	900
910	1.181	1.183	1.186	1.189	1.192	1.194	1.197	1.200	1.203	1.205	1.208	910
920	1.208	1.211	1.214	1.216	1.219	1.222	1.225	1.228	1.230	1.233	1.236	920
930	1.236	1.239	1.241	1.244	1.247	1.250	1.253	1.255	1.258	1.261	1.264	930
940	1.264	1.267	1.270	1.272	1.275	1.278	1.281	1.284	1.287	1.289	1.292	940
950	1.292	1.295	1.298	1.301	1.304	1.307	1.309	1.312	1.315	1.318	1.321	950
960	1.321	1.324	1.327	1.330	1.332	1.335	1.338	1.341	1.344	1.347	1.350	960
970	1.350	1.353	1.356	1.359	1.361	1.364	1.367	1.370	1.373	1.376	1.379	970
980	1.379	1.382	1.385	1.388	1.391	1.394	1.397	1.400	1.403	1.406	1.409	980
990	1.409	1.411	1.414	1.417	1.420	1.423	1.426	1.429	1.432	1.435	1.438	990
1000	1.438	1.441	1.444	1.447	1.450	1.453	1.456	1.459	1.462	1.465	1.468	1000
1010	1.468	1.471	1.474	1.477	1.480	1.483	1.487	1.490	1.493	1.496	1.499	1010
1020	1.499	1.502	1.505	1.508	1.511	1.514	1.517	1.520	1.523	1.526	1.529	1020
1030	1.529	1.532	1.536	1.539	1.542	1.545	1.548	1.551	1.554	1.557	1.560	1030
1040	1.560	1.563	1.566	1.570	1.573	1.576	1.579	1.582	1.585	1.588	1.591	1040
1050	1.591	1.595	1.598	1.601	1.604	1.607	1.610	1.613	1.617	1.620	1.623	1050
1060	1.623	1.626	1.629	1.632	1.636	1.639	1.642	1.645	1.648	1.652	1.655	1060
1070	1.655	1.658	1.661	1.664	1.668	1.671	1.674	1.677	1.680	1.684	1.687	1070
1080	1.687	1.690	1.693	1.696	1.700	1.703	1.706	1.709	1.713	1.716	1.719	1080
1090	1.719	1.722	1.726	1.729	1.732	1.735	1.739	1.742	1.745	1.748	1.752	1090
1100	1.752	1.755	1.758	1.762	1.765	1.768	1.771	1.775	1.778	1.781	1.785	1100
1110	1.785	1.788	1.791	1.795	1.798	1.801	1.804	1.808	1.811	1.814	1.818	1110
1120	1.818	1.821	1.824	1.828	1.831	1.834	1.838	1.841	1.844	1.848	1.851	1120
1130	1.851	1.855	1.858	1.861	1.865	1.868	1.871	1.875	1.878	1.882	1.885	1130
1140	1.885	1.888	1.892	1.895	1.898	1.902	1.905	1.909	1.912	1.915	1.919	1140
1150	1.919	1.922	1.926	1.929	1.933	1.936	1.939	1.943	1.946	1.950	1.953	1150
1160	1.953	1.957	1.960	1.963	1.967	1.970	1.974	1.977	1.981	1.984	1.988	1160
1170	1.988	1.991	1.995	1.998	2.002	2.005	2.009	2.012	2.015	2.019	2.022	1170
1180	2.022	2.026	2.029	2.033	2.036	2.040	2.043	2.047	2.051	2.054	2.058	1180
1190	2.058	2.061	2.065	2.068	2.072	2.075	2.079	2.082	2.086	2.089	2.093	1190
1200	2.093	2.096	2.100	2.104	2.107	2.111	2.114	2.118	2.121	2.125	2.128	1200
1210	2.128	2.132	2.136	2.139	2.143	2.146	2.150	2.154	2.157	2.161	2.164	1210
1220	2.164	2.168	2.172	2.175	2.179	2.182	2.186	2.190	2.193	2.197	2.201	1220
1230	2.201	2.204	2.208	2.211	2.215	2.219	2.222	2.226	2.230	2.233	2.237	1230
1240	2.237	2.241	2.244	2.248	2.252	2.255	2.259	2.263	2.266	2.270	2.274	1240
1250	2.274	2.277	2.281	2.285	2.288	2.292	2.296	2.299	2.303	2.307	2.311	1250
1260	2.311	2.314	2.318	2.322	2.325	2.329	2.333	2.337	2.340	2.344	2.348	1260
1270	2.348	2.351	2.355	2.359	2.363	2.366	2.370	2.374	2.378	2.381	2.385	1270
1280	2.385	2.389	2.393	2.396	2.400	2.404	2.408	2.412	2.415	2.419	2.423	1280
1290	2.423	2.427	2.430	2.434	2.438	2.442	2.446	2.449	2.453	2.457	2.461	1290
1300	2.461	2.465	2.469	2.472	2.476	2.480	2.484	2.488	2.491	2.495	2.499	1300

Type B Thermocouple
Thermoelectric Voltage as a Function of Temperature (°F)
Reference Junctions at 32°F
THERMOELECTRIC VOLTAGE IN ABSOLUTE MILLIVOLTS

°F	0	1	2	3	4	5	6	7	8	9	10	°F
1310	2.499	2.503	2.507	2.511	2.515	2.518	2.522	2.526	2.530	2.534	2.538	1310
1320	2.538	2.542	2.545	2.549	2.553	2.557	2.561	2.565	2.569	2.573	2.576	1320
1330	2.576	2.580	2.584	2.588	2.592	2.596	2.600	2.604	2.608	2.612	2.615	1330
1340	2.615	2.619	2.623	2.627	2.631	2.635	2.639	2.643	2.647	2.651	2.655	1340
1350	2.655	2.659	2.663	2.667	2.670	2.674	2.678	2.682	2.686	2.690	2.694	1350
1360	2.694	2.698	2.702	2.706	2.710	2.714	2.718	2.722	2.726	2.730	2.734	1360
1370	2.734	2.738	2.742	2.746	2.750	2.754	2.758	2.762	2.766	2.770	2.774	1370
1380	2.774	2.778	2.782	2.786	2.790	2.794	2.798	2.802	2.806	2.810	2.814	1380
1390	2.814	2.818	2.822	2.826	2.830	2.835	2.839	2.843	2.847	2.851	2.855	1390
1400	2.855	2.859	2.863	2.867	2.871	2.875	2.879	2.883	2.887	2.892	2.896	1400
1410	2.896	2.900	2.904	2.908	2.912	2.916	2.920	2.924	2.928	2.933	2.937	1410
1420	2.937	2.941	2.945	2.949	2.953	2.957	2.961	2.966	2.970	2.974	2.978	1420
1430	2.978	2.982	2.986	2.990	2.995	2.999	3.003	3.007	3.011	3.015	3.019	1430
1440	3.019	3.024	3.028	3.032	3.036	3.040	3.045	3.049	3.053	3.057	3.061	1440
1450	3.061	3.065	3.070	3.074	3.078	3.082	3.086	3.091	3.095	3.099	3.103	1450
1460	3.103	3.107	3.112	3.116	3.120	3.124	3.129	3.133	3.137	3.141	3.146	1460
1470	3.146	3.150	3.154	3.158	3.163	3.167	3.171	3.175	3.180	3.184	3.188	1470
1480	3.188	3.192	3.197	3.201	3.205	3.209	3.214	3.218	3.222	3.227	3.231	1480
1490	3.231	3.235	3.239	3.244	3.248	3.252	3.257	3.261	3.265	3.269	3.274	1490
1500	3.274	3.278	3.282	3.287	3.291	3.295	3.300	3.304	3.308	3.313	3.317	1500
1510	3.317	3.321	3.326	3.330	3.334	3.339	3.343	3.347	3.352	3.356	3.361	1510
1520	3.361	3.365	3.369	3.374	3.378	3.382	3.387	3.391	3.395	3.400	3.404	1520
1530	3.404	3.409	3.413	3.417	3.422	3.426	3.431	3.435	3.439	3.444	3.448	1530
1540	3.448	3.453	3.457	3.461	3.466	3.470	3.475	3.479	3.484	3.488	3.492	1540
1550	3.492	3.497	3.501	3.506	3.510	3.515	3.519	3.523	3.528	3.532	3.537	1550
1560	3.537	3.541	3.546	3.550	3.555	3.559	3.564	3.568	3.573	3.577	3.581	1560
1570	3.581	3.586	3.590	3.595	3.599	3.604	3.608	3.613	3.617	3.622	3.626	1570
1580	3.626	3.631	3.635	3.640	3.644	3.649	3.653	3.658	3.662	3.667	3.672	1580
1590	3.672	3.676	3.681	3.685	3.690	3.694	3.699	3.703	3.708	3.712	3.717	1590
1600	3.717	3.721	3.726	3.731	3.735	3.740	3.744	3.749	3.753	3.758	3.762	1600
1610	3.762	3.767	3.772	3.776	3.781	3.785	3.790	3.795	3.799	3.804	3.808	1610
1620	3.808	3.813	3.818	3.822	3.827	3.831	3.836	3.841	3.845	3.850	3.854	1620
1630	3.854	3.859	3.864	3.868	3.873	3.877	3.882	3.887	3.891	3.896	3.901	1630
1640	3.901	3.905	3.910	3.915	3.919	3.924	3.929	3.933	3.938	3.943	3.947	1640
1650	3.947	3.952	3.957	3.961	3.966	3.971	3.975	3.980	3.985	3.989	3.994	1650
1660	3.994	3.999	4.003	4.008	4.013	4.017	4.022	4.027	4.031	4.036	4.041	1660
1670	4.041	4.046	4.050	4.055	4.060	4.064	4.069	4.074	4.079	4.083	4.088	1670
1680	4.088	4.093	4.098	4.102	4.107	4.112	4.117	4.121	4.126	4.131	4.136	1680
1690	4.136	4.140	4.145	4.150	4.155	4.159	4.164	4.169	4.174	4.178	4.183	1690
1700	4.183	4.188	4.193	4.198	4.202	4.207	4.212	4.217	4.221	4.226	4.231	1700
1710	4.231	4.236	4.241	4.245	4.250	4.255	4.260	4.265	4.269	4.274	4.279	1710
1720	4.279	4.284	4.289	4.294	4.298	4.303	4.308	4.313	4.318	4.323	4.327	1720
1730	4.327	4.332	4.337	4.342	4.347	4.352	4.357	4.361	4.366	4.371	4.376	1730
1740	4.376	4.381	4.386	4.391	4.395	4.400	4.405	4.410	4.415	4.420	4.425	1740
1750	4.425	4.430	4.435	4.439	4.444	4.449	4.454	4.459	4.464	4.469	4.474	1750
1760	4.474	4.479	4.484	4.488	4.493	4.498	4.503	4.508	4.513	4.518	4.523	1760
1770	4.523	4.528	4.533	4.538	4.543	4.548	4.552	4.557	4.562	4.567	4.572	1770
1780	4.572	4.577	4.582	4.587	4.592	4.597	4.602	4.607	4.612	4.617	4.622	1780
1790	4.622	4.627	4.632	4.637	4.642	4.647	4.652	4.657	4.662	4.667	4.672	1790
1800	4.672	4.677	4.682	4.687	4.692	4.697	4.702	4.707	4.712	4.717	4.722	1800
1810	4.722	4.727	4.732	4.737	4.742	4.747	4.752	4.757	4.762	4.767	4.772	1810
1820	4.772	4.777	4.782	4.787	4.792	4.797	4.802	4.807	4.812	4.817	4.823	1820
1830	4.823	4.828	4.833	4.838	4.843	4.848	4.853	4.858	4.863	4.868	4.873	1830
1840	4.873	4.878	4.883	4.888	4.894	4.899	4.904	4.909	4.914	4.919	4.924	1840
1850	4.924	4.929	4.934	4.939	4.945	4.950	4.955	4.960	4.965	4.970	4.975	1850
1860	4.975	4.980	4.985	4.991	4.996	5.001	5.006	5.011	5.016	5.021	5.027	1860
1870	5.027	5.032	5.037	5.042	5.047	5.052	5.057	5.063	5.068	5.073	5.078	1870
1880	5.078	5.083	5.088	5.094	5.099	5.104	5.109	5.114	5.119	5.125	5.130	1880
1890	5.130	5.135	5.140	5.145	5.150	5.156	5.161	5.166	5.171	5.176	5.182	1890
1900	5.182	5.187	5.192	5.197	5.202	5.208	5.213	5.218	5.223	5.229	5.234	1900
1910	5.234	5.239	5.244	5.249	5.255	5.260	5.265	5.270	5.276	5.281	5.286	1910
1920	5.286	5.291	5.297	5.302	5.307	5.312	5.318	5.323	5.328	5.333	5.339	1920
1930	5.339	5.344	5.349	5.354	5.360	5.365	5.370	5.376	5.381	5.386	5.391	1930
1940	5.391	5.397	5.402	5.407	5.413	5.418	5.423	5.428	5.434	5.439	5.444	1940
1950	5.444	5.450	5.455	5.460	5.466	5.471	5.476	5.482	5.487	5.492	5.497	1950
1960	5.497	5.503	5.508	5.513	5.519	5.524	5.529	5.535	5.540	5.545	5.551	1960
1970	5.551	5.556	5.561	5.567	5.572	5.578	5.583	5.588	5.594	5.599	5.604	1970

Type B Thermocouple
Thermoelectric Voltage as a Function of Temperature (°F)
Reference Junctions at 32°F
THERMOELECTRIC VOLTAGE IN ABSOLUTE MILLIVOLTS

°F	0	1	2	3	4	5	6	7	8	9	10	°F
1980	5.604	5.610	5.615	5.620	5.626	5.631	5.637	5.642	5.647	5.653	5.658	1980
1990	5.658	5.663	5.669	5.674	5.680	5.685	5.690	5.696	5.701	5.707	5.712	1990
2000	5.712	5.717	5.723	5.728	5.734	5.739	5.744	5.750	5.755	5.761	5.766	2000
2010	5.766	5.771	5.777	5.782	5.788	5.793	5.799	5.804	5.810	5.815	5.820	2010
2020	5.820	5.826	5.831	5.837	5.842	5.848	5.853	5.859	5.864	5.869	5.875	2020
2030	5.875	5.880	5.886	5.891	5.897	5.902	5.908	5.913	5.919	5.924	5.930	2030
2040	5.930	5.935	5.941	5.946	5.951	5.957	5.962	5.968	5.973	5.979	5.984	2040
2050	5.984	5.990	5.995	6.001	6.006	6.012	6.017	6.023	6.028	6.034	6.039	2050
2060	6.039	6.045	6.051	6.056	6.062	6.067	6.073	6.078	6.084	6.089	6.095	2060
2070	6.095	6.100	6.106	6.111	6.117	6.122	6.128	6.134	6.139	6.145	6.150	2070
2080	6.150	6.156	6.161	6.167	6.172	6.178	6.184	6.189	6.195	6.200	6.206	2080
2090	6.206	6.211	6.217	6.223	6.228	6.234	6.239	6.245	6.250	6.256	6.262	2090
2100	6.262	6.267	6.273	6.278	6.284	6.290	6.295	6.301	6.306	6.312	6.318	2100
2110	6.318	6.323	6.329	6.334	6.340	6.346	6.351	6.357	6.362	6.368	6.374	2110
2120	6.374	6.379	6.385	6.391	6.396	6.402	6.408	6.413	6.419	6.424	6.430	2120
2130	6.430	6.436	6.441	6.447	6.453	6.458	6.464	6.470	6.475	6.481	6.487	2130
2140	6.487	6.492	6.498	6.504	6.509	6.515	6.521	6.526	6.532	6.538	6.543	2140
2150	6.543	6.549	6.555	6.560	6.566	6.572	6.577	6.583	6.589	6.594	6.600	2150
2160	6.600	6.606	6.612	6.617	6.623	6.629	6.634	6.640	6.646	6.651	6.657	2160
2170	6.657	6.663	6.669	6.674	6.680	6.686	6.692	6.697	6.703	6.709	6.714	2170
2180	6.714	6.720	6.726	6.732	6.737	6.743	6.749	6.755	6.760	6.766	6.772	2180
2190	6.772	6.778	6.783	6.789	6.795	6.801	6.806	6.812	6.818	6.824	6.829	2190
2200	6.829	6.835	6.841	6.847	6.852	6.858	6.864	6.870	6.876	6.881	6.887	2200
2210	6.887	6.893	6.899	6.904	6.910	6.916	6.922	6.928	6.933	6.939	6.945	2210
2220	6.945	6.951	6.957	6.962	6.968	6.974	6.980	6.986	6.991	6.997	7.003	2220
2230	7.003	7.009	7.015	7.021	7.026	7.032	7.038	7.044	7.050	7.055	7.061	2230
2240	7.061	7.067	7.073	7.079	7.085	7.090	7.096	7.102	7.108	7.114	7.120	2240
2250	7.120	7.126	7.131	7.137	7.143	7.149	7.155	7.161	7.167	7.172	7.178	2250
2260	7.178	7.184	7.190	7.196	7.202	7.208	7.213	7.219	7.225	7.231	7.237	2260
2270	7.237	7.243	7.249	7.255	7.260	7.266	7.272	7.278	7.284	7.290	7.296	2270
2280	7.296	7.302	7.308	7.314	7.319	7.325	7.331	7.337	7.343	7.349	7.355	2280
2290	7.355	7.361	7.367	7.373	7.378	7.384	7.390	7.396	7.402	7.408	7.414	2290
2300	7.414	7.420	7.426	7.432	7.438	7.444	7.450	7.456	7.461	7.467	7.473	2300
2310	7.473	7.479	7.485	7.491	7.497	7.503	7.509	7.515	7.521	7.527	7.533	2310
2320	7.533	7.539	7.545	7.551	7.557	7.563	7.569	7.575	7.581	7.587	7.592	2320
2330	7.592	7.598	7.604	7.610	7.616	7.622	7.628	7.634	7.640	7.646	7.652	2330
2340	7.652	7.658	7.664	7.670	7.676	7.682	7.688	7.694	7.700	7.706	7.712	2340
2350	7.712	7.718	7.724	7.730	7.736	7.742	7.748	7.754	7.760	7.766	7.772	2350
2360	7.772	7.778	7.784	7.790	7.796	7.802	7.808	7.814	7.820	7.827	7.833	2360
2370	7.833	7.839	7.845	7.851	7.857	7.863	7.869	7.875	7.881	7.887	7.893	2370
2380	7.893	7.899	7.905	7.911	7.917	7.923	7.929	7.935	7.941	7.947	7.953	2380
2390	7.953	7.959	7.966	7.972	7.978	7.984	7.990	7.996	8.002	8.008	8.014	2390
2400	8.014	8.020	8.026	8.032	8.038	8.044	8.051	8.057	8.063	8.069	8.075	2400
2410	8.075	8.081	8.087	8.093	8.099	8.105	8.111	8.118	8.124	8.130	8.136	2410
2420	8.136	8.142	8.148	8.154	8.160	8.166	8.172	8.179	8.185	8.191	8.197	2420
2430	8.197	8.203	8.209	8.215	8.221	8.227	8.234	8.240	8.246	8.252	8.258	2430
2440	8.258	8.264	8.270	8.276	8.283	8.289	8.295	8.301	8.307	8.313	8.319	2440
2450	8.319	8.326	8.332	8.338	8.344	8.350	8.356	8.362	8.369	8.375	8.381	2450
2460	8.381	8.387	8.393	8.399	8.405	8.412	8.418	8.424	8.430	8.436	8.442	2460
2470	8.442	8.449	8.455	8.461	8.467	8.473	8.479	8.486	8.492	8.498	8.504	2470
2480	8.504	8.510	8.516	8.523	8.529	8.535	8.541	8.547	8.554	8.560	8.566	2480
2490	8.566	8.572	8.578	8.585	8.591	8.597	8.603	8.609	8.616	8.622	8.628	2490
2500	8.628	8.634	8.640	8.647	8.653	8.659	8.665	8.671	8.678	8.684	8.690	2500
2510	8.690	8.696	8.702	8.709	8.715	8.721	8.727	8.733	8.740	8.746	8.752	2510
2520	8.752	8.758	8.765	8.771	8.777	8.783	8.790	8.796	8.802	8.808	8.814	2520
2530	8.814	8.821	8.827	8.833	8.839	8.846	8.852	8.858	8.864	8.871	8.877	2530
2540	8.877	8.883	8.889	8.896	8.902	8.908	8.914	8.921	8.927	8.933	8.939	2540
2550	8.939	8.946	8.952	8.958	8.964	8.971	8.977	8.983	8.989	8.996	9.002	2550
2560	9.002	9.008	9.015	9.021	9.027	9.033	9.040	9.046	9.052	9.058	9.065	2560
2570	9.065	9.071	9.077	9.084	9.090	9.096	9.102	9.109	9.115	9.121	9.128	2570
2580	9.128	9.134	9.140	9.146	9.153	9.159	9.165	9.172	9.178	9.184	9.191	2580
2590	9.191	9.197	9.203	9.209	9.216	9.222	9.228	9.235	9.241	9.247	9.254	2590
2600	9.254	9.260	9.266	9.273	9.279	9.285	9.291	9.298	9.304	9.310	9.317	2600
2610	9.317	9.323	9.329	9.336	9.342	9.348	9.355	9.361	9.367	9.374	9.380	2610
2620	9.380	9.386	9.393	9.399	9.405	9.412	9.418	9.424	9.431	9.437	9.443	2620
2630	9.443	9.450	9.456	9.462	9.469	9.475	9.481	9.488	9.494	9.500	9.507	2630
2640	9.507	9.513	9.519	9.526	9.532	9.538	9.545	9.551	9.558	9.564	9.570	2640

Type B Thermocouple
Thermoelectric Voltage as a Function of Temperature (°F)
Reference Junctions at 32°F
THERMOELECTRIC VOLTAGE IN ABSOLUTE MILLIVOLTS

°F	0	1	2	3	4	5	6	7	8	9	10	°F
2650	9.570	9.577	9.583	9.589	9.596	9.602	9.608	9.615	9.621	9.627	9.634	2650
2660	9.634	9.640	9.647	9.653	9.659	9.666	9.672	9.678	9.685	9.691	9.697	2660
2670	9.697	9.704	9.710	9.717	9.723	9.729	9.736	9.742	9.748	9.755	9.761	2670
2680	9.761	9.768	9.774	9.780	9.787	9.793	9.800	9.806	9.812	9.819	9.825	2680
2690	9.825	9.831	9.838	9.844	9.851	9.857	9.863	9.870	9.876	9.883	9.889	2690
2700	9.889	9.895	9.902	9.908	9.915	9.921	9.927	9.934	9.940	9.947	9.953	2700
2710	9.953	9.959	9.966	9.972	9.979	9.985	9.991	9.998	10.004	10.011	10.017	2710
2720	10.017	10.023	10.030	10.036	10.043	10.049	10.056	10.062	10.068	10.075	10.081	2720
2730	10.081	10.088	10.094	10.100	10.107	10.113	10.120	10.126	10.133	10.139	10.145	2730
2740	10.145	10.152	10.158	10.165	10.171	10.178	10.184	10.190	10.197	10.203	10.210	2740
2750	10.210	10.216	10.223	10.229	10.235	10.242	10.248	10.255	10.261	10.268	10.274	2750
2760	10.274	10.280	10.287	10.293	10.300	10.306	10.313	10.319	10.325	10.332	10.338	2760
2770	10.338	10.345	10.351	10.358	10.364	10.371	10.377	10.383	10.390	10.396	10.403	2770
2780	10.403	10.409	10.416	10.422	10.429	10.435	10.441	10.448	10.454	10.461	10.467	2780
2790	10.467	10.474	10.480	10.487	10.493	10.500	10.506	10.512	10.519	10.525	10.532	2790
2800	10.532	10.538	10.545	10.551	10.558	10.564	10.571	10.577	10.584	10.590	10.596	2800
2810	10.596	10.603	10.609	10.616	10.622	10.629	10.635	10.642	10.648	10.655	10.661	2810
2820	10.661	10.668	10.674	10.680	10.687	10.693	10.700	10.706	10.713	10.719	10.726	2820
2830	10.726	10.732	10.739	10.745	10.752	10.758	10.765	10.771	10.778	10.784	10.790	2830
2840	10.790	10.797	10.803	10.810	10.816	10.823	10.829	10.836	10.842	10.849	10.855	2840
2850	10.855	10.862	10.868	10.875	10.881	10.888	10.894	10.901	10.907	10.914	10.920	2850
2860	10.920	10.926	10.933	10.939	10.946	10.952	10.959	10.965	10.972	10.978	10.985	2860
2870	10.985	10.991	10.998	11.004	11.011	11.017	11.024	11.030	11.037	11.043	11.050	2870
2880	11.050	11.056	11.063	11.069	11.076	11.082	11.089	11.095	11.102	11.108	11.115	2880
2890	11.115	11.121	11.128	11.134	11.141	11.147	11.154	11.160	11.166	11.173	11.179	2890
2900	11.179	11.186	11.192	11.199	11.205	11.212	11.218	11.225	11.231	11.238	11.244	2900
2910	11.244	11.251	11.257	11.264	11.270	11.277	11.283	11.290	11.296	11.303	11.309	2910
2920	11.309	11.316	11.322	11.329	11.335	11.342	11.348	11.355	11.361	11.368	11.374	2920
2930	11.374	11.381	11.387	11.394	11.400	11.407	11.413	11.420	11.426	11.433	11.439	2930
2940	11.439	11.446	11.452	11.459	11.465	11.472	11.478	11.485	11.491	11.498	11.504	2940
2950	11.504	11.511	11.517	11.524	11.530	11.537	11.543	11.550	11.556	11.563	11.569	2950
2960	11.569	11.576	11.582	11.589	11.595	11.602	11.608	11.615	11.621	11.628	11.634	2960
2970	11.634	11.641	11.647	11.654	11.660	11.667	11.673	11.680	11.686	11.693	11.699	2970
2980	11.699	11.706	11.712	11.719	11.725	11.732	11.738	11.745	11.751	11.758	11.764	2980
2990	11.764	11.771	11.777	11.784	11.790	11.797	11.803	11.810	11.816	11.823	11.829	2990
3000	11.829	11.836	11.842	11.849	11.855	11.862	11.868	11.875	11.881	11.888	11.894	3000
3010	11.894	11.901	11.907	11.914	11.920	11.927	11.933	11.940	11.946	11.953	11.959	3010
3020	11.959	11.966	11.972	11.979	11.985	11.992	11.998	12.005	12.011	12.018	12.024	3020
3030	12.024	12.031	12.037	12.044	12.050	12.057	12.063	12.070	12.076	12.083	12.089	3030
3040	12.089	12.096	12.102	12.109	12.115	12.121	12.128	12.134	12.141	12.147	12.154	3040
3050	12.154	12.160	12.167	12.173	12.180	12.186	12.193	12.199	12.206	12.212	12.219	3050
3060	12.219	12.225	12.232	12.238	12.245	12.251	12.258	12.264	12.271	12.277	12.284	3060
3070	12.284	12.290	12.297	12.303	12.310	12.316	12.323	12.329	12.336	12.342	12.349	3070
3080	12.349	12.355	12.362	12.368	12.374	12.381	12.387	12.394	12.400	12.407	12.413	3080
3090	12.413	12.420	12.426	12.433	12.439	12.446	12.452	12.459	12.465	12.472	12.478	3090
3100	12.478	12.485	12.491	12.498	12.504	12.511	12.517	12.523	12.530	12.536	12.543	3100
3110	12.543	12.549	12.556	12.562	12.569	12.575	12.582	12.588	12.595	12.601	12.608	3110
3120	12.608	12.614	12.621	12.627	12.633	12.640	12.646	12.653	12.659	12.666	12.672	3120
3130	12.672	12.679	12.685	12.692	12.698	12.705	12.711	12.718	12.724	12.730	12.737	3130
3140	12.737	12.743	12.750	12.756	12.763	12.769	12.776	12.782	12.789	12.795	12.801	3140
3150	12.801	12.808	12.814	12.821	12.827	12.834	12.840	12.847	12.853	12.860	12.866	3150
3160	12.866	12.872	12.879	12.885	12.892	12.898	12.905	12.911	12.918	12.924	12.930	3160
3170	12.930	12.937	12.943	12.950	12.956	12.963	12.969	12.976	12.982	12.988	12.995	3170
3180	12.995	13.001	13.008	13.014	13.021	13.027	13.034	13.040	13.046	13.053	13.059	3180
3190	13.059	13.066	13.072	13.079	13.085	13.091	13.098	13.104	13.111	13.117	13.124	3190
3200	13.124	13.130	13.136	13.143	13.149	13.156	13.162	13.169	13.175	13.181	13.188	3200
3210	13.188	13.194	13.201	13.207	13.213	13.220	13.226	13.233	13.239	13.246	13.252	3210
3220	13.252	13.258	13.265	13.271	13.278	13.284	13.290	13.297	13.303	13.310	13.316	3220
3230	13.316	13.322	13.329	13.335	13.342	13.348	13.354	13.361	13.367	13.374	13.380	3230
3240	13.380	13.387	13.393	13.399	13.406	13.412	13.418	13.425	13.431	13.438	13.444	3240
3250	13.444	13.450	13.457	13.463	13.470	13.476	13.482	13.489	13.495	13.502	13.508	3250
3260	13.508	13.514	13.521	13.527	13.533	13.540	13.546	13.553	13.559	13.565	13.572	3260
3270	13.572	13.578	13.585	13.591	13.597	13.604	13.610	13.616	13.623	13.629	13.635	3270
3280	13.635	13.642	13.648	13.655	13.661	13.667	13.674	13.680	13.686	13.693	13.699	3280
3290	13.699	13.706	13.712	13.718	13.725	13.731	13.737	13.744	13.750	13.756	13.763	3290
3300	13.763	13.769	13.775	13.782	13.788	13.794	13.801	13.807	13.814			3300

Type E Thermocouple
Thermoelectric Voltage as a Function of Temperature (°C)
Reference Junctions at 0°C
THERMOELECTRIC VOLTAGE IN ABSOLUTE MILLIVOLTS

°C	0	1	2	3	4	5	6	7	8	9	10	°C
−270	−9.835	−9.833	−9.831	−9.828	−9.825	−9.821	−9.817	−9.813	−9.808	−9.802	−9.797	−270
−260	−9.797	−9.791	−9.784	−9.777	−9.770	−9.762	−9.754	−9.746	−9.737	−9.728	−9.719	−260
−250	−9.719	−9.709	−9.699	−9.688	−9.677	−9.666	−9.654	−9.642	−9.630	−9.617	−9.604	−250
−240	−9.604	−9.591	−9.577	−9.563	−9.549	−9.534	−9.519	−9.503	−9.488	−9.472	−9.455	−240
−230	−9.455	−9.438	−9.421	−9.404	−9.386	−9.368	−9.350	−9.332	−9.313	−9.293	−9.274	−230
−220	−9.274	−9.254	−9.234	−9.214	−9.193	−9.172	−9.151	−9.129	−9.107	−9.085	−9.063	−220
−210	−9.063	−9.040	−9.017	−8.994	−8.971	−8.947	−8.923	−8.899	−8.874	−8.850	−8.824	−210
−200	−8.824	−8.799	−8.774	−8.748	−8.722	−8.696	−8.669	−8.642	−8.615	−8.588	−8.561	−200
−190	−8.561	−8.533	−8.505	−8.477	−8.449	−8.420	−8.391	−8.362	−8.333	−8.303	−8.273	−190
−180	−8.273	−8.243	−8.213	−8.183	−8.152	−8.121	−8.090	−8.058	−8.027	−7.995	−7.963	−180
−170	−7.963	−7.931	−7.898	−7.866	−7.833	−7.800	−7.767	−7.733	−7.699	−7.665	−7.631	−170
−160	−7.631	−7.597	−7.562	−7.528	−7.493	−7.458	−7.422	−7.387	−7.351	−7.315	−7.279	−160
−150	−7.279	−7.243	−7.206	−7.169	−7.132	−7.095	−7.058	−7.020	−6.983	−6.945	−6.907	−150
−140	−6.907	−6.869	−6.830	−6.792	−6.753	−6.714	−6.675	−6.635	−6.596	−6.556	−6.516	−140
−130	−6.516	−6.476	−6.436	−6.395	−6.354	−6.314	−6.273	−6.231	−6.190	−6.149	−6.107	−130
−120	−6.107	−6.065	−6.023	−5.981	−5.938	−5.896	−5.853	−5.810	−5.767	−5.724	−5.680	−120
−110	−5.680	−5.637	−5.593	−5.549	−5.505	−5.460	−5.416	−5.371	−5.327	−5.282	−5.237	−110
−100	−5.237	−5.191	−5.146	−5.100	−5.055	−5.009	−4.963	−4.916	−4.870	−4.824	−4.777	−100
−90	−4.777	−4.730	−4.683	−4.636	−4.588	−4.541	−4.493	−4.446	−4.398	−4.350	−4.301	−90
−80	−4.301	−4.253	−4.204	−4.156	−4.107	−4.058	−4.009	−3.959	−3.910	−3.860	−3.811	−80
−70	−3.811	−3.761	−3.711	−3.661	−3.610	−3.560	−3.509	−3.459	−3.408	−3.357	−3.306	−70
−60	−3.306	−3.254	−3.203	−3.152	−3.100	−3.048	−2.996	−2.944	−2.892	−2.839	−2.787	−60
−50	−2.787	−2.734	−2.681	−2.628	−2.575	−2.522	−2.469	−2.416	−2.362	−2.308	−2.254	−50
−40	−2.254	−2.200	−2.146	−2.092	−2.038	−1.983	−1.929	−1.874	−1.819	−1.764	−1.709	−40
−30	−1.709	−1.654	−1.599	−1.543	−1.487	−1.432	−1.376	−1.320	−1.264	−1.208	−1.151	−30
−20	−1.151	−1.095	−1.038	−0.982	−0.925	−0.868	−0.811	−0.754	−0.696	−0.639	−0.581	−20
−10	−0.581	−0.524	−0.466	−0.408	−0.350	−0.292	−0.234	−0.176	−0.117	−0.059	0.000	−10
0	0.000	0.059	0.118	0.176	0.235	0.295	0.354	0.413	0.472	0.532	0.591	0
10	0.591	0.651	0.711	0.770	0.830	0.890	0.950	1.011	1.071	1.131	1.192	10
20	1.192	1.252	1.313	1.373	1.434	1.495	1.556	1.617	1.678	1.739	1.801	20
30	1.801	1.862	1.924	1.985	2.047	2.109	2.171	2.233	2.295	2.357	2.419	30
40	2.419	2.482	2.544	2.607	2.669	2.732	2.795	2.858	2.921	2.984	3.047	40
50	3.047	3.110	3.173	3.237	3.300	3.364	3.428	3.491	3.555	3.619	3.683	50
60	3.683	3.748	3.812	3.876	3.941	4.005	4.070	4.134	4.199	4.264	4.329	60
70	4.329	4.394	4.459	4.524	4.590	4.655	4.720	4.786	4.852	4.917	4.983	70
80	4.983	5.049	5.115	5.181	5.247	5.314	5.380	5.446	5.513	5.579	5.646	80
90	5.646	5.713	5.780	5.846	5.913	5.981	6.048	6.115	6.182	6.250	6.317	90
100	6.317	6.385	6.452	6.520	6.588	6.656	6.724	6.792	6.860	6.928	6.996	100
110	6.996	7.064	7.133	7.201	7.270	7.339	7.407	7.476	7.545	7.614	7.683	110
120	7.683	7.752	7.821	7.890	7.960	8.029	8.099	8.168	8.238	8.307	8.377	120
130	8.377	8.447	8.517	8.587	8.657	8.727	8.797	8.867	8.938	9.008	9.078	130
140	9.078	9.149	9.220	9.290	9.361	9.432	9.503	9.573	9.644	9.715	9.787	140
150	9.787	9.858	9.929	10.000	10.072	10.143	10.215	10.286	10.358	10.429	10.501	150
160	10.501	10.573	10.645	10.717	10.789	10.861	10.933	11.005	11.077	11.150	11.222	160
170	11.222	11.294	11.367	11.439	11.512	11.585	11.657	11.730	11.803	11.876	11.949	170
180	11.949	12.022	12.095	12.168	12.241	12.314	12.387	12.461	12.534	12.608	12.681	180
190	12.681	12.755	12.828	12.902	12.975	13.049	13.123	13.197	13.271	13.345	13.419	190
200	13.419	13.493	13.567	13.641	13.715	13.789	13.864	13.938	14.012	14.087	14.161	200
210	14.161	14.236	14.310	14.385	14.460	14.534	14.609	14.684	14.759	14.834	14.909	210
220	14.909	14.984	15.059	15.134	15.209	15.284	15.359	15.435	15.510	15.585	15.661	220
230	15.661	15.736	15.812	15.887	15.963	16.038	16.114	16.190	16.266	16.341	16.417	230
240	16.417	16.493	16.569	16.645	16.721	16.797	16.873	16.949	17.025	17.101	17.178	240
250	17.178	17.254	17.330	17.406	17.483	17.559	17.636	17.712	17.789	17.865	17.942	250
260	17.942	18.018	18.095	18.172	18.248	18.325	18.402	18.479	18.556	18.633	18.710	260
270	18.710	18.787	18.864	18.941	19.018	19.095	19.172	19.249	19.326	19.404	19.481	270
280	19.481	19.558	19.636	19.713	19.790	19.868	19.945	20.023	20.100	20.178	20.256	280
290	20.256	20.333	20.411	20.488	20.566	20.644	20.722	20.800	20.877	20.955	21.033	290
300	21.033	21.111	21.189	21.267	21.345	21.423	21.501	21.579	21.657	21.735	21.814	300
310	21.814	21.892	21.970	22.048	22.127	22.205	22.283	22.362	22.440	22.518	22.597	310
320	22.597	22.675	22.754	22.832	22.911	22.989	23.068	23.147	23.225	23.304	23.383	320
330	23.383	23.461	23.540	23.619	23.698	23.777	23.855	23.934	24.013	24.092	24.171	330
340	24.171	24.250	24.329	24.408	24.487	24.566	24.645	24.724	24.803	24.882	24.961	340

Type E Thermocouple
Thermoelectric Voltage as a Function of Temperature (°C)
Reference Junctions at 0°C
THERMOELECTRIC VOLTAGE IN ABSOLUTE MILLIVOLTS

°C	0	1	2	3	4	5	6	7	8	9	10	°C
350	24.961	25.041	25.120	25.199	25.278	25.357	25.437	25.516	25.595	25.675	25.754	350
360	25.754	25.833	25.913	25.992	26.072	26.151	26.230	26.310	26.389	26.469	26.549	360
370	26.549	26.628	26.708	26.787	26.867	26.947	27.026	27.106	27.186	27.265	27.345	370
380	27.345	27.425	27.504	27.584	27.664	27.744	27.824	27.903	27.983	28.063	28.143	380
390	28.143	28.223	28.303	28.383	28.463	28.543	28.623	28.703	28.783	28.863	28.943	390
400	28.943	29.023	29.103	29.183	29.263	29.343	29.423	29.503	29.584	29.664	29.744	400
410	29.744	29.824	29.904	29.984	30.065	30.145	30.225	30.305	30.386	30.466	30.546	410
420	30.546	30.627	30.707	30.787	30.868	30.948	31.028	31.109	31.189	31.270	31.350	420
430	31.350	31.430	31.511	31.591	31.672	31.752	31.833	31.913	31.994	32.074	32.155	430
440	32.155	32.235	32.316	32.396	32.477	32.557	32.638	32.719	32.799	32.880	32.960	440
450	32.960	33.041	33.122	33.202	33.283	33.364	33.444	33.525	33.605	33.686	33.767	450
460	33.767	33.848	33.928	34.009	34.090	34.170	34.251	34.332	34.413	34.493	34.574	460
470	34.574	34.655	34.736	34.816	34.897	34.978	35.059	35.140	35.220	35.301	35.382	470
480	35.382	35.463	35.544	35.624	35.705	35.786	35.867	35.948	36.029	36.109	36.190	480
490	36.190	36.271	36.352	36.433	36.514	36.595	36.675	36.756	36.837	36.918	36.999	490
500	36.999	37.080	37.161	37.242	37.323	37.403	37.484	37.565	37.646	37.727	37.808	500
510	37.808	37.889	37.970	38.051	38.132	38.213	38.293	38.374	38.455	38.536	38.617	510
520	38.617	38.698	38.779	38.860	38.941	39.022	39.103	39.184	39.264	39.345	39.426	520
530	39.426	39.507	39.588	39.669	39.750	39.831	39.912	39.993	40.074	40.155	40.236	530
540	40.236	40.316	40.397	40.478	40.559	40.640	40.721	40.802	40.883	40.964	41.045	540
550	41.045	41.125	41.206	41.287	41.368	41.449	41.530	41.611	41.692	41.773	41.853	550
560	41.853	41.934	42.015	42.096	42.177	42.258	42.339	42.419	42.500	42.581	42.662	560
570	42.662	42.743	42.824	42.904	42.985	43.066	43.147	43.228	43.308	43.389	43.470	570
580	43.470	43.551	43.632	43.712	43.793	43.874	43.955	44.035	44.116	44.197	44.278	580
590	44.278	44.358	44.439	44.520	44.601	44.681	44.762	44.843	44.923	45.004	45.085	590
600	45.085	45.165	45.246	45.327	45.407	45.488	45.569	45.649	45.730	45.811	45.891	600
610	45.891	45.972	46.052	46.133	46.213	46.294	46.375	46.455	46.536	46.616	46.697	610
620	46.697	46.777	46.858	46.938	47.019	47.099	47.180	47.260	47.341	47.421	47.502	620
630	47.502	47.582	47.663	47.743	47.824	47.904	47.984	48.065	48.145	48.226	48.306	630
640	48.306	48.386	48.467	48.547	48.627	48.708	48.788	48.868	48.949	49.029	49.109	640
650	49.109	49.189	49.270	49.350	49.430	49.510	49.591	49.671	49.751	49.831	49.911	650
660	49.911	49.992	50.072	50.152	50.232	50.312	50.392	50.472	50.553	50.633	50.713	660
670	50.713	50.793	50.873	50.953	51.033	51.113	51.193	51.273	51.353	51.433	51.513	670
680	51.513	51.593	51.673	51.753	51.833	51.913	51.993	52.073	52.152	52.232	52.312	680
690	52.312	52.392	52.472	52.552	52.632	52.711	52.791	52.871	52.951	53.031	53.110	690
700	53.110	53.190	53.270	53.350	53.429	53.509	53.589	53.668	53.748	53.828	53.907	700
710	53.907	53.987	54.066	54.146	54.226	54.305	54.385	54.464	54.544	54.623	54.703	710
720	54.703	54.782	54.862	54.941	55.021	55.100	55.180	55.259	55.339	55.418	55.498	720
730	55.498	55.577	55.656	55.736	55.815	55.894	55.974	56.053	56.132	56.212	56.291	730
740	56.291	56.370	56.449	56.529	56.608	56.687	56.766	56.845	56.924	57.004	57.083	740
750	57.083	57.162	57.241	57.320	57.399	57.478	57.557	57.636	57.715	57.794	57.873	750
760	57.873	57.952	58.031	58.110	58.189	58.268	58.347	58.426	58.505	58.584	58.663	760
770	58.663	58.742	58.820	58.899	58.978	59.057	59.136	59.214	59.293	59.372	59.451	770
780	59.451	59.529	59.608	59.687	59.765	59.844	59.923	60.001	60.080	60.159	60.237	780
790	60.237	60.316	60.394	60.473	60.551	60.630	60.708	60.787	60.865	60.944	61.022	790
800	61.022	61.101	61.179	61.258	61.336	61.414	61.493	61.571	61.649	61.728	61.806	800
810	61.806	61.884	61.962	62.041	62.119	62.197	62.275	62.353	62.432	62.510	62.588	810
820	62.588	62.666	62.744	62.822	62.900	62.978	63.056	63.134	63.212	63.290	63.368	820
830	63.368	63.446	63.524	63.602	63.680	63.758	63.836	63.914	63.992	64.069	64.147	830
840	64.147	64.225	64.303	64.380	64.458	64.536	64.614	64.691	64.769	64.847	64.924	840
850	64.924	65.002	65.080	65.157	65.235	65.312	65.390	65.467	65.545	65.622	65.700	850
860	65.700	65.777	65.855	65.932	66.009	66.087	66.164	66.241	66.319	66.396	66.473	860
870	66.473	66.551	66.628	66.705	66.782	66.859	66.937	67.014	67.091	67.168	67.245	870
880	67.245	67.322	67.399	67.476	67.553	67.630	67.707	67.784	67.861	67.938	68.015	880
890	68.015	68.092	68.169	68.246	68.323	68.399	68.476	68.553	68.630	68.706	68.783	890
900	68.783	68.860	68.936	69.013	69.090	69.166	69.243	69.320	69.396	69.473	69.549	900
910	69.549	69.626	69.702	69.779	69.855	69.931	70.008	70.084	70.161	70.237	70.313	910
920	70.313	70.390	70.466	70.542	70.618	70.694	70.771	70.847	70.923	70.999	71.075	920
930	71.075	71.151	71.227	71.304	71.380	71.456	71.532	71.608	71.683	71.759	71.835	930
940	71.835	71.911	71.987	72.063	72.139	72.215	72.290	72.366	72.442	72.518	72.593	940
950	72.593	72.669	72.745	72.820	72.896	72.972	73.047	73.123	73.199	73.274	73.350	950
960	73.350	73.425	73.501	73.576	73.652	73.727	73.802	73.878	73.953	74.029	74.104	960
970	74.104	74.179	74.255	74.330	74.405	74.480	74.556	74.631	74.706	74.781	74.857	970
980	74.857	74.932	75.007	75.082	75.157	75.232	75.307	75.382	75.458	75.533	75.608	980
990	75.608	75.683	75.758	75.833	75.908	75.983	76.058	76.133	76.208	76.283	76.358	990

Type E Thermocouple
Thermoelectric Voltage as a Function of Temperature (°F)
Reference Junctions at 32°F
THERMOELECTRIC VOLTAGE IN ABSOLUTE MILLIVOLTS

°F	0	1	2	3	4	5	6	7	8	9	10	°F
−460							−9.835	−9.834	−9.833	−9.832	−9.830	−460
−450	−9.830	−9.829	−9.827	−9.825	−9.823	−9.821	−9.819	−9.817	−9.814	−9.812	−9.809	−450
−440	−9.809	−9.806	−9.803	−9.800	−9.797	−9.793	−9.790	−9.786	−9.783	−9.779	−9.775	−440
−430	−9.775	−9.771	−9.767	−9.762	−9.758	−9.753	−9.749	−9.744	−9.739	−9.734	−9.729	−430
−420	−9.729	−9.724	−9.719	−9.713	−9.708	−9.702	−9.696	−9.690	−9.684	−9.678	−9.672	−420
−410	−9.672	−9.666	−9.659	−9.653	−9.646	−9.639	−9.633	−9.626	−9.619	−9.611	−9.604	−410
−400	−9.604	−9.597	−9.589	−9.582	−9.574	−9.566	−9.558	−9.550	−9.542	−9.534	−9.526	−400
−390	−9.526	−9.517	−9.509	−9.500	−9.491	−9.482	−9.473	−9.464	−9.455	−9.446	−9.437	−390
−380	−9.437	−9.427	−9.418	−9.408	−9.398	−9.388	−9.378	−9.368	−9.358	−9.348	−9.338	−380
−370	−9.338	−9.327	−9.317	−9.306	−9.296	−9.285	−9.274	−9.263	−9.252	−9.241	−9.229	−370
−360	−9.229	−9.218	−9.207	−9.195	−9.184	−9.172	−9.160	−9.148	−9.136	−9.124	−9.112	−360
−350	−9.112	−9.100	−9.088	−9.075	−9.063	−9.050	−9.038	−9.025	−9.012	−8.999	−8.986	−350
−340	−8.986	−8.973	−8.960	−8.947	−8.934	−8.920	−8.907	−8.893	−8.880	−8.866	−8.852	−340
−330	−8.852	−8.838	−8.824	−8.810	−8.796	−8.782	−8.768	−8.754	−8.739	−8.725	−8.710	−330
−320	−8.710	−8.696	−8.681	−8.666	−8.651	−8.636	−8.621	−8.606	−8.591	−8.576	−8.561	−320
−310	−8.561	−8.545	−8.530	−8.514	−8.499	−8.483	−8.468	−8.452	−8.436	−8.420	−8.404	−310
−300	−8.404	−8.388	−8.372	−8.355	−8.339	−8.323	−8.306	−8.290	−8.273	−8.257	−8.240	−300
−290	−8.240	−8.223	−8.206	−8.189	−8.172	−8.155	−8.138	−8.121	−8.104	−8.086	−8.069	−290
−280	−8.069	−8.051	−8.034	−8.016	−7.999	−7.981	−7.963	−7.945	−7.927	−7.909	−7.891	−280
−270	−7.891	−7.873	−7.855	−7.837	−7.818	−7.800	−7.781	−7.763	−7.744	−7.726	−7.707	−270
−260	−7.707	−7.688	−7.669	−7.650	−7.631	−7.612	−7.593	−7.574	−7.555	−7.535	−7.516	−260
−250	−7.516	−7.497	−7.477	−7.458	−7.438	−7.418	−7.399	−7.379	−7.359	−7.339	−7.319	−250
−240	−7.319	−7.299	−7.279	−7.259	−7.239	−7.218	−7.198	−7.178	−7.157	−7.137	−7.116	−240
−230	−7.116	−7.095	−7.075	−7.054	−7.033	−7.012	−6.991	−6.970	−6.949	−6.928	−6.907	−230
−220	−6.907	−6.886	−6.864	−6.843	−6.822	−6.800	−6.779	−6.757	−6.735	−6.714	−6.692	−220
−210	−6.692	−6.670	−6.648	−6.626	−6.604	−6.582	−6.560	−6.538	−6.516	−6.494	−6.471	−210
−200	−6.471	−6.449	−6.427	−6.404	−6.382	−6.359	−6.336	−6.314	−6.291	−6.268	−6.245	−200
−190	−6.245	−6.222	−6.199	−6.176	−6.153	−6.130	−6.107	−6.084	−6.060	−6.037	−6.013	−190
−180	−6.013	−5.990	−5.967	−5.943	−5.919	−5.896	−5.872	−5.848	−5.824	−5.800	−5.776	−180
−170	−5.776	−5.752	−5.728	−5.704	−5.680	−5.656	−5.632	−5.607	−5.583	−5.559	−5.534	−170
−160	−5.534	−5.510	−5.485	−5.460	−5.436	−5.411	−5.386	−5.362	−5.337	−5.312	−5.287	−160
−150	−5.287	−5.262	−5.237	−5.212	−5.186	−5.161	−5.136	−5.111	−5.085	−5.060	−5.034	−150
−140	−5.034	−5.009	−4.983	−4.958	−4.932	−4.906	−4.880	−4.855	−4.829	−4.803	−4.777	−140
−130	−4.777	−4.751	−4.725	−4.699	−4.672	−4.646	−4.620	−4.594	−4.567	−4.541	−4.515	−130
−120	−4.515	−4.488	−4.462	−4.435	−4.408	−4.382	−4.355	−4.328	−4.301	−4.274	−4.248	−120
−110	−4.248	−4.221	−4.194	−4.167	−4.139	−4.112	−4.085	−4.058	−4.031	−4.003	−3.976	−110
−100	−3.976	−3.949	−3.921	−3.894	−3.866	−3.838	−3.811	−3.783	−3.755	−3.728	−3.700	−100
−90	−3.700	−3.672	−3.644	−3.616	−3.588	−3.560	−3.532	−3.504	−3.476	−3.447	−3.419	−90
−80	−3.419	−3.391	−3.363	−3.334	−3.306	−3.277	−3.249	−3.220	−3.192	−3.163	−3.134	−80
−70	−3.134	−3.106	−3.077	−3.048	−3.019	−2.990	−2.961	−2.932	−2.903	−2.874	−2.845	−70
−60	−2.845	−2.816	−2.787	−2.758	−2.728	−2.699	−2.670	−2.640	−2.611	−2.581	−2.552	−60
−50	−2.552	−2.522	−2.493	−2.463	−2.433	−2.404	−2.374	−2.344	−2.314	−2.284	−2.254	−50
−40	−2.254	−2.224	−2.194	−2.164	−2.134	−2.104	−2.074	−2.044	−2.014	−1.983	−1.953	−40
−30	−1.953	−1.923	−1.892	−1.862	−1.831	−1.801	−1.770	−1.740	−1.709	−1.678	−1.648	−30
−20	−1.648	−1.617	−1.586	−1.555	−1.525	−1.494	−1.463	−1.432	−1.401	−1.370	−1.339	−20
−10	−1.339	−1.308	−1.276	−1.245	−1.214	−1.183	−1.151	−1.120	−1.089	−1.057	−1.026	−10
0	−1.026	−0.994	−0.963	−0.931	−0.900	−0.868	−0.836	−0.805	−0.773	−0.741	−0.709	0
10	−0.709	−0.677	−0.645	−0.613	−0.581	−0.549	−0.517	−0.485	−0.453	−0.421	−0.389	10
20	−0.389	−0.357	−0.324	−0.292	−0.260	−0.227	−0.195	−0.163	−0.130	−0.098	−0.065	20
30	−0.065	−0.033	0.000	0.033	0.065	0.098	0.131	0.163	0.196	0.229	0.262	30
40	0.262	0.295	0.327	0.360	0.393	0.426	0.459	0.492	0.525	0.558	0.591	40
50	0.591	0.624	0.658	0.691	0.724	0.757	0.790	0.824	0.857	0.890	0.924	50
60	0.924	0.957	0.990	1.024	1.057	1.091	1.124	1.158	1.192	1.225	1.259	60
70	1.259	1.292	1.326	1.360	1.394	1.427	1.461	1.495	1.529	1.563	1.597	70
80	1.597	1.631	1.665	1.699	1.733	1.767	1.801	1.835	1.869	1.903	1.937	80
90	1.937	1.972	2.006	2.040	2.075	2.109	2.143	2.178	2.212	2.247	2.281	90
100	2.281	2.316	2.350	2.385	2.419	2.454	2.489	2.523	2.558	2.593	2.627	100
110	2.627	2.662	2.697	2.732	2.767	2.802	2.837	2.872	2.907	2.942	2.977	110
120	2.977	3.012	3.047	3.082	3.117	3.152	3.187	3.223	3.258	3.293	3.329	120
130	3.329	3.364	3.399	3.435	3.470	3.506	3.541	3.577	3.612	3.648	3.683	130

Type E Thermocouple
Thermoelectric Voltage as a Function of Temperature (°F)
Reference Junctions at 32°F
THERMOELECTRIC VOLTAGE IN ABSOLUTE MILLIVOLTS

°F	0	1	2	3	4	5	6	7	8	9	10	°F
140	3.683	3.719	3.755	3.790	3.826	3.862	3.898	3.933	3.969	4.005	4.041	140
150	4.041	4.077	4.113	4.149	4.185	4.221	4.257	4.293	4.329	4.365	4.401	150
160	4.401	4.437	4.474	4.510	4.546	4.582	4.619	4.655	4.691	4.728	4.764	160
170	4.764	4.801	4.837	4.874	4.910	4.947	4.983	5.020	5.056	5.093	5.130	170
180	5.130	5.166	5.203	5.240	5.277	5.314	5.350	5.387	5.424	5.461	5.498	180
190	5.498	5.535	5.572	5.609	5.646	5.683	5.720	5.757	5.794	5.832	5.869	190
200	5.869	5.906	5.943	5.981	6.018	6.055	6.092	6.130	6.167	6.205	6.242	200
210	6.242	6.280	6.317	6.355	6.392	6.430	6.467	6.505	6.543	6.580	6.618	210
220	6.618	6.656	6.693	6.731	6.769	6.807	6.845	6.882	6.920	6.958	6.996	220
230	6.996	7.034	7.072	7.110	7.148	7.186	7.224	7.262	7.300	7.339	7.377	230
240	7.377	7.415	7.453	7.491	7.530	7.568	7.606	7.645	7.683	7.721	7.760	240
250	7.760	7.798	7.837	7.875	7.914	7.952	7.991	8.029	8.068	8.106	8.145	250
260	8.145	8.184	8.222	8.261	8.300	8.338	8.377	8.416	8.455	8.494	8.532	260
270	8.532	8.571	8.610	8.649	8.688	8.727	8.766	8.805	8.844	8.883	8.922	270
280	8.922	8.961	9.000	9.039	9.078	9.118	9.157	9.196	9.235	9.274	9.314	280
290	9.314	9.353	9.392	9.432	9.471	9.510	9.550	9.589	9.629	9.668	9.708	290
300	9.708	9.747	9.787	9.826	9.866	9.905	9.945	9.984	10.024	10.064	10.103	300
310	10.103	10.143	10.183	10.223	10.262	10.302	10.342	10.382	10.421	10.461	10.501	310
320	10.501	10.541	10.581	10.621	10.661	10.701	10.741	10.781	10.821	10.861	10.901	320
330	10.901	10.941	10.981	11.021	11.061	11.101	11.142	11.182	11.222	11.262	11.302	330
340	11.302	11.343	11.383	11.423	11.464	11.504	11.544	11.585	11.625	11.665	11.706	340
350	11.706	11.746	11.787	11.827	11.868	11.908	11.949	11.989	12.030	12.070	12.111	350
360	12.111	12.152	12.192	12.233	12.273	12.314	12.355	12.396	12.436	12.477	12.518	360
370	12.518	12.559	12.599	12.640	12.681	12.722	12.763	12.804	12.844	12.885	12.926	370
380	12.926	12.967	13.008	13.049	13.090	13.131	13.172	13.213	13.254	13.295	13.336	380
390	13.336	13.378	13.419	13.460	13.501	13.542	13.583	13.624	13.666	13.707	13.748	390
400	13.748	13.789	13.831	13.872	13.913	13.955	13.996	14.037	14.079	14.120	14.161	400
410	14.161	14.203	14.244	14.286	14.327	14.368	14.410	14.451	14.493	14.534	14.576	410
420	14.576	14.618	14.659	14.701	14.742	14.784	14.826	14.867	14.909	14.950	14.992	420
430	14.992	15.034	15.076	15.117	15.159	15.201	15.243	15.284	15.326	15.368	15.410	430
440	15.410	15.451	15.493	15.535	15.577	15.619	15.661	15.703	15.745	15.787	15.829	440
450	15.829	15.871	15.912	15.954	15.996	16.038	16.080	16.123	16.165	16.207	16.249	450
460	16.249	16.291	16.333	16.375	16.417	16.459	16.501	16.544	16.586	16.628	16.670	460
470	16.670	16.712	16.755	16.797	16.839	16.881	16.924	16.966	17.008	17.051	17.093	470
480	17.093	17.135	17.178	17.220	17.262	17.305	17.347	17.389	17.432	17.474	17.517	480
490	17.517	17.559	17.602	17.644	17.687	17.729	17.772	17.814	17.857	17.899	17.942	490
500	17.942	17.984	18.027	18.070	18.112	18.155	18.197	18.240	18.283	18.325	18.368	500
510	18.368	18.411	18.453	18.496	18.539	18.581	18.624	18.667	18.710	18.752	18.795	510
520	18.795	18.838	18.881	18.924	18.966	19.009	19.052	19.095	19.138	19.181	19.223	520
530	19.223	19.266	19.309	19.352	19.395	19.438	19.481	19.524	19.567	19.610	19.653	530
540	19.653	19.696	19.739	19.782	19.825	19.868	19.911	19.954	19.997	20.040	20.083	540
550	20.083	20.126	20.169	20.212	20.256	20.299	20.342	20.385	20.428	20.471	20.514	550
560	20.514	20.558	20.601	20.644	20.687	20.730	20.774	20.817	20.860	20.903	20.947	560
570	20.947	20.990	21.033	21.076	21.120	21.163	21.206	21.250	21.293	21.336	21.380	570
580	21.380	21.423	21.466	21.510	21.553	21.597	21.640	21.683	21.727	21.770	21.814	580
590	21.814	21.857	21.901	21.944	21.987	22.031	22.074	22.118	22.161	22.205	22.248	590
600	22.248	22.292	22.336	22.379	22.423	22.466	22.510	22.553	22.597	22.640	22.684	600
610	22.684	22.728	22.771	22.815	22.859	22.902	22.946	22.989	23.033	23.077	23.120	610
620	23.120	23.164	23.208	23.252	23.295	23.339	23.383	23.426	23.470	23.514	23.558	620
630	23.558	23.601	23.645	23.689	23.733	23.777	23.820	23.864	23.908	23.952	23.996	630
640	23.996	24.039	24.083	24.127	24.171	24.215	24.259	24.302	24.346	24.390	24.434	640
650	24.434	24.478	24.522	24.566	24.610	24.654	24.698	24.742	24.786	24.829	24.873	650
660	24.873	24.917	24.961	25.005	25.049	25.093	25.137	25.181	25.225	25.269	25.313	660
670	25.313	25.357	25.401	25.445	25.490	25.534	25.578	25.622	25.666	25.710	25.754	670
680	25.754	25.798	25.842	25.886	25.930	25.974	26.019	26.063	26.107	26.151	26.195	680
690	26.195	26.239	26.283	26.328	26.372	26.416	26.460	26.504	26.549	26.593	26.637	690
700	26.637	26.681	26.725	26.770	26.814	26.858	26.902	26.947	26.991	27.035	27.079	700
710	27.079	27.124	27.168	27.212	27.256	27.301	27.345	27.389	27.434	27.478	27.522	710
720	27.522	27.566	27.611	27.655	27.699	27.744	27.788	27.832	27.877	27.921	27.966	720
730	27.966	28.010	28.054	28.099	28.143	28.187	28.232	28.276	28.321	28.365	28.409	730

Type E Thermocouple
Thermoelectric Voltage as a Function of Temperature (°F)
Reference Junctions at 32°F
THERMOELECTRIC VOLTAGE IN ABSOLUTE MILLIVOLTS

°F	0	1	2	3	4	5	6	7	8	9	10	°F
740	28.409	28.454	28.498	28.543	28.587	28.632	28.676	28.720	28.765	28.809	28.854	740
750	28.854	28.898	28.943	28.987	29.032	29.076	29.121	29.165	29.210	29.254	29.299	750
760	29.299	29.343	29.388	29.432	29.477	29.521	29.566	29.610	29.655	29.699	29.744	760
770	29.744	29.788	29.833	29.878	29.922	29.967	30.011	30.056	30.100	30.145	30.190	770
780	30.190	30.234	30.279	30.323	30.368	30.412	30.457	30.502	30.546	30.591	30.636	780
790	30.636	30.680	30.725	30.769	30.814	30.859	30.903	30.948	30.993	31.037	31.082	790
800	31.082	31.127	31.171	31.216	31.261	31.305	31.350	31.395	31.439	31.484	31.529	800
810	31.529	31.573	31.618	31.663	31.707	31.752	31.797	31.842	31.886	31.931	31.976	810
820	31.976	32.020	32.065	32.110	32.155	32.199	32.244	32.289	32.334	32.378	32.423	820
830	32.423	32.468	32.513	32.557	32.602	32.647	32.692	32.736	32.781	32.826	32.871	830
840	32.871	32.916	32.960	33.005	33.050	33.095	33.140	33.184	33.229	33.274	33.319	840
850	33.319	33.364	33.408	33.453	33.498	33.543	33.588	33.632	33.677	33.722	33.767	850
860	33.767	33.812	33.857	33.901	33.946	33.991	34.036	34.081	34.126	34.170	34.215	860
870	34.215	34.260	34.305	34.350	34.395	34.440	34.484	34.529	34.574	34.619	34.664	870
880	34.664	34.709	34.754	34.798	34.843	34.888	34.933	34.978	35.023	35.068	35.113	880
890	35.113	35.157	35.202	35.247	35.292	35.337	35.382	35.427	35.472	35.517	35.562	890
900	35.562	35.606	35.651	35.696	35.741	35.786	35.831	35.876	35.921	35.966	36.011	900
910	36.011	36.056	36.100	36.145	36.190	36.235	36.280	36.325	36.370	36.415	36.460	910
920	36.460	36.505	36.550	36.595	36.640	36.684	36.729	36.774	36.819	36.864	36.909	920
930	36.909	36.954	36.999	37.044	37.089	37.134	37.179	37.224	37.269	37.314	37.358	930
940	37.358	37.403	37.448	37.493	37.538	37.583	37.628	37.673	37.718	37.763	37.808	940
950	37.808	37.853	37.898	37.943	37.988	38.033	38.078	38.123	38.168	38.213	38.257	950
960	38.257	38.302	38.347	38.392	38.437	38.482	38.527	38.572	38.617	38.662	38.707	960
970	38.707	38.752	38.797	38.842	38.887	38.932	38.977	39.022	39.067	39.112	39.157	970
980	39.157	39.202	39.247	39.291	39.336	39.381	39.426	39.471	39.516	39.561	39.606	980
990	39.606	39.651	39.696	39.741	39.786	39.831	39.876	39.921	39.966	40.011	40.056	990
1000	40.056	40.101	40.146	40.191	40.236	40.280	40.325	40.370	40.415	40.460	40.505	1000
1010	40.505	40.550	40.595	40.640	40.685	40.730	40.775	40.820	40.865	40.910	40.955	1010
1020	40.955	41.000	41.045	41.090	41.134	41.179	41.224	41.269	41.314	41.359	41.404	1020
1030	41.404	41.449	41.494	41.539	41.584	41.629	41.674	41.719	41.764	41.808	41.853	1030
1040	41.853	41.898	41.943	41.988	42.033	42.078	42.123	42.168	42.213	42.258	42.303	1040
1050	42.303	42.348	42.392	42.437	42.482	42.527	42.572	42.617	42.662	42.707	42.752	1050
1060	42.752	42.797	42.842	42.886	42.931	42.976	43.021	43.066	43.111	43.156	43.201	1060
1070	43.201	43.246	43.290	43.335	43.380	43.425	43.470	43.515	43.560	43.605	43.650	1070
1080	43.650	43.694	43.739	43.784	43.829	43.874	43.919	43.964	44.008	44.053	44.098	1080
1090	44.098	44.143	44.188	44.233	44.278	44.322	44.367	44.412	44.457	44.502	44.547	1090
1100	44.547	44.592	44.636	44.681	44.726	44.771	44.816	44.861	44.905	44.950	44.995	1100
1110	44.995	45.040	45.085	45.130	45.174	45.219	45.264	45.309	45.354	45.398	45.443	1110
1120	45.443	45.488	45.533	45.578	45.622	45.667	45.712	45.757	45.802	45.846	45.891	1120
1130	45.891	45.936	45.981	46.025	46.070	46.115	46.160	46.205	46.249	46.294	46.339	1130
1140	46.339	46.384	46.428	46.473	46.518	46.563	46.607	46.652	46.697	46.742	46.786	1140
1150	46.786	46.831	46.876	46.921	46.965	47.010	47.055	47.099	47.144	47.189	47.234	1150
1160	47.234	47.278	47.323	47.368	47.412	47.457	47.502	47.546	47.591	47.636	47.681	1160
1170	47.681	47.725	47.770	47.815	47.859	47.904	47.949	47.993	48.038	48.083	48.127	1170
1180	48.127	48.172	48.217	48.261	48.306	48.351	48.395	48.440	48.484	48.529	48.574	1180
1190	48.574	48.618	48.663	48.708	48.752	48.797	48.842	48.886	48.931	48.975	49.020	1190
1200	49.020	49.065	49.109	49.154	49.198	49.243	49.288	49.332	49.377	49.421	49.466	1200
1210	49.466	49.510	49.555	49.600	49.644	49.689	49.733	49.778	49.822	49.867	49.911	1210
1220	49.911	49.956	50.001	50.045	50.090	50.134	50.179	50.223	50.268	50.312	50.357	1220
1230	50.357	50.401	50.446	50.490	50.535	50.579	50.624	50.668	50.713	50.757	50.802	1230
1240	50.802	50.846	50.891	50.935	50.980	51.024	51.069	51.113	51.157	51.202	51.246	1240
1250	51.246	51.291	51.335	51.380	51.424	51.469	51.513	51.557	51.602	51.646	51.691	1250
1260	51.691	51.735	51.780	51.824	51.868	51.913	51.957	52.002	52.046	52.090	52.135	1260
1270	52.135	52.179	52.223	52.268	52.312	52.357	52.401	52.445	52.490	52.534	52.578	1270
1280	52.578	52.623	52.667	52.711	52.756	52.800	52.844	52.889	52.933	52.977	53.022	1280
1290	53.022	53.066	53.110	53.155	53.199	53.243	53.288	53.332	53.376	53.420	53.465	1290
1300	53.465	53.509	53.553	53.597	53.642	53.686	53.730	53.774	53.819	53.863	53.907	1300
1310	53.907	53.951	53.996	54.040	54.084	54.128	54.173	54.217	54.261	54.305	54.349	1310
1320	54.349	54.394	54.438	54.482	54.526	54.570	54.615	54.659	54.703	54.747	54.791	1320
1330	54.791	54.835	54.880	54.924	54.968	55.012	55.056	55.100	55.145	55.189	55.233	1330

Type E Thermocouple
Thermoelectric Voltage as a Function of Temperature (°F)
Reference Junctions at 32°F
THERMOELECTRIC VOLTAGE IN ABSOLUTE MILLIVOLTS

°F	0	1	2	3	4	5	6	7	8	9	10	°F
1340	55.233	55.277	55.321	55.365	55.409	55.453	55.498	55.542	55.586	55.630	55.674	1340
1350	55.674	55.718	55.762	55.806	55.850	55.894	55.938	55.982	56.026	56.071	56.115	1350
1360	56.115	56.159	56.203	56.247	56.291	56.335	56.379	56.423	56.467	56.511	56.555	1360
1370	56.555	56.599	56.643	56.687	56.731	56.775	56.819	56.863	56.907	56.951	56.995	1370
1380	56.995	57.039	57.083	57.127	57.171	57.215	57.259	57.303	57.346	57.390	57.434	1380
1390	57.434	57.478	57.522	57.566	57.610	57.654	57.698	57.742	57.786	57.830	57.873	1390
1400	57.873	57.917	57.961	58.005	58.049	58.093	58.137	58.181	58.224	58.268	58.312	1400
1410	58.312	58.356	58.400	58.444	58.487	58.531	58.575	58.619	58.663	58.707	58.750	1410
1420	58.750	58.794	58.838	58.882	58.926	58.969	59.013	59.057	59.101	59.144	59.188	1420
1430	59.188	59.232	59.276	59.319	59.363	59.407	59.451	59.494	59.538	59.582	59.626	1430
1440	59.626	59.669	59.713	59.757	59.800	59.844	59.888	59.932	59.975	60.019	60.063	1440
1450	60.063	60.106	60.150	60.194	60.237	60.281	60.325	60.368	60.412	60.455	60.499	1450
1460	60.499	60.543	60.586	60.630	60.674	60.717	60.761	60.804	60.848	60.892	60.935	1460
1470	60.935	60.979	61.022	61.066	61.109	61.153	61.197	61.240	61.284	61.327	61.371	1470
1480	61.371	61.414	61.458	61.501	61.545	61.588	61.632	61.675	61.719	61.762	61.806	1480
1490	61.806	61.849	61.893	61.936	61.980	62.023	62.067	62.110	62.154	62.197	62.240	1490
1500	62.240	62.284	62.327	62.371	62.414	62.458	62.501	62.544	62.588	62.631	62.675	1500
1510	62.675	62.718	62.761	62.805	62.848	62.892	62.935	62.978	63.022	63.065	63.108	1510
1520	63.108	63.152	63.195	63.238	63.282	63.325	63.368	63.412	63.455	63.498	63.542	1520
1530	63.542	63.585	63.628	63.671	63.715	63.758	63.801	63.844	63.888	63.931	63.974	1530
1540	63.974	64.017	64.061	64.104	64.147	64.190	64.234	64.277	64.320	64.363	64.406	1540
1550	64.406	64.450	64.493	64.536	64.579	64.622	64.665	64.709	64.752	64.795	64.838	1550
1560	64.838	64.881	64.924	64.967	65.011	65.054	65.097	65.140	65.183	65.226	65.269	1560
1570	65.269	65.312	65.355	65.398	65.441	65.484	65.528	65.571	65.614	65.657	65.700	1570
1580	65.700	65.743	65.786	65.829	65.872	65.915	65.958	66.001	66.044	66.087	66.130	1580
1590	66.130	66.173	66.216	66.259	66.302	66.345	66.387	66.430	66.473	66.516	66.559	1590
1600	66.559	66.602	66.645	66.688	66.731	66.774	66.817	66.859	66.902	66.945	66.988	1600
1610	66.988	67.031	67.074	67.117	67.159	67.202	67.245	67.288	67.331	67.374	67.416	1610
1620	67.416	67.459	67.502	67.545	67.588	67.630	67.673	67.716	67.759	67.801	67.844	1620
1630	67.844	67.887	67.930	67.972	68.015	68.058	68.101	68.143	68.186	68.229	68.271	1630
1640	68.271	68.314	68.357	68.399	68.442	68.485	68.527	68.570	68.613	68.655	68.698	1640
1650	68.698	68.740	68.783	68.826	68.868	68.911	68.953	68.996	69.039	69.081	69.124	1650
1660	69.124	69.166	69.209	69.251	69.294	69.337	69.379	69.422	69.464	69.507	69.549	1660
1670	69.549	69.592	69.634	69.677	69.719	69.762	69.804	69.847	69.889	69.931	69.974	1670
1680	69.974	70.016	70.059	70.101	70.144	70.186	70.228	70.271	70.313	70.356	70.398	1680
1690	70.398	70.440	70.483	70.525	70.567	70.610	70.652	70.694	70.737	70.779	70.821	1690
1700	70.821	70.864	70.906	70.948	70.991	71.033	71.075	71.118	71.160	71.202	71.244	1700
1710	71.244	71.287	71.329	71.371	71.413	71.456	71.498	71.540	71.582	71.624	71.667	1710
1720	71.667	71.709	71.751	71.793	71.835	71.878	71.920	71.962	72.004	72.046	72.088	1720
1730	72.088	72.130	72.173	72.215	72.257	72.299	72.341	72.383	72.425	72.467	72.509	1730
1740	72.509	72.551	72.593	72.635	72.678	72.720	72.762	72.804	72.846	72.888	72.930	1740
1750	72.930	72.972	73.014	73.056	73.098	73.140	73.182	73.224	73.266	73.308	73.350	1750
1760	73.350	73.392	73.434	73.475	73.517	73.559	73.601	73.643	73.685	73.727	73.769	1760
1770	73.769	73.811	73.853	73.895	73.936	73.978	74.020	74.062	74.104	74.146	74.188	1770
1780	74.188	74.229	74.271	74.313	74.355	74.397	74.439	74.480	74.522	74.564	74.606	1780
1790	74.606	74.648	74.689	74.731	74.773	74.815	74.857	74.898	74.940	74.982	75.024	1790
1800	75.024	75.065	75.107	75.149	75.191	75.232	75.274	75.316	75.357	75.399	75.441	1800
1810	75.441	75.483	75.524	75.566	75.608	75.649	75.691	75.733	75.774	75.816	75.858	1810
1820	75.858	75.899	75.941	75.983	76.024	76.066	76.108	76.149	76.191	76.233	76.274	1820
1830	76.274	76.316	76.358									1830

Appendix B

Type J Thermocouple
Thermoelectric Voltage as a Function of Temperature (°C)
Reference Junctions at 0°C

THERMOELECTRIC VOLTAGE IN ABSOLUTE MILLIVOLTS

°C	0	1	2	3	4	5	6	7	8	9	10	°C
−210	−8.096	−8.076	−8.057	−8.037	−8.017	−7.996	−7.976	−7.955	−7.934	−7.912	−7.890	−210
−200	−7.890	−7.868	−7.846	−7.824	−7.801	−7.778	−7.755	−7.731	−7.707	−7.683	−7.659	−200
−190	−7.659	−7.634	−7.609	−7.584	−7.559	−7.533	−7.508	−7.482	−7.455	−7.429	−7.402	−190
−180	−7.402	−7.375	−7.348	−7.321	−7.293	−7.265	−7.237	−7.209	−7.180	−7.151	−7.122	−180
−170	−7.122	−7.093	−7.064	−7.034	−7.004	−6.974	−6.944	−6.914	−6.883	−6.852	−6.821	−170
−160	−6.821	−6.790	−6.758	−6.727	−6.695	−6.663	−6.630	−6.598	−6.565	−6.532	−6.499	−160
−150	−6.499	−6.466	−6.433	−6.399	−6.365	−6.331	−6.297	−6.263	−6.228	−6.194	−6.159	−150
−140	−6.159	−6.124	−6.089	−6.053	−6.018	−5.982	−5.946	−5.910	−5.874	−5.837	−5.801	−140
−130	−5.801	−5.764	−5.727	−5.690	−5.653	−5.615	−5.578	−5.540	−5.502	−5.464	−5.426	−130
−120	−5.426	−5.388	−5.349	−5.311	−5.272	−5.233	−5.194	−5.155	−5.115	−5.076	−5.036	−120
−110	−5.036	−4.996	−4.956	−4.916	−4.876	−4.836	−4.795	−4.755	−4.714	−4.673	−4.632	−110
−100	−4.632	−4.591	−4.550	−4.508	−4.467	−4.425	−4.383	−4.341	−4.299	−4.257	−4.215	−100
−90	−4.215	−4.172	−4.130	−4.087	−4.044	−4.001	−3.958	−3.915	−3.872	−3.829	−3.785	−90
−80	−3.785	−3.742	−3.698	−3.654	−3.610	−3.566	−3.522	−3.478	−3.433	−3.389	−3.344	−80
−70	−3.344	−3.299	−3.255	−3.210	−3.165	−3.120	−3.074	−3.029	−2.984	−2.938	−2.892	−70
−60	−2.892	−2.847	−2.801	−2.755	−2.709	−2.663	−2.617	−2.570	−2.524	−2.478	−2.431	−60
−50	−2.431	−2.384	−2.338	−2.291	−2.244	−2.197	−2.150	−2.102	−2.055	−2.008	−1.960	−50
−40	−1.960	−1.913	−1.865	−1.818	−1.770	−1.722	−1.674	−1.626	−1.578	−1.530	−1.481	−40
−30	−1.481	−1.433	−1.385	−1.336	−1.288	−1.239	−1.190	−1.141	−1.093	−1.044	−0.995	−30
−20	−0.995	−0.945	−0.896	−0.847	−0.798	−0.748	−0.699	−0.650	−0.600	−0.550	−0.501	−20
−10	−0.501	−0.451	−0.401	−0.351	−0.301	−0.251	−0.201	−0.151	−0.101	−0.050	0.000	−10
0	0.000	0.050	0.101	0.151	0.202	0.253	0.303	0.354	0.405	0.456	0.507	0
10	0.507	0.558	0.609	0.660	0.711	0.762	0.813	0.865	0.916	0.967	1.019	10
20	1.019	1.070	1.122	1.174	1.225	1.277	1.329	1.381	1.432	1.484	1.536	20
30	1.536	1.588	1.640	1.693	1.745	1.797	1.849	1.901	1.954	2.006	2.058	30
40	2.058	2.111	2.163	2.216	2.268	2.321	2.374	2.426	2.479	2.532	2.585	40
50	2.585	2.638	2.691	2.743	2.796	2.849	2.902	2.956	3.009	3.062	3.115	50
60	3.115	3.168	3.221	3.275	3.328	3.381	3.435	3.488	3.542	3.595	3.649	60
70	3.649	3.702	3.756	3.809	3.863	3.917	3.971	4.024	4.078	4.132	4.186	70
80	4.186	4.239	4.293	4.347	4.401	4.455	4.509	4.563	4.617	4.671	4.725	80
90	4.725	4.780	4.834	4.888	4.942	4.996	5.050	5.105	5.159	5.213	5.268	90
100	5.268	5.322	5.376	5.431	5.485	5.540	5.594	5.649	5.703	5.758	5.812	100
110	5.812	5.867	5.921	5.976	6.031	6.085	6.140	6.195	6.249	6.304	6.359	110
120	6.359	6.414	6.468	6.523	6.578	6.633	6.688	6.742	6.797	6.852	6.907	120
130	6.907	6.962	7.017	7.072	7.127	7.182	7.237	7.292	7.347	7.402	7.457	130
140	7.457	7.512	7.567	7.622	7.677	7.732	7.787	7.843	7.898	7.953	8.008	140
150	8.008	8.063	8.118	8.174	8.229	8.284	8.339	8.394	8.450	8.505	8.560	150
160	8.560	8.616	8.671	8.726	8.781	8.837	8.892	8.947	9.003	9.058	9.113	160
170	9.113	9.169	9.224	9.279	9.335	9.390	9.446	9.501	9.556	9.612	9.667	170
180	9.667	9.723	9.778	9.834	9.889	9.944	10.000	10.055	10.111	10.166	10.222	180
190	10.222	10.277	10.333	10.388	10.444	10.499	10.555	10.610	10.666	10.721	10.777	190
200	10.777	10.832	10.888	10.943	10.999	11.054	11.110	11.165	11.221	11.276	11.332	200
210	11.332	11.387	11.443	11.498	11.554	11.609	11.665	11.720	11.776	11.831	11.887	210
220	11.887	11.943	11.998	12.054	12.109	12.165	12.220	12.276	12.331	12.387	12.442	220
230	12.442	12.498	12.553	12.609	12.664	12.720	12.776	12.831	12.887	12.942	12.998	230
240	12.998	13.053	13.109	13.164	13.220	13.275	13.331	13.386	13.442	13.497	13.553	240
250	13.553	13.608	13.664	13.719	13.775	13.830	13.886	13.941	13.997	14.052	14.108	250
260	14.108	14.163	14.219	14.274	14.330	14.385	14.441	14.496	14.552	14.607	14.663	260
270	14.663	14.718	14.774	14.829	14.885	14.940	14.995	15.051	15.106	15.162	15.217	270
280	15.217	15.273	15.328	15.383	15.439	15.494	15.550	15.605	15.661	15.716	15.771	280
290	15.771	15.827	15.882	15.938	15.993	16.048	16.104	16.159	16.214	16.270	16.325	290
300	16.325	16.380	16.436	16.491	16.547	16.602	16.657	16.713	16.768	16.823	16.879	300
310	16.879	16.934	16.989	17.044	17.100	17.155	17.210	17.266	17.321	17.376	17.432	310
320	17.432	17.487	17.542	17.597	17.653	17.708	17.763	17.818	17.874	17.929	17.984	320
330	17.984	18.039	18.095	18.150	18.205	18.260	18.316	18.371	18.426	18.481	18.537	330
340	18.537	18.592	18.647	18.702	18.757	18.813	18.868	18.923	18.978	19.033	19.089	340
350	19.089	19.144	19.199	19.254	19.309	19.364	19.420	19.475	19.530	19.585	19.640	350
360	19.640	19.695	19.751	19.806	19.861	19.916	19.971	20.026	20.081	20.137	20.192	360
370	20.192	20.247	20.302	20.357	20.412	20.467	20.523	20.578	20.633	20.688	20.743	370
380	20.743	20.798	20.853	20.909	20.964	21.019	21.074	21.129	21.184	21.239	21.295	380

Type J Thermocouple
Thermoelectric Voltage as a Function of Temperature (°C)
Reference Junctions at 0°C
THERMOELECTRIC VOLTAGE IN ABSOLUTE MILLIVOLTS

°C	0	1	2	3	4	5	6	7	8	9	10	°C
390	21.295	21.350	21.405	21.460	21.515	21.570	21.625	21.680	21.736	21.791	21.846	390
400	21.846	21.901	21.956	22.011	22.066	22.122	22.177	22.232	22.287	22.342	22.397	400
410	22.397	22.453	22.508	22.563	22.618	22.673	22.728	22.784	22.839	22.894	22.949	410
420	22.949	23.004	23.060	23.115	23.170	23.225	23.280	23.336	23.391	23.446	23.501	420
430	23.501	23.556	23.612	23.667	23.722	23.777	23.833	23.888	23.943	23.999	24.054	430
440	24.054	24.109	24.164	24.220	24.275	24.330	24.386	24.441	24.496	24.552	24.607	440
450	24.607	24.662	24.718	24.773	24.829	24.884	24.939	24.995	25.050	25.106	25.161	450
460	25.161	25.217	25.272	25.327	25.383	25.438	25.494	25.549	25.605	25.661	25.716	460
470	25.716	25.772	25.827	25.883	25.938	25.994	26.050	26.105	26.161	26.216	26.272	470
480	26.272	26.328	26.383	26.439	26.495	26.551	26.606	26.662	26.718	26.774	26.829	480
490	26.829	26.885	26.941	26.997	27.053	27.109	27.165	27.220	27.276	27.332	27.388	490
500	27.388	27.444	27.500	27.556	27.612	27.668	27.724	27.780	27.836	27.893	27.949	500
510	27.949	28.005	28.061	28.117	28.173	28.230	28.286	28.342	28.398	28.455	28.511	510
520	28.511	28.567	28.624	28.680	28.736	28.793	28.849	28.906	28.962	29.019	29.075	520
530	29.075	29.132	29.188	29.245	29.301	29.358	29.415	29.471	29.528	29.585	29.642	530
540	29.642	29.698	29.755	29.812	29.869	29.926	29.983	30.039	30.096	30.153	30.210	540
550	30.210	30.267	30.324	30.381	30.439	30.496	30.553	30.610	30.667	30.724	30.782	550
560	30.782	30.839	30.896	30.954	31.011	31.068	31.126	31.183	31.241	31.298	31.356	560
570	31.356	31.413	31.471	31.528	31.586	31.644	31.702	31.759	31.817	31.875	31.933	570
580	31.933	31.991	32.048	32.106	32.164	32.222	32.280	32.338	32.396	32.455	32.513	580
590	32.513	32.571	32.629	32.687	32.746	32.804	32.862	32.921	32.979	33.038	33.096	590
600	33.096	33.155	33.213	33.272	33.330	33.389	33.448	33.506	33.565	33.624	33.683	600
610	33.683	33.742	33.800	33.859	33.918	33.977	34.036	34.095	34.155	34.214	34.273	610
620	34.273	34.332	34.391	34.451	34.510	34.569	34.629	34.688	34.748	34.807	34.867	620
630	34.867	34.926	34.986	35.046	35.105	35.165	35.225	35.285	35.344	35.404	35.464	630
640	35.464	35.524	35.584	35.644	35.704	35.764	35.825	35.885	35.945	36.005	36.066	640
650	36.066	36.126	36.186	36.247	36.307	36.368	36.428	36.489	36.549	36.610	36.671	650
660	36.671	36.732	36.792	36.853	36.914	36.975	37.036	37.097	37.158	37.219	37.280	660
670	37.280	37.341	37.402	37.463	37.525	37.586	37.647	37.709	37.770	37.831	37.893	670
680	37.893	37.954	38.016	38.078	38.139	38.201	38.262	38.324	38.386	38.448	38.510	680
690	38.510	38.572	38.633	38.695	38.757	38.819	38.882	38.944	39.006	39.068	39.130	690
700	39.130	39.192	39.255	39.317	39.379	39.442	39.504	39.567	39.629	39.692	39.754	700
710	39.754	39.817	39.880	39.942	40.005	40.068	40.131	40.193	40.256	40.319	40.382	710
720	40.382	40.445	40.508	40.571	40.634	40.697	40.760	40.823	40.886	40.950	41.013	720
730	41.013	41.076	41.139	41.203	41.266	41.329	41.393	41.456	41.520	41.583	41.647	730
740	41.647	41.710	41.774	41.837	41.901	41.965	42.028	42.092	42.156	42.219	42.283	740
750	42.283	42.347	42.411	42.475	42.538	42.602	42.666	42.730	42.794	42.858	42.922	750

Type J Thermocouple
Thermoelectric Voltage as a Function of Temperature (°F)
Reference Junctions at 32°F
THERMOELECTRIC VOLTAGE IN ABSOLUTE MILLIVOLTS

°F	0	1	2	3	4	5	6	7	8	9	10	°F
−350					−8.096	−8.085	−8.074	−8.063	−8.052	−8.041	−8.030	−350
−340	−8.030	−8.019	−8.008	−7.996	−7.985	−7.973	−7.962	−7.950	−7.938	−7.927	−7.915	−340
−330	−7.915	−7.903	−7.890	−7.878	−7.866	−7.854	−7.841	−7.829	−7.816	−7.803	−7.791	−330
−320	−7.791	−7.778	−7.765	−7.752	−7.739	−7.726	−7.712	−7.699	−7.686	−7.672	−7.659	−320
−310	−7.659	−7.645	−7.631	−7.618	−7.604	−7.590	−7.576	−7.562	−7.548	−7.533	−7.519	−310
−300	−7.519	−7.505	−7.490	−7.476	−7.461	−7.447	−7.432	−7.417	−7.402	−7.387	−7.372	−300
−290	−7.372	−7.357	−7.342	−7.327	−7.311	−7.296	−7.281	−7.265	−7.250	−7.234	−7.218	−290
−280	−7.218	−7.202	−7.187	−7.171	−7.155	−7.139	−7.122	−7.106	−7.090	−7.074	−7.057	−280
−270	−7.057	−7.041	−7.024	−7.008	−6.991	−6.974	−6.958	−6.941	−6.924	−6.907	−6.890	−270
−260	−6.890	−6.873	−6.856	−6.838	−6.821	−6.804	−6.786	−6.769	−6.751	−6.734	−6.716	−260
−250	−6.716	−6.698	−6.680	−6.663	−6.645	−6.627	−6.609	−6.591	−6.572	−6.554	−6.536	−250
−240	−6.536	−6.518	−6.499	−6.481	−6.462	−6.444	−6.425	−6.407	−6.388	−6.369	−6.350	−240
−230	−6.350	−6.331	−6.312	−6.293	−6.274	−6.255	−6.236	−6.217	−6.198	−6.178	−6.159	−230
−220	−6.159	−6.139	−6.120	−6.100	−6.081	−6.061	−6.041	−6.022	−6.002	−5.982	−5.962	−220
−210	−5.962	−5.942	−5.922	−5.902	−5.882	−5.861	−5.841	−5.821	−5.801	−5.780	−5.760	−210
−200	−5.760	−5.739	−5.719	−5.698	−5.678	−5.657	−5.636	−5.615	−5.594	−5.574	−5.553	−200
−190	−5.553	−5.532	−5.511	−5.490	−5.468	−5.447	−5.426	−5.405	−5.383	−5.362	−5.341	−190
−180	−5.341	−5.319	−5.298	−5.276	−5.255	−5.233	−5.211	−5.190	−5.168	−5.146	−5.124	−180
−170	−5.124	−5.102	−5.080	−5.058	−5.036	−5.014	−4.992	−4.970	−4.948	−4.925	−4.903	−170
−160	−4.903	−4.881	−4.858	−4.836	−4.813	−4.791	−4.768	−4.746	−4.723	−4.700	−4.678	−160
−150	−4.678	−4.655	−4.632	−4.609	−4.586	−4.563	−4.540	−4.517	−4.494	−4.471	−4.448	−150
−140	−4.448	−4.425	−4.402	−4.379	−4.355	−4.332	−4.309	−4.285	−4.262	−4.238	−4.215	−140
−130	−4.215	−4.191	−4.168	−4.144	−4.120	−4.097	−4.073	−4.049	−4.025	−4.001	−3.978	−130
−120	−3.978	−3.954	−3.930	−3.906	−3.882	−3.858	−3.833	−3.809	−3.785	−3.761	−3.737	−120
−110	−3.737	−3.712	−3.688	−3.664	−3.639	−3.615	−3.590	−3.566	−3.541	−3.517	−3.492	−110
−100	−3.492	−3.468	−3.443	−3.418	−3.394	−3.369	−3.344	−3.319	−3.294	−3.270	−3.245	−100
−90	−3.245	−3.220	−3.195	−3.170	−3.145	−3.120	−3.094	−3.069	−3.044	−3.019	−2.994	−90
−80	−2.994	−2.968	−2.943	−2.918	−2.892	−2.867	−2.842	−2.816	−2.791	−2.765	−2.740	−80
−70	−2.740	−2.714	−2.689	−2.663	−2.637	−2.612	−2.586	−2.560	−2.534	−2.509	−2.483	−70
−60	−2.483	−2.457	−2.431	−2.405	−2.379	−2.353	−2.327	−2.301	−2.275	−2.249	−2.223	−60
−50	−2.223	−2.197	−2.171	−2.144	−2.118	−2.092	−2.066	−2.039	−2.013	−1.987	−1.960	−50
−40	−1.960	−1.934	−1.908	−1.881	−1.855	−1.828	−1.802	−1.775	−1.748	−1.722	−1.695	−40
−30	−1.695	−1.669	−1.642	−1.615	−1.589	−1.562	−1.535	−1.508	−1.481	−1.455	−1.428	−30
−20	−1.428	−1.401	−1.374	−1.347	−1.320	−1.293	−1.266	−1.239	−1.212	−1.185	−1.158	−20
−10	−1.158	−1.131	−1.103	−1.076	−1.049	−1.022	−0.995	−0.967	−0.940	−0.913	−0.885	−10
0	−0.885	−0.858	−0.831	−0.803	−0.776	−0.748	−0.721	−0.694	−0.666	−0.639	−0.611	0
10	−0.611	−0.583	−0.556	−0.528	−0.501	−0.473	−0.445	−0.418	−0.390	−0.362	−0.334	10
20	−0.334	−0.307	−0.279	−0.251	−0.223	−0.195	−0.168	−0.140	−0.112	−0.084	−0.056	20
30	−0.056	−0.028	0.000	0.028	0.056	0.084	0.112	0.140	0.168	0.196	0.224	30
40	0.224	0.253	0.281	0.309	0.337	0.365	0.394	0.422	0.450	0.478	0.507	40
50	0.507	0.535	0.563	0.592	0.620	0.648	0.677	0.705	0.734	0.762	0.791	50
60	0.791	0.819	0.848	0.876	0.905	0.933	0.962	0.990	1.019	1.048	1.076	60
70	1.076	1.105	1.134	1.162	1.191	1.220	1.248	1.277	1.306	1.335	1.363	70
80	1.363	1.392	1.421	1.450	1.479	1.507	1.536	1.565	1.594	1.623	1.652	80
90	1.652	1.681	1.710	1.739	1.768	1.797	1.826	1.855	1.884	1.913	1.942	90
100	1.942	1.971	2.000	2.029	2.058	2.088	2.117	2.146	2.175	2.204	2.233	100
110	2.233	2.263	2.292	2.321	2.350	2.380	2.409	2.438	2.467	2.497	2.526	110
120	2.526	2.555	2.585	2.614	2.644	2.673	2.702	2.732	2.761	2.791	2.820	120
130	2.820	2.849	2.879	2.908	2.938	2.967	2.997	3.026	3.056	3.085	3.115	130
140	3.115	3.145	3.174	3.204	3.233	3.263	3.293	3.322	3.352	3.381	3.411	140
150	3.411	3.441	3.470	3.500	3.530	3.560	3.589	3.619	3.649	3.678	3.708	150
160	3.708	3.738	3.768	3.798	3.827	3.857	3.887	3.917	3.947	3.976	4.006	160
170	4.006	4.036	4.066	4.096	4.126	4.156	4.186	4.216	4.245	4.275	4.305	170
180	4.305	4.335	4.365	4.395	4.425	4.455	4.485	4.515	4.545	4.575	4.605	180
190	4.605	4.635	4.665	4.695	4.725	4.755	4.786	4.816	4.846	4.876	4.906	190
200	4.906	4.936	4.966	4.996	5.026	5.057	5.087	5.117	5.147	5.177	5.207	200
210	5.207	5.238	5.268	5.298	5.328	5.358	5.389	5.419	5.449	5.479	5.509	210
220	5.509	5.540	5.570	5.600	5.630	5.661	5.691	5.721	5.752	5.782	5.812	220
230	5.812	5.843	5.873	5.903	5.934	5.964	5.994	6.025	6.055	6.085	6.116	230
240	6.116	6.146	6.176	6.207	6.237	6.268	6.298	6.328	6.359	6.389	6.420	240

Type J Thermocouple
Thermoelectric Voltage as a Function of Temperature (°F)
Reference Junctions at 32°F
THERMOELECTRIC VOLTAGE IN ABSOLUTE MILLIVOLTS

°F	0	1	2	3	4	5	6	7	8	9	10	°F
250	6.420	6.450	6.481	6.511	6.541	6.572	6.602	6.633	6.663	6.694	6.724	250
260	6.724	6.755	6.785	6.816	6.846	6.877	6.907	6.938	6.968	6.999	7.029	260
270	7.029	7.060	7.090	7.121	7.151	7.182	7.212	7.243	7.274	7.304	7.335	270
280	7.335	7.365	7.396	7.426	7.457	7.488	7.518	7.549	7.579	7.610	7.641	280
290	7.641	7.671	7.702	7.732	7.763	7.794	7.824	7.855	7.885	7.916	7.947	290
300	7.947	7.977	8.008	8.039	8.069	8.100	8.131	8.161	8.192	8.223	8.253	300
310	8.253	8.284	8.315	8.345	8.376	8.407	8.437	8.468	8.499	8.530	8.560	310
320	8.560	8.591	8.622	8.652	8.683	8.714	8.745	8.775	8.806	8.837	8.867	320
330	8.867	8.898	8.929	8.960	8.990	9.021	9.052	9.083	9.113	9.144	9.175	330
340	9.175	9.206	9.236	9.267	9.298	9.329	9.359	9.390	9.421	9.452	9.483	340
350	9.483	9.513	9.544	9.575	9.606	9.636	9.667	9.698	9.729	9.760	9.790	350
360	9.790	9.821	9.852	9.883	9.914	9.944	9.975	10.006	10.037	10.068	10.098	360
370	10.098	10.129	10.160	10.191	10.222	10.252	10.283	10.314	10.345	10.376	10.407	370
380	10.407	10.437	10.468	10.499	10.530	10.561	10.592	10.622	10.653	10.684	10.715	380
390	10.715	10.746	10.777	10.807	10.838	10.869	10.900	10.931	10.962	10.992	11.023	390
400	11.023	11.054	11.085	11.116	11.147	11.177	11.208	11.239	11.270	11.301	11.332	400
410	11.332	11.363	11.393	11.424	11.455	11.486	11.517	11.548	11.578	11.609	11.640	410
420	11.640	11.671	11.702	11.733	11.764	11.794	11.825	11.856	11.887	11.918	11.949	420
430	11.949	11.980	12.010	12.041	12.072	12.103	12.134	12.165	12.196	12.226	12.257	430
440	12.257	12.288	12.319	12.350	12.381	12.411	12.442	12.473	12.504	12.535	12.566	440
450	12.566	12.597	12.627	12.658	12.689	12.720	12.751	12.782	12.813	12.843	12.874	450
460	12.874	12.905	12.936	12.967	12.998	13.029	13.059	13.090	13.121	13.152	13.183	460
470	13.183	13.214	13.244	13.275	13.306	13.337	13.368	13.399	13.430	13.460	13.491	470
480	13.491	13.522	13.553	13.584	13.615	13.645	13.676	13.707	13.738	13.769	13.800	480
490	13.800	13.830	13.861	13.892	13.923	13.954	13.985	14.015	14.046	14.077	14.108	490
500	14.108	14.139	14.170	14.200	14.231	14.262	14.293	14.324	14.355	14.385	14.416	500
510	14.416	14.447	14.478	14.509	14.539	14.570	14.601	14.632	14.663	14.694	14.724	510
520	14.724	14.755	14.786	14.817	14.848	14.878	14.909	14.940	14.971	15.002	15.032	520
530	15.032	15.063	15.094	15.125	15.156	15.186	15.217	15.248	15.279	15.310	15.340	530
540	15.340	15.371	15.402	15.433	15.464	15.494	15.525	15.556	15.587	15.617	15.648	540
550	15.648	15.679	15.710	15.741	15.771	15.802	15.833	15.864	15.894	15.925	15.956	550
560	15.956	15.987	16.018	16.048	16.079	16.110	16.141	16.171	16.202	16.233	16.264	560
570	16.264	16.294	16.325	16.356	16.387	16.417	16.448	16.479	16.510	16.540	16.571	570
580	16.571	16.602	16.633	16.663	16.694	16.725	16.756	16.786	16.817	16.848	16.879	580
590	16.879	16.909	16.940	16.971	17.001	17.032	17.063	17.094	17.124	17.155	17.186	590
600	17.186	17.217	17.247	17.278	17.309	17.339	17.370	17.401	17.432	17.462	17.493	600
610	17.493	17.524	17.554	17.585	17.616	17.646	17.677	17.708	17.739	17.769	17.800	610
620	17.800	17.831	17.861	17.892	17.923	17.953	17.984	18.015	18.046	18.076	18.107	620
630	18.107	18.138	18.168	18.199	18.230	18.260	18.291	18.322	18.352	18.383	18.414	630
640	18.414	18.444	18.475	18.506	18.537	18.567	18.598	18.629	18.659	18.690	18.721	640
650	18.721	18.751	18.782	18.813	18.843	18.874	18.905	18.935	18.966	18.997	19.027	650
660	19.027	19.058	19.089	19.119	19.150	19.180	19.211	19.242	19.272	19.303	19.334	660
670	19.334	19.364	19.395	19.426	19.456	19.487	19.518	19.548	19.579	19.610	19.640	670
680	19.640	19.671	19.702	19.732	19.763	19.793	19.824	19.855	19.885	19.916	19.947	680
690	19.947	19.977	20.008	20.039	20.069	20.100	20.131	20.161	20.192	20.222	20.253	690
700	20.253	20.284	20.314	20.345	20.376	20.406	20.437	20.467	20.498	20.529	20.559	700
710	20.559	20.590	20.621	20.651	20.682	20.713	20.743	20.774	20.804	20.835	20.866	710
720	20.866	20.896	20.927	20.958	20.988	21.019	21.049	21.080	21.111	21.141	21.172	720
730	21.172	21.203	21.233	21.264	21.295	21.325	21.356	21.386	21.417	21.448	21.478	730
740	21.478	21.509	21.540	21.570	21.601	21.631	21.662	21.693	21.723	21.754	21.785	740
750	21.785	21.815	21.846	21.877	21.907	21.938	21.968	21.999	22.030	22.060	22.091	750
760	22.091	22.122	22.152	22.183	22.214	22.244	22.275	22.305	22.336	22.367	22.397	760
770	22.397	22.428	22.459	22.489	22.520	22.551	22.581	22.612	22.643	22.673	22.704	770
780	22.704	22.735	22.765	22.796	22.826	22.857	22.888	22.918	22.949	22.980	23.010	780
790	23.010	23.041	23.072	23.102	23.133	23.164	23.194	23.225	23.256	23.286	23.317	790
800	23.317	23.348	23.378	23.409	23.440	23.471	23.501	23.532	23.563	23.593	23.624	800
810	23.624	23.655	23.685	23.716	23.747	23.777	23.808	23.839	23.870	23.900	23.931	810
820	23.931	23.962	23.992	24.023	24.054	24.085	24.115	24.146	24.177	24.207	24.238	820
830	24.238	24.269	24.300	24.330	24.361	24.392	24.423	24.453	24.484	24.515	24.546	830
840	24.546	24.576	24.607	24.638	24.669	24.699	24.730	24.761	24.792	24.822	24.853	840

Type J Thermocouple
Thermoelectric Voltage as a Function of Temperature (°F)
Reference Junctions at 32°F

THERMOELECTRIC VOLTAGE IN ABSOLUTE MILLIVOLTS

°F	0	1	2	3	4	5	6	7	8	9	10	°F
850	24.853	24.884	24.915	24.946	24.976	25.007	25.038	25.069	25.099	25.130	25.161	850
860	25.161	25.192	25.223	25.254	25.284	25.315	25.346	25.377	25.408	25.438	25.469	860
870	25.469	25.500	25.531	25.562	25.593	25.623	25.654	25.685	25.716	25.747	25.778	870
880	25.778	25.809	25.840	25.870	25.901	25.932	25.963	25.994	26.025	26.056	26.087	880
890	26.087	26.118	26.148	26.179	26.210	26.241	26.272	26.303	26.334	26.365	26.396	890
900	26.396	26.427	26.458	26.489	26.520	26.551	26.582	26.613	26.644	26.675	26.705	900
910	26.705	26.736	26.767	26.798	26.829	26.860	26.891	26.922	26.954	26.985	27.016	910
920	27.016	27.047	27.078	27.109	27.140	27.171	27.202	27.233	27.264	27.295	27.326	920
930	27.326	27.357	27.388	27.419	27.450	27.482	27.513	27.544	27.575	27.606	27.637	930
940	27.637	27.668	27.699	27.731	27.762	27.793	27.824	27.855	27.886	27.917	27.949	940
950	27.949	27.980	28.011	28.042	28.073	28.105	28.136	28.167	28.198	28.230	28.261	950
960	28.261	28.292	28.323	28.355	28.386	28.417	28.448	28.480	28.511	28.542	28.573	960
970	28.573	28.605	28.636	28.667	28.699	28.730	28.761	28.793	28.824	28.855	28.887	970
980	28.887	28.918	28.950	28.981	29.012	29.044	29.075	29.107	29.138	29.169	29.201	980
990	29.201	29.232	29.264	29.295	29.327	29.358	29.390	29.421	29.452	29.484	29.515	990
1000	29.515	29.547	29.578	29.610	29.642	29.673	29.705	29.736	29.768	29.799	29.831	1000
1010	29.831	29.862	29.894	29.926	29.957	29.989	30.020	30.052	30.084	30.115	30.147	1010
1020	30.147	30.179	30.210	30.242	30.274	30.305	30.337	30.369	30.400	30.432	30.464	1020
1030	30.464	30.496	30.527	30.559	30.591	30.623	30.654	30.686	30.718	30.750	30.782	1030
1040	30.782	30.813	30.845	30.877	30.909	30.941	30.973	31.005	31.036	31.068	31.100	1040
1050	31.100	31.132	31.164	31.196	31.228	31.260	31.292	31.324	31.356	31.388	31.420	1050
1060	31.420	31.452	31.484	31.516	31.548	31.580	31.612	31.644	31.676	31.708	31.740	1060
1070	31.740	31.772	31.804	31.836	31.868	31.901	31.933	31.965	31.997	32.029	32.061	1070
1080	32.061	32.094	32.126	32.158	32.190	32.222	32.255	32.287	32.319	32.351	32.384	1080
1090	32.384	32.416	32.448	32.480	32.513	32.545	32.577	32.610	32.642	32.674	32.707	1090
1100	32.707	32.739	32.772	32.804	32.836	32.869	32.901	32.934	32.966	32.999	33.031	1100
1110	33.031	33.064	33.096	33.129	33.161	33.194	33.226	33.259	33.291	33.324	33.356	1110
1120	33.356	33.389	33.422	33.454	33.487	33.519	33.552	33.585	33.617	33.650	33.683	1120
1130	33.683	33.715	33.748	33.781	33.814	33.846	33.879	33.912	33.945	33.977	34.010	1130
1140	34.010	34.043	34.076	34.109	34.141	34.174	34.207	34.240	34.273	34.306	34.339	1140
1150	34.339	34.372	34.405	34.437	34.470	34.503	34.536	34.569	34.602	34.635	34.668	1150
1160	34.668	34.701	34.734	34.767	34.801	34.834	34.867	34.900	34.933	34.966	34.999	1160
1170	34.999	35.032	35.065	35.099	35.132	35.165	35.198	35.231	35.265	35.298	35.331	1170
1180	35.331	35.364	35.398	35.431	35.464	35.498	35.531	35.564	35.598	35.631	35.664	1180
1190	35.664	35.698	35.731	35.764	35.798	35.831	35.865	35.898	35.932	35.965	35.999	1190
1200	35.999	36.032	36.066	36.099	36.133	36.166	36.200	36.233	36.267	36.301	36.334	1200
1210	36.334	36.368	36.401	36.435	36.469	36.502	36.536	36.570	36.603	36.637	36.671	1210
1220	36.671	36.705	36.738	36.772	36.806	36.840	36.873	36.907	36.941	36.975	37.009	1220
1230	37.009	37.043	37.076	37.110	37.144	37.178	37.212	37.246	37.280	37.314	37.348	1230
1240	37.348	37.382	37.416	37.450	37.484	37.518	37.552	37.586	37.620	37.654	37.688	1240
1250	37.688	37.722	37.756	37.790	37.825	37.859	37.893	37.927	37.961	37.995	38.030	1250
1260	38.030	38.064	38.098	38.132	38.167	38.201	38.235	38.269	38.304	38.338	38.372	1260
1270	38.372	38.407	38.441	38.475	38.510	38.544	38.578	38.613	38.647	38.682	38.716	1270
1280	38.716	38.751	38.785	38.819	38.854	38.888	38.923	38.957	38.992	39.027	39.061	1280
1290	39.061	39.096	39.130	39.165	39.199	39.234	39.269	39.303	39.338	39.373	39.407	1290
1300	39.407	39.442	39.477	39.511	39.546	39.581	39.615	39.650	39.685	39.720	39.754	1300
1310	39.754	39.789	39.824	39.859	39.894	39.928	39.963	39.998	40.033	40.068	40.103	1310
1320	40.103	40.138	40.172	40.207	40.242	40.277	40.312	40.347	40.382	40.417	40.452	1320
1330	40.452	40.487	40.522	40.557	40.592	40.627	40.662	40.697	40.732	40.767	40.802	1330
1340	40.802	40.837	40.872	40.908	40.943	40.978	41.013	41.048	41.083	41.118	41.154	1340
1350	41.154	41.189	41.224	41.259	41.294	41.329	41.365	41.400	41.435	41.470	41.506	1350
1360	41.506	41.541	41.576	41.611	41.647	41.682	41.717	41.753	41.788	41.823	41.859	1360
1370	41.859	41.894	41.929	41.965	42.000	42.035	42.071	42.106	42.142	42.177	42.212	1370
1380	42.212	42.248	42.283	42.319	42.354	42.390	42.425	42.460	42.496	42.531	42.567	1380
1390	42.567	42.602	42.638	42.673	42.709	42.744	42.780	42.815	42.851	42.886	42.922	1390

Type K Thermocouple
Thermoelectric Voltage as a Function of Temperature (°C)
Reference Junctions at 0°C

THERMOELECTRIC VOLTAGE IN ABSOLUTE MILLIVOLTS

°C	0	1	2	3	4	5	6	7	8	9	10	°C
−270	−6.458	−6.457	−6.456	−6.455	−6.453	−6.452	−6.450	−6.448	−6.446	−6.444	−6.441	−270
−260	−6.441	−6.438	−6.435	−6.432	−6.429	−6.425	−6.421	−6.417	−6.413	−6.408	−6.404	−260
−250	−6.404	−6.399	−6.394	−6.388	−6.382	−6.377	−6.371	−6.364	−6.358	−6.351	−6.344	−250
−240	−6.344	−6.337	−6.329	−6.322	−6.314	−6.306	−6.297	−6.289	−6.280	−6.271	−6.262	−240
−230	−6.262	−6.253	−6.243	−6.233	−6.223	−6.213	−6.202	−6.192	−6.181	−6.170	−6.158	−230
−220	−6.158	−6.147	−6.135	−6.123	−6.111	−6.099	−6.087	−6.074	−6.061	−6.048	−6.035	−220
−210	−6.035	−6.021	−6.007	−5.994	−5.980	−5.965	−5.951	−5.936	−5.922	−5.907	−5.891	−210
−200	−5.891	−5.876	−5.860	−5.845	−5.829	−5.813	−5.796	−5.780	−5.763	−5.747	−5.730	−200
−190	−5.730	−5.712	−5.695	−5.678	−5.660	−5.642	−5.624	−5.606	−5.587	−5.569	−5.550	−190
−180	−5.550	−5.531	−5.512	−5.493	−5.474	−5.454	−5.434	−5.414	−5.394	−5.374	−5.354	−180
−170	−5.354	−5.333	−5.313	−5.292	−5.271	−5.249	−5.228	−5.207	−5.185	−5.163	−5.141	−170
−160	−5.141	−5.119	−5.097	−5.074	−5.051	−5.029	−5.006	−4.983	−4.959	−4.936	−4.912	−160
−150	−4.912	−4.889	−4.865	−4.841	−4.817	−4.792	−4.768	−4.743	−4.719	−4.694	−4.669	−150
−140	−4.669	−4.644	−4.618	−4.593	−4.567	−4.541	−4.515	−4.489	−4.463	−4.437	−4.410	−140
−130	−4.410	−4.384	−4.357	−4.330	−4.303	−4.276	−4.248	−4.221	−4.193	−4.166	−4.138	−130
−120	−4.138	−4.110	−4.082	−4.053	−4.025	−3.997	−3.968	−3.939	−3.910	−3.881	−3.852	−120
−110	−3.852	−3.823	−3.793	−3.764	−3.734	−3.704	−3.674	−3.644	−3.614	−3.584	−3.553	−110
−100	−3.553	−3.523	−3.492	−3.461	−3.430	−3.399	−3.368	−3.337	−3.305	−3.274	−3.242	−100
−90	−3.242	−3.211	−3.179	−3.147	−3.115	−3.082	−3.050	−3.018	−2.985	−2.953	−2.920	−90
−80	−2.920	−2.887	−2.854	−2.821	−2.788	−2.754	−2.721	−2.687	−2.654	−2.620	−2.586	−80
−70	−2.586	−2.552	−2.518	−2.484	−2.450	−2.416	−2.381	−2.347	−2.312	−2.277	−2.243	−70
−60	−2.243	−2.208	−2.173	−2.137	−2.102	−2.067	−2.032	−1.996	−1.961	−1.925	−1.889	−60
−50	−1.889	−1.853	−1.817	−1.781	−1.745	−1.709	−1.673	−1.636	−1.600	−1.563	−1.527	−50
−40	−1.527	−1.490	−1.453	−1.416	−1.379	−1.342	−1.305	−1.268	−1.231	−1.193	−1.156	−40
−30	−1.156	−1.118	−1.081	−1.043	−1.005	−0.968	−0.930	−0.892	−0.854	−0.816	−0.777	−30
−20	−0.777	−0.739	−0.701	−0.662	−0.624	−0.585	−0.547	−0.508	−0.469	−0.431	−0.392	−20
−10	−0.392	−0.353	−0.314	−0.275	−0.236	−0.197	−0.157	−0.118	−0.079	−0.039	0.000	−10
0	0.000	0.039	0.079	0.119	0.158	0.198	0.238	0.277	0.317	0.357	0.397	0
10	0.397	0.437	0.477	0.517	0.557	0.597	0.637	0.677	0.718	0.758	0.798	10
20	0.798	0.838	0.879	0.919	0.960	1.000	1.041	1.081	1.122	1.162	1.203	20
30	1.203	1.244	1.285	1.325	1.366	1.407	1.448	1.489	1.529	1.570	1.611	30
40	1.611	1.652	1.693	1.734	1.776	1.817	1.858	1.899	1.940	1.981	2.022	40
50	2.022	2.064	2.105	2.146	2.188	2.229	2.270	2.312	2.353	2.394	2.436	50
60	2.436	2.477	2.519	2.560	2.601	2.643	2.684	2.726	2.767	2.809	2.850	60
70	2.850	2.892	2.933	2.975	3.016	3.058	3.100	3.141	3.183	3.224	3.266	70
80	3.266	3.307	3.349	3.390	3.432	3.473	3.515	3.556	3.598	3.639	3.681	80
90	3.681	3.722	3.764	3.805	3.847	3.888	3.930	3.971	4.012	4.054	4.095	90
100	4.095	4.137	4.178	4.219	4.261	4.302	4.343	4.384	4.426	4.467	4.508	100
110	4.508	4.549	4.590	4.632	4.673	4.714	4.755	4.796	4.837	4.878	4.919	110
120	4.919	4.960	5.001	5.042	5.083	5.124	5.164	5.205	5.246	5.287	5.327	120
130	5.327	5.368	5.409	5.450	5.490	5.531	5.571	5.612	5.652	5.693	5.733	130
140	5.733	5.774	5.814	5.855	5.895	5.936	5.976	6.016	6.057	6.097	6.137	140
150	6.137	6.177	6.218	6.258	6.298	6.338	6.378	6.419	6.459	6.499	6.539	150
160	6.539	6.579	6.619	6.659	6.699	6.739	6.779	6.819	6.859	6.899	6.939	160
170	6.939	6.979	7.019	7.059	7.099	7.139	7.179	7.219	7.259	7.299	7.338	170
180	7.338	7.378	7.418	7.458	7.498	7.538	7.578	7.618	7.658	7.697	7.737	180
190	7.737	7.777	7.817	7.857	7.897	7.937	7.977	8.017	8.057	8.097	8.137	190
200	8.137	8.177	8.216	8.256	8.296	8.336	8.376	8.416	8.456	8.497	8.537	200
210	8.537	8.577	8.617	8.657	8.697	8.737	8.777	8.817	8.857	8.898	8.938	210
220	8.938	8.978	9.018	9.058	9.099	9.139	9.179	9.220	9.260	9.300	9.341	220
230	9.341	9.381	9.421	9.462	9.502	9.543	9.583	9.624	9.664	9.705	9.745	230
240	9.745	9.786	9.826	9.867	9.907	9.948	9.989	10.029	10.070	10.111	10.151	240
250	10.151	10.192	10.233	10.274	10.315	10.355	10.396	10.437	10.478	10.519	10.560	250
260	10.560	10.600	10.641	10.682	10.723	10.764	10.805	10.846	10.887	10.928	10.969	260
270	10.969	11.010	11.051	11.093	11.134	11.175	11.216	11.257	11.298	11.339	11.381	270
280	11.381	11.422	11.463	11.504	11.546	11.587	11.628	11.669	11.711	11.752	11.793	280
290	11.793	11.835	11.876	11.918	11.959	12.000	12.042	12.083	12.125	12.166	12.207	290
300	12.207	12.249	12.290	12.332	12.373	12.415	12.456	12.498	12.539	12.581	12.623	300
310	12.623	12.664	12.706	12.747	12.789	12.831	12.872	12.914	12.955	12.997	13.039	310
320	13.039	13.080	13.122	13.164	13.205	13.247	13.289	13.331	13.372	13.414	13.456	320

Type K Thermocouple
Thermoelectric Voltage as a Function of Temperature (°C)
Reference Junctions at 0°C
THERMOELECTRIC VOLTAGE IN ABSOLUTE MILLIVOLTS

°C	0	1	2	3	4	5	6	7	8	9	10	°C
330	13.456	13.497	13.539	13.581	13.623	13.665	13.706	13.748	13.790	13.832	13.874	330
340	13.874	13.915	13.957	13.999	14.041	14.083	14.125	14.167	14.208	14.250	14.292	340
350	14.292	14.334	14.376	14.418	14.460	14.502	14.544	14.586	14.628	14.670	14.712	350
360	14.712	14.754	14.796	14.838	14.880	14.922	14.964	15.006	15.048	15.090	15.132	360
370	15.132	15.174	15.216	15.258	15.300	15.342	15.384	15.426	15.468	15.510	15.552	370
380	15.552	15.594	15.636	15.679	15.721	15.763	15.805	15.847	15.889	15.931	15.974	380
390	15.974	16.016	16.058	16.100	16.142	16.184	16.227	16.269	16.311	16.353	16.395	390
400	16.395	16.438	16.480	16.522	16.564	16.607	16.649	16.691	16.733	16.776	16.818	400
410	16.818	16.860	16.902	16.945	16.987	17.029	17.072	17.114	17.156	17.199	17.241	410
420	17.241	17.283	17.326	17.368	17.410	17.453	17.495	17.537	17.580	17.622	17.664	420
430	17.664	17.707	17.749	17.792	17.834	17.876	17.919	17.961	18.004	18.046	18.088	430
440	18.088	18.131	18.173	18.216	18.258	18.301	18.343	18.385	18.428	18.470	18.513	440
450	18.513	18.555	18.598	18.640	18.683	18.725	18.768	18.810	18.853	18.895	18.938	450
460	18.938	18.980	19.023	19.065	19.108	19.150	19.193	19.235	19.278	19.320	19.363	460
470	19.363	19.405	19.448	19.490	19.533	19.576	19.618	19.661	19.703	19.746	19.788	470
480	19.788	19.831	19.873	19.916	19.959	20.001	20.044	20.086	20.129	20.172	20.214	480
490	20.214	20.257	20.299	20.342	20.385	20.427	20.470	20.512	20.555	20.598	20.640	490
500	20.640	20.683	20.725	20.768	20.811	20.853	20.896	20.938	20.981	21.024	21.066	500
510	21.066	21.109	21.152	21.194	21.237	21.280	21.322	21.365	21.407	21.450	21.493	510
520	21.493	21.535	21.578	21.621	21.663	21.706	21.749	21.791	21.834	21.876	21.919	520
530	21.919	21.962	22.004	22.047	22.090	22.132	22.175	22.218	22.260	22.303	22.346	530
540	22.346	22.388	22.431	22.473	22.516	22.559	22.601	22.644	22.687	22.729	22.772	540
550	22.772	22.815	22.857	22.900	22.942	22.985	23.028	23.070	23.113	23.156	23.198	550
560	23.198	23.241	23.284	23.326	23.369	23.411	23.454	23.497	23.539	23.582	23.624	560
570	23.624	23.667	23.710	23.752	23.795	23.837	23.880	23.923	23.965	24.008	24.050	570
580	24.050	24.093	24.136	24.178	24.221	24.263	24.306	24.348	24.391	24.434	24.476	580
590	24.476	24.519	24.561	24.604	24.646	24.689	24.731	24.774	24.817	24.859	24.902	590
600	24.902	24.944	24.987	25.029	25.072	25.114	25.157	25.199	25.242	25.284	25.327	600
610	25.327	25.369	25.412	25.454	25.497	25.539	25.582	25.624	25.666	25.709	25.751	610
620	25.751	25.794	25.836	25.879	25.921	25.964	26.006	26.048	26.091	26.133	26.176	620
630	26.176	26.218	26.260	26.303	26.345	26.387	26.430	26.472	26.515	26.557	26.599	630
640	26.599	26.642	26.684	26.726	26.769	26.811	26.853	26.896	26.938	26.980	27.022	640
650	27.022	27.065	27.107	27.149	27.192	27.234	27.276	27.318	27.361	27.403	27.445	650
660	27.445	27.487	27.529	27.572	27.614	27.656	27.698	27.740	27.783	27.825	27.867	660
670	27.867	27.909	27.951	27.993	28.035	28.078	28.120	28.162	28.204	28.246	28.288	670
680	28.288	28.330	28.372	28.414	28.456	28.498	28.540	28.583	28.625	28.667	28.709	680
690	28.709	28.751	28.793	28.835	28.877	28.919	28.961	29.002	29.044	29.086	29.128	690
700	29.128	29.170	29.212	29.254	29.296	29.338	29.380	29.422	29.464	29.505	29.547	700
710	29.547	29.589	29.631	29.673	29.715	29.756	29.798	29.840	29.882	29.924	29.965	710
720	29.965	30.007	30.049	30.091	30.132	30.174	30.216	30.257	30.299	30.341	30.383	720
730	30.383	30.424	30.466	30.508	30.549	30.591	30.632	30.674	30.716	30.757	30.799	730
740	30.799	30.840	30.882	30.924	30.965	31.007	31.048	31.090	31.131	31.173	31.214	740
750	31.214	31.256	31.297	31.339	31.380	31.422	31.463	31.504	31.546	31.587	31.629	750
760	31.629	31.670	31.712	31.753	31.794	31.836	31.877	31.918	31.960	32.001	32.042	760
770	32.042	32.084	32.125	32.166	32.207	32.249	32.290	32.331	32.372	32.414	32.455	770
780	32.455	32.496	32.537	32.578	32.619	32.661	32.702	32.743	32.784	32.825	32.866	780
790	32.866	32.907	32.948	32.990	33.031	33.072	33.113	33.154	33.195	33.236	33.277	790
800	33.277	33.318	33.359	33.400	33.441	33.482	33.523	33.564	33.604	33.645	33.686	800
810	33.686	33.727	33.768	33.809	33.850	33.891	33.931	33.972	34.013	34.054	34.095	810
820	34.095	34.136	34.176	34.217	34.258	34.299	34.339	34.380	34.421	34.461	34.502	820
830	34.502	34.543	34.583	34.624	34.665	34.705	34.746	34.787	34.827	34.868	34.909	830
840	34.909	34.949	34.990	35.030	35.071	35.111	35.152	35.192	35.233	35.273	35.314	840
850	35.314	35.354	35.395	35.435	35.476	35.516	35.557	35.597	35.637	35.678	35.718	850
860	35.718	35.758	35.799	35.839	35.880	35.920	35.960	36.000	36.041	36.081	36.121	860
870	36.121	36.162	36.202	36.242	36.282	36.323	36.363	36.403	36.443	36.483	36.524	870
880	36.524	36.564	36.604	36.644	36.684	36.724	36.764	36.804	36.844	36.885	36.925	880
890	36.925	36.965	37.005	37.045	37.085	37.125	37.165	37.205	37.245	37.285	37.325	890
900	37.325	37.365	37.405	37.445	37.484	37.524	37.564	37.604	37.644	37.684	37.724	900
910	37.724	37.764	37.803	37.843	37.883	37.923	37.963	38.002	38.042	38.082	38.122	910
920	38.122	38.162	38.201	38.241	38.281	38.320	38.360	38.400	38.439	38.479	38.519	920

Type K Thermocouple
Thermoelectric Voltage as a Function of Temperature (°C)
Reference Junctions at 0°C
THERMOELECTRIC VOLTAGE IN ABSOLUTE MILLIVOLTS

°C	0	1	2	3	4	5	6	7	8	9	10	°C
930	38.519	38.558	38.598	38.638	38.677	38.717	38.756	38.796	38.836	38.875	38.915	930
940	38.915	38.954	38.994	39.033	39.073	39.112	39.152	39.191	39.231	39.270	39.310	940
950	39.310	39.349	39.388	39.428	39.467	39.507	39.546	39.585	39.625	39.664	39.703	950
960	39.703	39.743	39.782	39.821	39.861	39.900	39.939	39.979	40.018	40.057	40.096	960
970	40.096	40.136	40.175	40.214	40.253	40.292	40.332	40.371	40.410	40.449	40.488	970
980	40.488	40.527	40.566	40.605	40.645	40.684	40.723	40.762	40.801	40.840	40.879	980
990	40.879	40.918	40.957	40.996	41.035	41.074	41.113	41.152	41.191	41.230	41.269	990
1000	41.269	41.308	41.347	41.385	41.424	41.463	41.502	41.541	41.580	41.619	41.657	1000
1010	41.657	41.696	41.735	41.774	41.813	41.851	41.890	41.929	41.968	42.006	42.045	1010
1020	42.045	42.084	42.123	42.161	42.200	42.239	42.277	42.316	42.355	42.393	42.432	1020
1030	42.432	42.470	42.509	42.548	42.586	42.625	42.663	42.702	42.740	42.779	42.817	1030
1040	42.817	42.856	42.894	42.933	42.971	43.010	43.048	43.087	43.125	43.164	43.202	1040
1050	43.202	43.240	43.279	43.317	43.356	43.394	43.432	43.471	43.509	43.547	43.585	1050
1060	43.585	43.624	43.662	43.700	43.739	43.777	43.815	43.853	43.891	43.930	43.968	1060
1070	43.968	44.006	44.044	44.082	44.121	44.159	44.197	44.235	44.273	44.311	44.349	1070
1080	44.349	44.387	44.425	44.463	44.501	44.539	44.577	44.615	44.653	44.691	44.729	1080
1090	44.729	44.767	44.805	44.843	44.881	44.919	44.957	44.995	45.033	45.070	45.108	1090
1100	45.108	45.146	45.184	45.222	45.260	45.297	45.335	45.373	45.411	45.448	45.486	1100
1110	45.486	45.524	45.561	45.599	45.637	45.675	45.712	45.750	45.787	45.825	45.863	1110
1120	45.863	45.900	45.938	45.975	46.013	46.051	46.088	46.126	46.163	46.201	46.238	1120
1130	46.238	46.275	46.313	46.350	46.388	46.425	46.463	46.500	46.537	46.575	46.612	1130
1140	46.612	46.649	46.687	46.724	46.761	46.799	46.836	46.873	46.910	46.948	46.985	1140
1150	46.985	47.022	47.059	47.096	47.134	47.171	47.208	47.245	47.282	47.319	47.356	1150
1160	47.356	47.393	47.430	47.468	47.505	47.542	47.579	47.616	47.653	47.689	47.726	1160
1170	47.726	47.763	47.800	47.837	47.874	47.911	47.948	47.985	48.021	48.058	48.095	1170
1180	48.095	48.132	48.169	48.205	48.242	48.279	48.316	48.352	48.389	48.426	48.462	1180
1190	48.462	48.499	48.536	48.572	48.609	48.645	48.682	48.718	48.755	48.792	48.828	1190
1200	48.828	48.865	48.901	48.937	48.974	49.010	49.047	49.083	49.120	49.156	49.192	1200
1210	49.192	49.229	49.265	49.301	49.338	49.374	49.410	49.446	49.483	49.519	49.555	1210
1220	49.555	49.591	49.627	49.663	49.700	49.736	49.772	49.808	49.844	49.880	49.916	1220
1230	49.916	49.952	49.988	50.024	50.060	50.096	50.132	50.168	50.204	50.240	50.276	1230
1240	50.276	50.311	50.347	50.383	50.419	50.455	50.491	50.526	50.562	50.598	50.633	1240
1250	50.633	50.669	50.705	50.741	50.776	50.812	50.847	50.883	50.919	50.954	50.990	1250
1260	50.990	51.025	51.061	51.096	51.132	51.167	51.203	51.238	51.274	51.309	51.344	1260
1270	51.344	51.380	51.415	51.450	51.486	51.521	51.556	51.592	51.627	51.662	51.697	1270
1280	51.697	51.733	51.768	51.803	51.838	51.873	51.908	51.943	51.979	52.014	52.049	1280
1290	52.049	52.084	52.119	52.154	52.189	52.224	52.259	52.294	52.329	52.364	52.398	1290
1300	52.398	52.433	52.468	52.503	52.538	52.573	52.608	52.642	52.677	52.712	52.747	1300
1310	52.747	52.781	52.816	52.851	52.886	52.920	52.955	52.989	53.024	53.059	53.093	1310
1320	53.093	53.128	53.162	53.197	53.232	53.266	53.301	53.335	53.370	53.404	53.439	1320
1330	53.439	53.473	53.507	53.542	53.576	53.611	53.645	53.679	53.714	53.748	53.782	1330
1340	53.782	53.817	53.851	53.885	53.920	53.954	53.988	54.022	54.057	54.091	54.125	1340
1350	54.125	54.159	54.193	54.228	54.262	54.296	54.330	54.364	54.398	54.432	54.466	1350
1360	54.466	54.501	54.535	54.569	54.603	54.637	54.671	54.705	54.739	54.773	54.807	1360
1370	54.807	54.841	54.875									1370

Type K Thermocouple
Thermoelectric Voltage as a Function of Temperature (°F)
Reference Junctions at 32°F

THERMOELECTRIC VOLTAGE IN ABSOLUTE MILLIVOLTS

°F	0	1	2	3	4	5	6	7	8	9	10	°F
−460							−6.458	−6.457	−6.457	−6.456	−6.456	−460
−450	−6.456	−6.455	−6.454	−6.454	−6.453	−6.452	−6.451	−6.450	−6.449	−6.448	−6.447	−450
−440	−6.447	−6.445	−6.444	−6.443	−6.441	−6.440	−6.438	−6.436	−6.435	−6.433	−6.431	−440
−430	−6.431	−6.429	−6.427	−6.425	−6.423	−6.421	−6.419	−6.416	−6.414	−6.411	−6.409	−430
−420	−6.409	−6.406	−6.404	−6.401	−6.398	−6.395	−6.392	−6.389	−6.386	−6.383	−6.380	−420
−410	−6.380	−6.377	−6.373	−6.370	−6.366	−6.363	−6.359	−6.355	−6.352	−6.348	−6.344	−410
−400	−6.344	−6.340	−6.336	−6.332	−6.328	−6.323	−6.319	−6.315	−6.310	−6.306	−6.301	−400
−390	−6.301	−6.296	−6.292	−6.287	−6.282	−6.277	−6.272	−6.267	−6.262	−6.257	−6.251	−390
−380	−6.251	−6.246	−6.241	−6.235	−6.230	−6.224	−6.219	−6.213	−6.207	−6.201	−6.195	−380
−370	−6.195	−6.189	−6.183	−6.177	−6.171	−6.165	−6.158	−6.152	−6.146	−6.139	−6.133	−370
−360	−6.133	−6.126	−6.119	−6.113	−6.106	−6.099	−6.092	−6.085	−6.078	−6.071	−6.064	−360
−350	−6.064	−6.057	−6.049	−6.042	−6.035	−6.027	−6.020	−6.012	−6.004	−5.997	−5.989	−350
−340	−5.989	−5.981	−5.973	−5.965	−5.957	−5.949	−5.941	−5.933	−5.925	−5.917	−5.908	−340
−330	−5.908	−5.900	−5.891	−5.883	−5.874	−5.866	−5.857	−5.848	−5.839	−5.831	−5.822	−330
−320	−5.822	−5.813	−5.804	−5.795	−5.786	−5.776	−5.767	−5.758	−5.748	−5.739	−5.730	−320
−310	−5.730	−5.720	−5.711	−5.701	−5.691	−5.682	−5.672	−5.662	−5.652	−5.642	−5.632	−310
−300	−5.632	−5.622	−5.612	−5.602	−5.592	−5.581	−5.571	−5.561	−5.550	−5.540	−5.529	−300
−290	−5.529	−5.519	−5.508	−5.497	−5.487	−5.476	−5.465	−5.454	−5.443	−5.432	−5.421	−290
−280	−5.421	−5.410	−5.399	−5.388	−5.376	−5.365	−5.354	−5.342	−5.331	−5.319	−5.308	−280
−270	−5.308	−5.296	−5.285	−5.273	−5.261	−5.249	−5.238	−5.226	−5.214	−5.202	−5.190	−270
−260	−5.190	−5.178	−5.165	−5.153	−5.141	−5.129	−5.116	−5.104	−5.092	−5.079	−5.067	−260
−250	−5.067	−5.054	−5.041	−5.029	−5.016	−5.003	−4.990	−4.978	−4.965	−4.952	−4.939	−250
−240	−4.939	−4.926	−4.912	−4.899	−4.886	−4.873	−4.860	−4.846	−4.833	−4.819	−4.806	−240
−230	−4.806	−4.792	−4.779	−4.765	−4.752	−4.738	−4.724	−4.710	−4.697	−4.683	−4.669	−230
−220	−4.669	−4.655	−4.641	−4.627	−4.613	−4.598	−4.584	−4.570	−4.556	−4.541	−4.527	−220
−210	−4.527	−4.512	−4.498	−4.484	−4.469	−4.454	−4.440	−4.425	−4.410	−4.396	−4.381	−210
−200	−4.381	−4.366	−4.351	−4.336	−4.321	−4.306	−4.291	−4.276	−4.261	−4.245	−4.230	−200
−190	−4.230	−4.215	−4.200	−4.184	−4.169	−4.153	−4.138	−4.122	−4.107	−4.091	−4.075	−190
−180	−4.075	−4.060	−4.044	−4.028	−4.012	−3.997	−3.981	−3.965	−3.949	−3.933	−3.917	−180
−170	−3.917	−3.901	−3.884	−3.868	−3.852	−3.836	−3.819	−3.803	−3.787	−3.770	−3.754	−170
−160	−3.754	−3.737	−3.721	−3.704	−3.688	−3.671	−3.654	−3.637	−3.621	−3.604	−3.587	−160
−150	−3.587	−3.570	−3.553	−3.536	−3.519	−3.502	−3.485	−3.468	−3.451	−3.434	−3.417	−150
−140	−3.417	−3.399	−3.382	−3.365	−3.347	−3.330	−3.312	−3.295	−3.277	−3.260	−3.242	−140
−130	−3.242	−3.225	−3.207	−3.189	−3.172	−3.154	−3.136	−3.118	−3.100	−3.082	−3.065	−130
−120	−3.065	−3.047	−3.029	−3.010	−2.992	−2.974	−2.956	−2.938	−2.920	−2.902	−2.883	−120
−110	−2.883	−2.865	−2.847	−2.828	−2.810	−2.791	−2.773	−2.754	−2.736	−2.717	−2.699	−110
−100	−2.699	−2.680	−2.661	−2.643	−2.624	−2.605	−2.586	−2.567	−2.549	−2.530	−2.511	−100
−90	−2.511	−2.492	−2.473	−2.454	−2.435	−2.416	−2.397	−2.377	−2.358	−2.339	−2.320	−90
−80	−2.320	−2.300	−2.281	−2.262	−2.243	−2.223	−2.204	−2.184	−2.165	−2.145	−2.126	−80
−70	−2.126	−2.106	−2.087	−2.067	−2.047	−2.028	−2.008	−1.988	−1.968	−1.949	−1.929	−70
−60	−1.929	−1.909	−1.889	−1.869	−1.849	−1.829	−1.809	−1.789	−1.769	−1.749	−1.729	−60
−50	−1.729	−1.709	−1.689	−1.669	−1.648	−1.628	−1.608	−1.588	−1.567	−1.547	−1.527	−50
−40	−1.527	−1.506	−1.486	−1.465	−1.445	−1.424	−1.404	−1.383	−1.363	−1.342	−1.322	−40
−30	−1.322	−1.301	−1.280	−1.260	−1.239	−1.218	−1.197	−1.177	−1.156	−1.135	−1.114	−30
−20	−1.114	−1.093	−1.072	−1.051	−1.031	−1.010	−0.989	−0.968	−0.946	−0.925	−0.904	−20
−10	−0.904	−0.883	−0.862	−0.841	−0.820	−0.799	−0.777	−0.756	−0.735	−0.714	−0.692	−10
0	−0.692	−0.671	−0.650	−0.628	−0.607	−0.585	−0.564	−0.543	−0.521	−0.500	−0.478	0
10	−0.478	−0.457	−0.435	−0.413	−0.392	−0.370	−0.349	−0.327	−0.305	−0.284	−0.262	10
20	−0.262	−0.240	−0.218	−0.197	−0.175	−0.153	−0.131	−0.109	−0.088	−0.066	−0.044	20
30	−0.044	−0.022	0.000	0.022	0.044	0.066	0.088	0.110	0.132	0.154	0.176	30
40	0.176	0.198	0.220	0.242	0.264	0.286	0.308	0.331	0.353	0.375	0.397	40
50	0.397	0.419	0.441	0.464	0.486	0.508	0.530	0.553	0.575	0.597	0.619	50
60	0.619	0.642	0.664	0.686	0.709	0.731	0.753	0.776	0.798	0.821	0.843	60
70	0.843	0.865	0.888	0.910	0.933	0.955	0.978	1.000	1.023	1.045	1.068	70
80	1.068	1.090	1.113	1.135	1.158	1.181	1.203	1.226	1.248	1.271	1.294	80
90	1.294	1.316	1.339	1.362	1.384	1.407	1.430	1.452	1.475	1.498	1.520	90
100	1.520	1.543	1.566	1.589	1.611	1.634	1.657	1.680	1.703	1.725	1.748	100
110	1.748	1.771	1.794	1.817	1.839	1.862	1.885	1.908	1.931	1.954	1.977	110
120	1.977	2.000	2.022	2.045	2.068	2.091	2.114	2.137	2.160	2.183	2.206	120
130	2.206	2.229	2.252	2.275	2.298	2.321	2.344	2.367	2.390	2.413	2.436	130

Type K Thermocouple
Thermoelectric Voltage as a Function of Temperature (°F)
Reference Junctions at 32°F
THERMOELECTRIC VOLTAGE IN ABSOLUTE MILLIVOLTS

°F	0	1	2	3	4	5	6	7	8	9	10	°F
140	2.436	2.459	2.482	2.505	2.528	2.551	2.574	2.597	2.620	2.643	2.666	140
150	2.666	2.689	2.712	2.735	2.758	2.781	2.804	2.827	2.850	2.873	2.896	150
160	2.896	2.920	2.943	2.966	2.989	3.012	3.035	3.058	3.081	3.104	3.127	160
170	3.127	3.150	3.173	3.196	3.220	3.243	3.266	3.289	3.312	3.335	3.358	170
180	3.358	3.381	3.404	3.427	3.450	3.473	3.496	3.519	3.543	3.566	3.589	180
190	3.589	3.612	3.635	3.658	3.681	3.704	3.727	3.750	3.773	3.796	3.819	190
200	3.819	3.842	3.865	3.888	3.911	3.934	3.957	3.980	4.003	4.026	4.049	200
210	4.049	4.072	4.095	4.118	4.141	4.164	4.187	4.210	4.233	4.256	4.279	210
220	4.279	4.302	4.325	4.348	4.371	4.394	4.417	4.439	4.462	4.485	4.508	220
230	4.508	4.531	4.554	4.577	4.600	4.622	4.645	4.668	4.691	4.714	4.737	230
240	4.737	4.759	4.782	4.805	4.828	4.851	4.873	4.896	4.919	4.942	4.964	240
250	4.964	4.987	5.010	5.033	5.055	5.078	5.101	5.124	5.146	5.169	5.192	250
260	5.192	5.214	5.237	5.260	5.282	5.305	5.327	5.350	5.373	5.395	5.418	260
270	5.418	5.440	5.463	5.486	5.508	5.531	5.553	5.576	5.598	5.621	5.643	270
280	5.643	5.666	5.688	5.711	5.733	5.756	5.778	5.801	5.823	5.846	5.868	280
290	5.868	5.891	5.913	5.936	5.958	5.980	6.003	6.025	6.048	6.070	6.092	290
300	6.092	6.115	6.137	6.160	6.182	6.204	6.227	6.249	6.271	6.294	6.316	300
310	6.316	6.338	6.361	6.383	6.405	6.428	6.450	6.472	6.494	6.517	6.539	310
320	6.539	6.561	6.583	6.606	6.628	6.650	6.672	6.695	6.717	6.739	6.761	320
330	6.761	6.784	6.806	6.828	6.850	6.873	6.895	6.917	6.939	6.961	6.984	330
340	6.984	7.006	7.028	7.050	7.072	7.094	7.117	7.139	7.161	7.183	7.205	340
350	7.205	7.228	7.250	7.272	7.294	7.316	7.338	7.361	7.383	7.405	7.427	350
360	7.427	7.449	7.471	7.494	7.516	7.538	7.560	7.582	7.604	7.627	7.649	360
370	7.649	7.671	7.693	7.715	7.737	7.760	7.782	7.804	7.826	7.848	7.870	370
380	7.870	7.893	7.915	7.937	7.959	7.981	8.003	8.026	8.048	8.070	8.092	380
390	8.092	8.114	8.137	8.159	8.181	8.203	8.225	8.248	8.270	8.292	8.314	390
400	8.314	8.336	8.359	8.381	8.403	8.425	8.448	8.470	8.492	8.514	8.537	400
410	8.537	8.559	8.581	8.603	8.626	8.648	8.670	8.692	8.715	8.737	8.759	410
420	8.759	8.782	8.804	8.826	8.849	8.871	8.893	8.916	8.938	8.960	8.983	420
430	8.983	9.005	9.027	9.050	9.072	9.094	9.117	9.139	9.161	9.184	9.206	430
440	9.206	9.229	9.251	9.273	9.296	9.318	9.341	9.363	9.385	9.408	9.430	440
450	9.430	9.453	9.475	9.498	9.520	9.543	9.565	9.588	9.610	9.633	9.655	450
460	9.655	9.678	9.700	9.723	9.745	9.768	9.790	9.813	9.835	9.858	9.880	460
470	9.880	9.903	9.926	9.948	9.971	9.993	10.016	10.038	10.061	10.084	10.106	470
480	10.106	10.129	10.151	10.174	10.197	10.219	10.242	10.265	10.287	10.310	10.333	480
490	10.333	10.355	10.378	10.401	10.423	10.446	10.469	10.491	10.514	10.537	10.560	490
500	10.560	10.582	10.605	10.628	10.650	10.673	10.696	10.719	10.741	10.764	10.787	500
510	10.787	10.810	10.833	10.855	10.878	10.901	10.924	10.947	10.969	10.992	11.015	510
520	11.015	11.038	11.061	11.083	11.106	11.129	11.152	11.175	11.198	11.221	11.243	520
530	11.243	11.266	11.289	11.312	11.335	11.358	11.381	11.404	11.426	11.449	11.472	530
540	11.472	11.495	11.518	11.541	11.564	11.587	11.610	11.633	11.656	11.679	11.702	540
550	11.702	11.725	11.748	11.770	11.793	11.816	11.839	11.862	11.885	11.908	11.931	550
560	11.931	11.954	11.977	12.000	12.023	12.046	12.069	12.092	12.115	12.138	12.161	560
570	12.161	12.184	12.207	12.230	12.254	12.277	12.300	12.323	12.346	12.369	12.392	570
580	12.392	12.415	12.438	12.461	12.484	12.507	12.530	12.553	12.576	12.599	12.623	580
590	12.623	12.646	12.669	12.692	12.715	12.738	12.761	12.784	12.807	12.831	12.854	590
600	12.854	12.877	12.900	12.923	12.946	12.969	12.992	13.016	13.039	13.062	13.085	600
610	13.085	13.108	13.131	13.154	13.178	13.201	13.224	13.247	13.270	13.293	13.317	610
620	13.317	13.340	13.363	13.386	13.409	13.433	13.456	13.479	13.502	13.525	13.549	620
630	13.549	13.572	13.595	13.618	13.641	13.665	13.688	13.711	13.734	13.757	13.781	630
640	13.781	13.804	13.827	13.850	13.874	13.897	13.920	13.943	13.967	13.990	14.013	640
650	14.013	14.036	14.060	14.083	14.106	14.129	14.153	14.176	14.199	14.222	14.246	650
660	14.246	14.269	14.292	14.316	14.339	14.362	14.385	14.409	14.432	14.455	14.479	660
670	14.479	14.502	14.525	14.548	14.572	14.595	14.618	14.642	14.665	14.688	14.712	670
680	14.712	14.735	14.758	14.782	14.805	14.828	14.852	14.875	14.898	14.922	14.945	680
690	14.945	14.968	14.992	15.015	15.038	15.062	15.085	15.108	15.132	15.155	15.178	690
700	15.178	15.202	15.225	15.248	15.272	15.295	15.318	15.342	15.365	15.389	15.412	700
710	15.412	15.435	15.459	15.482	15.505	15.529	15.552	15.576	15.599	15.622	15.646	710
720	15.646	15.669	15.693	15.716	15.739	15.763	15.786	15.810	15.833	15.856	15.880	720
730	15.880	15.903	15.927	15.950	15.974	15.997	16.020	16.044	16.067	16.091	16.114	730

Type K Thermocouple
Thermoelectric Voltage as a Function of Temperature (°F)
Reference Junctions at 32°F
THERMOELECTRIC VOLTAGE IN ABSOLUTE MILLIVOLTS

°F	0	1	2	3	4	5	6	7	8	9	10	°F
740	16.114	16.138	16.161	16.184	16.208	16.231	16.255	16.278	16.302	16.325	16.349	740
750	16.349	16.372	16.395	16.419	16.442	16.466	16.489	16.513	16.536	16.560	16.583	750
760	16.583	16.607	16.630	16.654	16.677	16.700	16.724	16.747	16.771	16.794	16.818	760
770	16.818	16.841	16.865	16.888	16.912	16.935	16.959	16.982	17.006	17.029	17.053	770
780	17.053	17.076	17.100	17.123	17.147	17.170	17.194	17.217	17.241	17.264	17.288	780
790	17.288	17.311	17.335	17.358	17.382	17.406	17.429	17.453	17.476	17.500	17.523	790
800	17.523	17.547	17.570	17.594	17.617	17.641	17.664	17.688	17.711	17.735	17.759	800
810	17.759	17.782	17.806	17.829	17.853	17.876	17.900	17.923	17.947	17.971	17.994	810
820	17.994	18.018	18.041	18.065	18.088	18.112	18.136	18.159	18.183	18.206	18.230	820
830	18.230	18.253	18.277	18.301	18.324	18.348	18.371	18.395	18.418	18.442	18.466	830
840	18.466	18.489	18.513	18.536	18.560	18.584	18.607	18.631	18.654	18.678	18.702	840
850	18.702	18.725	18.749	18.772	18.796	18.820	18.843	18.867	18.890	18.914	18.938	850
860	18.938	18.961	18.985	19.008	19.032	19.056	19.079	19.103	19.127	19.150	19.174	860
870	19.174	19.197	19.221	19.245	19.268	19.292	19.316	19.339	19.363	19.386	19.410	870
880	19.410	19.434	19.457	19.481	19.505	19.528	19.552	19.576	19.599	19.623	19.646	880
890	19.646	19.670	19.694	19.717	19.741	19.765	19.788	19.812	19.836	19.859	19.883	890
900	19.883	19.907	19.930	19.954	19.978	20.001	20.025	20.049	20.072	20.096	20.120	900
910	20.120	20.143	20.167	20.190	20.214	20.238	20.261	20.285	20.309	20.332	20.356	910
920	20.356	20.380	20.403	20.427	20.451	20.474	20.498	20.522	20.545	20.569	20.593	920
930	20.593	20.616	20.640	20.664	20.688	20.711	20.735	20.759	20.782	20.806	20.830	930
940	20.830	20.853	20.877	20.901	20.924	20.948	20.972	20.995	21.019	21.043	21.066	940
950	21.066	21.090	21.114	21.137	21.161	21.185	21.208	21.232	21.256	21.280	21.303	950
960	21.303	21.327	21.351	21.374	21.398	21.422	21.445	21.469	21.493	21.516	21.540	960
970	21.540	21.564	21.587	21.611	21.635	21.659	21.682	21.706	21.730	21.753	21.777	970
980	21.777	21.801	21.824	21.848	21.872	21.895	21.919	21.943	21.966	21.990	22.014	980
990	22.014	22.038	22.061	22.085	22.109	22.132	22.156	22.180	22.203	22.227	22.251	990
1000	22.251	22.274	22.298	22.322	22.346	22.369	22.393	22.417	22.440	22.464	22.488	1000
1010	22.488	22.511	22.535	22.559	22.582	22.606	22.630	22.654	22.677	22.701	22.725	1010
1020	22.725	22.748	22.772	22.796	22.819	22.843	22.867	22.890	22.914	22.938	22.961	1020
1030	22.961	22.985	23.009	23.032	23.056	23.080	23.104	23.127	23.151	23.175	23.198	1030
1040	23.198	23.222	23.246	23.269	23.293	23.317	23.340	23.364	23.388	23.411	23.435	1040
1050	23.435	23.459	23.482	23.506	23.530	23.553	23.577	23.601	23.624	23.648	23.672	1050
1060	23.672	23.695	23.719	23.743	23.766	23.790	23.814	23.837	23.861	23.885	23.908	1060
1070	23.908	23.932	23.956	23.979	24.003	24.027	24.050	24.074	24.098	24.121	24.145	1070
1080	24.145	24.169	24.192	24.216	24.240	24.263	24.287	24.311	24.334	24.358	24.382	1080
1090	24.382	24.405	24.429	24.453	24.476	24.500	24.523	24.547	24.571	24.594	24.618	1090
1100	24.618	24.642	24.665	24.689	24.713	24.736	24.760	24.783	24.807	24.831	24.854	1100
1110	24.854	24.878	24.902	24.925	24.949	24.972	24.996	25.020	25.043	25.067	25.091	1110
1120	25.091	25.114	25.138	25.161	25.185	25.209	25.232	25.256	25.279	25.303	25.327	1120
1130	25.327	25.350	25.374	25.397	25.421	25.445	25.468	25.492	25.515	25.539	25.563	1130
1140	25.563	25.586	25.610	25.633	25.657	25.681	25.704	25.728	25.751	25.775	25.799	1140
1150	25.799	25.822	25.846	25.869	25.893	25.916	25.940	25.964	25.987	26.011	26.034	1150
1160	26.034	26.058	26.081	26.105	26.128	26.152	26.176	26.199	26.223	26.246	26.270	1160
1170	26.270	26.293	26.317	26.340	26.364	26.387	26.411	26.435	26.458	26.482	26.505	1170
1180	26.505	26.529	26.552	26.576	26.599	26.623	26.646	26.670	26.693	26.717	26.740	1180
1190	26.740	26.764	26.787	26.811	26.834	26.858	26.881	26.905	26.928	26.952	26.975	1190
1200	26.975	26.999	27.022	27.046	27.069	27.093	27.116	27.140	27.163	27.187	27.210	1200
1210	27.210	27.234	27.257	27.281	27.304	27.328	27.351	27.375	27.398	27.422	27.445	1210
1220	27.445	27.468	27.492	27.515	27.539	27.562	27.586	27.609	27.633	27.656	27.679	1220
1230	27.679	27.703	27.726	27.750	27.773	27.797	27.820	27.843	27.867	27.890	27.914	1230
1240	27.914	27.937	27.961	27.984	28.007	28.031	28.054	28.078	28.101	28.124	28.148	1240
1250	28.148	28.171	28.195	28.218	28.241	28.265	28.288	28.311	28.335	28.358	28.382	1250
1260	28.382	28.405	28.428	28.452	28.475	28.498	28.522	28.545	28.569	28.592	28.615	1260
1270	28.615	28.639	28.662	28.685	28.709	28.732	28.755	28.779	28.802	28.825	28.849	1270
1280	28.849	28.872	28.895	28.919	28.942	28.965	28.988	29.012	29.035	29.058	29.082	1280
1290	29.082	29.105	29.128	29.152	29.175	29.198	29.221	29.245	29.268	29.291	29.315	1290
1300	29.315	29.338	29.361	29.384	29.408	29.431	29.454	29.477	29.501	29.524	29.547	1300
1310	29.547	29.570	29.594	29.617	29.640	29.663	29.687	29.710	29.733	29.756	29.780	1310
1320	29.780	29.803	29.826	29.849	29.872	29.896	29.919	29.942	29.965	29.989	30.012	1320
1330	30.012	30.035	30.058	30.081	30.104	30.128	30.151	30.174	30.197	30.220	30.244	1330

Type K Thermocouple
Thermoelectric Voltage as a Function of Temperature (°F)
Reference Junctions at 32°F
THERMOELECTRIC VOLTAGE IN ABSOLUTE MILLIVOLTS

°F	0	1	2	3	4	5	6	7	8	9	10	°F
1340	30.244	30.267	30.290	30.313	30.336	30.359	30.383	30.406	30.429	30.452	30.475	1340
1350	30.475	30.498	30.521	30.545	30.568	30.591	30.614	30.637	30.660	30.683	30.706	1350
1360	30.706	30.730	30.753	30.776	30.799	30.822	30.845	30.868	30.891	30.914	30.937	1360
1370	30.937	30.961	30.984	31.007	31.030	31.053	31.076	31.099	31.122	31.145	31.168	1370
1380	31.168	31.191	31.214	31.237	31.260	31.283	31.306	31.329	31.353	31.376	31.399	1380
1390	31.399	31.422	31.445	31.468	31.491	31.514	31.537	31.560	31.583	31.606	31.629	1390
1400	31.629	31.652	31.675	31.698	31.721	31.744	31.767	31.790	31.813	31.836	31.859	1400
1410	31.859	31.882	31.905	31.927	31.950	31.973	31.996	32.019	32.042	32.065	32.088	1410
1420	32.088	32.111	32.134	32.157	32.180	32.203	32.226	32.249	32.272	32.294	32.317	1420
1430	32.317	32.340	32.363	32.386	32.409	32.432	32.455	32.478	32.501	32.523	32.546	1430
1440	32.546	32.569	32.592	32.615	32.638	32.661	32.683	32.706	32.729	32.752	32.775	1440
1450	32.775	32.798	32.821	32.843	32.866	32.889	32.912	32.935	32.958	32.980	33.003	1450
1460	33.003	33.026	33.049	33.072	33.094	33.117	33.140	33.163	33.186	33.208	33.231	1460
1470	33.231	33.254	33.277	33.300	33.322	33.345	33.368	33.391	33.413	33.436	33.459	1470
1480	33.459	33.482	33.504	33.527	33.550	33.573	33.595	33.618	33.641	33.664	33.686	1480
1490	33.686	33.709	33.732	33.754	33.777	33.800	33.823	33.845	33.868	33.891	33.913	1490
1500	33.913	33.936	33.959	33.981	34.004	34.027	34.049	34.072	34.095	34.117	34.140	1500
1510	34.140	34.163	34.185	34.208	34.231	34.253	34.276	34.299	34.321	34.344	34.366	1510
1520	34.366	34.389	34.412	34.434	34.457	34.480	34.502	34.525	34.547	34.570	34.593	1520
1530	34.593	34.615	34.638	34.660	34.683	34.705	34.728	34.751	34.773	34.796	34.818	1530
1540	34.818	34.841	34.863	34.886	34.909	34.931	34.954	34.976	34.999	35.021	35.044	1540
1550	35.044	35.066	35.089	35.111	35.134	35.156	35.179	35.201	35.224	35.246	35.269	1550
1560	35.269	35.291	35.314	35.336	35.359	35.381	35.404	35.426	35.449	35.471	35.494	1560
1570	35.494	35.516	35.539	35.561	35.583	35.606	35.628	35.651	35.673	35.696	35.718	1570
1580	35.718	35.741	35.763	35.785	35.808	35.830	35.853	35.875	35.897	35.920	35.942	1580
1590	35.942	35.965	35.987	36.009	36.032	36.054	36.077	36.099	36.121	36.144	36.166	1590
1600	36.166	36.188	36.211	36.233	36.256	36.278	36.300	36.323	36.345	36.367	36.390	1600
1610	36.390	36.412	36.434	36.457	36.479	36.501	36.524	36.546	36.568	36.590	36.613	1610
1620	36.613	36.635	36.657	36.680	36.702	36.724	36.746	36.769	36.791	36.813	36.836	1620
1630	36.836	36.858	36.880	36.902	36.925	36.947	36.969	36.991	37.014	37.036	37.058	1630
1640	37.058	37.080	37.103	37.125	37.147	37.169	37.191	37.214	37.236	37.258	37.280	1640
1650	37.280	37.303	37.325	37.347	37.369	37.391	37.413	37.436	37.458	37.480	37.502	1650
1660	37.502	37.524	37.547	37.569	37.591	37.613	37.635	37.657	37.679	37.702	37.724	1660
1670	37.724	37.746	37.768	37.790	37.812	37.834	37.857	37.879	37.901	37.923	37.945	1670
1680	37.945	37.967	37.989	38.011	38.033	38.055	38.078	38.100	38.122	38.144	38.166	1680
1690	38.166	38.188	38.210	38.232	38.254	38.276	38.298	38.320	38.342	38.364	38.387	1690
1700	38.387	38.409	38.431	38.453	38.475	38.497	38.519	38.541	38.563	38.585	38.607	1700
1710	38.607	38.629	38.651	38.673	38.695	38.717	38.739	38.761	38.783	38.805	38.827	1710
1720	38.827	38.849	38.871	38.893	38.915	38.937	38.959	38.981	39.003	39.024	39.046	1720
1730	39.046	39.068	39.090	39.112	39.134	39.156	39.178	39.200	39.222	39.244	39.266	1730
1740	39.266	39.288	39.310	39.331	39.353	39.375	39.397	39.419	39.441	39.463	39.485	1740
1750	39.485	39.507	39.529	39.550	39.572	39.594	39.616	39.638	39.660	39.682	39.703	1750
1760	39.703	39.725	39.747	39.769	39.791	39.813	39.835	39.856	39.878	39.900	39.922	1760
1770	39.922	39.944	39.965	39.987	40.009	40.031	40.053	40.075	40.096	40.118	40.140	1770
1780	40.140	40.162	40.183	40.205	40.227	40.249	40.271	40.292	40.314	40.336	40.358	1780
1790	40.358	40.379	40.401	40.423	40.445	40.466	40.488	40.510	40.532	40.553	40.575	1790
1800	40.575	40.597	40.619	40.640	40.662	40.684	40.705	40.727	40.749	40.770	40.792	1800
1810	40.792	40.814	40.836	40.857	40.879	40.901	40.922	40.944	40.966	40.987	41.009	1810
1820	41.009	41.031	41.052	41.074	41.096	41.117	41.139	41.161	41.182	41.204	41.225	1820
1830	41.225	41.247	41.269	41.290	41.312	41.334	41.355	41.377	41.398	41.420	41.442	1830
1840	41.442	41.463	41.485	41.506	41.528	41.550	41.571	41.593	41.614	41.636	41.657	1840
1850	41.657	41.679	41.701	41.722	41.744	41.765	41.787	41.808	41.830	41.851	41.873	1850
1860	41.873	41.895	41.916	41.938	41.959	41.981	42.002	42.024	42.045	42.067	42.088	1860
1870	42.088	42.110	42.131	42.153	42.174	42.196	42.217	42.239	42.260	42.282	42.303	1870
1880	42.303	42.325	42.346	42.367	42.389	42.410	42.432	42.453	42.475	42.496	42.518	1880
1890	42.518	42.539	42.560	42.582	42.603	42.625	42.646	42.668	42.689	42.710	42.732	1890
1900	42.732	42.753	42.775	42.796	42.817	42.839	42.860	42.882	42.903	42.924	42.946	1900
1910	42.946	42.967	42.989	43.010	43.031	43.053	43.074	43.095	43.117	43.138	43.159	1910
1920	43.159	43.181	43.202	43.223	43.245	43.266	43.287	43.309	43.330	43.351	43.373	1920
1930	43.373	43.394	43.415	43.436	43.458	43.479	43.500	43.522	43.543	43.564	43.585	1930

Type K Thermocouple
Thermoelectric Voltage as a Function of Temperature (°F)
Reference Junctions at 32°F
THERMOELECTRIC VOLTAGE IN ABSOLUTE MILLIVOLTS

°F	0	1	2	3	4	5	6	7	8	9	10	°F
1940	43.585	43.607	43.628	43.649	43.671	43.692	43.713	43.734	43.756	43.777	43.798	1940
1950	43.798	43.819	43.841	43.862	43.883	43.904	43.925	43.947	43.968	43.989	44.010	1950
1960	44.010	44.031	44.053	44.074	44.095	44.116	44.137	44.159	44.180	44.201	44.222	1960
1970	44.222	44.243	44.265	44.286	44.307	44.328	44.349	44.370	44.391	44.413	44.434	1970
1980	44.434	44.455	44.476	44.497	44.518	44.539	44.560	44.582	44.603	44.624	44.645	1980
1990	44.645	44.666	44.687	44.708	44.729	44.750	44.771	44.793	44.814	44.835	44.856	1990
2000	44.856	44.877	44.898	44.919	44.940	44.961	44.982	45.003	45.024	45.045	45.066	2000
2010	45.066	45.087	45.108	45.129	45.150	45.171	45.192	45.213	45.234	45.255	45.276	2010
2020	45.276	45.297	45.318	45.339	45.360	45.381	45.402	45.423	45.444	45.465	45.486	2020
2030	45.486	45.507	45.528	45.549	45.570	45.591	45.612	45.633	45.654	45.675	45.695	2030
2040	45.695	45.716	45.737	45.758	45.779	45.800	45.821	45.842	45.863	45.884	45.904	2040
2050	45.904	45.925	45.946	45.967	45.988	46.009	46.030	46.051	46.071	46.092	46.113	2050
2060	46.113	46.134	46.155	46.176	46.196	46.217	46.238	46.259	46.280	46.300	46.321	2060
2070	46.321	46.342	46.363	46.384	46.404	46.425	46.446	46.467	46.488	46.508	46.529	2070
2080	46.529	46.550	46.571	46.591	46.612	46.633	46.654	46.674	46.695	46.716	46.737	2080
2090	46.737	46.757	46.778	46.799	46.819	46.840	46.861	46.881	46.902	46.923	46.944	2090
2100	46.944	46.964	46.985	47.006	47.026	47.047	47.068	47.088	47.109	47.130	47.150	2100
2110	47.150	47.171	47.191	47.212	47.233	47.253	47.274	47.295	47.315	47.336	47.356	2110
2120	47.356	47.377	47.398	47.418	47.439	47.459	47.480	47.500	47.521	47.542	47.562	2120
2130	47.562	47.583	47.603	47.624	47.644	47.665	47.685	47.706	47.726	47.747	47.767	2130
2140	47.767	47.788	47.808	47.829	47.849	47.870	47.890	47.911	47.931	47.952	47.972	2140
2150	47.972	47.993	48.013	48.034	48.054	48.075	48.095	48.116	48.136	48.156	48.177	2150
2160	48.177	48.197	48.218	48.238	48.258	48.279	48.299	48.320	48.340	48.360	48.381	2160
2170	48.381	48.401	48.422	48.442	48.462	48.483	48.503	48.523	48.544	48.564	48.584	2170
2180	48.584	48.605	48.625	48.645	48.666	48.686	48.706	48.727	48.747	48.767	48.787	2180
2190	48.787	48.808	48.828	48.848	48.869	48.889	48.909	48.929	48.950	48.970	48.990	2190
2200	48.990	49.010	49.031	49.051	49.071	49.091	49.111	49.132	49.152	49.172	49.192	2200
2210	49.192	49.212	49.233	49.253	49.273	49.293	49.313	49.333	49.354	49.374	49.394	2210
2220	49.394	49.414	49.434	49.454	49.474	49.495	49.515	49.535	49.555	49.575	49.595	2220
2230	49.595	49.615	49.635	49.655	49.675	49.696	49.716	49.736	49.756	49.776	49.796	2230
2240	49.796	49.816	49.836	49.856	49.876	49.896	49.916	49.936	49.956	49.976	49.996	2240
2250	49.996	50.016	50.036	50.056	50.076	50.096	50.116	50.136	50.156	50.176	50.196	2250
2260	50.196	50.216	50.236	50.256	50.276	50.296	50.315	50.335	50.355	50.375	50.395	2260
2270	50.395	50.415	50.435	50.455	50.475	50.494	50.514	50.534	50.554	50.574	50.594	2270
2280	50.594	50.614	50.633	50.653	50.673	50.693	50.713	50.733	50.752	50.772	50.792	2280
2290	50.792	50.812	50.832	50.851	50.871	50.891	50.911	50.930	50.950	50.970	50.990	2290
2300	50.990	51.009	51.029	51.049	51.069	51.088	51.108	51.128	51.148	51.167	51.187	2300
2310	51.187	51.207	51.226	51.246	51.266	51.285	51.305	51.325	51.344	51.364	51.384	2310
2320	51.384	51.403	51.423	51.443	51.462	51.482	51.501	51.521	51.541	51.560	51.580	2320
2330	51.580	51.599	51.619	51.639	51.658	51.678	51.697	51.717	51.736	51.756	51.776	2330
2340	51.776	51.795	51.815	51.834	51.854	51.873	51.893	51.912	51.932	51.951	51.971	2340
2350	51.971	51.990	52.010	52.029	52.049	52.068	52.088	52.107	52.127	52.146	52.165	2350
2360	52.165	52.185	52.204	52.224	52.243	52.263	52.282	52.301	52.321	52.340	52.360	2360
2370	52.360	52.379	52.398	52.418	52.437	52.457	52.476	52.495	52.515	52.534	52.553	2370
2380	52.553	52.573	52.592	52.611	52.631	52.650	52.669	52.689	52.708	52.727	52.747	2380
2390	52.747	52.766	52.785	52.805	52.824	52.843	52.862	52.882	52.901	52.920	52.939	2390
2400	52.939	52.959	52.978	52.997	53.016	53.036	53.055	53.074	53.093	53.113	53.132	2400
2410	53.132	53.151	53.170	53.189	53.209	53.228	53.247	53.266	53.285	53.304	53.324	2410
2420	53.324	53.343	53.362	53.381	53.400	53.419	53.439	53.458	53.477	53.496	53.515	2420
2430	53.515	53.534	53.553	53.572	53.592	53.611	53.630	53.649	53.668	53.687	53.706	2430
2440	53.706	53.725	53.744	53.763	53.782	53.801	53.821	53.840	53.859	53.878	53.897	2440
2450	53.897	53.916	53.935	53.954	53.973	53.992	54.011	54.030	54.049	54.068	54.087	2450
2460	54.087	54.106	54.125	54.144	54.163	54.182	54.201	54.220	54.239	54.258	54.277	2460
2470	54.277	54.296	54.315	54.334	54.353	54.372	54.391	54.410	54.429	54.447	54.466	2470
2480	54.466	54.485	54.504	54.523	54.542	54.561	54.580	54.599	54.618	54.637	54.656	2480
2490	54.656	54.675	54.694	54.712	54.731	54.750	54.769	54.788	54.807	54.826	54.845	2490
2500	54.845	54.864	54.882									2500

Type N Thermocouple
Thermoelectric Voltage as a Function of Temperature (°C)
Reference Junctions at 0°C
THERMOELECTRIC VOLTAGE IN ABSOLUTE MILLIVOLTS

°C	0	1	2	3	4	5	6	7	8	9	10	°C
−270	−4.345	−4.345	−4.344	−4.344	−4.343	−4.342	−4.341	−4.340	−4.339	−4.337	−4.336	−270
−260	−4.336	−4.334	−4.332	−4.330	−4.328	−4.326	−4.324	−4.321	−4.319	−4.316	−4.313	−260
−250	−4.313	−4.310	−4.307	−4.304	−4.300	−4.297	−4.293	−4.289	−4.285	−4.281	−4.277	−250
−240	−4.277	−4.273	−4.268	−4.263	−4.259	−4.254	−4.248	−4.243	−4.238	−4.232	−4.227	−240
−230	−4.227	−4.221	−4.215	−4.209	−4.202	−4.196	−4.189	−4.183	−4.176	−4.169	−4.162	−230
−220	−4.162	−4.155	−4.147	−4.140	−4.132	−4.124	−4.116	−4.108	−4.100	−4.091	−4.083	−220
−210	−4.083	−4.074	−4.066	−4.057	−4.048	−4.038	−4.029	−4.020	−4.010	−4.000	−3.990	−210
−200	−3.990	−3.980	−3.970	−3.960	−3.950	−3.939	−3.928	−3.918	−3.907	−3.896	−3.884	−200
−190	−3.884	−3.873	−3.862	−3.850	−3.838	−3.827	−3.815	−3.803	−3.790	−3.778	−3.766	−190
−180	−3.766	−3.753	−3.740	−3.727	−3.715	−3.701	−3.688	−3.675	−3.661	−3.648	−3.634	−180
−170	−3.634	−3.620	−3.607	−3.592	−3.578	−3.564	−3.550	−3.535	−3.521	−3.506	−3.491	−170
−160	−3.491	−3.476	−3.461	−3.446	−3.430	−3.415	−3.399	−3.384	−3.368	−3.352	−3.336	−160
−150	−3.336	−3.320	−3.304	−3.288	−3.271	−3.255	−3.238	−3.221	−3.204	−3.187	−3.170	−150
−140	−3.170	−3.153	−3.136	−3.118	−3.101	−3.083	−3.066	−3.048	−3.030	−3.012	−2.994	−140
−130	−2.994	−2.976	−2.957	−2.939	−2.921	−2.902	−2.883	−2.864	−2.846	−2.827	−2.807	−130
−120	−2.807	−2.788	−2.769	−2.750	−2.730	−2.711	−2.691	−2.671	−2.651	−2.632	−2.612	−120
−110	−2.612	−2.591	−2.571	−2.551	−2.531	−2.510	−2.490	−2.469	−2.448	−2.427	−2.407	−110
−100	−2.407	−2.386	−2.365	−2.343	−2.322	−2.301	−2.280	−2.258	−2.237	−2.215	−2.193	−100
−90	−2.193	−2.171	−2.150	−2.128	−2.106	−2.084	−2.061	−2.039	−2.017	−1.995	−1.972	−90
−80	−1.972	−1.950	−1.927	−1.904	−1.882	−1.859	−1.836	−1.813	−1.790	−1.767	−1.744	−80
−70	−1.744	−1.721	−1.697	−1.674	−1.651	−1.627	−1.604	−1.580	−1.556	−1.533	−1.509	−70
−60	−1.509	−1.485	−1.461	−1.437	−1.413	−1.389	−1.365	−1.341	−1.317	−1.293	−1.268	−60
−50	−1.268	−1.244	−1.220	−1.195	−1.171	−1.146	−1.121	−1.097	−1.072	−1.047	−1.023	−50
−40	−1.023	−0.998	−0.973	−0.948	−0.923	−0.898	−0.873	−0.848	−0.823	−0.797	−0.772	−40
−30	−0.772	−0.747	−0.722	−0.696	−0.671	−0.646	−0.620	−0.595	−0.569	−0.544	−0.518	−30
−20	−0.518	−0.492	−0.467	−0.441	−0.415	−0.390	−0.364	−0.338	−0.312	−0.286	−0.260	−20
−10	−0.260	−0.234	−0.208	−0.183	−0.157	−0.130	−0.104	−0.078	−0.052	−0.026	0.000	−10
0	0.000	0.026	0.052	0.078	0.104	0.130	0.156	0.182	0.208	0.234	0.261	0
10	0.261	0.287	0.313	0.340	0.366	0.392	0.419	0.445	0.472	0.498	0.525	10
20	0.525	0.551	0.578	0.605	0.632	0.658	0.685	0.712	0.739	0.766	0.793	20
30	0.793	0.820	0.847	0.874	0.901	0.928	0.955	0.982	1.010	1.037	1.064	30
40	1.064	1.092	1.119	1.146	1.174	1.201	1.229	1.256	1.284	1.312	1.339	40
50	1.339	1.367	1.395	1.423	1.451	1.479	1.506	1.534	1.562	1.591	1.619	50
60	1.619	1.647	1.675	1.703	1.731	1.760	1.788	1.816	1.845	1.873	1.902	60
70	1.902	1.930	1.959	1.987	2.016	2.045	2.073	2.102	2.131	2.160	2.188	70
80	2.188	2.217	2.246	2.275	2.304	2.333	2.362	2.392	2.421	2.450	2.479	80
90	2.479	2.508	2.538	2.567	2.596	2.626	2.655	2.685	2.714	2.744	2.774	90
100	2.774	2.803	2.833	2.863	2.892	2.922	2.952	2.982	3.012	3.042	3.072	100
110	3.072	3.102	3.132	3.162	3.192	3.222	3.252	3.283	3.313	3.343	3.374	110
120	3.374	3.404	3.434	3.465	3.495	3.526	3.557	3.587	3.618	3.648	3.679	120
130	3.679	3.710	3.741	3.772	3.802	3.833	3.864	3.895	3.926	3.957	3.988	130
140	3.988	4.019	4.050	4.082	4.113	4.144	4.175	4.207	4.238	4.269	4.301	140
150	4.301	4.332	4.364	4.395	4.427	4.458	4.490	4.521	4.553	4.585	4.617	150
160	4.617	4.648	4.680	4.712	4.744	4.776	4.808	4.840	4.872	4.904	4.936	160
170	4.936	4.968	5.000	5.032	5.064	5.097	5.129	5.161	5.193	5.226	5.258	170
180	5.258	5.290	5.323	5.355	5.388	5.420	5.453	5.486	5.518	5.551	5.584	180
190	5.584	5.616	5.649	5.682	5.715	5.747	5.780	5.813	5.846	5.879	5.912	190
200	5.912	5.945	5.978	6.011	6.044	6.077	6.110	6.144	6.177	6.210	6.243	200
210	6.243	6.277	6.310	6.343	6.377	6.410	6.443	6.477	6.510	6.544	6.577	210
220	6.577	6.611	6.645	6.678	6.712	6.745	6.779	6.813	6.847	6.880	6.914	220
230	6.914	6.948	6.982	7.016	7.050	7.084	7.118	7.152	7.186	7.220	7.254	230
240	7.254	7.288	7.322	7.356	7.390	7.424	7.458	7.493	7.527	7.561	7.596	240
250	7.596	7.630	7.664	7.699	7.733	7.767	7.802	7.836	7.871	7.905	7.940	250
260	7.940	7.975	8.009	8.044	8.078	8.113	8.148	8.182	8.217	8.252	8.287	260
270	8.287	8.321	8.356	8.391	8.426	8.461	8.496	8.531	8.566	8.601	8.636	270
280	8.636	8.671	8.706	8.741	8.776	8.811	8.846	8.881	8.916	8.952	8.987	280
290	8.987	9.022	9.057	9.093	9.128	9.163	9.198	9.234	9.269	9.305	9.340	290
300	9.340	9.375	9.411	9.446	9.482	9.517	9.553	9.589	9.624	9.660	9.695	300
310	9.695	9.731	9.767	9.802	9.838	9.874	9.909	9.945	9.981	10.017	10.053	310
320	10.053	10.088	10.124	10.160	10.196	10.232	10.268	10.304	10.340	10.376	10.412	320

Type N Thermocouple
Thermoelectric Voltage as a Function of Temperature (°C)
Reference Junctions at 0°C
THERMOELECTRIC VOLTAGE IN ABSOLUTE MILLIVOLTS

°C	0	1	2	3	4	5	6	7	8	9	10	°C
330	10.412	10.448	10.484	10.520	10.556	10.592	10.628	10.664	10.700	10.736	10.772	330
340	10.772	10.809	10.845	10.881	10.917	10.954	10.990	11.026	11.062	11.099	11.135	340
350	11.135	11.171	11.208	11.244	11.281	11.317	11.354	11.390	11.426	11.463	11.499	350
360	11.499	11.536	11.572	11.609	11.646	11.682	11.719	11.755	11.792	11.829	11.865	360
370	11.865	11.902	11.939	11.975	12.012	12.049	12.086	12.122	12.159	12.196	12.233	370
380	12.233	12.270	12.306	12.343	12.380	12.417	12.454	12.491	12.528	12.565	12.602	380
390	12.602	12.639	12.676	12.713	12.750	12.787	12.824	12.861	12.898	12.935	12.972	390
400	12.972	13.009	13.046	13.084	13.121	13.158	13.195	13.232	13.269	13.307	13.344	400
410	13.344	13.381	13.418	13.456	13.493	13.530	13.568	13.605	13.642	13.680	13.717	410
420	13.717	13.754	13.792	13.829	13.867	13.904	13.942	13.979	14.017	14.054	14.091	420
430	14.091	14.129	14.167	14.204	14.242	14.279	14.317	14.354	14.392	14.430	14.467	430
440	14.467	14.505	14.542	14.580	14.618	14.655	14.693	14.731	14.769	14.806	14.844	440
450	14.844	14.882	14.919	14.957	14.995	15.033	15.071	15.108	15.146	15.184	15.222	450
460	15.222	15.260	15.298	15.336	15.373	15.411	15.449	15.487	15.525	15.563	15.601	460
470	15.601	15.639	15.677	15.715	15.753	15.791	15.829	15.867	15.905	15.943	15.981	470
480	15.981	16.019	16.057	16.095	16.133	16.172	16.210	16.248	16.286	16.324	16.362	480
490	16.362	16.400	16.439	16.477	16.515	16.553	16.591	16.630	16.668	16.706	16.744	490
500	16.744	16.783	16.821	16.859	16.897	16.936	16.974	17.012	17.051	17.089	17.127	500
510	17.127	17.166	17.204	17.243	17.281	17.319	17.358	17.396	17.434	17.473	17.511	510
520	17.511	17.550	17.588	17.627	17.665	17.704	17.742	17.781	17.819	17.858	17.896	520
530	17.896	17.935	17.973	18.012	18.050	18.089	18.127	18.166	18.204	18.243	18.282	530
540	18.282	18.320	18.359	18.397	18.436	18.475	18.513	18.552	18.591	18.629	18.668	540
550	18.668	18.707	18.745	18.784	18.823	18.861	18.900	18.939	18.977	19.016	19.055	550
560	19.055	19.094	19.132	19.171	19.210	19.249	19.287	19.326	19.365	19.404	19.443	560
570	19.443	19.481	19.520	19.559	19.598	19.637	19.676	19.714	19.753	19.792	19.831	570
580	19.831	19.870	19.909	19.948	19.986	20.025	20.064	20.103	20.142	20.181	20.220	580
590	20.220	20.259	20.298	20.337	20.376	20.415	20.453	20.492	20.531	20.570	20.609	590
600	20.609	20.648	20.687	20.726	20.765	20.804	20.843	20.882	20.921	20.960	20.999	600
610	20.999	21.038	21.077	21.116	21.155	21.195	21.234	21.273	21.312	21.351	21.390	610
620	21.390	21.429	21.468	21.507	21.546	21.585	21.624	21.663	21.702	21.742	21.781	620
630	21.781	21.820	21.859	21.898	21.937	21.976	22.015	22.055	22.094	22.133	22.172	630
640	22.172	22.211	22.250	22.289	22.329	22.368	22.407	22.446	22.485	22.524	22.564	640
650	22.564	22.603	22.642	22.681	22.720	22.760	22.799	22.838	22.877	22.916	22.956	650
660	22.956	22.995	23.034	23.073	23.112	23.152	23.191	23.230	23.269	23.309	23.348	660
670	23.348	23.387	23.426	23.466	23.505	23.544	23.583	23.623	23.662	23.701	23.740	670
680	23.740	23.780	23.819	23.858	23.897	23.937	23.976	24.015	24.054	24.094	24.133	680
690	24.133	24.172	24.212	24.251	24.290	24.329	24.369	24.408	24.447	24.487	24.526	690
700	24.526	24.565	24.604	24.644	24.683	24.722	24.762	24.801	24.840	24.879	24.919	700
710	24.919	24.958	24.997	25.037	25.076	25.115	25.155	25.194	25.233	25.273	25.312	710
720	25.312	25.351	25.391	25.430	25.469	25.508	25.548	25.587	25.626	25.666	25.705	720
730	25.705	25.744	25.784	25.823	25.862	25.902	25.941	25.980	26.020	26.059	26.098	730
740	26.098	26.138	26.177	26.216	26.255	26.295	26.334	26.373	26.413	26.452	26.491	740
750	26.491	26.531	26.570	26.609	26.649	26.688	26.727	26.767	26.806	26.845	26.885	750
760	26.885	26.924	26.963	27.002	27.042	27.081	27.120	27.160	27.199	27.238	27.278	760
770	27.278	27.317	27.356	27.396	27.435	27.474	27.513	27.553	27.592	27.631	27.671	770
780	27.671	27.710	27.749	27.788	27.828	27.867	27.906	27.946	27.985	28.024	28.063	780
790	28.063	28.103	28.142	28.181	28.221	28.260	28.299	28.338	28.378	28.417	28.456	790
800	28.456	28.495	28.535	28.574	28.613	28.652	28.692	28.731	28.770	28.809	28.849	800
810	28.849	28.888	28.927	28.966	29.006	29.045	29.084	29.123	29.163	29.202	29.241	810
820	29.241	29.280	29.319	29.359	29.398	29.437	29.476	29.516	29.555	29.594	29.633	820
830	29.633	29.672	29.712	29.751	29.790	29.829	29.868	29.908	29.947	29.986	30.025	830
840	30.025	30.064	30.103	30.143	30.182	30.221	30.260	30.299	30.338	30.378	30.417	840
850	30.417	30.456	30.495	30.534	30.573	30.612	30.652	30.691	30.730	30.769	30.808	850
860	30.808	30.847	30.886	30.925	30.964	31.004	31.043	31.082	31.121	31.160	31.199	860
870	31.199	31.238	31.277	31.316	31.355	31.394	31.434	31.473	31.512	31.551	31.590	870
880	31.590	31.629	31.668	31.707	31.746	31.785	31.824	31.863	31.902	31.941	31.980	880
890	31.980	32.019	32.058	32.097	32.136	32.175	32.214	32.253	32.292	32.331	32.370	890
900	32.370	32.409	32.448	32.487	32.526	32.565	32.604	32.643	32.682	32.721	32.760	900
910	32.760	32.799	32.838	32.877	32.916	32.955	32.993	33.032	33.071	33.110	33.149	910
920	33.149	33.188	33.227	33.266	33.305	33.344	33.382	33.421	33.460	33.499	33.538	920

Type N Thermocouple
Thermoelectric Voltage as a Function of Temperature (°C)
Reference Junctions at 0°C
THERMOELECTRIC VOLTAGE IN ABSOLUTE MILLIVOLTS

°C	0	1	2	3	4	5	6	7	8	9	10	°C
930	33.538	33.577	33.616	33.655	33.693	33.732	33.771	33.810	33.849	33.888	33.926	930
940	33.926	33.965	34.004	34.043	34.082	34.121	34.159	34.198	34.237	34.276	34.315	940
950	34.315	34.353	34.392	34.431	34.470	34.508	34.547	34.586	34.625	34.663	34.702	950
960	34.702	34.741	34.780	34.818	34.857	34.896	34.935	34.973	35.012	35.051	35.089	960
970	35.089	35.128	35.167	35.205	35.244	35.283	35.321	35.360	35.399	35.437	35.476	970
980	35.476	35.515	35.553	35.592	35.631	35.669	35.708	35.747	35.785	35.824	35.862	980
990	35.862	35.901	35.940	35.978	36.017	36.055	36.094	36.132	36.171	36.210	36.248	990
1000	36.248	36.287	36.325	36.364	36.402	36.441	36.479	36.518	36.556	36.595	36.633	1000
1010	36.633	36.672	36.710	36.749	36.787	36.826	36.864	36.903	36.941	36.980	37.018	1010
1020	37.018	37.057	37.095	37.134	37.172	37.210	37.249	37.287	37.326	37.364	37.402	1020
1030	37.402	37.441	37.479	37.518	37.556	37.594	37.633	37.671	37.710	37.748	37.786	1030
1040	37.786	37.825	37.863	37.901	37.940	37.978	38.016	38.055	38.093	38.131	38.169	1040
1050	38.169	38.208	38.246	38.284	38.323	38.361	38.399	38.437	38.476	38.514	38.552	1050
1060	38.552	38.590	38.628	38.667	38.705	38.743	38.781	38.819	38.858	38.896	38.934	1060
1070	38.934	38.972	39.010	39.049	39.087	39.125	39.163	39.201	39.239	39.277	39.315	1070
1080	39.315	39.354	39.392	39.430	39.468	39.506	39.544	39.582	39.620	39.658	39.696	1080
1090	39.696	39.734	39.772	39.810	39.848	39.886	39.924	39.962	40.000	40.038	40.076	1090
1100	40.076	40.114	40.152	40.190	40.228	40.266	40.304	40.342	40.380	40.418	40.456	1100
1110	40.456	40.494	40.532	40.570	40.607	40.645	40.683	40.721	40.759	40.797	40.835	1110
1120	40.835	40.872	40.910	40.948	40.986	41.024	41.062	41.099	41.137	41.175	41.213	1120
1130	41.213	41.250	41.288	41.326	41.364	41.401	41.439	41.477	41.515	41.552	41.590	1130
1140	41.590	41.628	41.665	41.703	41.741	41.778	41.816	41.854	41.891	41.929	41.966	1140
1150	41.966	42.004	42.042	42.079	42.117	42.154	42.192	42.229	42.267	42.305	42.342	1150
1160	42.342	42.380	42.417	42.455	42.492	42.530	42.567	42.605	42.642	42.680	42.717	1160
1170	42.717	42.754	42.792	42.829	42.867	42.904	42.941	42.979	43.016	43.054	43.091	1170
1180	43.091	43.128	43.166	43.203	43.240	43.278	43.315	43.352	43.389	43.427	43.464	1180
1190	43.464	43.501	43.538	43.576	43.613	43.650	43.687	43.725	43.762	43.799	43.836	1190
1200	43.836	43.873	43.910	43.948	43.985	44.022	44.059	44.096	44.133	44.170	44.207	1200
1210	44.207	44.244	44.281	44.318	44.355	44.393	44.430	44.467	44.504	44.541	44.577	1210
1220	44.577	44.614	44.651	44.688	44.725	44.762	44.799	44.836	44.873	44.910	44.947	1220
1230	44.947	44.984	45.020	45.057	45.094	45.131	45.168	45.204	45.241	45.278	45.315	1230
1240	45.315	45.352	45.388	45.425	45.462	45.498	45.535	45.572	45.609	45.645	45.682	1240
1250	45.682	45.719	45.755	45.792	45.828	45.865	45.902	45.938	45.975	46.011	46.048	1250
1260	46.048	46.085	46.121	46.158	46.194	46.231	46.267	46.304	46.340	46.377	46.413	1260
1270	46.413	46.449	46.486	46.522	46.559	46.595	46.631	46.668	46.704	46.741	46.777	1270
1280	46.777	46.813	46.850	46.886	46.922	46.959	46.995	47.031	47.067	47.104	47.140	1280
1290	47.140	47.176	47.212	47.249	47.285	47.321	47.357	47.393	47.430	47.466	47.502	1290

Type N Thermocouple
Thermoelectric Voltage as a Function of Temperature (°F)
Reference Junctions at 32°F
THERMOELECTRIC VOLTAGE IN ABSOLUTE MILLIVOLTS

°F	0	1	2	3	4	5	6	7	8	9	10	°F
−460							−4.345	−4.345	−4.345	−4.344	−4.344	−460
−450	−4.344	−4.344	−4.343	−4.343	−4.343	−4.342	−4.341	−4.341	−4.340	−4.340	−4.339	−450
−440	−4.339	−4.338	−4.337	−4.337	−4.336	−4.335	−4.334	−4.333	−4.332	−4.331	−4.330	−440
−430	−4.330	−4.329	−4.327	−4.326	−4.325	−4.324	−4.322	−4.321	−4.319	−4.318	−4.316	−430
−420	−4.316	−4.315	−4.313	−4.312	−4.310	−4.308	−4.306	−4.305	−4.303	−4.301	−4.299	−420
−410	−4.299	−4.297	−4.295	−4.293	−4.291	−4.289	−4.286	−4.284	−4.282	−4.279	−4.277	−410
−400	−4.277	−4.275	−4.272	−4.270	−4.267	−4.264	−4.262	−4.259	−4.256	−4.254	−4.251	−400
−390	−4.251	−4.248	−4.245	−4.242	−4.239	−4.236	−4.233	−4.230	−4.227	−4.223	−4.220	−390
−380	−4.220	−4.217	−4.213	−4.210	−4.207	−4.203	−4.200	−4.196	−4.192	−4.189	−4.185	−380
−370	−4.185	−4.181	−4.177	−4.174	−4.170	−4.166	−4.162	−4.158	−4.154	−4.150	−4.145	−370
−360	−4.145	−4.141	−4.137	−4.133	−4.128	−4.124	−4.120	−4.115	−4.111	−4.106	−4.102	−360
−350	−4.102	−4.097	−4.092	−4.088	−4.083	−4.078	−4.073	−4.068	−4.064	−4.059	−4.054	−350
−340	−4.054	−4.049	−4.043	−4.038	−4.033	−4.028	−4.023	−4.017	−4.012	−4.007	−4.001	−340
−330	−4.001	−3.996	−3.990	−3.985	−3.979	−3.974	−3.968	−3.962	−3.957	−3.951	−3.945	−330
−320	−3.945	−3.939	−3.933	−3.927	−3.921	−3.915	−3.909	−3.903	−3.897	−3.891	−3.884	−320
−310	−3.884	−3.878	−3.872	−3.865	−3.859	−3.853	−3.846	−3.840	−3.833	−3.827	−3.820	−310
−300	−3.820	−3.813	−3.807	−3.800	−3.793	−3.786	−3.779	−3.772	−3.766	−3.759	−3.752	−300
−290	−3.752	−3.745	−3.737	−3.730	−3.723	−3.716	−3.709	−3.701	−3.694	−3.687	−3.679	−290
−280	−3.679	−3.672	−3.664	−3.657	−3.649	−3.642	−3.634	−3.627	−3.619	−3.611	−3.603	−280
−270	−3.603	−3.596	−3.588	−3.580	−3.572	−3.564	−3.556	−3.548	−3.540	−3.532	−3.524	−270
−260	−3.524	−3.516	−3.507	−3.499	−3.491	−3.483	−3.474	−3.466	−3.458	−3.449	−3.441	−260
−250	−3.441	−3.432	−3.424	−3.415	−3.406	−3.398	−3.389	−3.380	−3.372	−3.363	−3.354	−250
−240	−3.354	−3.345	−3.336	−3.327	−3.318	−3.309	−3.300	−3.291	−3.282	−3.273	−3.264	−240
−230	−3.264	−3.255	−3.245	−3.236	−3.227	−3.217	−3.208	−3.199	−3.189	−3.180	−3.170	−230
−220	−3.170	−3.161	−3.151	−3.142	−3.132	−3.122	−3.113	−3.103	−3.093	−3.083	−3.074	−220
−210	−3.074	−3.064	−3.054	−3.044	−3.034	−3.024	−3.014	−3.004	−2.994	−2.984	−2.974	−210
−200	−2.974	−2.964	−2.953	−2.943	−2.933	−2.923	−2.912	−2.902	−2.892	−2.881	−2.871	−200
−190	−2.871	−2.860	−2.850	−2.839	−2.829	−2.818	−2.807	−2.797	−2.786	−2.775	−2.765	−190
−180	−2.765	−2.754	−2.743	−2.732	−2.722	−2.711	−2.700	−2.689	−2.678	−2.667	−2.656	−180
−170	−2.656	−2.645	−2.634	−2.623	−2.612	−2.600	−2.589	−2.578	−2.567	−2.555	−2.544	−170
−160	−2.544	−2.533	−2.521	−2.510	−2.499	−2.487	−2.476	−2.464	−2.453	−2.441	−2.430	−160
−150	−2.430	−2.418	−2.407	−2.395	−2.383	−2.372	−2.360	−2.348	−2.336	−2.325	−2.313	−150
−140	−2.313	−2.301	−2.289	−2.277	−2.265	−2.253	−2.241	−2.229	−2.217	−2.205	−2.193	−140
−130	−2.193	−2.181	−2.169	−2.157	−2.145	−2.133	−2.120	−2.108	−2.096	−2.084	−2.071	−130
−120	−2.071	−2.059	−2.047	−2.034	−2.022	−2.009	−1.997	−1.985	−1.972	−1.960	−1.947	−120
−110	−1.947	−1.935	−1.922	−1.909	−1.897	−1.884	−1.871	−1.859	−1.846	−1.833	−1.821	−110
−100	−1.821	−1.808	−1.795	−1.782	−1.770	−1.757	−1.744	−1.731	−1.718	−1.705	−1.692	−100
−90	−1.692	−1.679	−1.666	−1.653	−1.640	−1.627	−1.614	−1.601	−1.588	−1.575	−1.562	−90
−80	−1.562	−1.549	−1.535	−1.522	−1.509	−1.496	−1.483	−1.469	−1.456	−1.443	−1.429	−80
−70	−1.429	−1.416	−1.403	−1.389	−1.376	−1.363	−1.349	−1.336	−1.322	−1.309	−1.295	−70
−60	−1.295	−1.282	−1.268	−1.255	−1.241	−1.228	−1.214	−1.201	−1.187	−1.173	−1.160	−60
−50	−1.160	−1.146	−1.132	−1.119	−1.105	−1.091	−1.078	−1.064	−1.050	−1.036	−1.023	−50
−40	−1.023	−1.009	−0.995	−0.981	−0.967	−0.953	−0.940	−0.926	−0.912	−0.898	−0.884	−40
−30	−0.884	−0.870	−0.856	−0.842	−0.828	−0.814	−0.800	−0.786	−0.772	−0.758	−0.744	−30
−20	−0.744	−0.730	−0.716	−0.702	−0.688	−0.674	−0.660	−0.646	−0.631	−0.617	−0.603	−20
−10	−0.603	−0.589	−0.575	−0.561	−0.546	−0.532	−0.518	−0.504	−0.489	−0.475	−0.461	−10
0	−0.461	−0.447	−0.432	−0.418	−0.404	−0.390	−0.375	−0.361	−0.347	−0.332	−0.318	0
10	−0.318	−0.303	−0.289	−0.275	−0.260	−0.246	−0.232	−0.217	−0.203	−0.188	−0.174	10
20	−0.174	−0.159	−0.145	−0.130	−0.116	−0.102	−0.087	−0.073	−0.058	−0.044	−0.029	20
30	−0.029	−0.015	0.000	0.014	0.029	0.043	0.058	0.072	0.087	0.101	0.115	30
40	0.115	0.130	0.144	0.159	0.173	0.188	0.202	0.217	0.232	0.246	0.261	40
50	0.261	0.275	0.290	0.304	0.319	0.334	0.348	0.363	0.378	0.392	0.407	50
60	0.407	0.422	0.436	0.451	0.466	0.481	0.495	0.510	0.525	0.540	0.554	60
70	0.554	0.569	0.584	0.599	0.614	0.629	0.643	0.658	0.673	0.688	0.703	70
80	0.703	0.718	0.733	0.748	0.763	0.778	0.793	0.808	0.823	0.838	0.853	80
90	0.853	0.868	0.883	0.898	0.913	0.928	0.943	0.958	0.973	0.988	1.003	90
100	1.003	1.019	1.034	1.049	1.064	1.079	1.095	1.110	1.125	1.140	1.156	100
110	1.156	1.171	1.186	1.201	1.217	1.232	1.247	1.263	1.278	1.293	1.309	110
120	1.309	1.324	1.339	1.355	1.370	1.386	1.401	1.417	1.432	1.448	1.463	120
130	1.463	1.479	1.494	1.510	1.525	1.541	1.556	1.572	1.587	1.603	1.619	130

Type N Thermocouple
Thermoelectric Voltage as a Function of Temperature (°F)
Reference Junctions at 32°F
THERMOELECTRIC VOLTAGE IN ABSOLUTE MILLIVOLTS

°F	0	1	2	3	4	5	6	7	8	9	10	°F
140	1.619	1.634	1.650	1.666	1.681	1.697	1.713	1.728	1.744	1.760	1.775	140
150	1.775	1.791	1.807	1.823	1.838	1.854	1.870	1.886	1.902	1.917	1.933	150
160	1.933	1.949	1.965	1.981	1.997	2.013	2.029	2.045	2.060	2.076	2.092	160
170	2.092	2.108	2.124	2.140	2.156	2.172	2.188	2.204	2.221	2.237	2.253	170
180	2.253	2.269	2.285	2.301	2.317	2.333	2.349	2.366	2.382	2.398	2.414	180
190	2.414	2.430	2.447	2.463	2.479	2.495	2.512	2.528	2.544	2.561	2.577	190
200	2.577	2.593	2.610	2.626	2.642	2.659	2.675	2.691	2.708	2.724	2.741	200
210	2.741	2.757	2.774	2.790	2.807	2.823	2.840	2.856	2.873	2.889	2.906	210
220	2.906	2.922	2.939	2.955	2.972	2.989	3.005	3.022	3.038	3.055	3.072	220
230	3.072	3.088	3.105	3.122	3.139	3.155	3.172	3.189	3.205	3.222	3.239	230
240	3.239	3.256	3.273	3.289	3.306	3.323	3.340	3.357	3.374	3.391	3.407	240
250	3.407	3.424	3.441	3.458	3.475	3.492	3.509	3.526	3.543	3.560	3.577	250
260	3.577	3.594	3.611	3.628	3.645	3.662	3.679	3.696	3.713	3.730	3.748	260
270	3.748	3.765	3.782	3.799	3.816	3.833	3.850	3.868	3.885	3.902	3.919	270
280	3.919	3.936	3.954	3.971	3.988	4.005	4.023	4.040	4.057	4.075	4.092	280
290	4.092	4.109	4.127	4.144	4.161	4.179	4.196	4.214	4.231	4.248	4.266	290
300	4.266	4.283	4.301	4.318	4.336	4.353	4.371	4.388	4.406	4.423	4.441	300
310	4.441	4.458	4.476	4.493	4.511	4.529	4.546	4.564	4.581	4.599	4.617	310
320	4.617	4.634	4.652	4.670	4.687	4.705	4.723	4.740	4.758	4.776	4.794	320
330	4.794	4.811	4.829	4.847	4.865	4.882	4.900	4.918	4.936	4.954	4.971	330
340	4.971	4.989	5.007	5.025	5.043	5.061	5.079	5.097	5.114	5.132	5.150	340
350	5.150	5.168	5.186	5.204	5.222	5.240	5.258	5.276	5.294	5.312	5.330	350
360	5.330	5.348	5.366	5.384	5.402	5.420	5.439	5.457	5.475	5.493	5.511	360
370	5.511	5.529	5.547	5.565	5.584	5.602	5.620	5.638	5.656	5.674	5.693	370
380	5.693	5.711	5.729	5.747	5.766	5.784	5.802	5.820	5.839	5.857	5.875	380
390	5.875	5.894	5.912	5.930	5.949	5.967	5.985	6.004	6.022	6.040	6.059	390
400	6.059	6.077	6.096	6.114	6.132	6.151	6.169	6.188	6.206	6.225	6.243	400
410	6.243	6.262	6.280	6.299	6.317	6.336	6.354	6.373	6.391	6.410	6.429	410
420	6.429	6.447	6.466	6.484	6.503	6.521	6.540	6.559	6.577	6.596	6.615	420
430	6.615	6.633	6.652	6.671	6.689	6.708	6.727	6.745	6.764	6.783	6.802	430
440	6.802	6.820	6.839	6.858	6.877	6.895	6.914	6.933	6.952	6.971	6.989	440
450	6.989	7.008	7.027	7.046	7.065	7.084	7.102	7.121	7.140	7.159	7.178	450
460	7.178	7.197	7.216	7.235	7.254	7.273	7.291	7.310	7.329	7.348	7.367	460
470	7.367	7.386	7.405	7.424	7.443	7.462	7.481	7.500	7.519	7.538	7.557	470
480	7.557	7.576	7.596	7.615	7.634	7.653	7.672	7.691	7.710	7.729	7.748	480
490	7.748	7.767	7.787	7.806	7.825	7.844	7.863	7.882	7.902	7.921	7.940	490
500	7.940	7.959	7.978	7.998	8.017	8.036	8.055	8.075	8.094	8.113	8.132	500
510	8.132	8.152	8.171	8.190	8.209	8.229	8.248	8.267	8.287	8.306	8.325	510
520	8.325	8.345	8.364	8.383	8.403	8.422	8.441	8.461	8.480	8.500	8.519	520
530	8.519	8.538	8.558	8.577	8.597	8.616	8.636	8.655	8.675	8.694	8.713	530
540	8.713	8.733	8.752	8.772	8.791	8.811	8.830	8.850	8.870	8.889	8.909	540
550	8.909	8.928	8.948	8.967	8.987	9.006	9.026	9.046	9.065	9.085	9.104	550
560	9.104	9.124	9.144	9.163	9.183	9.202	9.222	9.242	9.261	9.281	9.301	560
570	9.301	9.320	9.340	9.360	9.379	9.399	9.419	9.439	9.458	9.478	9.498	570
580	9.498	9.517	9.537	9.557	9.577	9.596	9.616	9.636	9.656	9.676	9.695	580
590	9.695	9.715	9.735	9.755	9.775	9.794	9.814	9.834	9.854	9.874	9.894	590
600	9.894	9.913	9.933	9.953	9.973	9.993	10.013	10.033	10.053	10.072	10.092	600
610	10.092	10.112	10.132	10.152	10.172	10.192	10.212	10.232	10.252	10.272	10.292	610
620	10.292	10.312	10.332	10.352	10.372	10.392	10.412	10.432	10.452	10.472	10.492	620
630	10.492	10.512	10.532	10.552	10.572	10.592	10.612	10.632	10.652	10.672	10.692	630
640	10.692	10.712	10.732	10.752	10.772	10.793	10.813	10.833	10.853	10.873	10.893	640
650	10.893	10.913	10.933	10.954	10.974	10.994	11.014	11.034	11.054	11.075	11.095	650
660	11.095	11.115	11.135	11.155	11.176	11.196	11.216	11.236	11.256	11.277	11.297	660
670	11.297	11.317	11.337	11.358	11.378	11.398	11.418	11.439	11.459	11.479	11.499	670
680	11.499	11.520	11.540	11.560	11.581	11.601	11.621	11.642	11.662	11.682	11.703	680
690	11.703	11.723	11.743	11.764	11.784	11.804	11.825	11.845	11.865	11.886	11.906	690
700	11.906	11.926	11.947	11.967	11.988	12.008	12.028	12.049	12.069	12.090	12.110	700
710	12.110	12.131	12.151	12.171	12.192	12.212	12.233	12.253	12.274	12.294	12.315	710
720	12.315	12.335	12.356	12.376	12.397	12.417	12.438	12.458	12.479	12.499	12.520	720
730	12.520	12.540	12.561	12.581	12.602	12.622	12.643	12.663	12.684	12.705	12.725	730

Type N Thermocouple
Thermoelectric Voltage as a Function of Temperature (°F)
Reference Junctions at 32°F
THERMOELECTRIC VOLTAGE IN ABSOLUTE MILLIVOLTS

°F	0	1	2	3	4	5	6	7	8	9	10	°F
740	12.725	12.746	12.766	12.787	12.807	12.828	12.849	12.869	12.890	12.910	12.931	740
750	12.931	12.952	12.972	12.993	13.013	13.034	13.055	13.075	13.096	13.117	13.137	750
760	13.137	13.158	13.179	13.199	13.220	13.241	13.261	13.282	13.303	13.323	13.344	760
770	13.344	13.365	13.385	13.406	13.427	13.447	13.468	13.489	13.510	13.530	13.551	770
780	13.551	13.572	13.593	13.613	13.634	13.655	13.676	13.696	13.717	13.738	13.759	780
790	13.759	13.779	13.800	13.821	13.842	13.863	13.883	13.904	13.925	13.946	13.967	790
800	13.967	13.987	14.008	14.029	14.050	14.071	14.091	14.112	14.133	14.154	14.175	800
810	14.175	14.196	14.217	14.237	14.258	14.279	14.300	14.321	14.342	14.363	14.384	810
820	14.384	14.404	14.425	14.446	14.467	14.488	14.509	14.530	14.551	14.572	14.593	820
830	14.593	14.614	14.634	14.655	14.676	14.697	14.718	14.739	14.760	14.781	14.802	830
840	14.802	14.823	14.844	14.865	14.886	14.907	14.928	14.949	14.970	14.991	15.012	840
850	15.012	15.033	15.054	15.075	15.096	15.117	15.138	15.159	15.180	15.201	15.222	850
860	15.222	15.243	15.264	15.285	15.306	15.327	15.348	15.369	15.390	15.411	15.432	860
870	15.432	15.453	15.475	15.496	15.517	15.538	15.559	15.580	15.601	15.622	15.643	870
880	15.643	15.664	15.685	15.707	15.728	15.749	15.770	15.791	15.812	15.833	15.854	880
890	15.854	15.875	15.897	15.918	15.939	15.960	15.981	16.002	16.023	16.045	16.066	890
900	16.066	16.087	16.108	16.129	16.150	16.172	16.193	16.214	16.235	16.256	16.278	900
910	16.278	16.299	16.320	16.341	16.362	16.383	16.405	16.426	16.447	16.468	16.490	910
920	16.490	16.511	16.532	16.553	16.574	16.596	16.617	16.638	16.659	16.681	16.702	920
930	16.702	16.723	16.744	16.766	16.787	16.808	16.829	16.851	16.872	16.893	16.915	930
940	16.915	16.936	16.957	16.978	17.000	17.021	17.042	17.064	17.085	17.106	17.127	940
950	17.127	17.149	17.170	17.191	17.213	17.234	17.255	17.277	17.298	17.319	17.341	950
960	17.341	17.362	17.383	17.405	17.426	17.447	17.469	17.490	17.511	17.533	17.554	960
970	17.554	17.575	17.597	17.618	17.639	17.661	17.682	17.704	17.725	17.746	17.768	970
980	17.768	17.789	17.811	17.832	17.853	17.875	17.896	17.917	17.939	17.960	17.982	980
990	17.982	18.003	18.025	18.046	18.067	18.089	18.110	18.132	18.153	18.174	18.196	990
1000	18.196	18.217	18.239	18.260	18.282	18.303	18.325	18.346	18.367	18.389	18.410	1000
1010	18.410	18.432	18.453	18.475	18.496	18.518	18.539	18.561	18.582	18.603	18.625	1010
1020	18.625	18.646	18.668	18.689	18.711	18.732	18.754	18.775	18.797	18.818	18.840	1020
1030	18.840	18.861	18.883	18.904	18.926	18.947	18.969	18.990	19.012	19.033	19.055	1030
1040	19.055	19.076	19.098	19.120	19.141	19.163	19.184	19.206	19.227	19.249	19.270	1040
1050	19.270	19.292	19.313	19.335	19.356	19.378	19.400	19.421	19.443	19.464	19.486	1050
1060	19.486	19.507	19.529	19.550	19.572	19.594	19.615	19.637	19.658	19.680	19.701	1060
1070	19.701	19.723	19.745	19.766	19.788	19.809	19.831	19.853	19.874	19.896	19.917	1070
1080	19.917	19.939	19.961	19.982	20.004	20.025	20.047	20.069	20.090	20.112	20.133	1080
1090	20.133	20.155	20.177	20.198	20.220	20.241	20.263	20.285	20.306	20.328	20.350	1090
1100	20.350	20.371	20.393	20.415	20.436	20.458	20.479	20.501	20.523	20.544	20.566	1100
1110	20.566	20.588	20.609	20.631	20.653	20.674	20.696	20.718	20.739	20.761	20.783	1110
1120	20.783	20.804	20.826	20.848	20.869	20.891	20.913	20.934	20.956	20.978	20.999	1120
1130	20.999	21.021	21.043	21.064	21.086	21.108	21.129	21.151	21.173	21.195	21.216	1130
1140	21.216	21.238	21.260	21.281	21.303	21.325	21.346	21.368	21.390	21.411	21.433	1140
1150	21.433	21.455	21.477	21.498	21.520	21.542	21.563	21.585	21.607	21.629	21.650	1150
1160	21.650	21.672	21.694	21.716	21.737	21.759	21.781	21.802	21.824	21.846	21.868	1160
1170	21.868	21.889	21.911	21.933	21.955	21.976	21.998	22.020	22.041	22.063	22.085	1170
1180	22.085	22.107	22.128	22.150	22.172	22.194	22.215	22.237	22.259	22.281	22.302	1180
1190	22.302	22.324	22.346	22.368	22.390	22.411	22.433	22.455	22.477	22.498	22.520	1190
1200	22.520	22.542	22.564	22.585	22.607	22.629	22.651	22.672	22.694	22.716	22.738	1200
1210	22.738	22.760	22.781	22.803	22.825	22.847	22.868	22.890	22.912	22.934	22.956	1210
1220	22.956	22.977	22.999	23.021	23.043	23.064	23.086	23.108	23.130	23.152	23.173	1220
1230	23.173	23.195	23.217	23.239	23.261	23.282	23.304	23.326	23.348	23.370	23.391	1230
1240	23.391	23.413	23.435	23.457	23.479	23.500	23.522	23.544	23.566	23.588	23.609	1240
1250	23.609	23.631	23.653	23.675	23.697	23.718	23.740	23.762	23.784	23.806	23.828	1250
1260	23.828	23.849	23.871	23.893	23.915	23.937	23.958	23.980	24.002	24.024	24.046	1260
1270	24.046	24.068	24.089	24.111	24.133	24.155	24.177	24.198	24.220	24.242	24.264	1270
1280	24.264	24.286	24.308	24.329	24.351	24.373	24.395	24.417	24.439	24.460	24.482	1280
1290	24.482	24.504	24.526	24.548	24.569	24.591	24.613	24.635	24.657	24.679	24.700	1290
1300	24.700	24.722	24.744	24.766	24.788	24.810	24.831	24.853	24.875	24.897	24.919	1300
1310	24.919	24.941	24.962	24.984	25.006	25.028	25.050	25.072	25.093	25.115	25.137	1310
1320	25.137	25.159	25.181	25.203	25.225	25.246	25.268	25.290	25.312	25.334	25.356	1320
1330	25.356	25.377	25.399	25.421	25.443	25.465	25.487	25.508	25.530	25.552	25.574	1330

Type N Thermocouple
Thermoelectric Voltage as a Function of Temperature (°F)
Reference Junctions at 32°F
THERMOELECTRIC VOLTAGE IN ABSOLUTE MILLIVOLTS

°F	0	1	2	3	4	5	6	7	8	9	10	°F
1340	25.574	25.596	25.618	25.640	25.661	25.683	25.705	25.727	25.749	25.771	25.792	1340
1350	25.792	25.814	25.836	25.858	25.880	25.902	25.923	25.945	25.967	25.989	26.011	1350
1360	26.011	26.033	26.055	26.076	26.098	26.120	26.142	26.164	26.186	26.207	26.229	1360
1370	26.229	26.251	26.273	26.295	26.317	26.338	26.360	26.382	26.404	26.426	26.448	1370
1380	26.448	26.470	26.491	26.513	26.535	26.557	26.579	26.601	26.622	26.644	26.666	1380
1390	26.666	26.688	26.710	26.732	26.753	26.775	26.797	26.819	26.841	26.863	26.885	1390
1400	26.885	26.906	26.928	26.950	26.972	26.994	27.016	27.037	27.059	27.081	27.103	1400
1410	27.103	27.125	27.147	27.168	27.190	27.212	27.234	27.256	27.278	27.299	27.321	1410
1420	27.321	27.343	27.365	27.387	27.409	27.430	27.452	27.474	27.496	27.518	27.540	1420
1430	27.540	27.561	27.583	27.605	27.627	27.649	27.671	27.692	27.714	27.736	27.758	1430
1440	27.758	27.780	27.802	27.823	27.845	27.867	27.889	27.911	27.933	27.954	27.976	1440
1450	27.976	27.998	28.020	28.042	28.063	28.085	28.107	28.129	28.151	28.173	28.194	1450
1460	28.194	28.216	28.238	28.260	28.282	28.303	28.325	28.347	28.369	28.391	28.413	1460
1470	28.413	28.434	28.456	28.478	28.500	28.522	28.543	28.565	28.587	28.609	28.631	1470
1480	28.631	28.652	28.674	28.696	28.718	28.740	28.761	28.783	28.805	28.827	28.849	1480
1490	28.849	28.871	28.892	28.914	28.936	28.958	28.980	29.001	29.023	29.045	29.067	1490
1500	29.067	29.088	29.110	29.132	29.154	29.176	29.197	29.219	29.241	29.263	29.285	1500
1510	29.285	29.306	29.328	29.350	29.372	29.394	29.415	29.437	29.459	29.481	29.502	1510
1520	29.502	29.524	29.546	29.568	29.590	29.611	29.633	29.655	29.677	29.699	29.720	1520
1530	29.720	29.742	29.764	29.786	29.807	29.829	29.851	29.873	29.894	29.916	29.938	1530
1540	29.938	29.960	29.982	30.003	30.025	30.047	30.069	30.090	30.112	30.134	30.156	1540
1550	30.156	30.177	30.199	30.221	30.243	30.264	30.286	30.308	30.330	30.351	30.373	1550
1560	30.373	30.395	30.417	30.438	30.460	30.482	30.504	30.525	30.547	30.569	30.591	1560
1570	30.591	30.612	30.634	30.656	30.678	30.699	30.721	30.743	30.765	30.786	30.808	1570
1580	30.808	30.830	30.851	30.873	30.895	30.917	30.938	30.960	30.982	31.004	31.025	1580
1590	31.025	31.047	31.069	31.090	31.112	31.134	31.156	31.177	31.199	31.221	31.242	1590
1600	31.242	31.264	31.286	31.308	31.329	31.351	31.373	31.394	31.416	31.438	31.460	1600
1610	31.460	31.481	31.503	31.525	31.546	31.568	31.590	31.611	31.633	31.655	31.677	1610
1620	31.677	31.698	31.720	31.742	31.763	31.785	31.807	31.828	31.850	31.872	31.893	1620
1630	31.893	31.915	31.937	31.958	31.980	32.002	32.023	32.045	32.067	32.089	32.110	1630
1640	32.110	32.132	32.154	32.175	32.197	32.219	32.240	32.262	32.284	32.305	32.327	1640
1650	32.327	32.349	32.370	32.392	32.413	32.435	32.457	32.478	32.500	32.522	32.543	1650
1660	32.543	32.565	32.587	32.608	32.630	32.652	32.673	32.695	32.717	32.738	32.760	1660
1670	32.760	32.781	32.803	32.825	32.846	32.868	32.890	32.911	32.933	32.955	32.976	1670
1680	32.976	32.998	33.019	33.041	33.063	33.084	33.106	33.127	33.149	33.171	33.192	1680
1690	33.192	33.214	33.236	33.257	33.279	33.300	33.322	33.344	33.365	33.387	33.408	1690
1700	33.408	33.430	33.452	33.473	33.495	33.516	33.538	33.560	33.581	33.603	33.624	1700
1710	33.624	33.646	33.668	33.689	33.711	33.732	33.754	33.775	33.797	33.819	33.840	1710
1720	33.840	33.862	33.883	33.905	33.926	33.948	33.970	33.991	34.013	34.034	34.056	1720
1730	34.056	34.077	34.099	34.121	34.142	34.164	34.185	34.207	34.228	34.250	34.271	1730
1740	34.271	34.293	34.315	34.336	34.358	34.379	34.401	34.422	34.444	34.465	34.487	1740
1750	34.487	34.508	34.530	34.551	34.573	34.595	34.616	34.638	34.659	34.681	34.702	1750
1760	34.702	34.724	34.745	34.767	34.788	34.810	34.831	34.853	34.874	34.896	34.917	1760
1770	34.917	34.939	34.960	34.982	35.003	35.025	35.046	35.068	35.089	35.111	35.132	1770
1780	35.132	35.154	35.175	35.197	35.218	35.240	35.261	35.283	35.304	35.326	35.347	1780
1790	35.347	35.369	35.390	35.412	35.433	35.455	35.476	35.498	35.519	35.540	35.562	1790
1800	35.562	35.583	35.605	35.626	35.648	35.669	35.691	35.712	35.734	35.755	35.777	1800
1810	35.777	35.798	35.819	35.841	35.862	35.884	35.905	35.927	35.948	35.970	35.991	1810
1820	35.991	36.012	36.034	36.055	36.077	36.098	36.120	36.141	36.162	36.184	36.205	1820
1830	36.205	36.227	36.248	36.270	36.291	36.312	36.334	36.355	36.377	36.398	36.419	1830
1840	36.419	36.441	36.462	36.484	36.505	36.526	36.548	36.569	36.591	36.612	36.633	1840
1850	36.633	36.655	36.676	36.698	36.719	36.740	36.762	36.783	36.805	36.826	36.847	1850
1860	36.847	36.869	36.890	36.911	36.933	36.954	36.975	36.997	37.018	37.040	37.061	1860
1870	37.061	37.082	37.104	37.125	37.146	37.168	37.189	37.210	37.232	37.253	37.274	1870
1880	37.274	37.296	37.317	37.338	37.360	37.381	37.402	37.424	37.445	37.466	37.488	1880
1890	37.488	37.509	37.530	37.552	37.573	37.594	37.616	37.637	37.658	37.680	37.701	1890
1900	37.701	37.722	37.744	37.765	37.786	37.808	37.829	37.850	37.871	37.893	37.914	1900
1910	37.914	37.935	37.957	37.978	37.999	38.020	38.042	38.063	38.084	38.106	38.127	1910
1920	38.127	38.148	38.169	38.191	38.212	38.233	38.254	38.276	38.297	38.318	38.340	1920
1930	38.340	38.361	38.382	38.403	38.425	38.446	38.467	38.488	38.510	38.531	38.552	1930

Type N Thermocouple
Thermoelectric Voltage as a Function of Temperature (°F)
Reference Junctions at 32°F
THERMOELECTRIC VOLTAGE IN ABSOLUTE MILLIVOLTS

°F	0	1	2	3	4	5	6	7	8	9	10	°F
1940	38.552	38.573	38.594	38.616	38.637	38.658	38.679	38.701	38.722	38.743	38.764	1940
1950	38.764	38.786	38.807	38.828	38.849	38.870	38.892	38.913	38.934	38.955	38.976	1950
1960	38.976	38.998	39.019	39.040	39.061	39.082	39.104	39.125	39.146	39.167	39.188	1960
1970	39.188	39.210	39.231	39.252	39.273	39.294	39.315	39.337	39.358	39.379	39.400	1970
1980	39.400	39.421	39.442	39.464	39.485	39.506	39.527	39.548	39.569	39.591	39.612	1980
1990	39.612	39.633	39.654	39.675	39.696	39.717	39.739	39.760	39.781	39.802	39.823	1990
2000	39.823	39.844	39.865	39.886	39.908	39.929	39.950	39.971	39.992	40.013	40.034	2000
2010	40.034	40.055	40.076	40.097	40.119	40.140	40.161	40.182	40.203	40.224	40.245	2010
2020	40.245	40.266	40.287	40.308	40.329	40.351	40.372	40.393	40.414	40.435	40.456	2020
2030	40.456	40.477	40.498	40.519	40.540	40.561	40.582	40.603	40.624	40.645	40.666	2030
2040	40.666	40.687	40.708	40.729	40.750	40.772	40.793	40.814	40.835	40.856	40.877	2040
2050	40.877	40.898	40.919	40.940	40.961	40.982	41.003	41.024	41.045	41.066	41.087	2050
2060	41.087	41.108	41.129	41.150	41.171	41.192	41.213	41.234	41.255	41.276	41.297	2060
2070	41.297	41.318	41.338	41.359	41.380	41.401	41.422	41.443	41.464	41.485	41.506	2070
2080	41.506	41.527	41.548	41.569	41.590	41.611	41.632	41.653	41.674	41.695	41.716	2080
2090	41.716	41.736	41.757	41.778	41.799	41.820	41.841	41.862	41.883	41.904	41.925	2090
2100	41.925	41.946	41.966	41.987	42.008	42.029	42.050	42.071	42.092	42.113	42.134	2100
2110	42.134	42.154	42.175	42.196	42.217	42.238	42.259	42.280	42.300	42.321	42.342	2110
2120	42.342	42.363	42.384	42.405	42.425	42.446	42.467	42.488	42.509	42.530	42.550	2120
2130	42.550	42.571	42.592	42.613	42.634	42.655	42.675	42.696	42.717	42.738	42.759	2130
2140	42.759	42.779	42.800	42.821	42.842	42.862	42.883	42.904	42.925	42.946	42.966	2140
2150	42.966	42.987	43.008	43.029	43.049	43.070	43.091	43.112	43.132	43.153	43.174	2150
2160	43.174	43.195	43.215	43.236	43.257	43.278	43.298	43.319	43.340	43.360	43.381	2160
2170	43.381	43.402	43.423	43.443	43.464	43.485	43.505	43.526	43.547	43.567	43.588	2170
2180	43.588	43.609	43.629	43.650	43.671	43.692	43.712	43.733	43.754	43.774	43.795	2180
2190	43.795	43.815	43.836	43.857	43.877	43.898	43.919	43.939	43.960	43.981	44.001	2190
2200	44.001	44.022	44.042	44.063	44.084	44.104	44.125	44.146	44.166	44.187	44.207	2200
2210	44.207	44.228	44.248	44.269	44.290	44.310	44.331	44.351	44.372	44.393	44.413	2210
2220	44.413	44.434	44.454	44.475	44.495	44.516	44.536	44.557	44.577	44.598	44.619	2220
2230	44.619	44.639	44.660	44.680	44.701	44.721	44.742	44.762	44.783	44.803	44.824	2230
2240	44.824	44.844	44.865	44.885	44.906	44.926	44.947	44.967	44.988	45.008	45.029	2240
2250	45.029	45.049	45.069	45.090	45.110	45.131	45.151	45.172	45.192	45.213	45.233	2250
2260	45.233	45.254	45.274	45.294	45.315	45.335	45.356	45.376	45.396	45.417	45.437	2260
2270	45.437	45.458	45.478	45.498	45.519	45.539	45.560	45.580	45.600	45.621	45.641	2270
2280	45.641	45.662	45.682	45.702	45.723	45.743	45.763	45.784	45.804	45.824	45.845	2280
2290	45.845	45.865	45.885	45.906	45.926	45.946	45.967	45.987	46.007	46.028	46.048	2290
2300	46.048	46.068	46.089	46.109	46.129	46.149	46.170	46.190	46.210	46.231	46.251	2300
2310	46.251	46.271	46.291	46.312	46.332	46.352	46.372	46.393	46.413	46.433	46.453	2310
2320	46.453	46.474	46.494	46.514	46.534	46.555	46.575	46.595	46.615	46.636	46.656	2320
2330	46.656	46.676	46.696	46.716	46.737	46.757	46.777	46.797	46.817	46.838	46.858	2330
2340	46.858	46.878	46.898	46.918	46.938	46.959	46.979	46.999	47.019	47.039	47.059	2340
2350	47.059	47.079	47.100	47.120	47.140	47.160	47.180	47.200	47.220	47.241	47.261	2350
2360	47.261	47.281	47.301	47.321	47.341	47.361	47.381	47.401	47.421	47.442	47.462	2360
2370	47.462	47.482	47.502									2370

Type R Thermocouple
Thermoelectric Voltage as a Function of Temperature (°C)
Reference Junctions at 0°C
THERMOELECTRIC VOLTAGE IN ABSOLUTE MILLIVOLTS

°C	0	1	2	3	4	5	6	7	8	9	10	°C
−50	−0.226	−0.223	−0.219	−0.215	−0.211	−0.207	−0.204	−0.200	−0.196	−0.192	−0.188	−50
−40	−0.188	−0.184	−0.180	−0.175	−0.171	−0.167	−0.163	−0.158	−0.154	−0.150	−0.145	−40
−30	−0.145	−0.141	−0.137	−0.132	−0.128	−0.123	−0.119	−0.114	−0.109	−0.105	−0.100	−30
−20	−0.100	−0.095	−0.091	−0.086	−0.081	−0.076	−0.071	−0.066	−0.061	−0.056	−0.051	−20
−10	−0.051	−0.046	−0.041	−0.036	−0.031	−0.026	−0.021	−0.016	−0.011	−0.005	0.000	−10
0	0.000	0.005	0.011	0.016	0.021	0.027	0.032	0.038	0.043	0.049	0.054	0
10	0.054	0.060	0.065	0.071	0.077	0.082	0.088	0.094	0.100	0.105	0.111	10
20	0.111	0.117	0.123	0.129	0.135	0.141	0.147	0.152	0.158	0.165	0.171	20
30	0.171	0.177	0.183	0.189	0.195	0.201	0.207	0.214	0.220	0.226	0.232	30
40	0.232	0.239	0.245	0.251	0.258	0.264	0.271	0.277	0.283	0.290	0.296	40
0	0.296	0.303	0.310	0.316	0.323	0.329	0.336	0.343	0.349	0.356	0.363	50
60	0.363	0.369	0.376	0.383	0.390	0.397	0.403	0.410	0.417	0.424	0.431	60
70	0.431	0.438	0.445	0.452	0.459	0.466	0.473	0.480	0.487	0.494	0.501	70
80	0.501	0.508	0.515	0.523	0.530	0.537	0.544	0.552	0.559	0.566	0.573	80
90	0.573	0.581	0.588	0.595	0.603	0.610	0.617	0.625	0.632	0.640	0.647	90
100	0.647	0.655	0.662	0.670	0.677	0.685	0.692	0.700	0.708	0.715	0.723	100
110	0.723	0.730	0.738	0.746	0.754	0.761	0.769	0.777	0.784	0.792	0.800	110
120	0.800	0.808	0.816	0.824	0.831	0.839	0.847	0.855	0.863	0.871	0.879	120
130	0.879	0.887	0.895	0.903	0.911	0.919	0.927	0.935	0.943	0.951	0.959	130
140	0.959	0.967	0.975	0.983	0.992	1.000	1.008	1.016	1.024	1.032	1.041	140
150	1.041	1.049	1.057	1.065	1.074	1.082	1.090	1.099	1.107	1.115	1.124	150
160	1.124	1.132	1.140	1.149	1.157	1.166	1.174	1.183	1.191	1.200	1.208	160
170	1.208	1.217	1.225	1.234	1.242	1.251	1.259	1.268	1.276	1.285	1.294	170
180	1.294	1.302	1.311	1.319	1.328	1.337	1.345	1.354	1.363	1.372	1.380	180
190	1.380	1.389	1.398	1.407	1.415	1.424	1.433	1.442	1.450	1.459	1.468	190
200	1.468	1.477	1.486	1.495	1.504	1.512	1.521	1.530	1.539	1.548	1.557	200
210	1.557	1.566	1.575	1.584	1.593	1.602	1.611	1.620	1.629	1.638	1.647	210
220	1.647	1.656	1.665	1.674	1.683	1.692	1.702	1.711	1.720	1.729	1.738	220
230	1.738	1.747	1.756	1.766	1.775	1.784	1.793	1.802	1.812	1.821	1.830	230
240	1.830	1.839	1.849	1.858	1.867	1.876	1.886	1.895	1.904	1.914	1.923	240
250	1.923	1.932	1.942	1.951	1.960	1.970	1.979	1.988	1.998	2.007	2.017	250
260	2.017	2.026	2.036	2.045	2.054	2.064	2.073	2.083	2.092	2.102	2.111	260
270	2.111	2.121	2.130	2.140	2.149	2.159	2.169	2.178	2.188	2.197	2.207	270
280	2.207	2.216	2.226	2.236	2.245	2.255	2.264	2.274	2.284	2.293	2.303	280
290	2.303	2.313	2.322	2.332	2.342	2.351	2.361	2.371	2.381	2.390	2.400	290
300	2.400	2.410	2.420	2.429	2.439	2.449	2.459	2.468	2.478	2.488	2.498	300
310	2.498	2.508	2.517	2.527	2.537	2.547	2.557	2.567	2.577	2.586	2.596	310
320	2.596	2.606	2.616	2.626	2.636	2.646	2.656	2.666	2.676	2.685	2.695	320
330	2.695	2.705	2.715	2.725	2.735	2.745	2.755	2.765	2.775	2.785	2.795	330
340	2.795	2.805	2.815	2.825	2.835	2.845	2.855	2.866	2.876	2.886	2.896	340
350	2.896	2.906	2.916	2.926	2.936	2.946	2.956	2.966	2.977	2.987	2.997	350
360	2.997	3.007	3.017	3.027	3.037	3.048	3.058	3.068	3.078	3.088	3.099	360
370	3.099	3.109	3.119	3.129	3.139	3.150	3.160	3.170	3.180	3.191	3.201	370
380	3.201	3.211	3.221	3.232	3.242	3.252	3.263	3.273	3.283	3.293	3.304	380
390	3.304	3.314	3.324	3.335	3.345	3.355	3.366	3.376	3.386	3.397	3.407	390
400	3.407	3.418	3.428	3.438	3.449	3.459	3.470	3.480	3.490	3.501	3.511	400
410	3.511	3.522	3.532	3.543	3.553	3.563	3.574	3.584	3.595	3.605	3.616	410
420	3.616	3.626	3.637	3.647	3.658	3.668	3.679	3.689	3.700	3.710	3.721	420
430	3.721	3.731	3.742	3.752	3.763	3.774	3.784	3.795	3.805	3.816	3.826	430
440	3.826	3.837	3.848	3.858	3.869	3.879	3.890	3.901	3.911	3.922	3.933	440
450	3.933	3.943	3.954	3.964	3.975	3.986	3.996	4.007	4.018	4.028	4.039	450
460	4.039	4.050	4.061	4.071	4.082	4.093	4.103	4.114	4.125	4.136	4.146	460
470	4.146	4.157	4.168	4.178	4.189	4.200	4.211	4.222	4.232	4.243	4.254	470
480	4.254	4.265	4.275	4.286	4.297	4.308	4.319	4.329	4.340	4.351	4.362	480
490	4.362	4.373	4.384	4.394	4.405	4.416	4.427	4.438	4.449	4.460	4.471	490
500	4.471	4.481	4.492	4.503	4.514	4.525	4.536	4.547	4.558	4.569	4.580	500
510	4.580	4.591	4.601	4.612	4.623	4.634	4.645	4.656	4.667	4.678	4.689	510
520	4.689	4.700	4.711	4.722	4.733	4.744	4.755	4.766	4.777	4.788	4.799	520
530	4.799	4.810	4.821	4.832	4.843	4.854	4.865	4.876	4.888	4.899	4.910	530
540	4.910	4.921	4.932	4.943	4.954	4.965	4.976	4.987	4.998	5.009	5.021	540

Type R Thermocouple
Thermoelectric Voltage as a Function of Temperature (°C)
Reference Junctions at 0°C
THERMOELECTRIC VOLTAGE IN ABSOLUTE MILLIVOLTS

°C	0	1	2	3	4	5	6	7	8	9	10	°C
550	5.021	5.032	5.043	5.054	5.065	5.076	5.087	5.099	5.110	5.121	5.132	550
560	5.132	5.143	5.154	5.166	5.177	5.188	5.199	5.210	5.221	5.233	5.244	560
570	5.244	5.255	5.266	5.278	5.289	5.300	5.311	5.322	5.334	5.345	5.356	570
580	5.356	5.368	5.379	5.390	5.401	5.413	5.424	5.435	5.446	5.458	5.469	580
590	5.469	5.480	5.492	5.503	5.514	5.526	5.537	5.548	5.560	5.571	5.582	590
600	5.582	5.594	5.605	5.616	5.628	5.639	5.650	5.662	5.673	5.685	5.696	600
610	5.696	5.707	5.719	5.730	5.742	5.753	5.764	5.776	5.787	5.799	5.810	610
620	5.810	5.821	5.833	5.844	5.856	5.867	5.879	5.890	5.902	5.913	5.925	620
630	5.925	5.936	5.948	5.959	5.971	5.982	5.994	6.005	6.017	6.028	6.040	630
640	6.040	6.051	6.063	6.074	6.086	6.098	6.109	6.121	6.132	6.144	6.155	640
650	6.155	6.167	6.179	6.190	6.202	6.213	6.225	6.237	6.248	6.260	6.272	650
660	6.272	6.283	6.295	6.307	6.318	6.330	6.342	6.353	6.365	6.377	6.388	660
670	6.388	6.400	6.412	6.423	6.435	6.447	6.458	6.470	6.482	6.494	6.505	670
680	6.505	6.517	6.529	6.541	6.552	6.564	6.576	6.588	6.599	6.611	6.623	680
690	6.623	6.635	6.647	6.658	6.670	6.682	6.694	6.706	6.718	6.729	6.741	690
700	6.741	6.753	6.765	6.777	6.789	6.800	6.812	6.824	6.836	6.848	6.860	700
710	6.860	6.872	6.884	6.895	6.907	6.919	6.931	6.943	6.955	6.967	6.979	710
720	6.979	6.991	7.003	7.015	7.027	7.039	7.051	7.063	7.074	7.086	7.098	720
730	7.098	7.110	7.122	7.134	7.146	7.158	7.170	7.182	7.194	7.206	7.218	730
740	7.218	7.231	7.243	7.255	7.267	7.279	7.291	7.303	7.315	7.327	7.339	740
750	7.339	7.351	7.363	7.375	7.387	7.399	7.412	7.424	7.436	7.448	7.460	750
760	7.460	7.472	7.484	7.496	7.509	7.521	7.533	7.545	7.557	7.569	7.582	760
770	7.582	7.594	7.606	7.618	7.630	7.642	7.655	7.667	7.679	7.691	7.703	770
780	7.703	7.716	7.728	7.740	7.752	7.765	7.777	7.789	7.801	7.814	7.826	780
790	7.826	7.838	7.850	7.863	7.875	7.887	7.900	7.912	7.924	7.937	7.949	790
800	7.949	7.961	7.973	7.986	7.998	8.010	8.023	8.035	8.047	8.060	8.072	800
810	8.072	8.085	8.097	8.109	8.122	8.134	8.146	8.159	8.171	8.184	8.196	810
820	8.196	8.208	8.221	8.233	8.246	8.258	8.271	8.283	8.295	8.308	8.320	820
830	8.320	8.333	8.345	8.358	8.370	8.383	8.395	8.408	8.420	8.433	8.445	830
840	8.445	8.458	8.470	8.483	8.495	8.508	8.520	8.533	8.545	8.558	8.570	840
850	8.570	8.583	8.595	8.608	8.621	8.633	8.646	8.658	8.671	8.683	8.696	850
860	8.696	8.709	8.721	8.734	8.746	8.759	8.772	8.784	8.797	8.810	8.822	860
870	8.822	8.835	8.847	8.860	8.873	8.885	8.898	8.911	8.923	8.936	8.949	870
880	8.949	8.961	8.974	8.987	9.000	9.012	9.025	9.038	9.050	9.063	9.076	880
890	9.076	9.089	9.101	9.114	9.127	9.140	9.152	9.165	9.178	9.191	9.203	890
900	9.203	9.216	9.229	9.242	9.254	9.267	9.280	9.293	9.306	9.319	9.331	900
910	9.331	9.344	9.357	9.370	9.383	9.395	9.408	9.421	9.434	9.447	9.460	910
920	9.460	9.473	9.485	9.498	9.511	9.524	9.537	9.550	9.563	9.576	9.589	920
930	9.589	9.602	9.614	9.627	9.640	9.653	9.666	9.679	9.692	9.705	9.718	930
940	9.718	9.731	9.744	9.757	9.770	9.783	9.796	9.809	9.822	9.835	9.848	940
950	9.848	9.861	9.874	9.887	9.900	9.913	9.926	9.939	9.952	9.965	9.978	950
960	9.978	9.991	10.004	10.017	10.030	10.043	10.056	10.069	10.082	10.095	10.109	960
970	10.109	10.122	10.135	10.148	10.161	10.174	10.187	10.200	10.213	10.227	10.240	970
980	10.240	10.253	10.266	10.279	10.292	10.305	10.319	10.332	10.345	10.358	10.371	980
990	10.371	10.384	10.398	10.411	10.424	10.437	10.450	10.464	10.477	10.490	10.503	990
1000	10.503	10.516	10.530	10.543	10.556	10.569	10.583	10.596	10.609	10.622	10.636	1000
1010	10.636	10.649	10.662	10.675	10.689	10.702	10.715	10.729	10.742	10.755	10.768	1010
1020	10.768	10.782	10.795	10.808	10.822	10.835	10.848	10.862	10.875	10.888	10.902	1020
1030	10.902	10.915	10.928	10.942	10.955	10.968	10.982	10.995	11.009	11.022	11.035	1030
1040	11.035	11.049	11.062	11.076	11.089	11.102	11.116	11.129	11.143	11.156	11.170	1040
1050	11.170	11.183	11.196	11.210	11.223	11.237	11.250	11.264	11.277	11.291	11.304	1050
1060	11.304	11.318	11.331	11.345	11.358	11.372	11.385	11.399	11.412	11.426	11.439	1060
1070	11.439	11.453	11.466	11.480	11.493	11.507	11.520	11.534	11.547	11.561	11.574	1070
1080	11.574	11.588	11.602	11.615	11.629	11.642	11.656	11.669	11.683	11.697	11.710	1080
1090	11.710	11.724	11.737	11.751	11.765	11.778	11.792	11.805	11.819	11.833	11.846	1090
1100	11.846	11.860	11.874	11.887	11.901	11.914	11.928	11.942	11.955	11.969	11.983	1100
1110	11.983	11.996	12.010	12.024	12.037	12.051	12.065	12.078	12.092	12.106	12.119	1110
1120	12.119	12.133	12.147	12.161	12.174	12.188	12.202	12.215	12.229	12.243	12.257	1120
1130	12.257	12.270	12.284	12.298	12.311	12.325	12.339	12.353	12.366	12.380	12.394	1130
1140	12.394	12.408	12.421	12.435	12.449	12.463	12.476	12.490	12.504	12.518	12.532	1140

Type R Thermocouple
Thermoelectric Voltage as a Function of Temperature (°C)
Reference Junctions at 0°C

THERMOELECTRIC VOLTAGE IN ABSOLUTE MILLIVOLTS

°C	0	1	2	3	4	5	6	7	8	9	10	°C
1150	12.532	12.545	12.559	12.573	12.587	12.600	12.614	12.628	12.642	12.656	12.669	1150
1160	12.669	12.683	12.697	12.711	12.725	12.739	12.752	12.766	12.780	12.794	12.808	1160
1170	12.808	12.822	12.835	12.849	12.863	12.877	12.891	12.905	12.918	12.932	12.946	1170
1180	12.946	12.960	12.974	12.988	13.002	13.016	13.029	13.043	13.057	13.071	13.085	1180
1190	13.085	13.099	13.113	13.127	13.140	13.154	13.168	13.182	13.196	13.210	13.224	1190
1200	13.224	13.238	13.252	13.266	13.280	13.293	13.307	13.321	13.335	13.349	13.363	1200
1210	13.363	13.377	13.391	13.405	13.419	13.433	13.447	13.461	13.475	13.489	13.502	1210
1220	13.502	13.516	13.530	13.544	13.558	13.572	13.586	13.600	13.614	13.628	13.642	1220
1230	13.642	13.656	13.670	13.684	13.698	13.712	13.726	13.740	13.754	13.768	13.782	1230
1240	13.782	13.796	13.810	13.824	13.838	13.852	13.866	13.880	13.894	13.908	13.922	1240
1250	13.922	13.936	13.950	13.964	13.978	13.992	14.006	14.020	14.034	14.048	14.062	1250
1260	14.062	14.076	14.090	14.104	14.118	14.132	14.146	14.160	14.174	14.188	14.202	1260
1270	14.202	14.216	14.230	14.244	14.258	14.272	14.286	14.301	14.315	14.329	14.343	1270
1280	14.343	14.357	14.371	14.385	14.399	14.413	14.427	14.441	14.455	14.469	14.483	1280
1290	14.483	14.497	14.511	14.525	14.539	14.554	14.568	14.582	14.596	14.610	14.624	1290
1300	14.624	14.638	14.652	14.666	14.680	14.694	14.708	14.722	14.737	14.751	14.765	1300
1310	14.765	14.779	14.793	14.807	14.821	14.835	14.849	14.863	14.877	14.891	14.906	1310
1320	14.906	14.920	14.934	14.948	14.962	14.976	14.990	15.004	15.018	15.032	15.047	1320
1330	15.047	15.061	15.075	15.089	15.103	15.117	15.131	15.145	15.159	15.173	15.188	1330
1340	15.188	15.202	15.216	15.230	15.244	15.258	15.272	15.286	15.300	15.315	15.329	1340
1350	15.329	15.343	15.357	15.371	15.385	15.399	15.413	15.427	15.442	15.456	15.470	1350
1360	15.470	15.484	15.498	15.512	15.526	15.540	15.555	15.569	15.583	15.597	15.611	1360
1370	15.611	15.625	15.639	15.653	15.667	15.682	15.696	15.710	15.724	15.738	15.752	1370
1380	15.752	15.766	15.780	15.795	15.809	15.823	15.837	15.851	15.865	15.879	15.893	1380
1390	15.893	15.908	15.922	15.936	15.950	15.964	15.978	15.992	16.006	16.021	16.035	1390
1400	16.035	16.049	16.063	16.077	16.091	16.105	16.119	16.134	16.148	16.162	16.176	1400
1410	16.176	16.190	16.204	16.218	16.232	16.247	16.261	16.275	16.289	16.303	16.317	1410
1420	16.317	16.331	16.345	16.360	16.374	16.388	16.402	16.416	16.430	16.444	16.458	1420
1430	16.458	16.472	16.487	16.501	16.515	16.529	16.543	16.557	16.571	16.585	16.599	1430
1440	16.599	16.614	16.628	16.642	16.656	16.670	16.684	16.698	16.712	16.726	16.741	1440
1450	16.741	16.755	16.769	16.783	16.797	16.811	16.825	16.839	16.853	16.867	16.882	1450
1460	16.882	16.896	16.910	16.924	16.938	16.952	16.966	16.980	16.994	17.008	17.022	1460
1470	17.022	17.037	17.051	17.065	17.079	17.093	17.107	17.121	17.135	17.149	17.163	1470
1480	17.163	17.177	17.192	17.206	17.220	17.234	17.248	17.262	17.276	17.290	17.304	1480
1490	17.304	17.318	17.332	17.346	17.360	17.374	17.388	17.403	17.417	17.431	17.445	1490
1500	17.445	17.459	17.473	17.487	17.501	17.515	17.529	17.543	17.557	17.571	17.585	1500
1510	17.585	17.599	17.613	17.627	17.641	17.655	17.669	17.684	17.698	17.712	17.726	1510
1520	17.726	17.740	17.754	17.768	17.782	17.796	17.810	17.824	17.838	17.852	17.866	1520
1530	17.866	17.880	17.894	17.908	17.922	17.936	17.950	17.964	17.978	17.992	18.006	1530
1540	18.006	18.020	18.034	18.048	18.062	18.076	18.090	18.104	18.118	18.132	18.146	1540
1550	18.146	18.160	18.174	18.188	18.202	18.216	18.230	18.244	18.258	18.272	18.286	1550
1560	18.286	18.299	18.313	18.327	18.341	18.355	18.369	18.383	18.397	18.411	18.425	1560
1570	18.425	18.439	18.453	18.467	18.481	18.495	18.509	18.523	18.537	18.550	18.564	1570
1580	18.564	18.578	18.592	18.606	18.620	18.634	18.648	18.662	18.676	18.690	18.703	1580
1590	18.703	18.717	18.731	18.745	18.759	18.773	18.787	18.801	18.815	18.828	18.842	1590
1600	18.842	18.856	18.870	18.884	18.898	18.912	18.926	18.939	18.953	18.967	18.981	1600
1610	18.981	18.995	19.009	19.023	19.036	19.050	19.064	19.078	19.092	19.106	19.119	1610
1620	19.119	19.133	19.147	19.161	19.175	19.188	19.202	19.216	19.230	19.244	19.257	1620
1630	19.257	19.271	19.285	19.299	19.313	19.326	19.340	19.354	19.368	19.382	19.395	1630
1640	19.395	19.409	19.423	19.437	19.450	19.464	19.478	19.492	19.505	19.519	19.533	1640
1650	19.533	19.547	19.560	19.574	19.588	19.602	19.615	19.629	19.643	19.656	19.670	1650
1660	19.670	19.684	19.698	19.711	19.725	19.739	19.752	19.766	19.780	19.793	19.807	1660
1670	19.807	19.821	19.834	19.848	19.862	19.875	19.889	19.903	19.916	19.930	19.944	1670
1680	19.944	19.957	19.971	19.985	19.998	20.012	20.025	20.039	20.053	20.066	20.080	1680
1690	20.080	20.093	20.107	20.120	20.134	20.148	20.161	20.175	20.188	20.202	20.215	1690
1700	20.215	20.229	20.242	20.256	20.269	20.283	20.296	20.309	20.323	20.336	20.350	1700
1710	20.350	20.363	20.377	20.390	20.403	20.417	20.430	20.443	20.457	20.470	20.483	1710
1720	20.483	20.497	20.510	20.523	20.537	20.550	20.563	20.576	20.590	20.603	20.616	1720
1730	20.616	20.629	20.642	20.656	20.669	20.682	20.695	20.708	20.721	20.734	20.748	1730
1740	20.748	20.761	20.774	20.787	20.800	20.813	20.826	20.839	20.852	20.865	20.878	1740
1750	20.878	20.891	20.904	20.916	20.929	20.942	20.955	20.968	20.981	20.994	21.006	1750
1760	21.006	21.019	21.032	21.045	21.057	21.070	21.083	21.096	21.108			1760

Type R Thermocouple
Thermoelectric Voltage as a Function of Temperature (°F)
Reference Junctions at 32°F
THERMOELECTRIC VOLTAGE IN ABSOLUTE MILLIVOLTS

°F	0	1	2	3	4	5	6	7	8	9	10	°F
-60			-0.226	-0.224	-0.222	-0.220	-0.218	-0.216	-0.214	-0.212	-0.210	-60
-50	-0.210	-0.207	-0.205	-0.203	-0.201	-0.199	-0.197	-0.194	-0.192	-0.190	-0.188	-50
-40	-0.188	-0.185	-0.183	-0.181	-0.179	-0.176	-0.174	-0.172	-0.169	-0.167	-0.165	-40
-30	-0.165	-0.162	-0.160	-0.158	-0.155	-0.153	-0.150	-0.148	-0.145	-0.143	-0.141	-30
-20	-0.141	-0.138	-0.136	-0.133	-0.131	-0.128	-0.126	-0.123	-0.121	-0.118	-0.116	-20
-10	-0.116	-0.113	-0.110	-0.108	-0.105	-0.103	-0.100	-0.097	-0.095	-0.092	-0.089	-10
0	-0.089	-0.087	-0.084	-0.082	-0.079	-0.076	-0.073	-0.071	-0.068	-0.065	-0.063	0
10	-0.063	-0.060	-0.057	-0.054	-0.051	-0.049	-0.046	-0.043	-0.040	-0.037	-0.035	10
20	-0.035	-0.032	-0.029	-0.026	-0.023	-0.020	-0.017	-0.015	-0.012	-0.009	-0.006	20
30	-0.006	-0.003	0.000	0.003	0.006	0.009	0.012	0.015	0.018	0.021	0.024	30
40	0.024	0.027	0.030	0.033	0.036	0.039	0.042	0.045	0.048	0.051	0.054	40
50	0.054	0.057	0.060	0.064	0.067	0.070	0.073	0.076	0.079	0.082	0.086	50
60	0.086	0.089	0.092	0.095	0.098	0.101	0.105	0.108	0.111	0.114	0.118	60
70	0.118	0.121	0.124	0.127	0.131	0.134	0.137	0.141	0.144	0.147	0.150	70
80	0.150	0.154	0.157	0.161	0.164	0.167	0.171	0.174	0.177	0.181	0.184	80
90	0.184	0.188	0.191	0.194	0.198	0.201	0.205	0.208	0.212	0.215	0.218	90
100	0.218	0.222	0.225	0.229	0.232	0.236	0.239	0.243	0.246	0.250	0.253	100
110	0.253	0.257	0.261	0.264	0.268	0.271	0.275	0.278	0.282	0.286	0.289	110
120	0.289	0.293	0.296	0.300	0.304	0.307	0.311	0.315	0.318	0.322	0.326	120
130	0.326	0.329	0.333	0.337	0.340	0.344	0.348	0.351	0.355	0.359	0.363	130
140	0.363	0.366	0.370	0.374	0.378	0.381	0.385	0.389	0.393	0.397	0.400	140
150	0.400	0.404	0.408	0.412	0.416	0.419	0.423	0.427	0.431	0.435	0.439	150
160	0.439	0.443	0.446	0.450	0.454	0.458	0.462	0.466	0.470	0.474	0.478	160
170	0.478	0.482	0.485	0.489	0.493	0.497	0.501	0.505	0.509	0.513	0.517	170
180	0.517	0.521	0.525	0.529	0.533	0.537	0.541	0.545	0.549	0.553	0.557	180
190	0.557	0.561	0.565	0.569	0.573	0.577	0.581	0.586	0.590	0.594	0.598	190
200	0.598	0.602	0.606	0.610	0.614	0.618	0.622	0.627	0.631	0.635	0.639	200
210	0.639	0.643	0.647	0.651	0.656	0.660	0.664	0.668	0.672	0.676	0.681	210
220	0.681	0.685	0.689	0.693	0.697	0.702	0.706	0.710	0.714	0.719	0.723	220
230	0.723	0.727	0.731	0.736	0.740	0.744	0.748	0.753	0.757	0.761	0.766	230
240	0.766	0.770	0.774	0.778	0.783	0.787	0.791	0.796	0.800	0.804	0.809	240
250	0.809	0.813	0.817	0.822	0.826	0.830	0.835	0.839	0.844	0.848	0.852	250
260	0.852	0.857	0.861	0.866	0.870	0.874	0.879	0.883	0.888	0.892	0.897	260
270	0.897	0.901	0.905	0.910	0.914	0.919	0.923	0.928	0.932	0.937	0.941	270
280	0.941	0.946	0.950	0.955	0.959	0.964	0.968	0.973	0.977	0.982	0.986	280
290	0.986	0.991	0.995	1.000	1.004	1.009	1.013	1.018	1.022	1.027	1.032	290
300	1.032	1.036	1.041	1.045	1.050	1.054	1.059	1.064	1.068	1.073	1.077	300
310	1.077	1.082	1.087	1.091	1.096	1.101	1.105	1.110	1.114	1.119	1.124	310
320	1.124	1.128	1.133	1.138	1.142	1.147	1.152	1.156	1.161	1.166	1.170	320
330	1.170	1.175	1.180	1.184	1.189	1.194	1.199	1.203	1.208	1.213	1.217	330
340	1.217	1.222	1.227	1.232	1.236	1.241	1.246	1.251	1.255	1.260	1.265	340
350	1.265	1.270	1.274	1.279	1.284	1.289	1.294	1.298	1.303	1.308	1.313	350
360	1.313	1.318	1.322	1.327	1.332	1.337	1.342	1.346	1.351	1.356	1.361	360
370	1.361	1.366	1.371	1.375	1.380	1.385	1.390	1.395	1.400	1.405	1.409	370
380	1.409	1.414	1.419	1.424	1.429	1.434	1.439	1.444	1.449	1.453	1.458	380
390	1.458	1.463	1.468	1.473	1.478	1.483	1.488	1.493	1.498	1.503	1.508	390
400	1.508	1.512	1.517	1.522	1.527	1.532	1.537	1.542	1.547	1.552	1.557	400
410	1.557	1.562	1.567	1.572	1.577	1.582	1.587	1.592	1.597	1.602	1.607	410
420	1.607	1.612	1.617	1.622	1.627	1.632	1.637	1.642	1.647	1.652	1.657	420
430	1.657	1.662	1.667	1.672	1.677	1.682	1.687	1.692	1.698	1.703	1.708	430
440	1.708	1.713	1.718	1.723	1.728	1.733	1.738	1.743	1.748	1.753	1.758	440
450	1.758	1.764	1.769	1.774	1.779	1.784	1.789	1.794	1.799	1.804	1.810	450
460	1.810	1.815	1.820	1.825	1.830	1.835	1.840	1.845	1.851	1.856	1.861	460
470	1.861	1.866	1.871	1.876	1.882	1.887	1.892	1.897	1.902	1.907	1.913	470
480	1.913	1.918	1.923	1.928	1.933	1.938	1.944	1.949	1.954	1.959	1.964	480
490	1.964	1.970	1.975	1.980	1.985	1.991	1.996	2.001	2.006	2.011	2.017	490
500	2.017	2.022	2.027	2.032	2.038	2.043	2.048	2.053	2.059	2.064	2.069	500
510	2.069	2.074	2.080	2.085	2.090	2.095	2.101	2.106	2.111	2.117	2.122	510
520	2.122	2.127	2.132	2.138	2.143	2.148	2.154	2.159	2.164	2.170	2.175	520
530	2.175	2.180	2.186	2.191	2.196	2.201	2.207	2.212	2.217	2.223	2.228	530

Type R Thermocouple
Thermoelectric Voltage as a Function of Temperature (°F)
Reference Junctions at 32°F
THERMOELECTRIC VOLTAGE IN ABSOLUTE MILLIVOLTS

°F	0	1	2	3	4	5	6	7	8	9	10	°F
540	2.228	2.233	2.239	2.244	2.249	2.255	2.260	2.266	2.271	2.276	2.282	540
550	2.282	2.287	2.292	2.298	2.303	2.308	2.314	2.319	2.325	2.330	2.335	550
560	2.335	2.341	2.346	2.351	2.357	2.362	2.368	2.373	2.378	2.384	2.389	560
570	2.389	2.395	2.400	2.405	2.411	2.416	2.422	2.427	2.433	2.438	2.443	570
580	2.443	2.449	2.454	2.460	2.465	2.471	2.476	2.481	2.487	2.492	2.498	580
590	2.498	2.503	2.509	2.514	2.520	2.525	2.531	2.536	2.541	2.547	2.552	590
600	2.552	2.558	2.563	2.569	2.574	2.580	2.585	2.591	2.596	2.602	2.607	600
610	2.607	2.613	2.618	2.624	2.629	2.635	2.640	2.646	2.651	2.657	2.662	610
620	2.662	2.668	2.673	2.679	2.684	2.690	2.695	2.701	2.706	2.712	2.718	620
630	2.718	2.723	2.729	2.734	2.740	2.745	2.751	2.756	2.762	2.767	2.773	630
640	2.773	2.779	2.784	2.790	2.795	2.801	2.806	2.812	2.818	2.823	2.829	640
650	2.829	2.834	2.840	2.845	2.851	2.857	2.862	2.868	2.873	2.879	2.885	650
660	2.885	2.890	2.896	2.901	2.907	2.913	2.918	2.924	2.929	2.935	2.941	660
670	2.941	2.946	2.952	2.957	2.963	2.969	2.974	2.980	2.986	2.991	2.997	670
680	2.997	3.002	3.008	3.014	3.019	3.025	3.031	3.036	3.042	3.048	3.053	680
690	3.053	3.059	3.065	3.070	3.076	3.082	3.087	3.093	3.099	3.104	3.110	690
700	3.110	3.116	3.121	3.127	3.133	3.138	3.144	3.150	3.155	3.161	3.167	700
710	3.167	3.172	3.178	3.184	3.189	3.195	3.201	3.207	3.212	3.218	3.224	710
720	3.224	3.229	3.235	3.241	3.247	3.252	3.258	3.264	3.269	3.275	3.281	720
730	3.281	3.287	3.292	3.298	3.304	3.309	3.315	3.321	3.327	3.332	3.338	730
740	3.338	3.344	3.350	3.355	3.361	3.367	3.373	3.378	3.384	3.390	3.396	740
750	3.396	3.401	3.407	3.413	3.419	3.424	3.430	3.436	3.442	3.448	3.453	750
760	3.453	3.459	3.465	3.471	3.476	3.482	3.488	3.494	3.500	3.505	3.511	760
770	3.511	3.517	3.523	3.529	3.534	3.540	3.546	3.552	3.558	3.563	3.569	770
780	3.569	3.575	3.581	3.587	3.592	3.598	3.604	3.610	3.616	3.622	3.627	780
790	3.627	3.633	3.639	3.645	3.651	3.657	3.662	3.668	3.674	3.680	3.686	790
800	3.686	3.692	3.697	3.703	3.709	3.715	3.721	3.727	3.733	3.738	3.744	800
810	3.744	3.750	3.756	3.762	3.768	3.774	3.779	3.785	3.791	3.797	3.803	810
820	3.803	3.809	3.815	3.821	3.826	3.832	3.838	3.844	3.850	3.856	3.862	820
830	3.862	3.868	3.874	3.879	3.885	3.891	3.897	3.903	3.909	3.915	3.921	830
840	3.921	3.927	3.933	3.938	3.944	3.950	3.956	3.962	3.968	3.974	3.980	840
850	3.980	3.986	3.992	3.998	4.004	4.009	4.015	4.021	4.027	4.033	4.039	850
860	4.039	4.045	4.051	4.057	4.063	4.069	4.075	4.081	4.087	4.093	4.099	860
870	4.099	4.105	4.110	4.116	4.122	4.128	4.134	4.140	4.146	4.152	4.158	870
880	4.158	4.164	4.170	4.176	4.182	4.188	4.194	4.200	4.206	4.212	4.218	880
890	4.218	4.224	4.230	4.236	4.242	4.248	4.254	4.260	4.266	4.272	4.278	890
900	4.278	4.284	4.290	4.296	4.302	4.308	4.314	4.320	4.326	4.332	4.338	900
910	4.338	4.344	4.350	4.356	4.362	4.368	4.374	4.380	4.386	4.392	4.398	910
920	4.398	4.404	4.410	4.416	4.422	4.428	4.434	4.440	4.446	4.452	4.458	920
930	4.458	4.465	4.471	4.477	4.483	4.489	4.495	4.501	4.507	4.513	4.519	930
940	4.519	4.525	4.531	4.537	4.543	4.549	4.555	4.561	4.567	4.574	4.580	940
950	4.580	4.586	4.592	4.598	4.604	4.610	4.616	4.622	4.628	4.634	4.640	950
960	4.640	4.647	4.653	4.659	4.665	4.671	4.677	4.683	4.689	4.695	4.701	960
970	4.701	4.707	4.714	4.720	4.726	4.732	4.738	4.744	4.750	4.756	4.762	970
980	4.762	4.769	4.775	4.781	4.787	4.793	4.799	4.805	4.811	4.818	4.824	980
990	4.824	4.830	4.836	4.842	4.848	4.854	4.860	4.867	4.873	4.879	4.885	990
1000	4.885	4.891	4.897	4.904	4.910	4.916	4.922	4.928	4.934	4.940	4.947	1000
1010	4.947	4.953	4.959	4.965	4.971	4.977	4.984	4.990	4.996	5.002	5.008	1010
1020	5.008	5.014	5.021	5.027	5.033	5.039	5.045	5.052	5.058	5.064	5.070	1020
1030	5.070	5.076	5.082	5.089	5.095	5.101	5.107	5.113	5.120	5.126	5.132	1030
1040	5.132	5.138	5.144	5.151	5.157	5.163	5.169	5.175	5.182	5.188	5.194	1040
1050	5.194	5.200	5.207	5.213	5.219	5.225	5.231	5.238	5.244	5.250	5.256	1050
1060	5.256	5.263	5.269	5.275	5.281	5.288	5.294	5.300	5.306	5.313	5.319	1060
1070	5.319	5.325	5.331	5.337	5.344	5.350	5.356	5.362	5.369	5.375	5.381	1070
1080	5.381	5.388	5.394	5.400	5.406	5.413	5.419	5.425	5.431	5.438	5.444	1080
1090	5.444	5.450	5.456	5.463	5.469	5.475	5.482	5.488	5.494	5.500	5.507	1090
1100	5.507	5.513	5.519	5.526	5.532	5.538	5.544	5.551	5.557	5.563	5.570	1100
1110	5.570	5.576	5.582	5.589	5.595	5.601	5.607	5.614	5.620	5.626	5.633	1110
1120	5.633	5.639	5.645	5.652	5.658	5.664	5.671	5.677	5.683	5.690	5.696	1120
1130	5.696	5.702	5.709	5.715	5.721	5.728	5.734	5.740	5.747	5.753	5.759	1130

Type R Thermocouple
Thermoelectric Voltage as a Function of Temperature (°F)
Reference Junctions at 32°F
THERMOELECTRIC VOLTAGE IN ABSOLUTE MILLIVOLTS

°F	0	1	2	3	4	5	6	7	8	9	10	°F
1140	5.759	5.766	5.772	5.778	5.785	5.791	5.797	5.804	5.810	5.816	5.823	1140
1150	5.823	5.829	5.835	5.842	5.848	5.855	5.861	5.867	5.874	5.880	5.886	1150
1160	5.886	5.893	5.899	5.905	5.912	5.918	5.925	5.931	5.937	5.944	5.950	1160
1170	5.950	5.957	5.963	5.969	5.976	5.982	5.988	5.995	6.001	6.008	6.014	1170
1180	6.014	6.021	6.027	6.033	6.040	6.046	6.053	6.059	6.065	6.072	6.078	1180
1190	6.078	6.085	6.091	6.098	6.104	6.110	6.117	6.123	6.130	6.136	6.143	1190
1200	6.143	6.149	6.155	6.162	6.168	6.175	6.181	6.188	6.194	6.201	6.207	1200
1210	6.207	6.213	6.220	6.226	6.233	6.239	6.246	6.252	6.259	6.265	6.272	1210
1220	6.272	6.278	6.285	6.291	6.297	6.304	6.310	6.317	6.323	6.330	6.336	1220
1230	6.336	6.343	6.349	6.356	6.362	6.369	6.375	6.382	6.388	6.395	6.401	1230
1240	6.401	6.408	6.414	6.421	6.427	6.434	6.440	6.447	6.453	6.460	6.466	1240
1250	6.466	6.473	6.479	6.486	6.492	6.499	6.505	6.512	6.518	6.525	6.532	1250
1260	6.532	6.538	6.545	6.551	6.558	6.564	6.571	6.577	6.584	6.590	6.597	1260
1270	6.597	6.603	6.610	6.616	6.623	6.630	6.636	6.643	6.649	6.656	6.662	1270
1280	6.662	6.669	6.675	6.682	6.689	6.695	6.702	6.708	6.715	6.721	6.728	1280
1290	6.728	6.735	6.741	6.748	6.754	6.761	6.767	6.774	6.781	6.787	6.794	1290
1300	6.794	6.800	6.807	6.814	6.820	6.827	6.833	6.840	6.847	6.853	6.860	1300
1310	6.860	6.866	6.873	6.880	6.886	6.893	6.899	6.906	6.913	6.919	6.926	1310
1320	6.926	6.932	6.939	6.946	6.952	6.959	6.966	6.972	6.979	6.985	6.992	1320
1330	6.992	6.999	7.005	7.012	7.019	7.025	7.032	7.039	7.045	7.052	7.059	1330
1340	7.059	7.065	7.072	7.078	7.085	7.092	7.098	7.105	7.112	7.118	7.125	1340
1350	7.125	7.132	7.138	7.145	7.152	7.158	7.165	7.172	7.178	7.185	7.192	1350
1360	7.192	7.198	7.205	7.212	7.218	7.225	7.232	7.239	7.245	7.252	7.259	1360
1370	7.259	7.265	7.272	7.279	7.285	7.292	7.299	7.305	7.312	7.319	7.326	1370
1380	7.326	7.332	7.339	7.346	7.352	7.359	7.366	7.373	7.379	7.386	7.393	1380
1390	7.393	7.399	7.406	7.413	7.420	7.426	7.433	7.440	7.447	7.453	7.460	1390
1400	7.460	7.467	7.474	7.480	7.487	7.494	7.500	7.507	7.514	7.521	7.527	1400
1410	7.527	7.534	7.541	7.548	7.554	7.561	7.568	7.575	7.582	7.588	7.595	1410
1420	7.595	7.602	7.609	7.615	7.622	7.629	7.636	7.642	7.649	7.656	7.663	1420
1430	7.663	7.670	7.676	7.683	7.690	7.697	7.703	7.710	7.717	7.724	7.731	1430
1440	7.731	7.737	7.744	7.751	7.758	7.765	7.771	7.778	7.785	7.792	7.799	1440
1450	7.799	7.805	7.812	7.819	7.826	7.833	7.840	7.846	7.853	7.860	7.867	1450
1460	7.867	7.874	7.880	7.887	7.894	7.901	7.908	7.915	7.921	7.928	7.935	1460
1470	7.935	7.942	7.949	7.956	7.963	7.969	7.976	7.983	7.990	7.997	8.004	1470
1480	8.004	8.010	8.017	8.024	8.031	8.038	8.045	8.052	8.058	8.065	8.072	1480
1490	8.072	8.079	8.086	8.093	8.100	8.107	8.113	8.120	8.127	8.134	8.141	1490
1500	8.141	8.148	8.155	8.162	8.168	8.175	8.182	8.189	8.196	8.203	8.210	1500
1510	8.210	8.217	8.224	8.231	8.237	8.244	8.251	8.258	8.265	8.272	8.279	1510
1520	8.279	8.286	8.293	8.300	8.306	8.313	8.320	8.327	8.334	8.341	8.348	1520
1530	8.348	8.355	8.362	8.369	8.376	8.383	8.390	8.397	8.403	8.410	8.417	1530
1540	8.417	8.424	8.431	8.438	8.445	8.452	8.459	8.466	8.473	8.480	8.487	1540
1550	8.487	8.494	8.501	8.508	8.515	8.522	8.529	8.535	8.542	8.549	8.556	1550
1560	8.556	8.563	8.570	8.577	8.584	8.591	8.598	8.605	8.612	8.619	8.626	1560
1570	8.626	8.633	8.640	8.647	8.654	8.661	8.668	8.675	8.682	8.689	8.696	1570
1580	8.696	8.703	8.710	8.717	8.724	8.731	8.738	8.745	8.752	8.759	8.766	1580
1590	8.766	8.773	8.780	8.787	8.794	8.801	8.808	8.815	8.822	8.829	8.836	1590
1600	8.836	8.843	8.850	8.857	8.864	8.871	8.878	8.885	8.892	8.899	8.907	1600
1610	8.907	8.914	8.921	8.928	8.935	8.942	8.949	8.956	8.963	8.970	8.977	1610
1620	8.977	8.984	8.991	8.998	9.005	9.012	9.019	9.026	9.033	9.040	9.048	1620
1630	9.048	9.055	9.062	9.069	9.076	9.083	9.090	9.097	9.104	9.111	9.118	1630
1640	9.118	9.125	9.132	9.140	9.147	9.154	9.161	9.168	9.175	9.182	9.189	1640
1650	9.189	9.196	9.203	9.210	9.218	9.225	9.232	9.239	9.246	9.253	9.260	1650
1660	9.260	9.267	9.274	9.282	9.289	9.296	9.303	9.310	9.317	9.324	9.331	1660
1670	9.331	9.338	9.346	9.353	9.360	9.367	9.374	9.381	9.388	9.395	9.403	1670
1680	9.403	9.410	9.417	9.424	9.431	9.438	9.445	9.453	9.460	9.467	9.474	1680
1690	9.474	9.481	9.488	9.495	9.503	9.510	9.517	9.524	9.531	9.538	9.546	1690
1700	9.546	9.553	9.560	9.567	9.574	9.581	9.589	9.596	9.603	9.610	9.617	1700
1710	9.617	9.624	9.632	9.639	9.646	9.653	9.660	9.668	9.675	9.682	9.689	1710
1720	9.689	9.696	9.704	9.711	9.718	9.725	9.732	9.740	9.747	9.754	9.761	1720
1730	9.761	9.768	9.776	9.783	9.790	9.797	9.804	9.812	9.819	9.826	9.833	1730

Type R Thermocouple
Thermoelectric Voltage as a Function of Temperature (°F)
Reference Junctions at 32°F
THERMOELECTRIC VOLTAGE IN ABSOLUTE MILLIVOLTS

°F	0	1	2	3	4	5	6	7	8	9	10	°F
1740	9.833	9.840	9.848	9.855	9.862	9.869	9.877	9.884	9.891	9.898	9.906	1740
1750	9.906	9.913	9.920	9.927	9.934	9.942	9.949	9.956	9.963	9.971	9.978	1750
1760	9.978	9.985	9.992	10.000	10.007	10.014	10.021	10.029	10.036	10.043	10.050	1760
1770	10.050	10.058	10.065	10.072	10.079	10.087	10.094	10.101	10.109	10.116	10.123	1770
1780	10.123	10.130	10.138	10.145	10.152	10.159	10.167	10.174	10.181	10.189	10.196	1780
1790	10.196	10.203	10.210	10.218	10.225	10.232	10.240	10.247	10.254	10.262	10.269	1790
1800	10.269	10.276	10.283	10.291	10.298	10.305	10.313	10.320	10.327	10.335	10.342	1800
1810	10.342	10.349	10.357	10.364	10.371	10.379	10.386	10.393	10.400	10.408	10.415	1810
1820	10.415	10.422	10.430	10.437	10.444	10.452	10.459	10.466	10.474	10.481	10.488	1820
1830	10.488	10.496	10.503	10.511	10.518	10.525	10.533	10.540	10.547	10.555	10.562	1830
1840	10.562	10.569	10.577	10.584	10.591	10.599	10.606	10.613	10.621	10.628	10.636	1840
1850	10.636	10.643	10.650	10.658	10.665	10.672	10.680	10.687	10.695	10.702	10.709	1850
1860	10.709	10.717	10.724	10.731	10.739	10.746	10.754	10.761	10.768	10.776	10.783	1860
1870	10.783	10.791	10.798	10.805	10.813	10.820	10.828	10.835	10.842	10.850	10.857	1870
1880	10.857	10.865	10.872	10.879	10.887	10.894	10.902	10.909	10.917	10.924	10.931	1880
1890	10.931	10.939	10.946	10.954	10.961	10.968	10.976	10.983	10.991	10.998	11.006	1890
1900	11.006	11.013	11.021	11.028	11.035	11.043	11.050	11.058	11.065	11.073	11.080	1900
1910	11.080	11.088	11.095	11.102	11.110	11.117	11.125	11.132	11.140	11.147	11.155	1910
1920	11.155	11.162	11.170	11.177	11.184	11.192	11.199	11.207	11.214	11.222	11.229	1920
1930	11.229	11.237	11.244	11.252	11.259	11.267	11.274	11.282	11.289	11.297	11.304	1930
1940	11.304	11.312	11.319	11.327	11.334	11.342	11.349	11.357	11.364	11.372	11.379	1940
1950	11.379	11.387	11.394	11.402	11.409	11.417	11.424	11.432	11.439	11.447	11.454	1950
1960	11.454	11.462	11.469	11.477	11.484	11.492	11.499	11.507	11.514	11.522	11.529	1960
1970	11.529	11.537	11.544	11.552	11.559	11.567	11.574	11.582	11.590	11.597	11.605	1970
1980	11.605	11.612	11.620	11.627	11.635	11.642	11.650	11.657	11.665	11.672	11.680	1980
1990	11.680	11.688	11.695	11.703	11.710	11.718	11.725	11.733	11.740	11.748	11.756	1990
2000	11.756	11.763	11.771	11.778	11.786	11.793	11.801	11.808	11.816	11.824	11.831	2000
2010	11.831	11.839	11.846	11.854	11.861	11.869	11.877	11.884	11.892	11.899	11.907	2010
2020	11.907	11.914	11.922	11.930	11.937	11.945	11.952	11.960	11.968	11.975	11.983	2020
2030	11.983	11.990	11.998	12.005	12.013	12.021	12.028	12.036	12.043	12.051	12.059	2030
2040	12.059	12.066	12.074	12.081	12.089	12.097	12.104	12.112	12.119	12.127	12.135	2040
2050	12.135	12.142	12.150	12.157	12.165	12.173	12.180	12.188	12.196	12.203	12.211	2050
2060	12.211	12.218	12.226	12.234	12.241	12.249	12.257	12.264	12.272	12.279	12.287	2060
2070	12.287	12.295	12.302	12.310	12.318	12.325	12.333	12.340	12.348	12.356	12.363	2070
2080	12.363	12.371	12.379	12.386	12.394	12.402	12.409	12.417	12.424	12.432	12.440	2080
2090	12.440	12.447	12.455	12.463	12.470	12.478	12.486	12.493	12.501	12.509	12.516	2090
2100	12.516	12.524	12.532	12.539	12.547	12.555	12.562	12.570	12.577	12.585	12.593	2100
2110	12.593	12.600	12.608	12.616	12.623	12.631	12.639	12.646	12.654	12.662	12.669	2110
2120	12.669	12.677	12.685	12.693	12.700	12.708	12.716	12.723	12.731	12.739	12.746	2120
2130	12.746	12.754	12.762	12.769	12.777	12.785	12.792	12.800	12.808	12.815	12.823	2130
2140	12.823	12.831	12.838	12.846	12.854	12.862	12.869	12.877	12.885	12.892	12.900	2140
2150	12.900	12.908	12.915	12.923	12.931	12.938	12.946	12.954	12.962	12.969	12.977	2150
2160	12.977	12.985	12.992	13.000	13.008	13.016	13.023	13.031	13.039	13.046	13.054	2160
2170	13.054	13.062	13.069	13.077	13.085	13.093	13.100	13.108	13.116	13.123	13.131	2170
2180	13.131	13.139	13.147	13.154	13.162	13.170	13.178	13.185	13.193	13.201	13.208	2180
2190	13.208	13.216	13.224	13.232	13.239	13.247	13.255	13.263	13.270	13.278	13.286	2190
2200	13.286	13.293	13.301	13.309	13.317	13.324	13.332	13.340	13.348	13.355	13.363	2200
2210	13.363	13.371	13.379	13.386	13.394	13.402	13.409	13.417	13.425	13.433	13.440	2210
2220	13.440	13.448	13.456	13.464	13.471	13.479	13.487	13.495	13.502	13.510	13.518	2220
2230	13.518	13.526	13.533	13.541	13.549	13.557	13.564	13.572	13.580	13.588	13.595	2230
2240	13.595	13.603	13.611	13.619	13.627	13.634	13.642	13.650	13.658	13.665	13.673	2240
2250	13.673	13.681	13.689	13.696	13.704	13.712	13.720	13.727	13.735	13.743	13.751	2250
2260	13.751	13.759	13.766	13.774	13.782	13.790	13.797	13.805	13.813	13.821	13.828	2260
2270	13.828	13.836	13.844	13.852	13.860	13.867	13.875	13.883	13.891	13.898	13.906	2270
2280	13.906	13.914	13.922	13.930	13.937	13.945	13.953	13.961	13.968	13.976	13.984	2280
2290	13.984	13.992	14.000	14.007	14.015	14.023	14.031	14.039	14.046	14.054	14.062	2290
2300	14.062	14.070	14.078	14.085	14.093	14.101	14.109	14.116	14.124	14.132	14.140	2300
2310	14.140	14.148	14.155	14.163	14.171	14.179	14.187	14.194	14.202	14.210	14.218	2310
2320	14.218	14.226	14.233	14.241	14.249	14.257	14.265	14.272	14.280	14.288	14.296	2320
2330	14.296	14.304	14.311	14.319	14.327	14.335	14.343	14.350	14.358	14.366	14.374	2330

Appendix B

Type R Thermocouple
Thermoelectric Voltage as a Function of Temperature (°F)
Reference Junctions at 32°F

THERMOELECTRIC VOLTAGE IN ABSOLUTE MILLIVOLTS

°F	0	1	2	3	4	5	6	7	8	9	10	°F
2340	14.374	14.382	14.389	14.397	14.405	14.413	14.421	14.429	14.436	14.444	14.452	2340
2350	14.452	14.460	14.468	14.475	14.483	14.491	14.499	14.507	14.514	14.522	14.530	2350
2360	14.530	14.538	14.546	14.554	14.561	14.569	14.577	14.585	14.593	14.600	14.608	2360
2370	14.608	14.616	14.624	14.632	14.640	14.647	14.655	14.663	14.671	14.679	14.686	2370
2380	14.686	14.694	14.702	14.710	14.718	14.726	14.733	14.741	14.749	14.757	14.765	2380
2390	14.765	14.772	14.780	14.788	14.796	14.804	14.812	14.819	14.827	14.835	14.843	2390
2400	14.843	14.851	14.859	14.866	14.874	14.882	14.890	14.898	14.906	14.913	14.921	2400
2410	14.921	14.929	14.937	14.945	14.953	14.960	14.968	14.976	14.984	14.992	15.000	2410
2420	15.000	15.007	15.015	15.023	15.031	15.039	15.047	15.054	15.062	15.070	15.078	2420
2430	15.078	15.086	15.094	15.101	15.109	15.117	15.125	15.133	15.141	15.148	15.156	2430
2440	15.156	15.164	15.172	15.180	15.188	15.195	15.203	15.211	15.219	15.227	15.235	2440
2450	15.235	15.242	15.250	15.258	15.266	15.274	15.282	15.289	15.297	15.305	15.313	2450
2460	15.313	15.321	15.329	15.337	15.344	15.352	15.360	15.368	15.376	15.384	15.391	2460
2470	15.391	15.399	15.407	15.415	15.423	15.431	15.438	15.446	15.454	15.462	15.470	2470
2480	15.470	15.478	15.486	15.493	15.501	15.509	15.517	15.525	15.533	15.540	15.548	2480
2490	15.548	15.556	15.564	15.572	15.580	15.587	15.595	15.603	15.611	15.619	15.627	2490
2500	15.627	15.635	15.642	15.650	15.658	15.666	15.674	15.682	15.689	15.697	15.705	2500
2510	15.705	15.713	15.721	15.729	15.737	15.744	15.752	15.760	15.768	15.776	15.784	2510
2520	15.784	15.791	15.799	15.807	15.815	15.823	15.831	15.839	15.846	15.854	15.862	2520
2530	15.862	15.870	15.878	15.886	15.893	15.901	15.909	15.917	15.925	15.933	15.941	2530
2540	15.941	15.948	15.956	15.964	15.972	15.980	15.988	15.995	16.003	16.011	16.019	2540
2550	16.019	16.027	16.035	16.043	16.050	16.058	16.066	16.074	16.082	16.090	16.097	2550
2560	16.097	16.105	16.113	16.121	16.129	16.137	16.145	16.152	16.160	16.168	16.176	2560
2570	16.176	16.184	16.192	16.199	16.207	16.215	16.223	16.231	16.239	16.247	16.254	2570
2580	16.254	16.262	16.270	16.278	16.286	16.294	16.301	16.309	16.317	16.325	16.333	2580
2590	16.333	16.341	16.349	16.356	16.364	16.372	16.380	16.388	16.396	16.403	16.411	2590
2600	16.411	16.419	16.427	16.435	16.443	16.450	16.458	16.466	16.474	16.482	16.490	2600
2610	16.490	16.498	16.505	16.513	16.521	16.529	16.537	16.545	16.552	16.560	16.568	2610
2620	16.568	16.576	16.584	16.592	16.599	16.607	16.615	16.623	16.631	16.639	16.646	2620
2630	16.646	16.654	16.662	16.670	16.678	16.686	16.694	16.701	16.709	16.717	16.725	2630
2640	16.725	16.733	16.741	16.748	16.756	16.764	16.772	16.780	16.788	16.795	16.803	2640
2650	16.803	16.811	16.819	16.827	16.835	16.842	16.850	16.858	16.866	16.874	16.882	2650
2660	16.882	16.889	16.897	16.905	16.913	16.921	16.929	16.936	16.944	16.952	16.960	2660
2670	16.960	16.968	16.976	16.983	16.991	16.999	17.007	17.015	17.022	17.030	17.038	2670
2680	17.038	17.046	17.054	17.062	17.069	17.077	17.085	17.093	17.101	17.109	17.116	2680
2690	17.116	17.124	17.132	17.140	17.148	17.156	17.163	17.171	17.179	17.187	17.195	2690
2700	17.195	17.202	17.210	17.218	17.226	17.234	17.242	17.249	17.257	17.265	17.273	2700
2710	17.273	17.281	17.288	17.296	17.304	17.312	17.320	17.328	17.335	17.343	17.351	2710
2720	17.351	17.359	17.367	17.374	17.382	17.390	17.398	17.406	17.413	17.421	17.429	2720
2730	17.429	17.437	17.445	17.453	17.460	17.468	17.476	17.484	17.492	17.499	17.507	2730
2740	17.507	17.515	17.523	17.531	17.538	17.546	17.554	17.562	17.570	17.577	17.585	2740
2750	17.585	17.593	17.601	17.609	17.616	17.624	17.632	17.640	17.648	17.655	17.663	2750
2760	17.663	17.671	17.679	17.687	17.694	17.702	17.710	17.718	17.726	17.733	17.741	2760
2770	17.741	17.749	17.757	17.765	17.772	17.780	17.788	17.796	17.804	17.811	17.819	2770
2780	17.819	17.827	17.835	17.842	17.850	17.858	17.866	17.874	17.881	17.889	17.897	2780
2790	17.897	17.905	17.913	17.920	17.928	17.936	17.944	17.951	17.959	17.967	17.975	2790
2800	17.975	17.983	17.990	17.998	18.006	18.014	18.021	18.029	18.037	18.045	18.053	2800
2810	18.053	18.060	18.068	18.076	18.084	18.091	18.099	18.107	18.115	18.123	18.130	2810
2820	18.130	18.138	18.146	18.154	18.161	18.169	18.177	18.185	18.192	18.200	18.208	2820
2830	18.208	18.216	18.223	18.231	18.239	18.247	18.255	18.262	18.270	18.278	18.286	2830
2840	18.286	18.293	18.301	18.309	18.317	18.324	18.332	18.340	18.348	18.355	18.363	2840
2850	18.363	18.371	18.379	18.386	18.394	18.402	18.410	18.417	18.425	18.433	18.441	2850
2860	18.441	18.448	18.456	18.464	18.472	18.479	18.487	18.495	18.502	18.510	18.518	2860
2870	18.518	18.526	18.533	18.541	18.549	18.557	18.564	18.572	18.580	18.588	18.595	2870
2880	18.595	18.603	18.611	18.619	18.626	18.634	18.642	18.649	18.657	18.665	18.673	2880
2890	18.673	18.680	18.688	18.696	18.703	18.711	18.719	18.727	18.734	18.742	18.750	2890
2900	18.750	18.758	18.765	18.773	18.781	18.788	18.796	18.804	18.812	18.819	18.827	2900
2910	18.827	18.835	18.842	18.850	18.858	18.865	18.873	18.881	18.889	18.896	18.904	2910
2920	18.904	18.912	18.919	18.927	18.935	18.943	18.950	18.958	18.966	18.973	18.981	2920
2930	18.981	18.989	18.996	19.004	19.012	19.019	19.027	19.035	19.043	19.050	19.058	2930

Type R Thermocouple
Thermoelectric Voltage as a Function of Temperature (°F)
Reference Junctions at 32°F
THERMOELECTRIC VOLTAGE IN ABSOLUTE MILLIVOLTS

°F	0	1	2	3	4	5	6	7	8	9	10	°F
2940	19.058	19.066	19.073	19.081	19.089	19.096	19.104	19.112	19.119	19.127	19.135	2940
2950	19.135	19.142	19.150	19.158	19.165	19.173	19.181	19.188	19.196	19.204	19.211	2950
2960	19.211	19.219	19.227	19.234	19.242	19.250	19.257	19.265	19.273	19.280	19.288	2960
2970	19.288	19.296	19.303	19.311	19.319	19.326	19.334	19.342	19.349	19.357	19.365	2970
2980	19.365	19.372	19.380	19.388	19.395	19.403	19.411	19.418	19.426	19.434	19.441	2980
2990	19.441	19.449	19.457	19.464	19.472	19.479	19.487	19.495	19.502	19.510	19.518	2990
3000	19.518	19.525	19.533	19.541	19.548	19.556	19.563	19.571	19.579	19.586	19.594	3000
3010	19.594	19.602	19.609	19.617	19.624	19.632	19.640	19.647	19.655	19.663	19.670	3010
3020	19.670	19.678	19.685	19.693	19.701	19.708	19.716	19.723	19.731	19.739	19.746	3020
3030	19.746	19.754	19.761	19.769	19.777	19.784	19.792	19.800	19.807	19.815	19.822	3030
3040	19.822	19.830	19.837	19.845	19.853	19.860	19.868	19.875	19.883	19.891	19.898	3040
3050	19.898	19.906	19.913	19.921	19.929	19.936	19.944	19.951	19.959	19.966	19.974	3050
3060	19.974	19.982	19.989	19.997	20.004	20.012	20.019	20.027	20.034	20.042	20.050	3060
3070	20.050	20.057	20.065	20.072	20.080	20.087	20.095	20.102	20.110	20.117	20.125	3070
3080	20.125	20.132	20.140	20.148	20.155	20.163	20.170	20.178	20.185	20.193	20.200	3080
3090	20.200	20.208	20.215	20.223	20.230	20.238	20.245	20.253	20.260	20.268	20.275	3090
3100	20.275	20.283	20.290	20.297	20.305	20.312	20.320	20.327	20.335	20.342	20.350	3100
3110	20.350	20.357	20.365	20.372	20.380	20.387	20.394	20.402	20.409	20.417	20.424	3110
3120	20.424	20.432	20.439	20.446	20.454	20.461	20.469	20.476	20.483	20.491	20.498	3120
3130	20.498	20.506	20.513	20.520	20.528	20.535	20.543	20.550	20.557	20.565	20.572	3130
3140	20.572	20.579	20.587	20.594	20.601	20.609	20.616	20.623	20.631	20.638	20.645	3140
3150	20.645	20.653	20.660	20.667	20.675	20.682	20.689	20.697	20.704	20.711	20.718	3150
3160	20.718	20.726	20.733	20.740	20.748	20.755	20.762	20.769	20.777	20.784	20.791	3160
3170	20.791	20.798	20.806	20.813	20.820	20.827	20.834	20.842	20.849	20.856	20.863	3170
3180	20.863	20.870	20.878	20.885	20.892	20.899	20.906	20.914	20.921	20.928	20.935	3180
3190	20.935	20.942	20.949	20.956	20.964	20.971	20.978	20.985	20.992	20.999	21.006	3190
3200	21.006	21.013	21.021	21.028	21.035	21.042	21.049	21.056	21.063	21.070	21.077	3200
3210	21.077	21.084	21.091	21.098	21.105							3210

Type S Thermocouple
Thermoelectric Voltage as a Function of Temperature (°C)
Reference Junctions at 0°C
THERMOELECTRIC VOLTAGE IN ABSOLUTE MILLIVOLTS

°C	0	1	2	3	4	5	6	7	8	9	10	°C
−50	−0.236	−0.232	−0.228	−0.224	−0.220	−0.215	−0.211	−0.207	−0.203	−0.199	−0.194	−50
−40	−0.194	−0.190	−0.186	−0.181	−0.177	−0.173	−0.168	−0.164	−0.159	−0.155	−0.150	−40
−30	−0.150	−0.145	−0.141	−0.136	−0.132	−0.127	−0.122	−0.117	−0.112	−0.108	−0.103	−30
−20	−0.103	−0.098	−0.093	−0.088	−0.083	−0.078	−0.073	−0.068	−0.063	−0.058	−0.053	−20
−10	−0.053	−0.048	−0.042	−0.037	−0.032	−0.027	−0.021	−0.016	−0.011	−0.005	0.000	−10
0	0.000	0.005	0.011	0.016	0.022	0.027	0.033	0.038	0.044	0.050	0.055	0
10	0.055	0.061	0.067	0.072	0.078	0.084	0.090	0.095	0.101	0.107	0.113	10
20	0.113	0.119	0.125	0.131	0.137	0.142	0.148	0.154	0.161	0.167	0.173	20
30	0.173	0.179	0.185	0.191	0.197	0.203	0.210	0.216	0.222	0.228	0.235	30
40	0.235	0.241	0.247	0.254	0.260	0.266	0.273	0.279	0.286	0.292	0.299	40
50	0.299	0.305	0.312	0.318	0.325	0.331	0.338	0.345	0.351	0.358	0.365	50
60	0.365	0.371	0.378	0.385	0.391	0.398	0.405	0.412	0.419	0.425	0.432	60
70	0.432	0.439	0.446	0.453	0.460	0.467	0.474	0.481	0.488	0.495	0.502	70
80	0.502	0.509	0.516	0.523	0.530	0.537	0.544	0.551	0.558	0.566	0.573	80
90	0.573	0.580	0.587	0.594	0.602	0.609	0.616	0.623	0.631	0.638	0.645	90
100	0.645	0.653	0.660	0.667	0.675	0.682	0.690	0.697	0.704	0.712	0.719	100
110	0.719	0.727	0.734	0.742	0.749	0.757	0.764	0.772	0.780	0.787	0.795	110
120	0.795	0.802	0.810	0.818	0.825	0.833	0.841	0.848	0.856	0.864	0.872	120
130	0.872	0.879	0.887	0.895	0.903	0.910	0.918	0.926	0.934	0.942	0.950	130
140	0.950	0.957	0.965	0.973	0.981	0.989	0.997	1.005	1.013	1.021	1.029	140
150	1.029	1.037	1.045	1.053	1.061	1.069	1.077	1.085	1.093	1.101	1.109	150
160	1.109	1.117	1.125	1.133	1.141	1.149	1.158	1.166	1.174	1.182	1.190	160
170	1.190	1.198	1.207	1.215	1.223	1.231	1.240	1.248	1.256	1.264	1.273	170
180	1.273	1.281	1.289	1.297	1.306	1.314	1.322	1.331	1.339	1.347	1.356	180
190	1.356	1.364	1.373	1.381	1.389	1.398	1.406	1.415	1.423	1.432	1.440	190
200	1.440	1.448	1.457	1.465	1.474	1.482	1.491	1.499	1.508	1.516	1.525	200
210	1.525	1.534	1.542	1.551	1.559	1.568	1.576	1.585	1.594	1.602	1.611	210
220	1.611	1.620	1.628	1.637	1.645	1.654	1.663	1.671	1.680	1.689	1.698	220
230	1.698	1.706	1.715	1.724	1.732	1.741	1.750	1.759	1.767	1.776	1.785	230
240	1.785	1.794	1.802	1.811	1.820	1.829	1.838	1.846	1.855	1.864	1.873	240
250	1.873	1.882	1.891	1.899	1.908	1.917	1.926	1.935	1.944	1.953	1.962	250
260	1.962	1.971	1.979	1.988	1.997	2.006	2.015	2.024	2.033	2.042	2.051	260
270	2.051	2.060	2.069	2.078	2.087	2.096	2.105	2.114	2.123	2.132	2.141	270
280	2.141	2.150	2.159	2.168	2.177	2.186	2.195	2.204	2.213	2.222	2.232	280
290	2.232	2.241	2.250	2.259	2.268	2.277	2.286	2.295	2.304	2.314	2.323	290
300	2.323	2.332	2.341	2.350	2.359	2.368	2.378	2.387	2.396	2.405	2.414	300
310	2.414	2.424	2.433	2.442	2.451	2.460	2.470	2.479	2.488	2.497	2.506	310
320	2.506	2.516	2.525	2.534	2.543	2.553	2.562	2.571	2.581	2.590	2.599	320
330	2.599	2.608	2.618	2.627	2.636	2.646	2.655	2.664	2.674	2.683	2.692	330
340	2.692	2.702	2.711	2.720	2.730	2.739	2.748	2.758	2.767	2.776	2.786	340
350	2.786	2.795	2.805	2.814	2.823	2.833	2.842	2.852	2.861	2.870	2.880	350
360	2.880	2.889	2.899	2.908	2.917	2.927	2.936	2.946	2.955	2.965	2.974	360
370	2.974	2.984	2.993	3.003	3.012	3.022	3.031	3.041	3.050	3.059	3.069	370
380	3.069	3.078	3.088	3.097	3.107	3.117	3.126	3.136	3.145	3.155	3.164	380
390	3.164	3.174	3.183	3.193	3.202	3.212	3.221	3.231	3.241	3.250	3.260	390
400	3.260	3.269	3.279	3.288	3.298	3.308	3.317	3.327	3.336	3.346	3.356	400
410	3.356	3.365	3.375	3.384	3.394	3.404	3.413	3.423	3.433	3.442	3.452	410
420	3.452	3.462	3.471	3.481	3.491	3.500	3.510	3.520	3.529	3.539	3.549	420
430	3.549	3.558	3.568	3.578	3.587	3.597	3.607	3.616	3.626	3.636	3.645	430
440	3.645	3.655	3.665	3.675	3.684	3.694	3.704	3.714	3.723	3.733	3.743	440
450	3.743	3.752	3.762	3.772	3.782	3.791	3.801	3.811	3.821	3.831	3.840	450
460	3.840	3.850	3.860	3.870	3.879	3.889	3.899	3.909	3.919	3.928	3.938	460
470	3.938	3.948	3.958	3.968	3.977	3.987	3.997	4.007	4.017	4.027	4.036	470
480	4.036	4.046	4.056	4.066	4.076	4.086	4.095	4.105	4.115	4.125	4.135	480
490	4.135	4.145	4.155	4.164	4.174	4.184	4.194	4.204	4.214	4.224	4.234	490
500	4.234	4.243	4.253	4.263	4.273	4.283	4.293	4.303	4.313	4.323	4.333	500
510	4.333	4.343	4.352	4.362	4.372	4.382	4.392	4.402	4.412	4.422	4.432	510
520	4.432	4.442	4.452	4.462	4.472	4.482	4.492	4.502	4.512	4.522	4.532	520
530	4.532	4.542	4.552	4.562	4.572	4.582	4.592	4.602	4.612	4.622	4.632	530
540	4.632	4.642	4.652	4.662	4.672	4.682	4.692	4.702	4.712	4.722	4.732	540

Type S Thermocouple
Thermoelectric Voltage as a Function of Temperature (°C)
Reference Junctions at 0°C

THERMOELECTRIC VOLTAGE IN ABSOLUTE MILLIVOLTS

°C	0	1	2	3	4	5	6	7	8	9	10	°C
550	4.732	4.742	4.752	4.762	4.772	4.782	4.792	4.802	4.812	4.822	4.832	550
560	4.832	4.842	4.852	4.862	4.873	4.883	4.893	4.903	4.913	4.923	4.933	560
570	4.933	4.943	4.953	4.963	4.973	4.984	4.994	5.004	5.014	5.024	5.034	570
580	5.034	5.044	5.054	5.065	5.075	5.085	5.095	5.105	5.115	5.125	5.136	580
590	5.136	5.146	5.156	5.166	5.176	5.186	5.197	5.207	5.217	5.227	5.237	590
600	5.237	5.247	5.258	5.268	5.278	5.288	5.298	5.309	5.319	5.329	5.339	600
610	5.339	5.350	5.360	5.370	5.380	5.391	5.401	5.411	5.421	5.431	5.442	610
620	5.442	5.452	5.462	5.473	5.483	5.493	5.503	5.514	5.524	5.534	5.544	620
630	5.544	5.555	5.565	5.575	5.586	5.596	5.606	5.617	5.627	5.637	5.648	630
640	5.648	5.658	5.668	5.679	5.689	5.700	5.710	5.720	5.731	5.741	5.751	640
650	5.751	5.762	5.772	5.782	5.793	5.803	5.814	5.824	5.834	5.845	5.855	650
660	5.855	5.866	5.876	5.887	5.897	5.907	5.918	5.928	5.939	5.949	5.960	660
670	5.960	5.970	5.980	5.991	6.001	6.012	6.022	6.033	6.043	6.054	6.064	670
680	6.064	6.075	6.085	6.096	6.106	6.117	6.127	6.138	6.148	6.159	6.169	680
690	6.169	6.180	6.190	6.201	6.211	6.222	6.232	6.243	6.253	6.264	6.274	690
700	6.274	6.285	6.295	6.306	6.316	6.327	6.338	6.348	6.359	6.369	6.380	700
710	6.380	6.390	6.401	6.412	6.422	6.433	6.443	6.454	6.465	6.475	6.486	710
720	6.486	6.496	6.507	6.518	6.528	6.539	6.549	6.560	6.571	6.581	6.592	720
730	6.592	6.603	6.613	6.624	6.635	6.645	6.656	6.667	6.677	6.688	6.699	730
740	6.699	6.709	6.720	6.731	6.741	6.752	6.763	6.773	6.784	6.795	6.805	740
750	6.805	6.816	6.827	6.838	6.848	6.859	6.870	6.880	6.891	6.902	6.913	750
760	6.913	6.923	6.934	6.945	6.956	6.966	6.977	6.988	6.999	7.009	7.020	760
770	7.020	7.031	7.042	7.053	7.063	7.074	7.085	7.096	7.107	7.117	7.128	770
780	7.128	7.139	7.150	7.161	7.171	7.182	7.193	7.204	7.215	7.225	7.236	780
790	7.236	7.247	7.258	7.269	7.280	7.291	7.301	7.312	7.323	7.334	7.345	790
800	7.345	7.356	7.367	7.377	7.388	7.399	7.410	7.421	7.432	7.443	7.454	800
810	7.454	7.465	7.476	7.486	7.497	7.508	7.519	7.530	7.541	7.552	7.563	810
820	7.563	7.574	7.585	7.596	7.607	7.618	7.629	7.640	7.651	7.661	7.672	820
830	7.672	7.683	7.694	7.705	7.716	7.727	7.738	7.749	7.760	7.771	7.782	830
840	7.782	7.793	7.804	7.815	7.826	7.837	7.848	7.859	7.870	7.881	7.892	840
850	7.892	7.904	7.915	7.926	7.937	7.948	7.959	7.970	7.981	7.992	8.003	850
860	8.003	8.014	8.025	8.036	8.047	8.058	8.069	8.081	8.092	8.103	8.114	860
870	8.114	8.125	8.136	8.147	8.158	8.169	8.180	8.192	8.203	8.214	8.225	870
880	8.225	8.236	8.247	8.258	8.270	8.281	8.292	8.303	8.314	8.325	8.336	880
890	8.336	8.348	8.359	8.370	8.381	8.392	8.404	8.415	8.426	8.437	8.448	890
900	8.448	8.460	8.471	8.482	8.493	8.504	8.516	8.527	8.538	8.549	8.560	900
910	8.560	8.572	8.583	8.594	8.605	8.617	8.628	8.639	8.650	8.662	8.673	910
920	8.673	8.684	8.695	8.707	8.718	8.729	8.741	8.752	8.763	8.774	8.786	920
930	8.786	8.797	8.808	8.820	8.831	8.842	8.854	8.865	8.876	8.888	8.899	930
940	8.899	8.910	8.922	8.933	8.944	8.956	8.967	8.978	8.990	9.001	9.012	940
950	9.012	9.024	9.035	9.047	9.058	9.069	9.081	9.092	9.103	9.115	9.126	950
960	9.126	9.138	9.149	9.160	9.172	9.183	9.195	9.206	9.217	9.229	9.240	960
970	9.240	9.252	9.263	9.275	9.286	9.298	9.309	9.320	9.332	9.343	9.355	970
980	9.355	9.366	9.378	9.389	9.401	9.412	9.424	9.435	9.447	9.458	9.470	980
990	9.470	9.481	9.493	9.504	9.516	9.527	9.539	9.550	9.562	9.573	9.585	990
1000	9.585	9.596	9.608	9.619	9.631	9.642	9.654	9.665	9.677	9.689	9.700	1000
1010	9.700	9.712	9.723	9.735	9.746	9.758	9.770	9.781	9.793	9.804	9.816	1010
1020	9.816	9.828	9.839	9.851	9.862	9.874	9.886	9.897	9.909	9.920	9.932	1020
1030	9.932	9.944	9.955	9.967	9.979	9.990	10.002	10.013	10.025	10.037	10.048	1030
1040	10.048	10.060	10.072	10.083	10.095	10.107	10.118	10.130	10.142	10.154	10.165	1040
1050	10.165	10.177	10.189	10.200	10.212	10.224	10.235	10.247	10.259	10.271	10.282	1050
1060	10.282	10.294	10.306	10.318	10.329	10.341	10.353	10.364	10.376	10.388	10.400	1060
1070	10.400	10.411	10.423	10.435	10.447	10.459	10.470	10.482	10.494	10.506	10.517	1070
1080	10.517	10.529	10.541	10.553	10.565	10.576	10.588	10.600	10.612	10.624	10.635	1080
1090	10.635	10.647	10.659	10.671	10.683	10.694	10.706	10.718	10.730	10.742	10.754	1090
1100	10.754	10.765	10.777	10.789	10.801	10.813	10.825	10.836	10.848	10.860	10.872	1100
1110	10.872	10.884	10.896	10.908	10.919	10.931	10.943	10.955	10.967	10.979	10.991	1110
1120	10.991	11.003	11.014	11.026	11.038	11.050	11.062	11.074	11.086	11.098	11.110	1120
1130	11.110	11.121	11.133	11.145	11.157	11.169	11.181	11.193	11.205	11.217	11.229	1130
1140	11.229	11.241	11.252	11.264	11.276	11.288	11.300	11.312	11.324	11.336	11.348	1140

Type S Thermocouple
Thermoelectric Voltage as a Function of Temperature (°C)
Reference Junctions at 0°C
THERMOELECTRIC VOLTAGE IN ABSOLUTE MILLIVOLTS

°C	0	1	2	3	4	5	6	7	8	9	10	°C
1150	11.348	11.360	11.372	11.384	11.396	11.408	11.420	11.432	11.443	11.455	11.467	1150
1160	11.467	11.479	11.491	11.503	11.515	11.527	11.539	11.551	11.563	11.575	11.587	1160
1170	11.587	11.599	11.611	11.623	11.635	11.647	11.659	11.671	11.683	11.695	11.707	1170
1180	11.707	11.719	11.731	11.743	11.755	11.767	11.779	11.791	11.803	11.815	11.827	1180
1190	11.827	11.839	11.851	11.863	11.875	11.887	11.899	11.911	11.923	11.935	11.947	1190
1200	11.947	11.959	11.971	11.983	11.995	12.007	12.019	12.031	12.043	12.055	12.067	1200
1210	12.067	12.079	12.091	12.103	12.116	12.128	12.140	12.152	12.164	12.176	12.188	1210
1220	12.188	12.200	12.212	12.224	12.236	12.248	12.260	12.272	12.284	12.296	12.308	1220
1230	12.308	12.320	12.332	12.345	12.357	12.369	12.381	12.393	12.405	12.417	12.429	1230
1240	12.429	12.441	12.453	12.465	12.477	12.489	12.501	12.514	12.526	12.538	12.550	1240
1250	12.550	12.562	12.574	12.586	12.598	12.610	12.622	12.634	12.647	12.659	12.671	1250
1260	12.671	12.683	12.695	12.707	12.719	12.731	12.743	12.755	12.767	12.780	12.792	1260
1270	12.792	12.804	12.816	12.828	12.840	12.852	12.864	12.876	12.888	12.901	12.913	1270
1280	12.913	12.925	12.937	12.949	12.961	12.973	12.985	12.997	13.010	13.022	13.034	1280
1290	13.034	13.046	13.058	13.070	13.082	13.094	13.107	13.119	13.131	13.143	13.155	1290
1300	13.155	13.167	13.179	13.191	13.203	13.216	13.228	13.240	13.252	13.264	13.276	1300
1310	13.276	13.288	13.300	13.313	13.325	13.337	13.349	13.361	13.373	13.385	13.397	1310
1320	13.397	13.410	13.422	13.434	13.446	13.458	13.470	13.482	13.495	13.507	13.519	1320
1330	13.519	13.531	13.543	13.555	13.567	13.579	13.592	13.604	13.616	13.628	13.640	1330
1340	13.640	13.652	13.664	13.677	13.689	13.701	13.713	13.725	13.737	13.749	13.761	1340
1350	13.761	13.774	13.786	13.798	13.810	13.822	13.834	13.846	13.859	13.871	13.883	1350
1360	13.883	13.895	13.907	13.919	13.931	13.943	13.956	13.968	13.980	13.992	14.004	1360
1370	14.004	14.016	14.028	14.040	14.053	14.065	14.077	14.089	14.101	14.113	14.125	1370
1380	14.125	14.138	14.150	14.162	14.174	14.186	14.198	14.210	14.222	14.235	14.247	1380
1390	14.247	14.259	14.271	14.283	14.295	14.307	14.319	14.332	14.344	14.356	14.368	1390
1400	14.368	14.380	14.392	14.404	14.416	14.429	14.441	14.453	14.465	14.477	14.489	1400
1410	14.489	14.501	14.513	14.526	14.538	14.550	14.562	14.574	14.586	14.598	14.610	1410
1420	14.610	14.622	14.635	14.647	14.659	14.671	14.683	14.695	14.707	14.719	14.731	1420
1430	14.731	14.744	14.756	14.768	14.780	14.792	14.804	14.816	14.828	14.840	14.852	1430
1440	14.852	14.865	14.877	14.889	14.901	14.913	14.925	14.937	14.949	14.961	14.973	1440
1450	14.973	14.985	14.998	15.010	15.022	15.034	15.046	15.058	15.070	15.082	15.094	1450
1460	15.094	15.106	15.118	15.130	15.143	15.155	15.167	15.179	15.191	15.203	15.215	1460
1470	15.215	15.227	15.239	15.251	15.263	15.275	15.287	15.299	15.311	15.324	15.336	1470
1480	15.336	15.348	15.360	15.372	15.384	15.396	15.408	15.420	15.432	15.444	15.456	1480
1490	15.456	15.468	15.480	15.492	15.504	15.516	15.528	15.540	15.552	15.564	15.576	1490
1500	15.576	15.589	15.601	15.613	15.625	15.637	15.649	15.661	15.673	15.685	15.697	1500
1510	15.697	15.709	15.721	15.733	15.745	15.757	15.769	15.781	15.793	15.805	15.817	1510
1520	15.817	15.829	15.841	15.853	15.865	15.877	15.889	15.901	15.913	15.925	15.937	1520
1530	15.937	15.949	15.961	15.973	15.985	15.997	16.009	16.021	16.033	16.045	16.057	1530
1540	16.057	16.069	16.080	16.092	16.104	16.116	16.128	16.140	16.152	16.164	16.176	1540
1550	16.176	16.188	16.200	16.212	16.224	16.236	16.248	16.260	16.272	16.284	16.296	1550
1560	16.296	16.308	16.319	16.331	16.343	16.355	16.367	16.379	16.391	16.403	16.415	1560
1570	16.415	16.427	16.439	16.451	16.462	16.474	16.486	16.498	16.510	16.522	16.534	1570
1580	16.534	16.546	16.558	16.569	16.581	16.593	16.605	16.617	16.629	16.641	16.653	1580
1590	16.653	16.664	16.676	16.688	16.700	16.712	16.724	16.736	16.747	16.759	16.771	1590
1600	16.771	16.783	16.795	16.807	16.819	16.830	16.842	16.854	16.866	16.878	16.890	1600
1610	16.890	16.901	16.913	16.925	16.937	16.949	16.960	16.972	16.984	16.996	17.008	1610
1620	17.008	17.019	17.031	17.043	17.055	17.067	17.078	17.090	17.102	17.114	17.125	1620
1630	17.125	17.137	17.149	17.161	17.173	17.184	17.196	17.208	17.220	17.231	17.243	1630
1640	17.243	17.255	17.267	17.278	17.290	17.302	17.313	17.325	17.337	17.349	17.360	1640
1650	17.360	17.372	17.384	17.396	17.407	17.419	17.431	17.442	17.454	17.466	17.477	1650
1660	17.477	17.489	17.501	17.512	17.524	17.536	17.548	17.559	17.571	17.583	17.594	1660
1670	17.594	17.606	17.617	17.629	17.641	17.652	17.664	17.676	17.687	17.699	17.711	1670
1680	17.711	17.722	17.734	17.745	17.757	17.769	17.780	17.792	17.803	17.815	17.826	1680
1690	17.826	17.838	17.850	17.861	17.873	17.884	17.896	17.907	17.919	17.930	17.942	1690
1700	17.942	17.953	17.965	17.976	17.988	17.999	18.010	18.022	18.033	18.045	18.056	1700
1710	18.056	18.068	18.079	18.090	18.102	18.113	18.124	18.136	18.147	18.158	18.170	1710
1720	18.170	18.181	18.192	18.204	18.215	18.226	18.237	18.249	18.260	18.271	18.282	1720
1730	18.282	18.293	18.305	18.316	18.327	18.338	18.349	18.360	18.372	18.383	18.394	1730
1740	18.394	18.405	18.416	18.427	18.438	18.449	18.460	18.471	18.482	18.493	18.504	1740
1750	18.504	18.515	18.526	18.536	18.547	18.558	18.569	18.580	18.591	18.602	18.612	1750
1760	18.612	18.623	18.634	18.645	18.655	18.666	18.677	18.687	18.698			1760

Type S Thermocouple
Thermoelectric Voltage as a Function of Temperature (°F)
Reference Junctions at 32°F
THERMOELECTRIC VOLTAGE IN ABSOLUTE MILLIVOLTS

°F	0	1	2	3	4	5	6	7	8	9	10	°F
-60			-0.236	-0.233	-0.231	-0.229	-0.227	-0.225	-0.222	-0.220	-0.218	-60
-50	-0.218	-0.215	-0.213	-0.211	-0.209	-0.206	-0.204	-0.202	-0.199	-0.197	-0.194	-50
-40	-0.194	-0.192	-0.190	-0.187	-0.185	-0.182	-0.180	-0.178	-0.175	-0.173	-0.170	-40
-30	-0.170	-0.168	-0.165	-0.163	-0.160	-0.158	-0.155	-0.153	-0.150	-0.148	-0.145	-30
-20	-0.145	-0.142	-0.140	-0.137	-0.135	-0.132	-0.129	-0.127	-0.124	-0.122	-0.119	-20
-10	-0.119	-0.116	-0.114	-0.111	-0.108	-0.106	-0.103	-0.100	-0.097	-0.095	-0.092	-10
0	-0.092	-0.089	-0.086	-0.084	-0.081	-0.078	-0.075	-0.073	-0.070	-0.067	-0.064	0
10	-0.064	-0.061	-0.058	-0.056	-0.053	-0.050	-0.047	-0.044	-0.041	-0.038	-0.035	10
20	-0.035	-0.033	-0.030	-0.027	-0.024	-0.021	-0.018	-0.015	-0.012	-0.009	-0.006	20
30	-0.006	-0.003	0.000	0.003	0.006	0.009	0.012	0.015	0.018	0.021	0.024	30
40	0.024	0.027	0.030	0.033	0.037	0.040	0.043	0.046	0.049	0.052	0.055	40
50	0.055	0.058	0.062	0.065	0.068	0.071	0.074	0.077	0.081	0.084	0.087	50
60	0.087	0.090	0.093	0.097	0.100	0.103	0.106	0.110	0.113	0.116	0.119	60
70	0.119	0.123	0.126	0.129	0.133	0.136	0.139	0.142	0.146	0.149	0.152	70
80	0.152	0.156	0.159	0.163	0.166	0.169	0.173	0.176	0.179	0.183	0.186	80
90	0.186	0.190	0.193	0.197	0.200	0.203	0.207	0.210	0.214	0.217	0.221	90
100	0.221	0.224	0.228	0.231	0.235	0.238	0.242	0.245	0.249	0.252	0.256	100
110	0.256	0.259	0.263	0.266	0.270	0.274	0.277	0.281	0.284	0.288	0.291	110
120	0.291	0.295	0.299	0.302	0.306	0.309	0.313	0.317	0.320	0.324	0.328	120
130	0.328	0.331	0.335	0.339	0.342	0.346	0.350	0.353	0.357	0.361	0.365	130
140	0.365	0.368	0.372	0.376	0.379	0.383	0.387	0.391	0.394	0.398	0.402	140
150	0.402	0.406	0.409	0.413	0.417	0.421	0.425	0.428	0.432	0.436	0.440	150
160	0.440	0.444	0.448	0.451	0.455	0.459	0.463	0.467	0.471	0.474	0.478	160
170	0.478	0.482	0.486	0.490	0.494	0.498	0.502	0.506	0.510	0.513	0.517	170
180	0.517	0.521	0.525	0.529	0.533	0.537	0.541	0.545	0.549	0.553	0.557	180
190	0.557	0.561	0.565	0.569	0.573	0.577	0.581	0.585	0.589	0.593	0.597	190
200	0.597	0.601	0.605	0.609	0.613	0.617	0.621	0.625	0.629	0.633	0.637	200
210	0.637	0.641	0.645	0.649	0.653	0.658	0.662	0.666	0.670	0.674	0.678	210
220	0.678	0.682	0.686	0.690	0.695	0.699	0.703	0.707	0.711	0.715	0.719	220
230	0.719	0.724	0.728	0.732	0.736	0.740	0.744	0.749	0.753	0.757	0.761	230
240	0.761	0.765	0.770	0.774	0.778	0.782	0.786	0.791	0.795	0.799	0.803	240
250	0.803	0.808	0.812	0.816	0.820	0.824	0.829	0.833	0.837	0.842	0.846	250
260	0.846	0.850	0.854	0.859	0.863	0.867	0.872	0.876	0.880	0.884	0.889	260
270	0.889	0.893	0.897	0.902	0.906	0.910	0.915	0.919	0.923	0.928	0.932	270
280	0.932	0.936	0.941	0.945	0.950	0.954	0.958	0.963	0.967	0.971	0.976	280
290	0.976	0.980	0.985	0.989	0.993	0.998	1.002	1.007	1.011	1.015	1.020	290
300	1.020	1.024	1.029	1.033	1.038	1.042	1.046	1.051	1.055	1.060	1.064	300
310	1.064	1.069	1.073	1.078	1.082	1.087	1.091	1.095	1.100	1.104	1.109	310
320	1.109	1.113	1.118	1.122	1.127	1.131	1.136	1.140	1.145	1.149	1.154	320
330	1.154	1.158	1.163	1.168	1.172	1.177	1.181	1.186	1.190	1.195	1.199	330
340	1.199	1.204	1.208	1.213	1.218	1.222	1.227	1.231	1.236	1.240	1.245	340
350	1.245	1.250	1.254	1.259	1.263	1.268	1.273	1.277	1.282	1.286	1.291	350
360	1.291	1.296	1.300	1.305	1.309	1.314	1.319	1.323	1.328	1.333	1.337	360
370	1.337	1.342	1.347	1.351	1.356	1.360	1.365	1.370	1.374	1.379	1.384	370
380	1.384	1.388	1.393	1.398	1.402	1.407	1.412	1.417	1.421	1.426	1.431	380
390	1.431	1.435	1.440	1.445	1.449	1.454	1.459	1.464	1.468	1.473	1.478	390
400	1.478	1.482	1.487	1.492	1.497	1.501	1.506	1.511	1.516	1.520	1.525	400
410	1.525	1.530	1.535	1.539	1.544	1.549	1.554	1.558	1.563	1.568	1.573	410
420	1.573	1.577	1.582	1.587	1.592	1.597	1.601	1.606	1.611	1.616	1.620	420
430	1.620	1.625	1.630	1.635	1.640	1.644	1.649	1.654	1.659	1.664	1.669	430
440	1.669	1.673	1.678	1.683	1.688	1.693	1.698	1.702	1.707	1.712	1.717	440
450	1.717	1.722	1.727	1.731	1.736	1.741	1.746	1.751	1.756	1.761	1.765	450
460	1.765	1.770	1.775	1.780	1.785	1.790	1.795	1.799	1.804	1.809	1.814	460
470	1.814	1.819	1.824	1.829	1.834	1.839	1.843	1.848	1.853	1.858	1.863	470
480	1.863	1.868	1.873	1.878	1.883	1.888	1.893	1.898	1.902	1.907	1.912	480
490	1.912	1.917	1.922	1.927	1.932	1.937	1.942	1.947	1.952	1.957	1.962	490
500	1.962	1.967	1.972	1.977	1.981	1.986	1.991	1.996	2.001	2.006	2.011	500
510	2.011	2.016	2.021	2.026	2.031	2.036	2.041	2.046	2.051	2.056	2.061	510
520	2.061	2.066	2.071	2.076	2.081	2.086	2.091	2.096	2.101	2.106	2.111	520
530	2.111	2.116	2.121	2.126	2.131	2.136	2.141	2.146	2.151	2.156	2.161	530

Type S Thermocouple
Thermoelectric Voltage as a Function of Temperature (°F)
Reference Junctions at 32°F
THERMOELECTRIC VOLTAGE IN ABSOLUTE MILLIVOLTS

°F	0	1	2	3	4	5	6	7	8	9	10	°F
540	2.161	2.166	2.171	2.176	2.181	2.186	2.191	2.196	2.201	2.206	2.211	540
550	2.211	2.216	2.221	2.227	2.232	2.237	2.242	2.247	2.252	2.257	2.262	550
560	2.262	2.267	2.272	2.277	2.282	2.287	2.292	2.297	2.302	2.307	2.313	560
570	2.313	2.318	2.323	2.328	2.333	2.338	2.343	2.348	2.353	2.358	2.363	570
580	2.363	2.368	2.374	2.379	2.384	2.389	2.394	2.399	2.404	2.409	2.414	580
590	2.414	2.419	2.425	2.430	2.435	2.440	2.445	2.450	2.455	2.460	2.465	590
600	2.465	2.471	2.476	2.481	2.486	2.491	2.496	2.501	2.506	2.512	2.517	600
610	2.517	2.522	2.527	2.532	2.537	2.542	2.548	2.553	2.558	2.563	2.568	610
620	2.568	2.573	2.578	2.584	2.589	2.594	2.599	2.604	2.609	2.615	2.620	620
630	2.620	2.625	2.630	2.635	2.640	2.646	2.651	2.656	2.661	2.666	2.672	630
640	2.672	2.677	2.682	2.687	2.692	2.697	2.703	2.708	2.713	2.718	2.723	640
650	2.723	2.729	2.734	2.739	2.744	2.749	2.755	2.760	2.765	2.770	2.775	650
660	2.775	2.781	2.786	2.791	2.796	2.801	2.807	2.812	2.817	2.822	2.828	660
670	2.828	2.833	2.838	2.843	2.848	2.854	2.859	2.864	2.869	2.875	2.880	670
680	2.880	2.885	2.890	2.895	2.901	2.906	2.911	2.916	2.922	2.927	2.932	680
690	2.932	2.937	2.943	2.948	2.953	2.958	2.964	2.969	2.974	2.979	2.985	690
700	2.985	2.990	2.995	3.000	3.006	3.011	3.016	3.022	3.027	3.032	3.037	700
710	3.037	3.043	3.048	3.053	3.058	3.064	3.069	3.074	3.080	3.085	3.090	710
720	3.090	3.095	3.101	3.106	3.111	3.117	3.122	3.127	3.132	3.138	3.143	720
730	3.143	3.148	3.154	3.159	3.164	3.169	3.175	3.180	3.185	3.191	3.196	730
740	3.196	3.201	3.207	3.212	3.217	3.223	3.228	3.233	3.238	3.244	3.249	740
750	3.249	3.254	3.260	3.265	3.270	3.276	3.281	3.286	3.292	3.297	3.302	750
760	3.302	3.308	3.313	3.318	3.324	3.329	3.334	3.340	3.345	3.350	3.356	760
770	3.356	3.361	3.366	3.372	3.377	3.382	3.388	3.393	3.398	3.404	3.409	770
780	3.409	3.414	3.420	3.425	3.430	3.436	3.441	3.447	3.452	3.457	3.463	780
790	3.463	3.468	3.473	3.479	3.484	3.489	3.495	3.500	3.506	3.511	3.516	790
800	3.516	3.522	3.527	3.532	3.538	3.543	3.549	3.554	3.559	3.565	3.570	800
810	3.570	3.575	3.581	3.586	3.592	3.597	3.602	3.608	3.613	3.619	3.624	810
820	3.624	3.629	3.635	3.640	3.645	3.651	3.656	3.662	3.667	3.672	3.678	820
830	3.678	3.683	3.689	3.694	3.699	3.705	3.710	3.716	3.721	3.726	3.732	830
840	3.732	3.737	3.743	3.748	3.754	3.759	3.764	3.770	3.775	3.781	3.786	840
850	3.786	3.791	3.797	3.802	3.808	3.813	3.819	3.824	3.829	3.835	3.840	850
860	3.840	3.846	3.851	3.857	3.862	3.867	3.873	3.878	3.884	3.889	3.895	860
870	3.895	3.900	3.906	3.911	3.916	3.922	3.927	3.933	3.938	3.944	3.949	870
880	3.949	3.955	3.960	3.965	3.971	3.976	3.982	3.987	3.993	3.998	4.004	880
890	4.004	4.009	4.015	4.020	4.025	4.031	4.036	4.042	4.047	4.053	4.058	890
900	4.058	4.064	4.069	4.075	4.080	4.086	4.091	4.096	4.102	4.107	4.113	900
910	4.113	4.118	4.124	4.129	4.135	4.140	4.146	4.151	4.157	4.162	4.168	910
920	4.168	4.173	4.179	4.184	4.190	4.195	4.201	4.206	4.212	4.217	4.223	920
930	4.223	4.228	4.234	4.239	4.245	4.250	4.256	4.261	4.267	4.272	4.278	930
940	4.278	4.283	4.289	4.294	4.300	4.305	4.311	4.316	4.322	4.327	4.333	940
950	4.333	4.338	4.344	4.349	4.355	4.360	4.366	4.371	4.377	4.382	4.388	950
960	4.388	4.393	4.399	4.404	4.410	4.415	4.421	4.426	4.432	4.438	4.443	960
970	4.443	4.449	4.454	4.460	4.465	4.471	4.476	4.482	4.487	4.493	4.498	970
980	4.498	4.504	4.509	4.515	4.521	4.526	4.532	4.537	4.543	4.548	4.554	980
990	4.554	4.559	4.565	4.570	4.576	4.582	4.587	4.593	4.598	4.604	4.609	990
1000	4.609	4.615	4.620	4.626	4.632	4.637	4.643	4.648	4.654	4.659	4.665	1000
1010	4.665	4.670	4.676	4.682	4.687	4.693	4.698	4.704	4.709	4.715	4.721	1010
1020	4.721	4.726	4.732	4.737	4.743	4.748	4.754	4.760	4.765	4.771	4.776	1020
1030	4.776	4.782	4.788	4.793	4.799	4.804	4.810	4.815	4.821	4.827	4.832	1030
1040	4.832	4.838	4.843	4.849	4.855	4.860	4.866	4.871	4.877	4.883	4.888	1040
1050	4.888	4.894	4.899	4.905	4.911	4.916	4.922	4.927	4.933	4.939	4.944	1050
1060	4.944	4.950	4.956	4.961	4.967	4.972	4.978	4.984	4.989	4.995	5.000	1060
1070	5.000	5.006	5.012	5.017	5.023	5.029	5.034	5.040	5.045	5.051	5.057	1070
1080	5.057	5.062	5.068	5.074	5.079	5.085	5.090	5.096	5.102	5.107	5.113	1080
1090	5.113	5.119	5.124	5.130	5.136	5.141	5.147	5.153	5.158	5.164	5.169	1090
1100	5.169	5.175	5.181	5.186	5.192	5.198	5.203	5.209	5.215	5.220	5.226	1100
1110	5.226	5.232	5.237	5.243	5.249	5.254	5.260	5.266	5.271	5.277	5.283	1110
1120	5.283	5.288	5.294	5.300	5.305	5.311	5.317	5.322	5.328	5.334	5.339	1120
1130	5.339	5.345	5.351	5.356	5.362	5.368	5.373	5.379	5.385	5.391	5.396	1130

Type S Thermocouple
Thermoelectric Voltage as a Function of Temperature (°F)
Reference Junctions at 32°F
THERMOELECTRIC VOLTAGE IN ABSOLUTE MILLIVOLTS

°F	0	1	2	3	4	5	6	7	8	9	10	°F
1140	5.396	5.402	5.408	5.413	5.419	5.425	5.430	5.436	5.442	5.447	5.453	1140
1150	5.453	5.459	5.465	5.470	5.476	5.482	5.487	5.493	5.499	5.504	5.510	1150
1160	5.510	5.516	5.522	5.527	5.533	5.539	5.544	5.550	5.556	5.562	5.567	1160
1170	5.567	5.573	5.579	5.585	5.590	5.596	5.602	5.608	5.613	5.619	5.625	1170
1180	5.625	5.631	5.636	5.642	5.648	5.653	5.659	5.665	5.671	5.676	5.682	1180
1190	5.682	5.688	5.694	5.700	5.705	5.711	5.717	5.723	5.728	5.734	5.740	1190
1200	5.740	5.746	5.751	5.757	5.763	5.769	5.774	5.780	5.786	5.792	5.797	1200
1210	5.797	5.803	5.809	5.815	5.821	5.826	5.832	5.838	5.844	5.849	5.855	1210
1220	5.855	5.861	5.867	5.873	5.878	5.884	5.890	5.896	5.902	5.907	5.913	1220
1230	5.913	5.919	5.925	5.931	5.936	5.942	5.948	5.954	5.960	5.965	5.971	1230
1240	5.971	5.977	5.983	5.989	5.994	6.000	6.006	6.012	6.018	6.023	6.029	1240
1250	6.029	6.035	6.041	6.047	6.052	6.058	6.064	6.070	6.076	6.082	6.087	1250
1260	6.087	6.093	6.099	6.105	6.111	6.117	6.122	6.128	6.134	6.140	6.146	1260
1270	6.146	6.152	6.157	6.163	6.169	6.175	6.181	6.187	6.192	6.198	6.204	1270
1280	6.204	6.210	6.216	6.222	6.227	6.233	6.239	6.245	6.251	6.257	6.263	1280
1290	6.263	6.268	6.274	6.280	6.286	6.292	6.298	6.304	6.309	6.315	6.321	1290
1300	6.321	6.327	6.333	6.339	6.345	6.350	6.356	6.362	6.368	6.374	6.380	1300
1310	6.380	6.386	6.392	6.397	6.403	6.409	6.415	6.421	6.427	6.433	6.439	1310
1320	6.439	6.445	6.450	6.456	6.462	6.468	6.474	6.480	6.486	6.492	6.498	1320
1330	6.498	6.503	6.509	6.515	6.521	6.527	6.533	6.539	6.545	6.551	6.557	1330
1340	6.557	6.562	6.568	6.574	6.580	6.586	6.592	6.598	6.604	6.610	6.616	1340
1350	6.616	6.622	6.627	6.633	6.639	6.645	6.651	6.657	6.663	6.669	6.675	1350
1360	6.675	6.681	6.687	6.693	6.699	6.704	6.710	6.716	6.722	6.728	6.734	1360
1370	6.734	6.740	6.746	6.752	6.758	6.764	6.770	6.776	6.782	6.788	6.794	1370
1380	6.794	6.800	6.805	6.811	6.817	6.823	6.829	6.835	6.841	6.847	6.853	1380
1390	6.853	6.859	6.865	6.871	6.877	6.883	6.889	6.895	6.901	6.907	6.913	1390
1400	6.913	6.919	6.925	6.931	6.937	6.943	6.948	6.954	6.960	6.966	6.972	1400
1410	6.972	6.978	6.984	6.990	6.996	7.002	7.008	7.014	7.020	7.026	7.032	1410
1420	7.032	7.038	7.044	7.050	7.056	7.062	7.068	7.074	7.080	7.086	7.092	1420
1430	7.092	7.098	7.104	7.110	7.116	7.122	7.128	7.134	7.140	7.146	7.152	1430
1440	7.152	7.158	7.164	7.170	7.176	7.182	7.188	7.194	7.200	7.206	7.212	1440
1450	7.212	7.218	7.224	7.230	7.236	7.242	7.248	7.254	7.260	7.266	7.272	1450
1460	7.272	7.278	7.285	7.291	7.297	7.303	7.309	7.315	7.321	7.327	7.333	1460
1470	7.333	7.339	7.345	7.351	7.357	7.363	7.369	7.375	7.381	7.387	7.393	1470
1480	7.393	7.399	7.405	7.411	7.417	7.423	7.429	7.436	7.442	7.448	7.454	1480
1490	7.454	7.460	7.466	7.472	7.478	7.484	7.490	7.496	7.502	7.508	7.514	1490
1500	7.514	7.520	7.526	7.533	7.539	7.545	7.551	7.557	7.563	7.569	7.575	1500
1510	7.575	7.581	7.587	7.593	7.599	7.605	7.612	7.618	7.624	7.630	7.636	1510
1520	7.636	7.642	7.648	7.654	7.660	7.666	7.672	7.679	7.685	7.691	7.697	1520
1530	7.697	7.703	7.709	7.715	7.721	7.727	7.733	7.740	7.746	7.752	7.758	1530
1540	7.758	7.764	7.770	7.776	7.782	7.788	7.795	7.801	7.807	7.813	7.819	1540
1550	7.819	7.825	7.831	7.837	7.843	7.850	7.856	7.862	7.868	7.874	7.880	1550
1560	7.880	7.886	7.892	7.899	7.905	7.911	7.917	7.923	7.929	7.935	7.942	1560
1570	7.942	7.948	7.954	7.960	7.966	7.972	7.978	7.985	7.991	7.997	8.003	1570
1580	8.003	8.009	8.015	8.021	8.028	8.034	8.040	8.046	8.052	8.058	8.065	1580
1590	8.065	8.071	8.077	8.083	8.089	8.095	8.101	8.108	8.114	8.120	8.126	1590
1600	8.126	8.132	8.138	8.145	8.151	8.157	8.163	8.169	8.176	8.182	8.188	1600
1610	8.188	8.194	8.200	8.206	8.213	8.219	8.225	8.231	8.237	8.244	8.250	1610
1620	8.250	8.256	8.262	8.268	8.275	8.281	8.287	8.293	8.299	8.305	8.312	1620
1630	8.312	8.318	8.324	8.330	8.336	8.343	8.349	8.355	8.361	8.368	8.374	1630
1640	8.374	8.380	8.386	8.392	8.399	8.405	8.411	8.417	8.423	8.430	8.436	1640
1650	8.436	8.442	8.448	8.455	8.461	8.467	8.473	8.479	8.486	8.492	8.498	1650
1660	8.498	8.504	8.511	8.517	8.523	8.529	8.536	8.542	8.548	8.554	8.560	1660
1670	8.560	8.567	8.573	8.579	8.585	8.592	8.598	8.604	8.610	8.617	8.623	1670
1680	8.623	8.629	8.635	8.642	8.648	8.654	8.660	8.667	8.673	8.679	8.685	1680
1690	8.685	8.692	8.698	8.704	8.711	8.717	8.723	8.729	8.736	8.742	8.748	1690
1700	8.748	8.754	8.761	8.767	8.773	8.780	8.786	8.792	8.798	8.805	8.811	1700
1710	8.811	8.817	8.823	8.830	8.836	8.842	8.849	8.855	8.861	8.867	8.874	1710
1720	8.874	8.880	8.886	8.893	8.899	8.905	8.912	8.918	8.924	8.930	8.937	1720
1730	8.937	8.943	8.949	8.956	8.962	8.968	8.975	8.981	8.987	8.993	9.000	1730

Type S Thermocouple
Thermoelectric Voltage as a Function of Temperature (°F)
Reference Junctions at 32°F
THERMOELECTRIC VOLTAGE IN ABSOLUTE MILLIVOLTS

°F	0	1	2	3	4	5	6	7	8	9	10	°F
1740	9.000	9.006	9.012	9.019	9.025	9.031	9.038	9.044	9.050	9.057	9.063	1740
1750	9.063	9.069	9.076	9.082	9.088	9.095	9.101	9.107	9.114	9.120	9.126	1750
1760	9.126	9.133	9.139	9.145	9.152	9.158	9.164	9.171	9.177	9.183	9.190	1760
1770	9.190	9.196	9.202	9.209	9.215	9.221	9.228	9.234	9.240	9.247	9.253	1770
1780	9.253	9.259	9.266	9.272	9.278	9.285	9.291	9.298	9.304	9.310	9.317	1780
1790	9.317	9.323	9.329	9.336	9.342	9.348	9.355	9.361	9.368	9.374	9.380	1790
1800	9.380	9.387	9.393	9.399	9.406	9.412	9.419	9.425	9.431	9.438	9.444	1800
1810	9.444	9.450	9.457	9.463	9.470	9.476	9.482	9.489	9.495	9.502	9.508	1810
1820	9.508	9.514	9.521	9.527	9.533	9.540	9.546	9.553	9.559	9.565	9.572	1820
1830	9.572	9.578	9.585	9.591	9.598	9.604	9.610	9.617	9.623	9.630	9.636	1830
1840	9.636	9.642	9.649	9.655	9.662	9.668	9.674	9.681	9.687	9.694	9.700	1840
1850	9.700	9.707	9.713	9.719	9.726	9.732	9.739	9.745	9.752	9.758	9.764	1850
1860	9.764	9.771	9.777	9.784	9.790	9.797	9.803	9.809	9.816	9.822	9.829	1860
1870	9.829	9.835	9.842	9.848	9.855	9.861	9.867	9.874	9.880	9.887	9.893	1870
1880	9.893	9.900	9.906	9.913	9.919	9.926	9.932	9.938	9.945	9.951	9.958	1880
1890	9.958	9.964	9.971	9.977	9.984	9.990	9.997	10.003	10.010	10.016	10.023	1890
1900	10.023	10.029	10.036	10.042	10.048	10.055	10.061	10.068	10.074	10.081	10.087	1900
1910	10.087	10.094	10.100	10.107	10.113	10.120	10.126	10.133	10.139	10.146	10.152	1910
1920	10.152	10.159	10.165	10.172	10.178	10.185	10.191	10.198	10.204	10.211	10.217	1920
1930	10.217	10.224	10.230	10.237	10.243	10.250	10.256	10.263	10.269	10.276	10.282	1930
1940	10.282	10.289	10.295	10.302	10.308	10.315	10.321	10.328	10.334	10.341	10.348	1940
1950	10.348	10.354	10.361	10.367	10.374	10.380	10.387	10.393	10.400	10.406	10.413	1950
1960	10.413	10.419	10.426	10.432	10.439	10.445	10.452	10.459	10.465	10.472	10.478	1960
1970	10.478	10.485	10.491	10.498	10.504	10.511	10.517	10.524	10.531	10.537	10.544	1970
1980	10.544	10.550	10.557	10.563	10.570	10.576	10.583	10.589	10.596	10.603	10.609	1980
1990	10.609	10.616	10.622	10.629	10.635	10.642	10.648	10.655	10.662	10.668	10.675	1990
2000	10.675	10.681	10.688	10.694	10.701	10.708	10.714	10.721	10.727	10.734	10.740	2000
2010	10.740	10.747	10.754	10.760	10.767	10.773	10.780	10.786	10.793	10.800	10.806	2010
2020	10.806	10.813	10.819	10.826	10.832	10.839	10.846	10.852	10.859	10.865	10.872	2020
2030	10.872	10.879	10.885	10.892	10.898	10.905	10.912	10.918	10.925	10.931	10.938	2030
2040	10.938	10.944	10.951	10.958	10.964	10.971	10.977	10.984	10.991	10.997	11.004	2040
2050	11.004	11.010	11.017	11.024	11.030	11.037	11.043	11.050	11.057	11.063	11.070	2050
2060	11.070	11.076	11.083	11.090	11.096	11.103	11.110	11.116	11.123	11.129	11.136	2060
2070	11.136	11.143	11.149	11.156	11.162	11.169	11.176	11.182	11.189	11.196	11.202	2070
2080	11.202	11.209	11.215	11.222	11.229	11.235	11.242	11.248	11.255	11.262	11.268	2080
2090	11.268	11.275	11.282	11.288	11.295	11.301	11.308	11.315	11.321	11.328	11.335	2090
2100	11.335	11.341	11.348	11.355	11.361	11.368	11.374	11.381	11.388	11.394	11.401	2100
2110	11.401	11.408	11.414	11.421	11.428	11.434	11.441	11.447	11.454	11.461	11.467	2110
2120	11.467	11.474	11.481	11.487	11.494	11.501	11.507	11.514	11.521	11.527	11.534	2120
2130	11.534	11.541	11.547	11.554	11.560	11.567	11.574	11.580	11.587	11.594	11.600	2130
2140	11.600	11.607	11.614	11.620	11.627	11.634	11.640	11.647	11.654	11.660	11.667	2140
2150	11.667	11.674	11.680	11.687	11.694	11.700	11.707	11.714	11.720	11.727	11.734	2150
2160	11.734	11.740	11.747	11.754	11.760	11.767	11.774	11.780	11.787	11.794	11.800	2160
2170	11.800	11.807	11.814	11.820	11.827	11.834	11.840	11.847	11.854	11.860	11.867	2170
2180	11.867	11.874	11.880	11.887	11.894	11.900	11.907	11.914	11.920	11.927	11.934	2180
2190	11.934	11.940	11.947	11.954	11.960	11.967	11.974	11.980	11.987	11.994	12.001	2190
2200	12.001	12.007	12.014	12.021	12.027	12.034	12.041	12.047	12.054	12.061	12.067	2200
2210	12.067	12.074	12.081	12.087	12.094	12.101	12.107	12.114	12.121	12.128	12.134	2210
2220	12.134	12.141	12.148	12.154	12.161	12.168	12.174	12.181	12.188	12.194	12.201	2220
2230	12.201	12.208	12.215	12.221	12.228	12.235	12.241	12.248	12.255	12.261	12.268	2230
2240	12.268	12.275	12.282	12.288	12.295	12.302	12.308	12.315	12.322	12.328	12.335	2240
2250	12.335	12.342	12.349	12.355	12.362	12.369	12.375	12.382	12.389	12.395	12.402	2250
2260	12.402	12.409	12.416	12.422	12.429	12.436	12.442	12.449	12.456	12.463	12.469	2260
2270	12.469	12.476	12.483	12.489	12.496	12.503	12.510	12.516	12.523	12.530	12.536	2270
2280	12.536	12.543	12.550	12.557	12.563	12.570	12.577	12.583	12.590	12.597	12.604	2280
2290	12.604	12.610	12.617	12.624	12.630	12.637	12.644	12.651	12.657	12.664	12.671	2290
2300	12.671	12.677	12.684	12.691	12.698	12.704	12.711	12.718	12.724	12.731	12.738	2300
2310	12.738	12.745	12.751	12.758	12.765	12.771	12.778	12.785	12.792	12.798	12.805	2310
2320	12.805	12.812	12.819	12.825	12.832	12.839	12.845	12.852	12.859	12.866	12.872	2320
2330	12.872	12.879	12.886	12.893	12.899	12.906	12.913	12.919	12.926	12.933	12.940	2330

Type S Thermocouple
Thermoelectric Voltage as a Function of Temperature (°F)
Reference Junctions at 32°F
THERMOELECTRIC VOLTAGE IN ABSOLUTE MILLIVOLTS

°F	0	1	2	3	4	5	6	7	8	9	10	°F
2340	12.940	12.946	12.953	12.960	12.967	12.973	12.980	12.987	12.993	13.000	13.007	2340
2350	13.007	13.014	13.020	13.027	13.034	13.041	13.047	13.054	13.061	13.067	13.074	2350
2360	13.074	13.081	13.088	13.094	13.101	13.108	13.115	13.121	13.128	13.135	13.142	2360
2370	13.142	13.148	13.155	13.162	13.168	13.175	13.182	13.189	13.195	13.202	13.209	2370
2380	13.209	13.216	13.222	13.229	13.236	13.243	13.249	13.256	13.263	13.269	13.276	2380
2390	13.276	13.283	13.290	13.296	13.303	13.310	13.317	13.323	13.330	13.337	13.344	2390
2400	13.344	13.350	13.357	13.364	13.371	13.377	13.384	13.391	13.397	13.404	13.411	2400
2410	13.411	13.418	13.424	13.431	13.438	13.445	13.451	13.458	13.465	13.472	13.478	2410
2420	13.478	13.485	13.492	13.499	13.505	13.512	13.519	13.526	13.532	13.539	13.546	2420
2430	13.546	13.552	13.559	13.566	13.573	13.579	13.586	13.593	13.600	13.606	13.613	2430
2440	13.613	13.620	13.627	13.633	13.640	13.647	13.654	13.660	13.667	13.674	13.681	2440
2450	13.681	13.687	13.694	13.701	13.708	13.714	13.721	13.728	13.734	13.741	13.748	2450
2460	13.748	13.755	13.761	13.768	13.775	13.782	13.788	13.795	13.802	13.809	13.815	2460
2470	13.815	13.822	13.829	13.836	13.842	13.849	13.856	13.863	13.869	13.876	13.883	2470
2480	13.883	13.890	13.896	13.903	13.910	13.916	13.923	13.930	13.937	13.943	13.950	2480
2490	13.950	13.957	13.964	13.970	13.977	13.984	13.991	13.997	14.004	14.011	14.018	2490
2500	14.018	14.024	14.031	14.038	14.045	14.051	14.058	14.065	14.072	14.078	14.085	2500
2510	14.085	14.092	14.098	14.105	14.112	14.119	14.125	14.132	14.139	14.146	14.152	2510
2520	14.152	14.159	14.166	14.173	14.179	14.186	14.193	14.200	14.206	14.213	14.220	2520
2530	14.220	14.226	14.233	14.240	14.247	14.253	14.260	14.267	14.274	14.280	14.287	2530
2540	14.287	14.294	14.301	14.307	14.314	14.321	14.328	14.334	14.341	14.348	14.354	2540
2550	14.354	14.361	14.368	14.375	14.381	14.388	14.395	14.402	14.408	14.415	14.422	2550
2560	14.422	14.429	14.435	14.442	14.449	14.455	14.462	14.469	14.476	14.482	14.489	2560
2570	14.489	14.496	14.503	14.509	14.516	14.523	14.530	14.536	14.543	14.550	14.556	2570
2580	14.556	14.563	14.570	14.577	14.583	14.590	14.597	14.604	14.610	14.617	14.624	2580
2590	14.624	14.631	14.637	14.644	14.651	14.657	14.664	14.671	14.678	14.684	14.691	2590
2600	14.691	14.698	14.705	14.711	14.718	14.725	14.731	14.738	14.745	14.752	14.758	2600
2610	14.758	14.765	14.772	14.778	14.785	14.792	14.799	14.805	14.812	14.819	14.826	2610
2620	14.826	14.832	14.839	14.846	14.852	14.859	14.866	14.873	14.879	14.886	14.893	2620
2630	14.893	14.899	14.906	14.913	14.920	14.926	14.933	14.940	14.946	14.953	14.960	2630
2640	14.960	14.967	14.973	14.980	14.987	14.994	15.000	15.007	15.014	15.020	15.027	2640
2650	15.027	15.034	15.041	15.047	15.054	15.061	15.067	15.074	15.081	15.088	15.094	2650
2660	15.094	15.101	15.108	15.114	15.121	15.128	15.134	15.141	15.148	15.155	15.161	2660
2670	15.161	15.168	15.175	15.181	15.188	15.195	15.202	15.208	15.215	15.222	15.228	2670
2680	15.228	15.235	15.242	15.248	15.255	15.262	15.269	15.275	15.282	15.289	15.295	2680
2690	15.295	15.302	15.309	15.315	15.322	15.329	15.336	15.342	15.349	15.356	15.362	2690
2700	15.362	15.369	15.376	15.382	15.389	15.396	15.403	15.409	15.416	15.423	15.429	2700
2710	15.429	15.436	15.443	15.449	15.456	15.463	15.469	15.476	15.483	15.490	15.496	2710
2720	15.496	15.503	15.510	15.516	15.523	15.530	15.536	15.543	15.550	15.556	15.563	2720
2730	15.563	15.570	15.576	15.583	15.590	15.597	15.603	15.610	15.617	15.623	15.630	2730
2740	15.630	15.637	15.643	15.650	15.657	15.663	15.670	15.677	15.683	15.690	15.697	2740
2750	15.697	15.703	15.710	15.717	15.723	15.730	15.737	15.743	15.750	15.757	15.763	2750
2760	15.763	15.770	15.777	15.783	15.790	15.797	15.804	15.810	15.817	15.824	15.830	2760
2770	15.830	15.837	15.844	15.850	15.857	15.864	15.870	15.877	15.883	15.890	15.897	2770
2780	15.897	15.903	15.910	15.917	15.923	15.930	15.937	15.943	15.950	15.957	15.963	2780
2790	15.963	15.970	15.977	15.983	15.990	15.997	16.003	16.010	16.017	16.023	16.030	2790
2800	16.030	16.037	16.043	16.050	16.057	16.063	16.070	16.077	16.083	16.090	16.096	2800
2810	16.096	16.103	16.110	16.116	16.123	16.130	16.136	16.143	16.150	16.156	16.163	2810
2820	16.163	16.170	16.176	16.183	16.189	16.196	16.203	16.209	16.216	16.223	16.229	2820
2830	16.229	16.236	16.243	16.249	16.256	16.262	16.269	16.276	16.282	16.289	16.296	2830
2840	16.296	16.302	16.309	16.315	16.322	16.329	16.335	16.342	16.349	16.355	16.362	2840
2850	16.362	16.368	16.375	16.382	16.388	16.395	16.402	16.408	16.415	16.421	16.428	2850
2860	16.428	16.435	16.441	16.448	16.454	16.461	16.468	16.474	16.481	16.488	16.494	2860
2870	16.494	16.501	16.507	16.514	16.521	16.527	16.534	16.540	16.547	16.554	16.560	2870
2880	16.560	16.567	16.573	16.580	16.587	16.593	16.600	16.606	16.613	16.620	16.626	2880
2890	16.626	16.633	16.639	16.646	16.653	16.659	16.666	16.672	16.679	16.686	16.692	2890
2900	16.692	16.699	16.705	16.712	16.719	16.725	16.732	16.738	16.745	16.751	16.758	2900
2910	16.758	16.765	16.771	16.778	16.784	16.791	16.797	16.804	16.811	16.817	16.824	2910
2920	16.824	16.830	16.837	16.844	16.850	16.857	16.863	16.870	16.876	16.883	16.890	2920
2930	16.890	16.896	16.903	16.909	16.916	16.922	16.929	16.935	16.942	16.949	16.955	2930

Type S Thermocouple
Thermoelectric Voltage as a Function of Temperature (°F)
Reference Junctions at 32°F
THERMOELECTRIC VOLTAGE IN ABSOLUTE MILLIVOLTS

°F	0	1	2	3	4	5	6	7	8	9	10	°F
2940	16.955	16.962	16.968	16.975	16.981	16.988	16.995	17.001	17.008	17.014	17.021	2940
2950	17.021	17.027	17.034	17.040	17.047	17.053	17.060	17.067	17.073	17.080	17.086	2950
2960	17.086	17.093	17.099	17.106	17.112	17.119	17.125	17.132	17.139	17.145	17.152	2960
2970	17.152	17.158	17.165	17.171	17.178	17.184	17.191	17.197	17.204	17.210	17.217	2970
2980	17.217	17.223	17.230	17.237	17.243	17.250	17.256	17.263	17.269	17.276	17.282	2980
2990	17.282	17.289	17.295	17.302	17.308	17.315	17.321	17.328	17.334	17.341	17.347	2990
3000	17.347	17.354	17.360	17.367	17.373	17.380	17.386	17.393	17.399	17.406	17.412	3000
3010	17.412	17.419	17.425	17.432	17.438	17.445	17.451	17.458	17.464	17.471	17.477	3010
3020	17.477	17.484	17.490	17.497	17.503	17.510	17.516	17.523	17.529	17.536	17.542	3020
3030	17.542	17.549	17.555	17.562	17.568	17.575	17.581	17.588	17.594	17.601	17.607	3030
3040	17.607	17.614	17.620	17.627	17.633	17.639	17.646	17.652	17.659	17.665	17.672	3040
3050	17.672	17.678	17.685	17.691	17.698	17.704	17.711	17.717	17.723	17.730	17.736	3050
3060	17.736	17.743	17.749	17.756	17.762	17.769	17.775	17.781	17.788	17.794	17.801	3060
3070	17.801	17.807	17.814	17.820	17.826	17.833	17.839	17.846	17.852	17.859	17.865	3070
3080	17.865	17.871	17.878	17.884	17.891	17.897	17.903	17.910	17.916	17.923	17.929	3080
3090	17.929	17.935	17.942	17.948	17.954	17.961	17.967	17.974	17.980	17.986	17.993	3090
3100	17.993	17.999	18.005	18.012	18.018	18.024	18.031	18.037	18.043	18.050	18.056	3100
3110	18.056	18.063	18.069	18.075	18.081	18.088	18.094	18.100	18.107	18.113	18.119	3110
3120	18.119	18.126	18.132	18.138	18.145	18.151	18.157	18.163	18.170	18.176	18.182	3120
3130	18.182	18.189	18.195	18.201	18.207	18.214	18.220	18.226	18.232	18.239	18.245	3130
3140	18.245	18.251	18.257	18.264	18.270	18.276	18.282	18.289	18.295	18.301	18.307	3140
3150	18.307	18.313	18.320	18.326	18.332	18.338	18.344	18.351	18.357	18.363	18.369	3150
3160	18.369	18.375	18.381	18.388	18.394	18.400	18.406	18.412	18.418	18.424	18.431	3160
3170	18.431	18.437	18.443	18.449	18.455	18.461	18.467	18.473	18.479	18.486	18.492	3170
3180	18.492	18.498	18.504	18.510	18.516	18.522	18.528	18.534	18.540	18.546	18.552	3180
3190	18.552	18.558	18.564	18.570	18.576	18.582	18.588	18.594	18.600	18.606	18.612	3190
3200	18.612	18.618	18.624	18.630	18.636	18.642	18.648	18.654	18.660	18.666	18.672	3200
3210	18.672	18.678	18.684	18.690	18.696							3210

Type T Thermocouple
Thermoelectric Voltage as a Function of Temperature (°C)
Reference Junctions at 0°C

THERMOELECTRIC VOLTAGE IN ABSOLUTE MILLIVOLTS

°C	0	1	2	3	4	5	6	7	8	9	10	°C
−270	−6.258	−6.256	−6.255	−6.253	−6.251	−6.248	−6.245	−6.242	−6.239	−6.236	−6.232	−270
−260	−6.232	−6.228	−6.224	−6.219	−6.214	−6.209	−6.204	−6.198	−6.193	−6.187	−6.181	−260
−250	−6.181	−6.174	−6.167	−6.160	−6.153	−6.146	−6.138	−6.130	−6.122	−6.114	−6.105	−250
−240	−6.105	−6.096	−6.087	−6.078	−6.068	−6.059	−6.049	−6.039	−6.028	−6.018	−6.007	−240
−230	−6.007	−5.996	−5.985	−5.973	−5.962	−5.950	−5.938	−5.926	−5.914	−5.901	−5.889	−230
−220	−5.889	−5.876	−5.863	−5.850	−5.836	−5.823	−5.809	−5.795	−5.782	−5.767	−5.753	−220
−210	−5.753	−5.739	−5.724	−5.710	−5.695	−5.680	−5.665	−5.650	−5.634	−5.619	−5.603	−210
−200	−5.603	−5.587	−5.571	−5.555	−5.539	−5.522	−5.506	−5.489	−5.473	−5.456	−5.439	−200
−190	−5.439	−5.421	−5.404	−5.387	−5.369	−5.351	−5.333	−5.315	−5.297	−5.279	−5.261	−190
−180	−5.261	−5.242	−5.223	−5.205	−5.186	−5.167	−5.147	−5.128	−5.109	−5.089	−5.069	−180
−170	−5.069	−5.050	−5.030	−5.010	−4.989	−4.969	−4.948	−4.928	−4.907	−4.886	−4.865	−170
−160	−4.865	−4.844	−4.823	−4.801	−4.780	−4.758	−4.737	−4.715	−4.693	−4.670	−4.648	−160
−150	−4.648	−4.626	−4.603	−4.581	−4.558	−4.535	−4.512	−4.489	−4.466	−4.442	−4.419	−150
−140	−4.419	−4.395	−4.371	−4.347	−4.323	−4.299	−4.275	−4.251	−4.226	−4.202	−4.177	−140
−130	−4.177	−4.152	−4.127	−4.102	−4.077	−4.051	−4.026	−4.000	−3.974	−3.949	−3.923	−130
−120	−3.923	−3.897	−3.870	−3.844	−3.818	−3.791	−3.764	−3.737	−3.711	−3.684	−3.656	−120
−110	−3.656	−3.629	−3.602	−3.574	−3.547	−3.519	−3.491	−3.463	−3.435	−3.407	−3.378	−110
−100	−3.378	−3.350	−3.321	−3.293	−3.264	−3.235	−3.206	−3.177	−3.147	−3.118	−3.089	−100
−90	−3.089	−3.059	−3.029	−2.999	−2.970	−2.939	−2.909	−2.879	−2.849	−2.818	−2.788	−90
−80	−2.788	−2.757	−2.726	−2.695	−2.664	−2.633	−2.602	−2.570	−2.539	−2.507	−2.475	−80
−70	−2.475	−2.444	−2.412	−2.380	−2.348	−2.315	−2.283	−2.250	−2.218	−2.185	−2.152	−70
−60	−2.152	−2.120	−2.087	−2.053	−2.020	−1.987	−1.953	−1.920	−1.886	−1.853	−1.819	−60
−50	−1.819	−1.785	−1.751	−1.717	−1.682	−1.648	−1.614	−1.579	−1.544	−1.510	−1.475	−50
−40	−1.475	−1.440	−1.405	−1.370	−1.334	−1.299	−1.263	−1.228	−1.192	−1.157	−1.121	−40
−30	−1.121	−1.085	−1.049	−1.013	−0.976	−0.940	−0.903	−0.867	−0.830	−0.794	−0.757	−30
−20	−0.757	−0.720	−0.683	−0.646	−0.608	−0.571	−0.534	−0.496	−0.458	−0.421	−0.383	−20
−10	−0.383	−0.345	−0.307	−0.269	−0.231	−0.193	−0.154	−0.116	−0.077	−0.039	0.000	−10
0	0.000	0.039	0.078	0.117	0.156	0.195	0.234	0.273	0.312	0.351	0.391	0
10	0.391	0.430	0.470	0.510	0.549	0.589	0.629	0.669	0.709	0.749	0.789	10
20	0.789	0.830	0.870	0.911	0.951	0.992	1.032	1.073	1.114	1.155	1.196	20
30	1.196	1.237	1.279	1.320	1.361	1.403	1.444	1.486	1.528	1.569	1.611	30
40	1.611	1.653	1.695	1.738	1.780	1.822	1.865	1.907	1.950	1.992	2.035	40
50	2.035	2.078	2.121	2.164	2.207	2.250	2.294	2.337	2.380	2.424	2.467	50
60	2.467	2.511	2.555	2.599	2.643	2.687	2.731	2.775	2.819	2.864	2.908	60
70	2.908	2.953	2.997	3.042	3.087	3.131	3.176	3.221	3.266	3.312	3.357	70
80	3.357	3.402	3.447	3.493	3.538	3.584	3.630	3.676	3.721	3.767	3.813	80
90	3.813	3.859	3.906	3.952	3.998	4.044	4.091	4.137	4.184	4.231	4.277	90
100	4.277	4.324	4.371	4.418	4.465	4.512	4.559	4.607	4.654	4.701	4.749	100
110	4.749	4.796	4.844	4.891	4.939	4.987	5.035	5.083	5.131	5.179	5.227	110
120	5.227	5.275	5.324	5.372	5.420	5.469	5.517	5.566	5.615	5.663	5.712	120
130	5.712	5.761	5.810	5.859	5.908	5.957	6.007	6.056	6.105	6.155	6.204	130
140	6.204	6.254	6.303	6.353	6.403	6.452	6.502	6.552	6.602	6.652	6.702	140
150	6.702	6.753	6.803	6.853	6.903	6.954	7.004	7.055	7.106	7.156	7.207	150
160	7.207	7.258	7.309	7.360	7.411	7.462	7.513	7.564	7.615	7.666	7.718	160
170	7.718	7.769	7.821	7.872	7.924	7.975	8.027	8.079	8.131	8.183	8.235	170
180	8.235	8.287	8.339	8.391	8.443	8.495	8.548	8.600	8.652	8.705	8.757	180
190	8.757	8.810	8.863	8.915	8.968	9.021	9.074	9.127	9.180	9.233	9.286	190
200	9.286	9.339	9.392	9.446	9.499	9.553	9.606	9.659	9.713	9.767	9.820	200
210	9.820	9.874	9.928	9.982	10.036	10.090	10.144	10.198	10.252	10.306	10.360	210
220	10.360	10.414	10.469	10.523	10.578	10.632	10.687	10.741	10.796	10.851	10.905	220
230	10.905	10.960	11.015	11.070	11.125	11.180	11.235	11.290	11.345	11.401	11.456	230
240	11.456	11.511	11.566	11.622	11.677	11.733	11.788	11.844	11.900	11.956	12.011	240
250	12.011	12.067	12.123	12.179	12.235	12.291	12.347	12.403	12.459	12.515	12.572	250
260	12.572	12.628	12.684	12.741	12.797	12.854	12.910	12.967	13.024	13.080	13.137	260
270	13.137	13.194	13.251	13.307	13.364	13.421	13.478	13.535	13.592	13.650	13.707	270
280	13.707	13.764	13.821	13.879	13.936	13.993	14.051	14.108	14.166	14.223	14.281	280
290	14.281	14.339	14.396	14.454	14.512	14.570	14.628	14.686	14.744	14.802	14.860	290
300	14.860	14.918	14.976	15.034	15.092	15.151	15.209	15.267	15.326	15.384	15.443	300
310	15.443	15.501	15.560	15.619	15.677	15.736	15.795	15.853	15.912	15.971	16.030	310
320	16.030	16.089	16.148	16.207	16.266	16.325	16.384	16.444	16.503	16.562	16.621	320

Type T Thermocouple
Thermoelectric Voltage as a Function of Temperature (°C)
Reference Junctions at 0°C

THERMOELECTRIC VOLTAGE IN ABSOLUTE MILLIVOLTS

°C	0	1	2	3	4	5	6	7	8	9	10	°C
330	16.621	16.681	16.740	16.800	16.859	16.919	16.978	17.038	17.097	17.157	17.217	330
340	17.217	17.277	17.336	17.396	17.456	17.516	17.576	17.636	17.696	17.756	17.816	340
350	17.816	17.877	17.937	17.997	18.057	18.118	18.178	18.238	18.299	18.359	18.420	350
360	18.420	18.480	18.541	18.602	18.662	18.723	18.784	18.845	18.905	18.966	19.027	360
370	19.027	19.088	19.149	19.210	19.271	19.332	19.393	19.455	19.516	19.577	19.638	370
380	19.638	19.699	19.761	19.822	19.883	19.945	20.006	20.068	20.129	20.191	20.252	380
390	20.252	20.314	20.376	20.437	20.499	20.560	20.622	20.684	20.746	20.807	20.869	390

Type T Thermocouple
Thermoelectric Voltage as a Function of Temperature (°F)
Reference Junctions at 32°F
THERMOELECTRIC VOLTAGE IN ABSOLUTE MILLIVOLTS

°F	0	1	2	3	4	5	6	7	8	9	10	°F
−460							−6.258	−6.257	−6.256	−6.255	−6.254	−460
−450	−6.254	−6.253	−6.252	−6.251	−6.250	−6.248	−6.247	−6.245	−6.243	−6.242	−6.240	−450
−440	−6.240	−6.238	−6.236	−6.234	−6.232	−6.230	−6.227	−6.225	−6.223	−6.220	−6.217	−440
−430	−6.217	−6.215	−6.212	−6.209	−6.206	−6.203	−6.200	−6.197	−6.194	−6.191	−6.187	−430
−420	−6.187	−6.184	−6.181	−6.177	−6.173	−6.170	−6.166	−6.162	−6.158	−6.154	−6.150	−420
−410	−6.150	−6.146	−6.142	−6.137	−6.133	−6.128	−6.124	−6.119	−6.115	−6.110	−6.105	−410
−400	−6.105	−6.100	−6.095	−6.090	−6.085	−6.080	−6.075	−6.069	−6.064	−6.059	−6.053	−400
−390	−6.053	−6.048	−6.042	−6.036	−6.030	−6.025	−6.019	−6.013	−6.007	−6.001	−5.995	−390
−380	−5.995	−5.988	−5.982	−5.976	−5.969	−5.963	−5.957	−5.950	−5.943	−5.937	−5.930	−380
−370	−5.930	−5.923	−5.916	−5.910	−5.903	−5.896	−5.889	−5.881	−5.874	−5.867	−5.860	−370
−360	−5.860	−5.853	−5.845	−5.838	−5.830	−5.823	−5.815	−5.808	−5.800	−5.792	−5.785	−360
−350	−5.785	−5.777	−5.769	−5.761	−5.753	−5.745	−5.737	−5.729	−5.721	−5.713	−5.705	−350
−340	−5.705	−5.697	−5.688	−5.680	−5.672	−5.663	−5.655	−5.646	−5.638	−5.629	−5.620	−340
−330	−5.620	−5.612	−5.603	−5.594	−5.585	−5.576	−5.568	−5.559	−5.550	−5.541	−5.532	−330
−320	−5.532	−5.522	−5.513	−5.504	−5.495	−5.486	−5.476	−5.467	−5.457	−5.448	−5.439	−320
−310	−5.439	−5.429	−5.419	−5.410	−5.400	−5.390	−5.381	−5.371	−5.361	−5.351	−5.341	−310
−300	−5.341	−5.331	−5.321	−5.311	−5.301	−5.291	−5.281	−5.271	−5.261	−5.250	−5.240	−300
−290	−5.240	−5.230	−5.219	−5.209	−5.198	−5.188	−5.177	−5.167	−5.156	−5.145	−5.135	−290
−280	−5.135	−5.124	−5.113	−5.102	−5.091	−5.080	−5.069	−5.058	−5.047	−5.036	−5.025	−280
−270	−5.025	−5.014	−5.003	−4.992	−4.980	−4.969	−4.958	−4.946	−4.935	−4.923	−4.912	−270
−260	−4.912	−4.900	−4.889	−4.877	−4.865	−4.853	−4.842	−4.830	−4.818	−4.806	−4.794	−260
−250	−4.794	−4.782	−4.770	−4.758	−4.746	−4.734	−4.722	−4.710	−4.698	−4.685	−4.673	−250
−240	−4.673	−4.661	−4.648	−4.636	−4.623	−4.611	−4.598	−4.586	−4.573	−4.560	−4.548	−240
−230	−4.548	−4.535	−4.522	−4.509	−4.497	−4.484	−4.471	−4.458	−4.445	−4.432	−4.419	−230
−220	−4.419	−4.406	−4.392	−4.379	−4.366	−4.353	−4.339	−4.326	−4.313	−4.299	−4.286	−220
−210	−4.286	−4.272	−4.259	−4.245	−4.232	−4.218	−4.204	−4.191	−4.177	−4.163	−4.149	−210
−200	−4.149	−4.135	−4.121	−4.107	−4.093	−4.079	−4.065	−4.051	−4.037	−4.023	−4.009	−200
−190	−4.009	−3.994	−3.980	−3.966	−3.951	−3.937	−3.923	−3.908	−3.894	−3.879	−3.864	−190
−180	−3.864	−3.850	−3.835	−3.820	−3.806	−3.791	−3.776	−3.761	−3.746	−3.732	−3.717	−180
−170	−3.717	−3.702	−3.687	−3.671	−3.656	−3.641	−3.626	−3.611	−3.596	−3.580	−3.565	−170
−160	−3.565	−3.550	−3.534	−3.519	−3.503	−3.488	−3.472	−3.457	−3.441	−3.425	−3.410	−160
−150	−3.410	−3.394	−3.378	−3.362	−3.347	−3.331	−3.315	−3.299	−3.283	−3.267	−3.251	−150
−140	−3.251	−3.235	−3.219	−3.203	−3.186	−3.170	−3.154	−3.138	−3.121	−3.105	−3.089	−140
−130	−3.089	−3.072	−3.056	−3.039	−3.023	−3.006	−2.989	−2.973	−2.956	−2.939	−2.923	−130
−120	−2.923	−2.906	−2.889	−2.872	−2.855	−2.838	−2.822	−2.805	−2.788	−2.771	−2.753	−120
−110	−2.753	−2.736	−2.719	−2.702	−2.685	−2.667	−2.650	−2.633	−2.616	−2.598	−2.581	−110
−100	−2.581	−2.563	−2.546	−2.528	−2.511	−2.493	−2.475	−2.458	−2.440	−2.422	−2.405	−100
−90	−2.405	−2.387	−2.369	−2.351	−2.333	−2.315	−2.297	−2.279	−2.261	−2.243	−2.225	−90
−80	−2.225	−2.207	−2.189	−2.171	−2.152	−2.134	−2.116	−2.098	−2.079	−2.061	−2.042	−80
−70	−2.042	−2.024	−2.005	−1.987	−1.968	−1.950	−1.931	−1.912	−1.894	−1.875	−1.856	−70
−60	−1.856	−1.838	−1.819	−1.800	−1.781	−1.762	−1.743	−1.724	−1.705	−1.686	−1.667	−60
−50	−1.667	−1.648	−1.629	−1.610	−1.591	−1.571	−1.552	−1.533	−1.513	−1.494	−1.475	−50
−40	−1.475	−1.455	−1.436	−1.416	−1.397	−1.377	−1.358	−1.338	−1.319	−1.299	−1.279	−40
−30	−1.279	−1.260	−1.240	−1.220	−1.200	−1.180	−1.160	−1.141	−1.121	−1.101	−1.081	−30
−20	−1.081	−1.061	−1.041	−1.021	−1.000	−0.980	−0.960	−0.940	−0.920	−0.899	−0.879	−20
−10	−0.879	−0.859	−0.838	−0.818	−0.798	−0.777	−0.757	−0.736	−0.716	−0.695	−0.674	−10
0	−0.674	−0.654	−0.633	−0.613	−0.592	−0.571	−0.550	−0.529	−0.509	−0.488	−0.467	0
10	−0.467	−0.446	−0.425	−0.404	−0.383	−0.362	−0.341	−0.320	−0.299	−0.277	−0.256	10
20	−0.256	−0.235	−0.214	−0.193	−0.171	−0.150	−0.129	−0.107	−0.086	−0.064	−0.043	20
30	−0.043	−0.022	0.000	0.022	0.043	0.065	0.086	0.108	0.130	0.151	0.173	30
40	0.173	0.195	0.216	0.238	0.260	0.282	0.303	0.325	0.347	0.369	0.391	40
50	0.391	0.413	0.435	0.457	0.479	0.501	0.523	0.545	0.567	0.589	0.611	50
60	0.611	0.634	0.656	0.678	0.700	0.722	0.745	0.767	0.789	0.812	0.834	60
70	0.834	0.857	0.879	0.902	0.924	0.947	0.969	0.992	1.014	1.037	1.060	70
80	1.060	1.082	1.105	1.128	1.151	1.173	1.196	1.219	1.242	1.265	1.288	80
90	1.288	1.311	1.334	1.357	1.380	1.403	1.426	1.449	1.472	1.495	1.518	90
100	1.518	1.542	1.565	1.588	1.611	1.635	1.658	1.681	1.705	1.728	1.752	100
110	1.752	1.775	1.799	1.822	1.846	1.869	1.893	1.917	1.940	1.964	1.988	110
120	1.988	2.011	2.035	2.059	2.083	2.107	2.131	2.154	2.178	2.202	2.226	120
130	2.226	2.250	2.274	2.298	2.322	2.347	2.371	2.395	2.419	2.443	2.467	130

Type T Thermocouple
Thermoelectric Voltage as a Function of Temperature (°F)
Reference Junctions at 32°F
THERMOELECTRIC VOLTAGE IN ABSOLUTE MILLIVOLTS

°F	0	1	2	3	4	5	6	7	8	9	10	°F
140	2.467	2.492	2.516	2.540	2.565	2.589	2.613	2.638	2.662	2.687	2.711	140
150	2.711	2.736	2.760	2.785	2.809	2.834	2.859	2.883	2.908	2.933	2.958	150
160	2.958	2.982	3.007	3.032	3.057	3.082	3.107	3.131	3.156	3.181	3.206	160
170	3.206	3.231	3.256	3.281	3.307	3.332	3.357	3.382	3.407	3.432	3.458	170
180	3.458	3.483	3.508	3.533	3.559	3.584	3.609	3.635	3.660	3.686	3.711	180
190	3.711	3.737	3.762	3.788	3.813	3.839	3.864	3.890	3.916	3.941	3.967	190
200	3.967	3.993	4.019	4.044	4.070	4.096	4.122	4.148	4.174	4.199	4.225	200
210	4.225	4.251	4.277	4.303	4.329	4.355	4.381	4.408	4.434	4.460	4.486	210
220	4.486	4.512	4.538	4.565	4.591	4.617	4.643	4.670	4.696	4.722	4.749	220
230	4.749	4.775	4.801	4.828	4.854	4.881	4.907	4.934	4.960	4.987	5.014	230
240	5.014	5.040	5.067	5.093	5.120	5.147	5.174	5.200	5.227	5.254	5.281	240
250	5.281	5.307	5.334	5.361	5.388	5.415	5.442	5.469	5.496	5.523	5.550	250
260	5.550	5.577	5.604	5.631	5.658	5.685	5.712	5.739	5.767	5.794	5.821	260
270	5.821	5.848	5.875	5.903	5.930	5.957	5.985	6.012	6.039	6.067	6.094	270
280	6.094	6.122	6.149	6.177	6.204	6.232	6.259	6.287	6.314	6.342	6.369	280
290	6.369	6.397	6.425	6.452	6.480	6.508	6.536	6.563	6.591	6.619	6.647	290
300	6.647	6.675	6.702	6.730	6.758	6.786	6.814	6.842	6.870	6.898	6.926	300
310	6.926	6.954	6.982	7.010	7.038	7.066	7.094	7.122	7.151	7.179	7.207	310
320	7.207	7.235	7.263	7.292	7.320	7.348	7.377	7.405	7.433	7.462	7.490	320
330	7.490	7.518	7.547	7.575	7.604	7.632	7.661	7.689	7.718	7.746	7.775	330
340	7.775	7.804	7.832	7.861	7.889	7.918	7.947	7.975	8.004	8.033	8.062	340
350	8.062	8.090	8.119	8.148	8.177	8.206	8.235	8.264	8.292	8.321	8.350	350
360	8.350	8.379	8.408	8.437	8.466	8.495	8.524	8.553	8.583	8.612	8.641	360
370	8.641	8.670	8.699	8.728	8.757	8.787	8.816	8.845	8.874	8.904	8.933	370
380	8.933	8.962	8.992	9.021	9.050	9.080	9.109	9.139	9.168	9.198	9.227	380
390	9.227	9.257	9.286	9.316	9.345	9.375	9.404	9.434	9.464	9.493	9.523	390
400	9.523	9.553	9.582	9.612	9.642	9.671	9.701	9.731	9.761	9.791	9.820	400
410	9.820	9.850	9.880	9.910	9.940	9.970	10.000	10.030	10.060	10.090	10.120	410
420	10.120	10.150	10.180	10.210	10.240	10.270	10.300	10.330	10.360	10.390	10.420	420
430	10.420	10.451	10.481	10.511	10.541	10.572	10.602	10.632	10.662	10.693	10.723	430
440	10.723	10.753	10.784	10.814	10.845	10.875	10.905	10.936	10.966	10.997	11.027	440
450	11.027	11.058	11.088	11.119	11.149	11.180	11.211	11.241	11.272	11.302	11.333	450
460	11.333	11.364	11.394	11.425	11.456	11.487	11.517	11.548	11.579	11.610	11.640	460
470	11.640	11.671	11.702	11.733	11.764	11.795	11.826	11.856	11.887	11.918	11.949	470
480	11.949	11.980	12.011	12.042	12.073	12.104	12.135	12.166	12.198	12.229	12.260	480
490	12.260	12.291	12.322	12.353	12.384	12.416	12.447	12.478	12.509	12.540	12.572	490
500	12.572	12.603	12.634	12.666	12.697	12.728	12.760	12.791	12.822	12.854	12.885	500
510	12.885	12.917	12.948	12.979	13.011	13.042	13.074	13.105	13.137	13.168	13.200	510
520	13.200	13.232	13.263	13.295	13.326	13.358	13.390	13.421	13.453	13.485	13.516	520
530	13.516	13.548	13.580	13.611	13.643	13.675	13.707	13.739	13.770	13.802	13.834	530
540	13.834	13.866	13.898	13.930	13.961	13.993	14.025	14.057	14.089	14.121	14.153	540
550	14.153	14.185	14.217	14.249	14.281	14.313	14.345	14.377	14.409	14.441	14.474	550
560	14.474	14.506	14.538	14.570	14.602	14.634	14.666	14.699	14.731	14.763	14.795	560
570	14.795	14.828	14.860	14.892	14.924	14.957	14.989	15.021	15.054	15.086	15.118	570
580	15.118	15.151	15.183	15.216	15.248	15.280	15.313	15.345	15.378	15.410	15.443	580
590	15.443	15.475	15.508	15.540	15.573	15.605	15.638	15.671	15.703	15.736	15.769	590
600	15.769	15.801	15.834	15.866	15.899	15.932	15.965	15.997	16.030	16.063	16.096	600
610	16.096	16.128	16.161	16.194	16.227	16.259	16.292	16.325	16.358	16.391	16.424	610
620	16.424	16.457	16.490	16.523	16.555	16.588	16.621	16.654	16.687	16.720	16.753	620
630	16.753	16.786	16.819	16.852	16.886	16.919	16.952	16.985	17.018	17.051	17.084	630
640	17.084	17.117	17.150	17.184	17.217	17.250	17.283	17.316	17.350	17.383	17.416	640
650	17.416	17.450	17.483	17.516	17.549	17.583	17.616	17.649	17.683	17.716	17.750	650
660	17.750	17.783	17.816	17.850	17.883	17.917	17.950	17.984	18.017	18.051	18.084	660
670	18.084	18.118	18.151	18.185	18.218	18.252	18.285	18.319	18.353	18.386	18.420	670
680	18.420	18.454	18.487	18.521	18.555	18.588	18.622	18.656	18.689	18.723	18.757	680
690	18.757	18.791	18.824	18.858	18.892	18.926	18.960	18.993	19.027	19.061	19.095	690
700	19.095	19.129	19.163	19.197	19.230	19.264	19.298	19.332	19.366	19.400	19.434	700
710	19.434	19.468	19.502	19.536	19.570	19.604	19.638	19.672	19.706	19.740	19.774	710
720	19.774	19.808	19.843	19.877	19.911	19.945	19.979	20.013	20.047	20.081	20.116	720
730	20.116	20.150	20.184	20.218	20.252	20.287	20.321	20.355	20.389	20.423	20.458	730
740	20.458	20.492	20.526	20.560	20.595	20.629	20.663	20.698	20.732	20.766	20.801	740
750	20.801	20.835	20.869									750

RTD Resistance vs Temperature (°C)
Platinum 100, Alpha = 0.00385

RESISTANCE IN OHMS

°C	0	1	2	3	4	5	6	7	8	9	10	°C
−200	18.49	18.93	19.36	19.79	20.22	20.65	21.08	21.51	21.94	22.37	22.80	−200
−190	22.80	23.23	23.66	24.09	24.52	24.94	25.37	25.80	26.23	26.65	27.08	−190
−180	27.08	27.50	27.93	28.35	28.78	29.20	29.63	30.05	30.47	30.90	31.32	−180
−170	31.32	31.74	32.16	32.59	33.01	33.43	33.85	34.27	34.69	35.11	35.53	−170
−160	35.53	35.95	36.37	36.79	37.21	37.63	38.04	38.46	38.88	39.30	39.71	−160
−150	39.71	40.13	40.55	40.96	41.38	41.79	42.21	42.63	43.04	43.45	43.87	−150
−140	43.87	44.28	44.70	45.11	45.52	45.94	46.35	46.76	47.18	47.59	48.00	−140
−130	48.00	48.41	48.82	49.23	49.64	50.06	50.47	50.88	51.29	51.70	52.11	−130
−120	52.11	52.52	52.92	53.33	53.74	54.15	54.56	54.97	55.38	55.78	56.19	−120
−110	56.19	56.60	57.00	57.41	57.82	58.22	58.63	59.04	59.44	59.85	60.25	−110
−100	60.25	60.66	61.06	61.47	61.87	62.28	62.68	63.09	63.49	63.90	64.30	−100
−90	64.30	64.70	65.11	65.51	65.91	66.31	66.72	67.12	67.52	67.92	68.33	−90
−80	68.33	68.73	69.13	69.53	69.93	70.33	70.73	71.13	71.53	71.93	72.33	−80
−70	72.33	72.73	73.13	73.53	73.93	74.33	74.73	75.13	75.53	75.93	76.33	−70
−60	76.33	76.73	77.13	77.52	77.92	78.32	78.72	79.11	79.51	79.91	80.31	−60
−50	80.31	80.70	81.10	81.50	81.89	82.29	82.69	83.08	83.48	83.88	84.27	−50
−40	84.27	84.67	85.06	85.46	85.85	86.25	86.64	87.04	87.43	87.83	88.22	−40
−30	88.22	88.62	89.01	89.40	89.80	90.19	90.59	90.98	91.37	91.77	92.16	−30
−20	92.16	92.55	92.95	93.34	93.73	94.12	94.52	94.91	95.30	95.69	96.09	−20
−10	96.09	96.48	96.87	97.26	97.65	98.04	98.44	98.83	99.22	99.61	100.00	−10
0	100.00	100.39	100.78	101.17	101.56	101.95	102.34	102.73	103.12	103.51	103.90	0
10	103.90	104.29	104.68	105.07	105.46	105.85	106.24	106.63	107.02	107.40	107.79	10
20	107.79	108.18	108.57	108.96	109.35	109.73	110.12	110.51	110.90	111.28	111.67	20
30	111.67	112.06	112.45	112.83	113.22	113.61	113.99	114.38	114.77	115.15	115.54	30
40	115.54	115.93	116.31	116.70	117.08	117.47	117.85	118.24	118.62	119.01	119.40	40
50	119.40	119.78	120.16	120.55	120.93	121.32	121.70	122.09	122.47	122.86	123.24	50
60	123.24	123.62	124.01	124.39	124.77	125.16	125.54	125.92	126.31	126.69	127.07	60
70	127.07	127.45	127.84	128.22	128.60	128.98	129.37	129.75	130.13	130.51	130.89	70
80	130.89	131.27	131.66	132.04	132.42	132.80	133.18	133.56	133.94	134.32	134.70	80
90	134.70	135.08	135.46	135.84	136.22	136.60	136.98	137.36	137.74	138.12	138.50	90
100	138.50	138.88	139.26	139.64	140.02	140.39	140.77	141.15	141.53	141.91	142.29	100
110	142.29	142.66	143.04	143.42	143.80	144.17	144.55	144.93	145.31	145.68	146.06	110
120	146.06	146.44	146.81	147.19	147.57	147.94	148.32	148.70	149.07	149.45	149.82	120
130	149.82	150.20	150.57	150.95	151.33	151.70	152.08	152.45	152.83	153.20	153.58	130
140	153.58	153.95	154.32	154.70	155.07	155.45	155.82	156.19	156.57	156.94	157.31	140
150	157.31	157.69	158.06	158.43	158.81	159.18	159.55	159.93	160.30	160.67	161.04	150
160	161.04	161.42	161.79	162.16	162.53	162.90	163.27	163.65	164.02	164.39	164.76	160
170	164.76	165.13	165.50	165.87	166.24	166.61	166.98	167.35	167.72	168.09	168.46	170
180	168.46	168.83	169.20	169.57	169.94	170.31	170.68	171.05	171.42	171.79	172.16	180
190	172.16	172.53	172.90	173.26	173.63	174.00	174.37	174.74	175.10	175.47	175.84	190
200	175.84	176.21	176.57	176.94	177.31	177.68	178.04	178.41	178.78	179.14	179.51	200
210	179.51	179.88	180.24	180.61	180.97	181.34	181.71	182.07	182.44	182.80	183.17	210
220	183.17	183.53	183.90	184.26	184.63	184.99	185.36	185.72	186.09	186.45	186.82	220
230	186.82	187.18	187.54	187.91	188.27	188.63	189.00	189.36	189.72	190.09	190.45	230
240	190.45	190.81	191.18	191.54	191.90	192.26	192.63	192.99	193.35	193.71	194.07	240
250	194.07	194.44	194.80	195.16	195.52	195.88	196.24	196.60	196.96	197.33	197.69	250
260	197.69	198.05	198.41	198.77	199.13	199.49	199.85	200.21	200.57	200.93	201.29	260
270	201.29	201.65	202.01	202.36	202.72	203.08	203.44	203.80	204.16	204.52	204.88	270
280	204.88	205.23	205.59	205.95	206.31	206.67	207.02	207.38	207.74	208.10	208.45	280
290	208.45	208.81	209.17	209.52	209.88	210.24	210.59	210.95	211.31	211.66	212.02	290
300	212.02	212.37	212.73	213.09	213.44	213.80	214.15	214.51	214.86	215.22	215.57	300
310	215.57	215.93	216.28	216.64	216.99	217.35	217.70	218.05	218.41	218.76	219.12	310
320	219.12	219.47	219.82	220.18	220.53	220.88	221.24	221.59	221.94	222.29	222.65	320
330	222.65	223.00	223.35	223.70	224.06	224.41	224.76	225.11	225.46	225.81	226.17	330
340	226.17	226.52	226.87	227.22	227.57	227.92	228.27	228.62	228.97	229.32	229.67	340
350	229.67	230.02	230.37	230.72	231.07	231.42	231.77	232.12	232.47	232.82	233.17	350
360	233.17	233.52	233.87	234.22	234.56	234.91	235.26	235.61	235.96	236.31	236.65	360
370	236.65	237.00	237.35	237.70	238.04	238.39	238.74	239.09	239.43	239.78	240.13	370
380	240.13	240.47	240.82	241.17	241.51	241.86	242.20	242.55	242.90	243.24	243.59	380
390	243.59	243.93	244.28	244.62	244.97	245.31	245.66	246.00	246.35	246.69	247.04	390

Appendix B

RTD Resistance vs Temperature (°C)
Platinum 100, Alpha = 0.00385

RESISTANCE IN OHMS

°C	0	1	2	3	4	5	6	7	8	9	10	°C
400	247.04	247.38	247.73	248.07	248.41	248.76	249.10	249.45	249.79	250.13	250.48	400
410	250.48	250.82	251.16	251.50	251.85	252.19	252.53	252.88	253.22	253.56	253.90	410
420	253.90	254.24	254.59	254.93	255.27	255.61	255.95	256.29	256.63	256.98	257.32	420
430	257.32	257.66	258.00	258.34	258.68	259.02	259.36	259.70	260.04	260.38	260.72	430
440	260.72	261.06	261.40	261.74	262.08	262.42	262.76	263.10	263.43	263.77	264.11	440
450	264.11	264.45	264.79	265.13	265.47	265.80	266.14	266.48	266.82	267.15	267.49	450
460	267.49	267.83	268.17	268.50	268.84	269.18	269.51	269.85	270.19	270.52	270.86	460
470	270.86	271.20	271.53	271.87	272.20	272.54	272.88	273.21	273.55	273.88	274.22	470
480	274.22	274.55	274.89	275.22	275.56	275.89	276.23	276.56	276.89	277.23	277.56	480
490	277.56	277.90	278.23	278.56	278.90	279.23	279.56	279.90	280.23	280.56	280.90	490
500	280.90	281.23	281.56	281.89	282.23	282.56	282.89	283.22	283.55	283.89	284.22	500
510	284.22	284.55	284.88	285.21	285.54	285.87	286.21	286.54	286.87	287.20	287.53	510
520	287.53	287.86	288.19	288.52	288.85	289.18	289.51	289.84	290.17	290.50	290.83	520
530	290.83	291.16	291.49	291.81	292.14	292.47	292.80	293.13	293.46	293.79	294.11	530
540	294.11	294.44	294.77	295.10	295.43	295.75	296.08	296.41	296.74	297.06	297.39	540
550	297.39	297.72	298.04	298.37	298.70	299.02	299.35	299.68	300.00	300.33	300.65	550
560	300.65	300.98	301.31	301.63	301.96	302.28	302.61	302.93	303.26	303.58	303.91	560
570	303.91	304.23	304.56	304.88	305.20	305.53	305.85	306.18	306.50	306.82	307.15	570
580	307.15	307.47	307.79	308.12	308.44	308.76	309.09	309.41	309.73	310.05	310.38	580
590	310.38	310.70	311.02	311.34	311.66	311.99	312.31	312.63	312.95	313.27	313.59	590
600	313.59	313.92	314.24	314.56	314.88	315.20	315.52	315.84	316.16	316.48	316.80	600
610	316.80	317.12	317.44	317.76	318.08	318.40	318.72	319.04	319.36	319.68	319.99	610
620	319.99	320.31	320.63	320.95	321.27	321.59	321.91	322.22	322.54	322.86	323.18	620
630	323.18	323.49	323.81	324.13	324.45	324.76	325.08	325.40	325.72	326.03	326.35	630
640	326.35	326.66	326.98	327.30	327.61	327.93	328.25	328.56	328.88	329.19	329.51	640
650	329.51	329.82	330.14	330.45	330.77	331.08	331.40	331.71	332.03	332.34	332.66	650
660	332.66	332.97	333.28	333.60	333.91	334.23	334.54	334.85	335.17	335.48	335.79	660
670	335.79	336.11	336.42	336.73	337.04	337.36	337.67	337.98	338.29	338.61	338.92	670
680	338.92	339.23	339.54	339.85	340.16	340.47	340.79	341.10	341.41	341.72	342.03	680
690	342.03	342.34	342.65	342.96	343.27	343.58	343.89	344.20	344.51	344.82	345.13	690
700	345.13	345.44	345.75	346.06	346.37	346.68	346.99	347.30	347.60	347.91	348.22	700
710	348.22	348.53	348.84	349.15	349.45	349.76	350.07	350.38	350.69	350.99	351.30	710
720	351.30	351.61	351.91	352.22	352.53	352.83	353.14	353.45	353.75	354.06	354.37	720
730	354.37	354.67	354.98	355.28	355.59	355.90	356.20	356.51	356.81	357.12	357.42	730
740	357.42	357.73	358.03	358.34	358.64	358.94	359.25	359.55	359.86	360.16	360.47	740
750	360.47	360.77	361.07	361.38	361.68	361.98	362.29	362.59	362.89	363.19	363.50	750
760	363.50	363.80	364.10	364.40	364.71	365.01	365.31	365.61	365.91	366.22	366.52	760
770	366.52	366.82	367.12	367.42	367.72	368.02	368.32	368.62	368.93	369.23	369.53	770
780	369.53	369.83	370.13	370.43	370.73	371.03	371.33	371.63	371.92	372.22	372.52	780
790	372.52	372.82	373.12	373.42	373.72	374.02	374.32	374.61	374.91	375.21	375.51	790
800	375.51	375.81	376.10	376.40	376.70	377.00	377.29	377.59	377.89	378.19	378.48	800
810	378.48	378.78	379.08	379.37	379.67	379.97	380.26	380.56	380.85	381.15	381.45	810
820	381.45	381.74	382.04	382.33	382.63	382.92	383.22	383.51	383.81	384.10	384.40	820
830	384.40	384.69	384.98	385.28	385.57	385.87	386.16	386.45	386.75	387.04	387.33	830
840	387.33	387.63	387.92	388.21	388.51	388.80	389.09	389.39	389.68	389.97	390.26	840

RTD Resistance vs Temperature (°F)
Platinum 100, Alpha = 0.00385

RESISTANCE IN OHMS

°F	0	1	2	3	4	5	6	7	8	9	10	°F
-330	18.01	18.25	18.49	18.73	18.97	19.21	19.45	19.69	19.93	20.17	20.41	-330
-320	20.41	20.65	20.89	21.13	21.37	21.61	21.85	22.09	22.33	22.56	22.80	-320
-310	22.80	23.04	23.28	23.52	23.76	23.99	24.23	24.47	24.71	24.94	25.18	-310
-300	25.18	25.42	25.66	25.89	26.13	26.37	26.60	26.84	27.08	27.31	27.55	-300
-290	27.55	27.79	28.02	28.26	28.50	28.73	28.97	29.20	29.44	29.67	29.91	-290
-280	29.91	30.14	30.38	30.62	30.85	31.09	31.32	31.55	31.79	32.02	32.26	-280
-270	32.26	32.49	32.73	32.96	33.20	33.43	33.66	33.90	34.13	34.36	34.60	-270
-260	34.60	34.83	35.06	35.30	35.53	35.76	36.00	36.23	36.46	36.70	36.93	-260
-250	36.93	37.16	37.39	37.63	37.86	38.09	38.32	38.55	38.79	39.02	39.25	-250
-240	39.25	39.48	39.71	39.95	40.18	40.41	40.64	40.87	41.10	41.33	41.56	-240
-230	41.56	41.79	42.03	42.26	42.49	42.72	42.95	43.18	43.41	43.64	43.87	-230
-220	43.87	44.10	44.33	44.56	44.79	45.02	45.25	45.48	45.71	45.94	46.17	-220
-210	46.17	46.40	46.63	46.85	47.08	47.31	47.54	47.77	48.00	48.23	48.46	-210
-200	48.46	48.69	48.91	49.14	49.37	49.60	49.83	50.06	50.28	50.51	50.74	-200
-190	50.74	50.97	51.20	51.42	51.65	51.88	52.11	52.33	52.56	52.79	53.02	-190
-180	53.02	53.24	53.47	53.70	53.92	54.15	54.38	54.60	54.83	55.06	55.28	-180
-170	55.28	55.51	55.74	55.96	56.19	56.42	56.64	56.87	57.10	57.32	57.55	-170
-160	57.55	57.77	58.00	58.22	58.45	58.68	58.90	59.13	59.35	59.58	59.80	-160
-150	59.80	60.03	60.25	60.48	60.70	60.93	61.15	61.38	61.60	61.83	62.05	-150
-140	62.05	62.28	62.50	62.73	62.95	63.18	63.40	63.63	63.85	64.07	64.30	-140
-130	64.30	64.52	64.75	64.97	65.19	65.42	65.64	65.87	66.09	66.31	66.54	-130
-120	66.54	66.76	66.98	67.21	67.43	67.66	67.88	68.10	68.33	68.55	68.77	-120
-110	68.77	68.99	69.22	69.44	69.66	69.89	70.11	70.33	70.55	70.78	71.00	-110
-100	71.00	71.22	71.45	71.67	71.89	72.11	72.33	72.56	72.78	73.00	73.22	-100
-90	73.22	73.45	73.67	73.89	74.11	74.33	74.56	74.78	75.00	75.22	75.44	-90
-80	75.44	75.66	75.89	76.11	76.33	76.55	76.77	76.99	77.21	77.43	77.66	-80
-70	77.66	77.88	78.10	78.32	78.54	78.76	78.98	79.20	79.42	79.64	79.87	-70
-60	79.87	80.09	80.31	80.53	80.75	80.97	81.19	81.41	81.63	81.85	82.07	-60
-50	82.07	82.29	82.51	82.73	82.95	83.17	83.39	83.61	83.83	84.05	84.27	-50
-40	84.27	84.49	84.71	84.93	85.15	85.37	85.59	85.81	86.03	86.25	86.47	-40
-30	86.47	86.69	86.91	87.13	87.35	87.56	87.78	88.00	88.22	88.44	88.66	-30
-20	88.66	88.88	89.10	89.32	89.54	89.76	89.97	90.19	90.41	90.63	90.85	-20
-10	90.85	91.07	91.29	91.50	91.72	91.94	92.16	92.38	92.60	92.82	93.03	-10
0	93.03	93.25	93.47	93.69	93.91	94.12	94.34	94.56	94.78	95.00	95.21	0
10	95.21	95.43	95.65	95.87	96.09	96.30	96.52	96.74	96.96	97.17	97.39	10
20	97.39	97.61	97.83	98.04	98.26	98.48	98.70	98.91	99.13	99.35	99.57	20
30	99.57	99.78	100.00	100.22	100.43	100.65	100.87	101.09	101.30	101.52	101.74	30
40	101.74	101.95	102.17	102.39	102.60	102.82	103.04	103.25	103.47	103.69	103.90	40
50	103.90	104.12	104.34	104.55	104.77	104.98	105.20	105.42	105.63	105.85	106.07	50
60	106.07	106.28	106.50	106.71	106.93	107.15	107.36	107.58	107.79	108.01	108.22	60
70	108.22	108.44	108.66	108.87	109.09	109.30	109.52	109.73	109.95	110.16	110.38	70
80	110.38	110.60	110.81	111.03	111.24	111.46	111.67	111.89	112.10	112.32	112.53	80
90	112.53	112.75	112.96	113.18	113.39	113.61	113.82	114.04	114.25	114.47	114.68	90
100	114.68	114.90	115.11	115.32	115.54	115.75	115.97	116.18	116.40	116.61	116.83	100
110	116.83	117.04	117.25	117.47	117.68	117.90	118.11	118.33	118.54	118.75	118.97	110
120	118.97	119.18	119.40	119.61	119.82	120.04	120.25	120.46	120.68	120.89	121.11	120
130	121.11	121.32	121.53	121.75	121.96	122.17	122.39	122.60	122.81	123.03	123.24	130
140	123.24	123.45	123.67	123.88	124.09	124.31	124.52	124.73	124.94	125.16	125.37	140
150	125.37	125.58	125.80	126.01	126.22	126.43	126.65	126.86	127.07	127.28	127.50	150
160	127.50	127.71	127.92	128.13	128.35	128.56	128.77	128.98	129.20	129.41	129.62	160
170	129.62	129.83	130.04	130.26	130.47	130.68	130.89	131.10	131.32	131.53	131.74	170
180	131.74	131.95	132.16	132.38	132.59	132.80	133.01	133.22	133.43	133.65	133.86	180
190	133.86	134.07	134.28	134.49	134.70	134.91	135.12	135.34	135.55	135.76	135.97	190
200	135.97	136.18	136.39	136.60	136.81	137.02	137.24	137.45	137.66	137.87	138.08	200
210	138.08	138.29	138.50	138.71	138.92	139.13	139.34	139.55	139.76	139.97	140.18	210
220	140.18	140.39	140.60	140.82	141.03	141.24	141.45	141.66	141.87	142.08	142.29	220
230	142.29	142.50	142.71	142.92	143.13	143.34	143.55	143.76	143.97	144.17	144.38	230
240	144.38	144.59	144.80	145.01	145.22	145.43	145.64	145.85	146.06	146.27	146.48	240
250	146.48	146.69	146.90	147.11	147.32	147.53	147.73	147.94	148.15	148.36	148.57	250
260	148.57	148.78	148.99	149.20	149.41	149.61	149.82	150.03	150.24	150.45	150.66	260

Appendix B

RTD Resistance vs Temperature (°F)
Platinum 100, Alpha = 0.00385

RESISTANCE IN OHMS

°F	0	1	2	3	4	5	6	7	8	9	10	°F
270	150.66	150.87	151.08	151.28	151.49	151.70	151.91	152.12	152.33	152.53	152.74	270
280	152.74	152.95	153.16	153.37	153.58	153.78	153.99	154.20	154.41	154.62	154.82	280
290	154.82	155.03	155.24	155.45	155.65	155.86	156.07	156.28	156.48	156.69	156.90	290
300	156.90	157.11	157.31	157.52	157.73	157.94	158.14	158.35	158.56	158.77	158.97	300
310	158.97	159.18	159.39	159.59	159.80	160.01	160.22	160.42	160.63	160.84	161.04	310
320	161.04	161.25	161.46	161.66	161.87	162.08	162.28	162.49	162.70	162.90	163.11	320
330	163.11	163.32	163.52	163.73	163.93	164.14	164.35	164.55	164.76	164.97	165.17	330
340	165.17	165.38	165.58	165.79	166.00	166.20	166.41	166.61	166.82	167.03	167.23	340
350	167.23	167.44	167.64	167.85	168.05	168.26	168.46	168.67	168.88	169.08	169.29	350
360	169.29	169.49	169.70	169.90	170.11	170.31	170.52	170.72	170.93	171.13	171.34	360
370	171.34	171.54	171.75	171.95	172.16	172.36	172.57	172.77	172.98	173.18	173.39	370
380	173.39	173.59	173.80	174.00	174.20	174.41	174.61	174.82	175.02	175.23	175.43	380
390	175.43	175.64	175.84	176.04	176.25	176.45	176.66	176.86	177.06	177.27	177.47	390
400	177.47	177.68	177.88	178.08	178.29	178.49	178.70	178.90	179.10	179.31	179.51	400
410	179.51	179.71	179.92	180.12	180.32	180.53	180.73	180.93	181.14	181.34	181.54	410
420	181.54	181.75	181.95	182.15	182.36	182.56	182.76	182.97	183.17	183.37	183.57	420
430	183.57	183.78	183.98	184.18	184.39	184.59	184.79	184.99	185.20	185.40	185.60	430
440	185.60	185.80	186.01	186.21	186.41	186.61	186.82	187.02	187.22	187.42	187.62	440
450	187.62	187.83	188.03	188.23	188.43	188.63	188.84	189.04	189.24	189.44	189.64	450
460	189.64	189.85	190.05	190.25	190.45	190.65	190.85	191.06	191.26	191.46	191.66	460
470	191.66	191.86	192.06	192.26	192.47	192.67	192.87	193.07	193.27	193.47	193.67	470
480	193.67	193.87	194.07	194.28	194.48	194.68	194.88	195.08	195.28	195.48	195.68	480
490	195.68	195.88	196.08	196.28	196.48	196.68	196.88	197.09	197.29	197.49	197.69	490
500	197.69	197.89	198.09	198.29	198.49	198.69	198.89	199.09	199.29	199.49	199.69	500
510	199.69	199.89	200.09	200.29	200.49	200.69	200.89	201.09	201.29	201.49	201.69	510
520	201.69	201.89	202.09	202.28	202.48	202.68	202.88	203.08	203.28	203.48	203.68	520
530	203.68	203.88	204.08	204.28	204.48	204.68	204.88	205.07	205.27	205.47	205.67	530
540	205.67	205.87	206.07	206.27	206.47	206.67	206.86	207.06	207.26	207.46	207.66	540
550	207.66	207.86	208.06	208.25	208.45	208.65	208.85	209.05	209.25	209.44	209.64	550
560	209.64	209.84	210.04	210.24	210.44	210.63	210.83	211.03	211.23	211.43	211.62	560
570	211.62	211.82	212.02	212.22	212.41	212.61	212.81	213.01	213.20	213.40	213.60	570
580	213.60	213.80	213.99	214.19	214.39	214.59	214.78	214.98	215.18	215.38	215.57	580
590	215.57	215.77	215.97	216.16	216.36	216.56	216.76	216.95	217.15	217.35	217.54	590
600	217.54	217.74	217.94	218.13	218.33	218.53	218.72	218.92	219.12	219.31	219.51	600
610	219.51	219.70	219.90	220.10	220.29	220.49	220.69	220.88	221.08	221.27	221.47	610
620	221.47	221.67	221.86	222.06	222.25	222.45	222.65	222.84	223.04	223.23	223.43	620
630	223.43	223.63	223.82	224.02	224.21	224.41	224.60	224.80	224.99	225.19	225.38	630
640	225.38	225.58	225.78	225.97	226.17	226.36	226.56	226.75	226.95	227.14	227.34	640
650	227.34	227.53	227.73	227.92	228.12	228.31	228.51	228.70	228.89	229.09	229.28	650
660	229.28	229.48	229.67	229.87	230.06	230.26	230.45	230.65	230.84	231.03	231.23	660
670	231.23	231.42	231.62	231.81	232.01	232.20	232.39	232.59	232.78	232.98	233.17	670
680	233.17	233.36	233.56	233.75	233.94	234.14	234.33	234.53	234.72	234.91	235.11	680
690	235.11	235.30	235.49	235.69	235.88	236.07	236.27	236.46	236.65	236.85	237.04	690
700	237.04	237.23	237.43	237.62	237.81	238.01	238.20	238.39	238.58	238.78	238.97	700
710	238.97	239.16	239.36	239.55	239.74	239.93	240.13	240.32	240.51	240.70	240.90	710
720	240.90	241.09	241.28	241.47	241.67	241.86	242.05	242.24	242.44	242.63	242.82	720
730	242.82	243.01	243.20	243.40	243.59	243.78	243.97	244.16	244.36	244.55	244.74	730
740	244.74	244.93	245.12	245.31	245.51	245.70	245.89	246.08	246.27	246.46	246.65	740
750	246.65	246.85	247.04	247.23	247.42	247.61	247.80	247.99	248.18	248.38	248.57	750
760	248.57	248.76	248.95	249.14	249.33	249.52	249.71	249.90	250.09	250.28	250.48	760
770	250.48	250.67	250.86	251.05	251.24	251.43	251.62	251.81	252.00	252.19	252.38	770
780	252.38	252.57	252.76	252.95	253.14	253.33	253.52	253.71	253.90	254.09	254.28	780
790	254.28	254.47	254.66	254.85	255.04	255.23	255.42	255.61	255.80	255.99	256.18	790
800	256.18	256.37	256.56	256.75	256.94	257.13	257.32	257.51	257.70	257.88	258.07	800
810	258.07	258.26	258.45	258.64	258.83	259.02	259.21	259.40	259.59	259.78	259.96	810
820	259.96	260.15	260.34	260.53	260.72	260.91	261.10	261.29	261.47	261.66	261.85	820
830	261.85	262.04	262.23	262.42	262.61	262.79	262.98	263.17	263.36	263.55	263.74	830
840	263.74	263.92	264.11	264.30	264.49	264.68	264.86	265.05	265.24	265.43	265.62	840
850	265.62	265.80	265.99	266.18	266.37	266.55	266.74	266.93	267.12	267.30	267.49	850
860	267.49	267.68	267.87	268.05	268.24	268.43	268.62	268.80	268.99	269.18	269.36	860
870	269.36	269.55	269.74	269.93	270.11	270.30	270.49	270.67	270.86	271.05	271.23	870
880	271.23	271.42	271.61	271.79	271.98	272.17	272.35	272.54	272.73	272.91	273.10	880
890	273.10	273.29	273.47	273.66	273.84	274.03	274.22	274.40	274.59	274.78	274.96	890
900	274.96	275.15	275.33	275.52	275.71	275.89	276.08	276.26	276.45	276.63	276.82	900
910	276.82	277.01	277.19	277.38	277.56	277.75	277.93	278.12	278.30	278.49	278.67	910

RTD Resistance vs Temperature (°F)
Platinum 100, Alpha = 0.00385

RESISTANCE IN OHMS

°F	0	1	2	3	4	5	6	7	8	9	10	°F
920	278.67	278.86	279.05	279.23	279.42	279.60	279.79	279.97	280.16	280.34	280.53	920
930	280.53	280.71	280.90	281.08	281.27	281.45	281.64	281.82	282.00	282.19	282.37	930
940	282.37	282.56	282.74	282.93	283.11	283.30	283.48	283.67	283.85	284.03	284.22	940
950	284.22	284.40	284.59	284.77	284.95	285.14	285.32	285.51	285.69	285.87	286.06	950
960	286.06	286.24	286.43	286.61	286.79	286.98	287.16	287.34	287.53	287.71	287.90	960
970	287.90	288.08	288.26	288.45	288.63	288.81	289.00	289.18	289.36	289.55	289.73	970
980	289.73	289.91	290.10	290.28	290.46	290.64	290.83	291.01	291.19	291.38	291.56	980
990	291.56	291.74	291.92	292.11	292.29	292.47	292.65	292.84	293.02	293.20	293.38	990
1000	293.38	293.57	293.75	293.93	294.11	294.30	294.48	294.66	294.84	295.03	295.21	1000
1010	295.21	295.39	295.57	295.75	295.94	296.12	296.30	296.48	296.66	296.84	297.03	1010
1020	297.03	297.21	297.39	297.57	297.75	297.93	298.12	298.30	298.48	298.66	298.84	1020
1030	298.84	299.02	299.20	299.39	299.57	299.75	299.93	300.11	300.29	300.47	300.65	1030
1040	300.65	300.84	301.02	301.20	301.38	301.56	301.74	301.92	302.10	302.28	302.46	1040
1050	302.46	302.64	302.82	303.00	303.18	303.37	303.55	303.73	303.91	304.09	304.27	1050
1060	304.27	304.45	304.63	304.81	304.99	305.17	305.35	305.53	305.71	305.89	306.07	1060
1070	306.07	306.25	306.43	306.61	306.79	306.97	307.15	307.33	307.51	307.69	307.87	1070
1080	307.87	308.05	308.22	308.40	308.58	308.76	308.94	309.12	309.30	309.48	309.66	1080
1090	309.66	309.84	310.02	310.20	310.38	310.56	310.73	310.91	311.09	311.27	311.45	1090
1100	311.45	311.63	311.81	311.99	312.17	312.34	312.52	312.70	312.88	313.06	313.24	1100
1110	313.24	313.42	313.59	313.77	313.95	314.13	314.31	314.49	314.66	314.84	315.02	1110
1120	315.02	315.20	315.38	315.55	315.73	315.91	316.09	316.27	316.44	316.62	316.80	1120
1130	316.80	316.98	317.16	317.33	317.51	317.69	317.87	318.04	318.22	318.40	318.58	1130
1140	318.58	318.75	318.93	319.11	319.29	319.46	319.64	319.82	319.99	320.17	320.35	1140
1150	320.35	320.53	320.70	320.88	321.06	321.23	321.41	321.59	321.76	321.94	322.12	1150
1160	322.12	322.29	322.47	322.65	322.82	323.00	323.18	323.35	323.53	323.71	323.88	1160
1170	323.88	324.06	324.24	324.41	324.59	324.76	324.94	325.12	325.29	325.47	325.64	1170
1180	325.64	325.82	326.00	326.17	326.35	326.52	326.70	326.88	327.05	327.23	327.40	1180
1190	327.40	327.58	327.75	327.93	328.11	328.28	328.46	328.63	328.81	328.98	329.16	1190
1200	329.16	329.33	329.51	329.68	329.86	330.03	330.21	330.38	330.56	330.73	330.91	1200
1210	330.91	331.08	331.26	331.43	331.61	331.78	331.96	332.13	332.31	332.48	332.66	1210
1220	332.66	332.83	333.00	333.18	333.35	333.53	333.70	333.88	334.05	334.23	334.40	1220
1230	334.40	334.57	334.75	334.92	335.10	335.27	335.44	335.62	335.79	335.97	336.14	1230
1240	336.14	336.31	336.49	336.66	336.84	337.01	337.18	337.36	337.53	337.70	337.88	1240
1250	337.88	338.05	338.22	338.40	338.57	338.74	338.92	339.09	339.26	339.44	339.61	1250
1260	339.61	339.78	339.96	340.13	340.30	340.47	340.65	340.82	340.99	341.17	341.34	1260
1270	341.34	341.51	341.68	341.86	342.03	342.20	342.38	342.55	342.72	342.89	343.07	1270
1280	343.07	343.24	343.41	343.58	343.75	343.93	344.10	344.27	344.44	344.62	344.79	1280
1290	344.79	344.96	345.13	345.30	345.48	345.65	345.82	345.99	346.16	346.33	346.51	1290
1300	346.51	346.68	346.85	347.02	347.19	347.36	347.54	347.71	347.88	348.05	348.22	1300
1310	348.22	348.39	348.56	348.74	348.91	349.08	349.25	349.42	349.59	349.76	349.93	1310
1320	349.93	350.10	350.28	350.45	350.62	350.79	350.96	351.13	351.30	351.47	351.64	1320
1330	351.64	351.81	351.98	352.15	352.32	352.49	352.66	352.83	353.01	353.18	353.35	1330
1340	353.35	353.52	353.69	353.86	354.03	354.20	354.37	354.54	354.71	354.88	355.05	1340
1350	355.05	355.22	355.39	355.56	355.73	355.90	356.07	356.24	356.40	356.57	356.74	1350
1360	356.74	356.91	357.08	357.25	357.42	357.59	357.76	357.93	358.10	358.27	358.44	1360
1370	358.44	358.61	358.78	358.94	359.11	359.28	359.45	359.62	359.79	359.96	360.13	1370
1380	360.13	360.30	360.47	360.63	360.80	360.97	361.14	361.31	361.48	361.65	361.81	1380
1390	361.81	361.98	362.15	362.32	362.49	362.66	362.82	362.99	363.16	363.33	363.50	1390
1400	363.50	363.67	363.83	364.00	364.17	364.34	364.51	364.67	364.84	365.01	365.18	1400
1410	365.18	365.34	365.51	365.68	365.85	366.01	366.18	366.35	366.52	366.68	366.85	1410
1420	366.85	367.02	367.19	367.35	367.52	367.69	367.86	368.02	368.19	368.36	368.52	1420
1430	368.52	368.69	368.86	369.03	369.19	369.36	369.53	369.69	369.86	370.03	370.19	1430
1440	370.19	370.36	370.53	370.69	370.86	371.03	371.19	371.36	371.53	371.69	371.86	1440
1450	371.86	372.02	372.19	372.36	372.52	372.69	372.86	373.02	373.19	373.35	373.52	1450
1460	373.52	373.69	373.85	374.02	374.18	374.35	374.51	374.68	374.85	375.01	375.18	1460
1470	375.18	375.34	375.51	375.67	375.84	376.01	376.17	376.34	376.50	376.67	376.83	1470
1480	376.83	377.00	377.16	377.33	377.49	377.66	377.82	377.99	378.15	378.32	378.48	1480
1490	378.48	378.65	378.81	378.98	379.14	379.31	379.47	379.64	379.80	379.97	380.13	1490
1500	380.13	380.29	380.46	380.62	380.79	380.95	381.12	381.28	381.45	381.61	381.77	1500
1510	381.77	381.94	382.10	382.27	382.43	382.59	382.76	382.92	383.09	383.25	383.41	1510
1520	383.41	383.58	383.74	383.90	384.07	384.23	384.40	384.56	384.72	384.89	385.05	1520
1530	385.05	385.21	385.38	385.54	385.70	385.87	386.03	386.19	386.36	386.52	386.68	1530
1540	386.68	386.85	387.01	387.17	387.33	387.50	387.66	387.82	387.99	388.15	388.31	1540
1550	388.31	388.47	388.64	388.80	388.96	389.13	389.29	389.45	389.61	389.78	389.94	1550
1560	389.94	390.10	390.26									1560

RTD Resistance vs Temperature (°C)
Platinum 100, Alpha = 0.00392

RESISTANCE IN OHMS

°C	0	1	2	3	4	5	6	7	8	9	10	°C
−200	17.14	17.57	18.00	18.43	18.87	19.30	19.73	20.17	20.60	21.03	21.47	−200
−190	21.47	21.90	22.33	22.77	23.20	23.63	24.07	24.50	24.93	25.37	25.80	−190
−180	25.80	26.23	26.66	27.10	27.53	27.96	28.39	28.82	29.26	29.69	30.12	−180
−170	30.12	30.55	30.98	31.41	31.84	32.27	32.70	33.13	33.56	33.99	34.41	−170
−160	34.41	34.84	35.27	35.70	36.12	36.55	36.98	37.40	37.83	38.26	38.68	−160
−150	38.68	39.11	39.53	39.95	40.38	40.80	41.22	41.65	42.07	42.49	42.91	−150
−140	42.91	43.34	43.76	44.18	44.60	45.02	45.44	45.86	46.28	46.70	47.12	−140
−130	47.12	47.53	47.95	48.37	48.79	49.20	49.62	50.04	50.45	50.87	51.29	−130
−120	51.29	51.70	52.12	52.53	52.95	53.36	53.78	54.19	54.61	55.02	55.44	−120
−110	55.44	55.85	56.26	56.68	57.09	57.50	57.92	58.33	58.75	59.16	59.57	−110
−100	59.57	59.98	60.39	60.80	61.22	61.63	62.04	62.45	62.86	63.27	63.68	−100
−90	63.68	64.09	64.50	64.91	65.32	65.73	66.14	66.55	66.95	67.36	67.77	−90
−80	67.77	68.18	68.59	69.00	69.40	69.81	70.22	70.63	71.03	71.44	71.85	−80
−70	71.85	72.26	72.66	73.07	73.48	73.88	74.29	74.69	75.10	75.50	75.91	−70
−60	75.91	76.32	76.72	77.13	77.53	77.94	78.34	78.74	79.15	79.55	79.96	−60
−50	79.96	80.36	80.77	81.17	81.57	81.98	82.38	82.78	83.19	83.59	83.99	−50
−40	83.99	84.39	84.80	85.20	85.60	86.00	86.41	86.81	87.21	87.61	88.01	−40
−30	88.01	88.41	88.81	89.22	89.62	90.02	90.42	90.82	91.22	91.62	92.02	−30
−20	92.02	92.42	92.82	93.22	93.62	94.02	94.42	94.82	95.22	95.62	96.02	−20
−10	96.02	96.42	96.81	97.21	97.61	98.01	98.41	98.81	99.21	99.60	100.00	−10
0	100.00	100.40	100.79	101.19	101.59	101.99	102.38	102.78	103.18	103.57	103.97	0
10	103.97	104.37	104.76	105.16	105.56	105.95	106.35	106.74	107.14	107.54	107.93	10
20	107.93	108.33	108.72	109.12	109.51	109.91	110.30	110.69	111.09	111.48	111.88	20
30	111.88	112.27	112.66	113.06	113.45	113.85	114.24	114.63	115.02	115.42	115.81	30
40	115.81	116.20	116.60	116.99	117.38	117.77	118.16	118.56	118.95	119.34	119.73	40
50	119.73	120.12	120.51	120.91	121.30	121.69	122.08	122.47	122.86	123.25	123.64	50
60	123.64	124.03	124.42	124.81	125.20	125.59	125.98	126.37	126.76	127.15	127.54	60
70	127.54	127.93	128.32	128.70	129.09	129.48	129.87	130.26	130.65	131.04	131.42	70
80	131.42	131.81	132.20	132.59	132.97	133.36	133.75	134.14	134.52	134.91	135.30	80
90	135.30	135.68	136.07	136.46	136.84	137.23	137.62	138.00	138.39	138.77	139.16	90
100	139.16	139.55	139.93	140.32	140.70	141.09	141.47	141.86	142.24	142.63	143.01	100
110	143.01	143.39	143.78	144.16	144.55	144.93	145.31	145.70	146.08	146.46	146.85	110
120	146.85	147.23	147.61	148.00	148.38	148.76	149.14	149.53	149.91	150.29	150.67	120
130	150.67	151.06	151.44	151.82	152.20	152.58	152.96	153.34	153.73	154.11	154.49	130
140	154.49	154.87	155.25	155.63	156.01	156.39	156.77	157.15	157.53	157.91	158.29	140
150	158.29	158.67	159.05	159.43	159.81	160.19	160.57	160.94	161.32	161.70	162.08	150
160	162.08	162.46	162.84	163.22	163.59	163.97	164.35	164.73	165.10	165.48	165.86	160
170	165.86	166.24	166.61	166.99	167.37	167.75	168.12	168.50	168.87	169.25	169.63	170
180	169.63	170.00	170.38	170.76	171.13	171.51	171.88	172.26	172.63	173.01	173.38	180
190	173.38	173.76	174.13	174.51	174.88	175.26	175.63	176.01	176.38	176.75	177.13	190
200	177.13	177.50	177.88	178.25	178.62	179.00	179.37	179.74	180.11	180.49	180.86	200
210	180.86	181.23	181.61	181.98	182.35	182.72	183.09	183.47	183.84	184.21	184.58	210
220	184.58	184.95	185.32	185.70	186.07	186.44	186.81	187.18	187.55	187.92	188.29	220
230	188.29	188.66	189.03	189.40	189.77	190.14	190.51	190.88	191.25	191.62	191.99	230
240	191.99	192.36	192.72	193.09	193.46	193.83	194.20	194.57	194.94	195.30	195.67	240
250	195.67	196.04	196.41	196.78	197.14	197.51	197.88	198.25	198.61	198.98	199.35	250
260	199.35	199.71	200.08	200.45	200.81	201.18	201.55	201.91	202.28	202.64	203.01	260
270	203.01	203.38	203.74	204.11	204.47	204.84	205.20	205.57	205.93	206.30	206.66	270
280	206.66	207.03	207.39	207.75	208.12	208.48	208.85	209.21	209.57	209.94	210.30	280
290	210.30	210.66	211.03	211.39	211.75	212.12	212.48	212.84	213.20	213.57	213.93	290
300	213.93	214.29	214.65	215.01	215.38	215.74	216.10	216.46	216.82	217.18	217.54	300
310	217.54	217.90	218.27	218.63	218.99	219.35	219.71	220.07	220.43	220.79	221.15	310
320	221.15	221.51	221.87	222.23	222.59	222.95	223.30	223.66	224.02	224.38	224.74	320
330	224.74	225.10	225.46	225.82	226.17	226.53	226.89	227.25	227.61	227.96	228.32	330
340	228.32	228.68	229.04	229.39	229.75	230.11	230.46	230.82	231.18	231.53	231.89	340
350	231.89	232.25	232.60	232.96	233.31	233.67	234.03	234.38	234.74	235.09	235.45	350
360	235.45	235.80	236.16	236.51	236.87	237.22	237.58	237.93	238.29	238.64	238.99	360
370	238.99	239.35	239.70	240.06	240.41	240.76	241.12	241.47	241.82	242.18	242.53	370
380	242.53	242.88	243.23	243.59	243.94	244.29	244.64	244.99	245.35	245.70	246.05	380
390	246.05	246.40	246.75	247.10	247.46	247.81	248.16	248.51	248.86	249.21	249.56	390

RTD Resistance vs Temperature (°C)
Platinum 100, Alpha = 0.00392

RESISTANCE IN OHMS

°C	0	1	2	3	4	5	6	7	8	9	10	°C
400	249.56	249.91	250.26	250.61	250.96	251.31	251.66	252.01	252.36	252.71	253.06	400
410	253.06	253.41	253.76	254.11	254.46	254.81	255.15	255.50	255.85	256.20	256.55	410
420	256.55	256.90	257.24	257.59	257.94	258.29	258.63	258.98	259.33	259.68	260.02	420
430	260.02	260.37	260.72	261.06	261.41	261.76	262.10	262.45	262.79	263.14	263.49	430
440	263.49	263.83	264.18	264.52	264.87	265.21	265.56	265.90	266.25	266.59	266.94	440
450	266.94	267.28	267.63	267.97	268.32	268.66	269.00	269.35	269.69	270.03	270.38	450
460	270.38	270.72	271.06	271.41	271.75	272.09	272.44	272.78	273.12	273.46	273.81	460
470	273.81	274.15	274.49	274.83	275.17	275.52	275.86	276.20	276.54	276.88	277.22	470
480	277.22	277.56	277.90	278.25	278.59	278.93	279.27	279.61	279.95	280.29	280.63	480
490	280.63	280.97	281.31	281.65	281.99	282.32	282.66	283.00	283.34	283.68	284.02	490
500	284.02	284.36	284.70	285.04	285.37	285.71	286.05	286.39	286.72	287.06	287.40	500
510	287.40	287.74	288.07	288.41	288.75	289.09	289.42	289.76	290.10	290.43	290.77	510
520	290.77	291.10	291.44	291.78	292.11	292.45	292.78	293.12	293.45	293.79	294.12	520
530	294.12	294.46	294.79	295.13	295.46	295.80	296.13	296.47	296.80	297.13	297.47	530
540	297.47	297.80	298.14	298.47	298.80	299.14	299.47	299.80	300.13	300.47	300.80	540
550	300.80	301.13	301.47	301.80	302.13	302.46	302.79	303.13	303.46	303.79	304.12	550
560	304.12	304.45	304.78	305.11	305.44	305.78	306.11	306.44	306.77	307.10	307.43	560
570	307.43	307.76	308.09	308.42	308.75	309.08	309.41	309.74	310.07	310.39	310.72	570
580	310.72	311.05	311.38	311.71	312.04	312.37	312.70	313.02	313.35	313.68	314.01	580
590	314.01	314.34	314.66	314.99	315.32	315.64	315.97	316.30	316.63	316.95	317.28	590
600	317.28	317.61	317.93	318.26	318.58	318.91	319.24	319.56	319.89	320.21	320.54	600
610	320.54	320.86	321.19	321.51	321.84	322.16	322.49	322.81	323.14	323.46	323.78	610
620	323.78	324.11	324.43	324.76	325.08	325.40	325.73	326.05	326.37	326.70	327.02	620
630	327.02	327.34	327.66	327.99	328.31	328.63	328.95	329.28	329.60	329.92	330.24	630
640	330.24	330.56	330.88	331.20	331.53	331.85	332.17	332.49	332.81	333.13		640

RTD Resistance vs Temperature (°F)
Platinum 100, Alpha = 0.00392

RESISTANCE IN OHMS

°F	0	1	2	3	4	5	6	7	8	9	10	°F
−330			17.14	17.38	17.62	17.86	18.10	18.34	18.58	18.82	19.06	−330
−320	19.06	19.30	19.54	19.78	20.02	20.26	20.50	20.75	20.99	21.23	21.47	−320
−310	21.47	21.71	21.95	22.19	22.43	22.67	22.91	23.15	23.39	23.63	23.87	−310
−300	23.87	24.12	24.36	24.60	24.84	25.08	25.32	25.56	25.80	26.04	26.28	−300
−290	26.28	26.52	26.76	27.00	27.24	27.48	27.72	27.96	28.20	28.44	28.68	−290
−280	28.68	28.92	29.16	29.40	29.64	29.88	30.12	30.36	30.60	30.84	31.07	−280
−270	31.07	31.31	31.55	31.79	32.03	32.27	32.51	32.75	32.99	33.22	33.46	−270
−260	33.46	33.70	33.94	34.18	34.41	34.65	34.89	35.13	35.37	35.60	35.84	−260
−250	35.84	36.08	36.31	36.55	36.79	37.03	37.26	37.50	37.74	37.97	38.21	−250
−240	38.21	38.44	38.68	38.92	39.15	39.39	39.62	39.86	40.10	40.33	40.57	−240
−230	40.57	40.80	41.04	41.27	41.51	41.74	41.98	42.21	42.45	42.68	42.91	−230
−220	42.91	43.15	43.38	43.62	43.85	44.08	44.32	44.55	44.79	45.02	45.25	−220
−210	45.25	45.49	45.72	45.95	46.18	46.42	46.65	46.88	47.12	47.35	47.58	−210
−200	47.58	47.81	48.04	48.28	48.51	48.74	48.97	49.20	49.44	49.67	49.90	−200
−190	49.90	50.13	50.36	50.59	50.82	51.06	51.29	51.52	51.75	51.98	52.21	−190
−180	52.21	52.44	52.67	52.90	53.13	53.36	53.59	53.82	54.05	54.28	54.52	−180
−170	54.52	54.75	54.98	55.21	55.44	55.67	55.90	56.13	56.36	56.59	56.82	−170
−160	56.82	57.05	57.28	57.50	57.73	57.96	58.19	58.42	58.65	58.88	59.11	−160
−150	59.11	59.34	59.57	59.80	60.03	60.26	60.48	60.71	60.94	61.17	61.40	−150
−140	61.40	61.63	61.86	62.08	62.31	62.54	62.77	63.00	63.22	63.45	63.68	−140
−130	63.68	63.91	64.14	64.36	64.59	64.82	65.05	65.27	65.50	65.73	65.96	−130
−120	65.96	66.18	66.41	66.64	66.86	67.09	67.32	67.55	67.77	68.00	68.23	−120
−110	68.23	68.45	68.68	68.91	69.13	69.36	69.59	69.81	70.04	70.27	70.49	−110
−100	70.49	70.72	70.94	71.17	71.40	71.62	71.85	72.07	72.30	72.53	72.75	−100
−90	72.75	72.98	73.20	73.43	73.66	73.88	74.11	74.33	74.56	74.78	75.01	−90
−80	75.01	75.23	75.46	75.68	75.91	76.14	76.36	76.59	76.81	77.04	77.26	−80
−70	77.26	77.49	77.71	77.94	78.16	78.39	78.61	78.83	79.06	79.28	79.51	−70
−60	79.51	79.73	79.96	80.18	80.41	80.63	80.85	81.08	81.30	81.53	81.75	−60
−50	81.75	81.98	82.20	82.42	82.65	82.87	83.10	83.32	83.54	83.77	83.99	−50
−40	83.99	84.21	84.44	84.66	84.89	85.11	85.33	85.56	85.78	86.00	86.23	−40
−30	86.23	86.45	86.67	86.90	87.12	87.34	87.57	87.79	88.01	88.24	88.46	−30
−20	88.46	88.68	88.90	89.13	89.35	89.57	89.80	90.02	90.24	90.46	90.69	−20
−10	90.69	90.91	91.13	91.35	91.58	91.80	92.02	92.24	92.47	92.69	92.91	−10
0	92.91	93.13	93.35	93.58	93.80	94.02	94.24	94.46	94.69	94.91	95.13	0
10	95.13	95.35	95.57	95.80	96.02	96.24	96.46	96.68	96.90	97.12	97.35	10
20	97.35	97.57	97.79	98.01	98.23	98.45	98.67	98.90	99.12	99.34	99.56	20
30	99.56	99.78	100.00	100.22	100.44	100.66	100.88	101.10	101.32	101.55	101.77	30
40	101.77	101.99	102.21	102.43	102.65	102.87	103.09	103.31	103.53	103.75	103.97	40
50	103.97	104.19	104.41	104.63	104.85	105.07	105.29	105.51	105.73	105.95	106.17	50
60	106.17	106.39	106.61	106.83	107.05	107.27	107.49	107.71	107.93	108.15	108.37	60
70	108.37	108.59	108.81	109.03	109.25	109.47	109.69	109.91	110.12	110.34	110.56	70
80	110.56	110.78	111.00	111.22	111.44	111.66	111.88	112.10	112.31	112.53	112.75	80
90	112.75	112.97	113.19	113.41	113.63	113.85	114.06	114.28	114.50	114.72	114.94	90
100	114.94	115.16	115.37	115.59	115.81	116.03	116.25	116.46	116.68	116.90	117.12	100
110	117.12	117.34	117.55	117.77	117.99	118.21	118.43	118.64	118.86	119.08	119.30	110
120	119.30	119.51	119.73	119.95	120.17	120.38	120.60	120.82	121.04	121.25	121.47	120
130	121.47	121.69	121.90	122.12	122.34	122.56	122.77	122.99	123.21	123.42	123.64	130
140	123.64	123.86	124.07	124.29	124.51	124.72	124.94	125.16	125.37	125.59	125.81	140
150	125.81	126.02	126.24	126.46	126.67	126.89	127.11	127.32	127.54	127.75	127.97	150
160	127.97	128.19	128.40	128.62	128.83	129.05	129.27	129.48	129.70	129.91	130.13	160
170	130.13	130.35	130.56	130.78	130.99	131.21	131.42	131.64	131.85	132.07	132.29	170
180	132.29	132.50	132.72	132.93	133.15	133.36	133.58	133.79	134.01	134.22	134.44	180
190	134.44	134.65	134.87	135.08	135.30	135.51	135.73	135.94	136.16	136.37	136.59	190
200	136.59	136.80	137.02	137.23	137.45	137.66	137.87	138.09	138.30	138.52	138.73	200
210	138.73	138.95	139.16	139.37	139.59	139.80	140.02	140.23	140.44	140.66	140.87	210
220	140.87	141.09	141.30	141.51	141.73	141.94	142.16	142.37	142.58	142.80	43.01	220
230	143.01	143.22	143.44	143.65	143.86	144.08	144.29	144.50	144.72	144.93	145.14	230
240	145.14	145.36	145.57	145.78	146.00	146.21	146.42	146.64	146.85	147.06	147.27	240
250	147.27	147.49	147.70	147.91	148.12	148.34	148.55	148.76	148.97	149.19	149.40	250
260	149.40	149.61	149.82	150.04	150.25	150.46	150.67	150.89	151.10	151.31	151.52	260

RTD Resistance vs Temperature (°F)
Platinum 100, Alpha = 0.00392

RESISTANCE IN OHMS

°F	0	1	2	3	4	5	6	7	8	9	10	°F
270	151.52	151.73	151.95	152.16	152.37	152.58	152.79	153.01	153.22	153.43	153.64	270
280	153.64	153.85	154.06	154.28	154.49	154.70	154.91	155.12	155.33	155.55	155.76	280
290	155.76	155.97	156.18	156.39	156.60	156.81	157.02	157.24	157.45	157.66	157.87	290
300	157.87	158.08	158.29	158.50	158.71	158.92	159.13	159.34	159.55	159.77	159.98	300
310	159.98	160.19	160.40	160.61	160.82	161.03	161.24	161.45	161.66	161.87	162.08	310
320	162.08	162.29	162.50	162.71	162.92	163.13	163.34	163.55	163.76	163.97	164.18	320
330	164.18	164.39	164.60	164.81	165.02	165.23	165.44	165.65	165.86	166.07	166.28	330
340	166.28	166.49	166.70	166.91	167.12	167.33	167.54	167.75	167.95	168.16	168.37	340
350	168.37	168.58	168.79	169.00	169.21	169.42	169.63	169.84	170.05	170.25	170.46	350
360	170.46	170.67	170.88	171.09	171.30	171.51	171.72	171.92	172.13	172.34	172.55	360
370	172.55	172.76	172.97	173.18	173.38	173.59	173.80	174.01	174.22	174.42	174.63	370
380	174.63	174.84	175.05	175.26	175.47	175.67	175.88	176.09	176.30	176.50	176.71	380
390	176.71	176.92	177.13	177.34	177.54	177.75	177.96	178.17	178.37	178.58	178.79	390
400	178.79	179.00	179.20	179.41	179.62	179.82	180.03	180.24	180.45	180.65	180.86	400
410	180.86	181.07	181.27	181.48	181.69	181.90	182.10	182.31	182.52	182.72	182.93	410
420	182.93	183.14	183.34	183.55	183.76	183.96	184.17	184.38	184.58	184.79	184.99	420
430	184.99	185.20	185.41	185.61	185.82	186.03	186.23	186.44	186.64	186.85	187.06	430
440	187.06	187.26	187.47	187.67	187.88	188.08	188.29	188.50	188.70	188.91	189.11	440
450	189.11	189.32	189.52	189.73	189.93	190.14	190.35	190.55	190.76	190.96	191.17	450
460	191.17	191.37	191.58	191.78	191.99	192.19	192.40	192.60	192.81	193.01	193.22	460
470	193.22	193.42	193.63	193.83	194.04	194.24	194.44	194.65	194.85	195.06	195.26	470
480	195.26	195.47	195.67	195.88	196.08	196.29	196.49	196.69	196.90	197.10	197.31	480
490	197.31	197.51	197.72	197.92	198.12	198.33	198.53	198.74	198.94	199.14	199.35	490
500	199.35	199.55	199.75	199.96	200.16	200.37	200.57	200.77	200.98	201.18	201.38	500
510	201.38	201.59	201.79	201.99	202.20	202.40	202.60	202.81	203.01	203.21	203.42	510
520	203.42	203.62	203.82	204.03	204.23	204.43	204.63	204.84	205.04	205.24	205.45	520
530	205.45	205.65	205.85	206.05	206.26	206.46	206.66	206.86	207.07	207.27	207.47	530
540	207.47	207.67	207.88	208.08	208.28	208.48	208.68	208.89	209.09	209.29	209.49	540
550	209.49	209.69	209.90	210.10	210.30	210.50	210.70	210.91	211.11	211.31	211.51	550
560	211.51	211.71	211.91	212.12	212.32	212.52	212.72	212.92	213.12	213.32	213.53	560
570	213.53	213.73	213.93	214.13	214.33	214.53	214.73	214.93	215.13	215.34	215.54	570
580	215.54	215.74	215.94	216.14	216.34	216.54	216.74	216.94	217.14	217.34	217.54	580
590	217.54	217.74	217.94	218.15	218.35	218.55	218.75	218.95	219.15	219.35	219.55	590
600	219.55	219.75	219.95	220.15	220.35	220.55	220.75	220.95	221.15	221.35	221.55	600
610	221.55	221.75	221.95	222.15	222.35	222.55	222.75	222.95	223.15	223.34	223.54	610
620	223.54	223.74	223.94	224.14	224.34	224.54	224.74	224.94	225.14	225.34	225.54	620
630	225.54	225.74	225.94	226.13	226.33	226.53	226.73	226.93	227.13	227.33	227.53	630
640	227.53	227.73	227.92	228.12	228.32	228.52	228.72	228.92	229.12	229.31	229.51	640
650	229.51	229.71	229.91	230.11	230.31	230.50	230.70	230.90	231.10	231.30	231.49	650
660	231.49	231.69	231.89	232.09	232.29	232.48	232.68	232.88	233.08	233.28	233.47	660
670	233.47	233.67	233.87	234.07	234.26	234.46	234.66	234.86	235.05	235.25	235.45	670
680	235.45	235.65	235.84	236.04	236.24	236.43	236.63	236.83	237.03	237.22	237.42	680
690	237.42	237.62	237.81	238.01	238.21	238.40	238.60	238.80	238.99	239.19	239.39	690
700	239.39	239.58	239.78	239.98	240.17	240.37	240.57	240.76	240.96	241.15	241.35	700
710	241.35	241.55	241.74	241.94	242.14	242.33	242.53	242.72	242.92	243.12	243.31	710
720	243.31	243.51	243.70	243.90	244.09	244.29	244.49	244.68	244.88	245.07	245.27	720
730	245.27	245.46	245.66	245.85	246.05	246.25	246.44	246.64	246.83	247.03	247.22	730
740	247.22	247.42	247.61	247.81	248.00	248.20	248.39	248.59	248.78	248.98	249.17	740
750	249.17	249.37	249.56	249.76	249.95	250.15	250.34	250.53	250.73	250.92	251.12	750
760	251.12	251.31	251.51	251.70	251.89	252.09	252.28	252.48	252.67	252.87	253.06	760
770	253.06	253.25	253.45	253.64	253.84	254.03	254.22	254.42	254.61	254.81	255.00	770
780	255.00	255.19	255.39	255.58	255.77	255.97	256.16	256.35	256.55	256.74	256.93	780
790	256.93	257.13	257.32	257.51	257.71	257.90	258.09	258.29	258.48	258.67	258.87	790
800	258.87	259.06	259.25	259.44	259.64	259.83	260.02	260.22	260.41	260.60	260.79	800
810	260.79	260.99	261.18	261.37	261.56	261.76	261.95	262.14	262.33	262.53	262.72	810
820	262.72	262.91	263.10	263.29	263.49	263.68	263.87	264.06	264.25	264.45	264.64	820
830	264.64	264.83	265.02	265.21	265.41	265.60	265.79	265.98	266.17	266.36	266.56	830
840	266.56	266.75	266.94	267.13	267.32	267.51	267.70	267.89	268.09	268.28	268.47	840
850	268.47	268.66	268.85	269.04	269.23	269.42	269.61	269.81	270.00	270.19	270.38	850
860	270.38	270.57	270.76	270.95	271.14	271.33	271.52	271.71	271.90	272.09	272.28	860

RTD Resistance vs Temperature (°F)
Platinum 100, Alpha = 0.00392

RESISTANCE IN OHMS

°F	0	1	2	3	4	5	6	7	8	9	10	°F
870	272.28	272.47	272.67	272.86	273.05	273.24	273.43	273.62	273.81	274.00	274.19	870
880	274.19	274.38	274.57	274.76	274.95	275.14	275.33	275.52	275.71	275.90	276.09	880
890	276.09	276.27	276.46	276.65	276.84	277.03	277.22	277.41	277.60	277.79	277.98	890
900	277.98	278.17	278.36	278.55	278.74	278.93	279.12	279.30	279.49	279.68	279.87	900
910	279.87	280.06	280.25	280.44	280.63	280.82	281.00	281.19	281.38	281.57	281.76	910
920	281.76	281.95	282.14	282.32	282.51	282.70	282.89	283.08	283.27	283.45	283.64	920
930	283.64	283.83	284.02	284.21	284.40	284.58	284.77	284.96	285.15	285.34	285.52	930
940	285.52	285.71	285.90	286.09	286.27	286.46	286.65	286.84	287.02	287.21	287.40	940
950	287.40	287.59	287.77	287.96	288.15	288.34	288.52	288.71	288.90	289.09	289.27	950
960	289.27	289.46	289.65	289.83	290.02	290.21	290.39	290.58	290.77	290.95	291.14	960
970	291.14	291.33	291.51	291.70	291.89	292.07	292.26	292.45	292.63	292.82	293.01	970
980	293.01	293.19	293.38	293.57	293.75	293.94	294.12	294.31	294.50	294.68	294.87	980
990	294.87	295.05	295.24	295.43	295.61	295.80	295.98	296.17	296.35	296.54	296.73	990
1000	296.73	296.91	297.10	297.28	297.47	297.65	297.84	298.02	298.21	298.39	298.58	1000
1010	298.58	298.77	298.95	299.14	299.32	299.51	299.69	299.88	300.06	300.25	300.43	1010
1020	300.43	300.62	300.80	300.98	301.17	301.35	301.54	301.72	301.91	302.09	302.28	1020
1030	302.28	302.46	302.65	302.83	303.01	303.20	303.38	303.57	303.75	303.94	304.12	1030
1040	304.12	304.30	304.49	304.67	304.86	305.04	305.22	305.41	305.59	305.78	305.96	1040
1050	305.96	306.14	306.33	306.51	306.69	306.88	307.06	307.24	307.43	307.61	307.79	1050
1060	307.79	307.98	308.16	308.34	308.53	308.71	308.89	309.08	309.26	309.44	309.63	1060
1070	309.63	309.81	309.99	310.18	310.36	310.54	310.72	310.91	311.09	311.27	311.45	1070
1080	311.45	311.64	311.82	312.00	312.18	312.37	312.55	312.73	312.91	313.10	313.28	1080
1090	313.28	313.46	313.64	313.83	314.01	314.19	314.37	314.55	314.74	314.92	315.10	1090
1100	315.10	315.28	315.46	315.64	315.83	316.01	316.19	316.37	316.55	316.73	316.92	1100
1110	316.92	317.10	317.28	317.46	317.64	317.82	318.00	318.19	318.37	318.55	318.73	1110
1120	318.73	318.91	319.09	319.27	319.45	319.63	319.81	320.00	320.18	320.36	320.54	1120
1130	320.54	320.72	320.90	321.08	321.26	321.44	321.62	321.80	321.98	322.16	322.34	1130
1140	322.34	322.52	322.70	322.88	323.06	323.24	323.42	323.60	323.78	323.96	324.14	1140
1150	324.14	324.32	324.50	324.68	324.86	325.04	325.22	325.40	325.58	325.76	325.94	1150
1160	325.94	326.12	326.30	326.48	326.66	326.84	327.02	327.20	327.38	327.56	327.74	1160
1170	327.74	327.91	328.09	328.27	328.45	328.63	328.81	328.99	329.17	329.35	329.53	1170
1180	329.53	329.70	329.88	330.06	330.24	330.42	330.60	330.78	330.95	331.13	331.31	1180
1190	331.31	331.49	331.67	331.85	332.02	332.20	332.38	332.56	332.74	332.92	333.09	1190
1200	333.09											1200

RTD Resistance vs Temperature (°C)
Platinum 200, Alpha = 0.00385

RESISTANCE IN OHMS

°C	0	1	2	3	4	5	6	7	8	9	10	°C
−200	36.99	37.85	38.72	39.58	40.44	41.31	42.17	43.03	43.89	44.75	45.61	−200
−190	45.61	46.46	47.32	48.18	49.03	49.89	50.74	51.60	52.45	53.30	54.16	−190
−180	54.16	55.01	55.86	56.71	57.56	58.41	59.25	60.10	60.95	61.79	62.64	−180
−170	62.64	63.48	64.33	65.17	66.02	66.86	67.70	68.54	69.38	70.22	71.06	−170
−160	71.06	71.90	72.74	73.58	74.42	75.25	76.09	76.92	77.76	78.59	79.43	−160
−150	79.43	80.26	81.09	81.93	82.76	83.59	84.42	85.25	86.08	86.91	87.74	−150
−140	87.74	88.57	89.39	90.22	91.05	91.87	92.70	93.53	94.35	95.17	96.00	−140
−130	96.00	96.82	97.64	98.47	99.29	100.11	100.93	101.75	102.57	103.39	104.21	−130
−120	104.21	105.03	105.85	106.67	107.48	108.30	109.12	109.93	110.75	111.57	112.38	−120
−110	112.38	113.20	114.01	114.82	115.64	116.45	117.26	118.07	118.89	119.70	120.51	−110
−100	120.51	121.32	122.13	122.94	123.75	124.56	125.37	126.17	126.98	127.79	128.60	−100
−90	128.60	129.40	130.21	131.02	131.82	132.63	133.43	134.24	135.04	135.85	136.65	−90
−80	136.65	137.45	138.26	139.06	139.86	140.66	141.47	142.27	143.07	143.87	144.67	−80
−70	144.67	145.47	146.27	147.07	147.87	148.67	149.47	150.26	151.06	151.86	152.66	−70
−60	152.66	153.45	154.25	155.05	155.84	156.64	157.43	158.23	159.02	159.82	160.61	−60
−50	160.61	161.41	162.20	163.00	163.79	164.58	165.37	166.17	166.96	167.75	168.54	−50
−40	168.54	169.33	170.13	170.92	171.71	172.50	173.29	174.08	174.87	175.66	176.44	−40
−30	176.44	177.23	178.02	178.81	179.60	180.39	181.17	181.96	182.75	183.53	184.32	−30
−20	184.32	185.11	185.89	186.68	187.46	188.25	189.03	189.82	190.60	191.39	192.17	−20
−10	192.17	192.96	193.74	194.52	195.31	196.09	196.87	197.65	198.44	199.22	200.00	−10
0	200.00	200.78	201.56	202.34	203.12	203.91	204.69	205.47	206.25	207.03	207.80	0
10	207.80	208.58	209.36	210.14	210.92	211.70	212.48	213.25	214.03	214.81	215.59	10
20	215.59	216.36	217.14	217.92	218.69	219.47	220.24	221.02	221.79	222.57	223.34	20
30	223.34	224.12	224.89	225.67	226.44	227.21	227.99	228.76	229.53	230.31	231.08	30
40	231.08	231.85	232.62	233.39	234.17	234.94	235.71	236.48	237.25	238.02	238.79	40
50	238.79	239.56	240.33	241.10	241.87	242.64	243.41	244.17	244.94	245.71	246.48	50
60	246.48	247.25	248.01	248.78	249.55	250.31	251.08	251.85	252.61	253.38	254.14	60
70	254.14	254.91	255.67	256.44	257.20	257.97	258.73	259.50	260.26	261.02	261.79	70
80	261.79	262.55	263.31	264.07	264.84	265.60	266.36	267.12	267.88	268.64	269.40	80
90	269.40	270.17	270.93	271.69	272.45	273.21	273.96	274.72	275.48	276.24	277.00	90
100	277.00	277.76	278.52	279.27	280.03	280.79	281.55	282.30	283.06	283.82	284.57	100
110	284.57	285.33	286.08	286.84	287.59	288.35	289.10	289.86	290.61	291.37	292.12	110
120	292.12	292.88	293.63	294.38	295.13	295.89	296.64	297.39	298.14	298.90	299.65	120
130	299.65	300.40	301.15	301.90	302.65	303.40	304.15	304.90	305.65	306.40	307.15	130
140	307.15	307.90	308.65	309.40	310.14	310.89	311.64	312.39	313.14	313.88	314.63	140
150	314.63	315.38	316.12	316.87	317.62	318.36	319.11	319.85	320.60	321.34	322.09	150
160	322.09	322.83	323.57	324.32	325.06	325.81	326.55	327.29	328.03	328.78	329.52	160
170	329.52	330.26	331.00	331.74	332.49	333.23	333.97	334.71	335.45	336.19	336.93	170
180	336.93	337.67	338.41	339.15	339.89	340.63	341.36	342.10	342.84	343.58	344.32	180
190	344.32	345.05	345.79	346.53	347.26	348.00	348.74	349.47	350.21	350.94	351.68	190
200	351.68	352.41	353.15	353.88	354.62	355.35	356.09	356.82	357.55	358.29	359.02	200
210	359.02	359.75	360.48	361.22	361.95	362.68	363.41	364.14	364.87	365.61	366.34	210
220	366.34	367.07	367.80	368.53	369.26	369.99	370.72	371.44	372.17	372.90	373.63	220
230	373.63	374.36	375.09	375.81	376.54	377.27	378.00	378.72	379.45	380.18	380.90	230
240	380.90	381.63	382.35	383.08	383.80	384.53	385.25	385.98	386.70	387.42	388.15	240
250	388.15	388.87	389.60	390.32	391.04	391.76	392.49	393.21	393.93	394.65	395.37	250
260	395.37	396.09	396.81	397.54	398.26	398.98	399.70	400.42	401.14	401.85	402.57	260
270	402.57	403.29	404.01	404.73	405.45	406.17	406.88	407.60	408.32	409.03	409.75	270
280	409.75	410.47	411.18	411.90	412.62	413.33	414.05	414.76	415.48	416.19	416.91	280
290	416.91	417.62	418.33	419.05	419.76	420.47	421.19	421.90	422.61	423.33	424.04	290
300	424.04	424.75	425.46	426.17	426.88	427.59	428.31	429.02	429.73	430.44	431.15	300
310	431.15	431.86	432.56	433.27	433.98	434.69	435.40	436.11	436.82	437.52	438.23	310
320	438.23	438.94	439.65	440.35	441.06	441.76	442.47	443.18	443.88	444.59	445.29	320
330	445.29	446.00	446.70	447.41	448.11	448.81	449.52	450.22	450.93	451.63	452.33	330
340	452.33	453.03	453.74	454.44	455.14	455.84	456.54	457.24	457.95	458.65	459.35	340
350	459.35	460.05	460.75	461.45	462.15	462.85	463.54	464.24	464.94	465.64	466.34	350
360	466.34	467.04	467.73	468.43	469.13	469.83	470.52	471.22	471.92	472.61	473.31	360
370	473.31	474.00	474.70	475.39	476.09	476.78	477.48	478.17	478.87	479.56	480.25	370
380	480.25	480.95	481.64	482.33	483.03	483.72	484.41	485.10	485.79	486.48	487.18	380
390	487.18	487.87	488.56	489.25	489.94	490.63	491.32	492.01	492.70	493.39	494.08	390

Appendix B

RTD Resistance vs Temperature (°C)
Platinum 200, Alpha = 0.00385

RESISTANCE IN OHMS

°C	0	1	2	3	4	5	6	7	8	9	10	°C
400	494.08	494.76	495.45	496.14	496.83	497.52	498.20	498.89	499.58	500.26	500.95	400
410	500.95	501.64	502.32	503.01	503.70	504.38	505.07	505.75	506.44	507.12	507.80	410
420	507.80	508.49	509.17	509.86	510.54	511.22	511.90	512.59	513.27	513.95	514.63	420
430	514.63	515.32	516.00	516.68	517.36	518.04	518.72	519.40	520.08	520.76	521.44	430
440	521.44	522.12	522.80	523.48	524.16	524.83	525.51	526.19	526.87	527.55	528.22	440
450	528.22	528.90	529.58	530.25	530.93	531.61	532.28	532.96	533.63	534.31	534.98	450
460	534.98	535.66	536.33	537.01	537.68	538.36	539.03	539.70	540.38	541.05	541.72	460
470	541.72	542.39	543.07	543.74	544.41	545.08	545.75	546.42	547.09	547.76	548.43	470
480	548.43	549.10	549.77	550.44	551.11	551.78	552.45	553.12	553.79	554.46	555.12	480
490	555.12	555.79	556.46	557.13	557.79	558.46	559.13	559.79	560.46	561.13	561.79	490
500	561.79	562.46	563.12	563.79	564.45	565.12	565.78	566.45	567.11	567.77	568.44	500
510	568.44	569.10	569.76	570.42	571.09	571.75	572.41	573.07	573.73	574.40	575.06	510
520	575.06	575.72	576.38	577.04	577.70	578.36	579.02	579.68	580.34	581.00	581.65	520
530	581.65	582.31	582.97	583.63	584.29	584.94	585.60	586.26	586.92	587.57	588.23	530
540	588.23	588.89	589.54	590.20	590.85	591.51	592.16	592.82	593.47	594.13	594.78	540
550	594.78	595.43	596.09	596.74	597.39	598.05	598.70	599.35	600.00	600.66	601.31	550
560	601.31	601.96	602.61	603.26	603.91	604.56	605.21	605.86	606.51	607.16	607.81	560
570	607.81	608.46	609.11	609.76	610.41	611.06	611.70	612.35	613.00	613.65	614.29	570
580	614.29	614.94	615.59	616.23	616.88	617.53	618.17	618.82	619.46	620.11	620.75	580
590	620.75	621.40	622.04	622.69	623.33	623.97	624.62	625.26	625.90	626.55	627.19	590
600	627.19	627.83	628.47	629.11	629.76	630.40	631.04	631.68	632.32	632.96	633.60	600
610	633.60	634.24	634.88	635.52	636.16	636.80	637.44	638.07	638.71	639.35	639.99	610
620	639.99	640.63	641.26	641.90	642.54	643.17	643.81	644.45	645.08	645.72	646.35	620
630	646.35	646.99	647.62	648.26	648.89	649.53	650.16	650.80	651.43	652.06	652.70	630
640	652.70	653.33	653.96	654.59	655.23	655.86	656.49	657.12	657.75	658.38	659.02	640
650	659.02	659.65	660.28	660.91	661.54	662.17	662.80	663.43	664.05	664.68	665.31	650
660	665.31	665.94	666.57	667.20	667.82	668.45	669.08	669.71	670.33	670.96	671.58	660
670	671.58	672.21	672.84	673.46	674.09	674.71	675.34	675.96	676.59	677.21	677.83	670
680	677.83	678.46	679.08	679.70	680.33	680.95	681.57	682.19	682.82	683.44	684.06	680
690	684.06	684.68	685.30	685.92	686.54	687.16	687.78	688.40	689.02	689.64	690.26	690
700	690.26	690.88	691.50	692.12	692.74	693.36	693.97	694.59	695.21	695.83	696.44	700
710	696.44	697.06	697.68	698.29	698.91	699.52	700.14	700.76	701.37	701.99	702.60	710
720	702.60	703.21	703.83	704.44	705.06	705.67	706.28	706.90	707.51	708.12	708.73	720
730	708.73	709.35	709.96	710.57	711.18	711.79	712.40	713.01	713.62	714.23	714.84	730
740	714.84	715.45	716.06	716.67	717.28	717.89	718.50	719.11	719.72	720.32	720.93	740
750	720.93	721.54	722.15	722.75	723.36	723.97	724.57	725.18	725.78	726.39	726.99	750
760	726.99	727.60	728.20	728.81	729.41	730.02	730.62	731.23	731.83	732.43	733.04	760
770	733.04	733.64	734.24	734.84	735.44	736.05	736.65	737.25	737.85	738.45	739.05	770
780	739.05	739.65	740.25	740.85	741.45	742.05	742.65	743.25	743.85	744.45	745.05	780
790	745.05	745.64	746.24	746.84	747.44	748.04	748.63	749.23	749.83	750.42	751.02	790
800	751.02	751.61	752.21	752.80	753.40	753.99	754.59	755.18	755.78	756.37	756.97	800
810	756.97	757.56	758.15	758.75	759.34	759.93	760.52	761.12	761.71	762.30	762.89	810
820	762.89	763.48	764.07	764.66	765.25	765.84	766.43	767.02	767.61	768.20	768.79	820
830	768.79	769.38	769.97	770.56	771.15	771.73	772.32	772.91	773.50	774.08	774.67	830
840	774.67	775.26	775.84	776.43	777.01	777.60	778.19	778.77	779.36	779.94	780.52	840

RTD Resistance vs Temperature (°F)
Platinum 200, Alpha = 0.00385

RESISTANCE IN OHMS

°F	0	1	2	3	4	5	6	7	8	9	10	°F
-330	36.02	36.51	36.99	37.47	37.95	38.43	38.91	39.39	39.87	40.35	40.83	-330
-320	40.83	41.31	41.78	42.26	42.74	43.22	43.70	44.17	44.65	45.13	45.61	-320
-310	45.61	46.08	46.56	47.04	47.51	47.99	48.46	48.94	49.41	49.89	50.36	-310
-300	50.36	50.84	51.31	51.79	52.26	52.74	53.21	53.68	54.16	54.63	55.10	-300
-290	55.10	55.57	56.05	56.52	56.99	57.46	57.93	58.41	58.88	59.35	59.82	-290
-280	59.82	60.29	60.76	61.23	61.70	62.17	62.64	63.11	63.58	64.05	64.52	-280
-270	64.52	64.99	65.45	65.92	66.39	66.86	67.33	67.79	68.26	68.73	69.20	-270
-260	69.20	69.66	70.13	70.60	71.06	71.53	71.99	72.46	72.93	73.39	73.86	-260
-250	73.86	74.32	74.79	75.25	75.72	76.18	76.65	77.11	77.57	78.04	78.50	-250
-240	78.50	78.96	79.43	79.89	80.35	80.82	81.28	81.74	82.20	82.67	83.13	-240
-230	83.13	83.59	84.05	84.51	84.97	85.43	85.90	86.36	86.82	87.28	87.74	-230
-220	87.74	88.20	88.66	89.12	89.58	90.04	90.50	90.96	91.42	91.87	92.33	-220
-210	92.33	92.79	93.25	93.71	94.17	94.63	95.08	95.54	96.00	96.46	96.91	-210
-200	96.91	97.37	97.83	98.28	98.74	99.20	99.65	100.11	100.57	101.02	101.48	-200
-190	101.48	101.93	102.39	102.85	103.30	103.76	104.21	104.67	105.12	105.58	106.03	-190
-180	106.03	106.49	106.94	107.39	107.85	108.30	108.76	109.21	109.66	110.12	110.57	-180
-170	110.57	111.02	111.48	111.93	112.38	112.83	113.29	113.74	114.19	114.64	115.09	-170
-160	115.09	115.55	116.00	116.45	116.90	117.35	117.80	118.25	118.71	119.16	119.61	-160
-150	119.61	120.06	120.51	120.96	121.41	121.86	122.31	122.76	123.21	123.66	124.11	-150
-140	124.11	124.56	125.01	125.46	125.91	126.35	126.80	127.25	127.70	128.15	128.60	-140
-130	128.60	129.05	129.49	129.94	130.39	130.84	131.29	131.73	132.18	132.63	133.08	-130
-120	133.08	133.52	133.97	134.42	134.86	135.31	135.76	136.20	136.65	137.10	137.54	-120
-110	137.54	137.99	138.44	138.88	139.33	139.77	140.22	140.66	141.11	141.55	142.00	-110
-100	142.00	142.45	142.89	143.34	143.78	144.22	144.67	145.11	145.56	146.00	146.45	-100
-90	146.45	146.89	147.34	147.78	148.22	148.67	149.11	149.55	150.00	150.44	150.88	-90
-80	150.88	151.33	151.77	152.21	152.66	153.10	153.54	153.98	154.43	154.87	155.31	-80
-70	155.31	155.75	156.20	156.64	157.08	157.52	157.96	158.41	158.85	159.29	159.73	-70
-60	159.73	160.17	160.61	161.05	161.50	161.94	162.38	162.82	163.26	163.70	164.14	-60
-50	164.14	164.58	165.02	165.46	165.90	166.34	166.78	167.22	167.66	168.10	168.54	-50
-40	168.54	168.98	169.42	169.86	170.30	170.74	171.18	171.62	172.06	172.50	172.94	-40
-30	172.94	173.37	173.81	174.25	174.69	175.13	175.57	176.01	176.44	176.88	177.32	-30
-20	177.32	177.76	178.20	178.63	179.07	179.51	179.95	180.39	180.82	181.26	181.70	-20
-10	181.70	182.14	182.57	183.01	183.45	183.88	184.32	184.76	185.19	185.63	186.07	-10
0	186.07	186.50	186.94	187.38	187.81	188.25	188.69	189.12	189.56	189.99	190.43	0
10	190.43	190.87	191.30	191.74	192.17	192.61	193.04	193.48	193.91	194.35	194.78	10
20	194.78	195.22	195.65	196.09	196.52	196.96	197.39	197.83	198.26	198.70	199.13	20
30	199.13	199.57	200.00	200.43	200.87	201.30	201.74	202.17	202.60	203.04	203.47	30
40	203.47	203.91	204.34	204.77	205.21	205.64	206.07	206.51	206.94	207.37	207.80	40
50	207.80	208.24	208.67	209.10	209.54	209.97	210.40	210.83	211.27	211.70	212.13	50
60	212.13	212.56	212.99	213.43	213.86	214.29	214.72	215.15	215.59	216.02	216.45	60
70	216.45	216.88	217.31	217.74	218.17	218.61	219.04	219.47	219.90	220.33	220.76	70
80	220.76	221.19	221.62	222.05	222.48	222.91	223.34	223.77	224.20	224.63	225.06	80
90	225.06	225.49	225.92	226.35	226.78	227.21	227.64	228.07	228.50	228.93	229.36	90
100	229.36	229.79	230.22	230.65	231.08	231.51	231.94	232.37	232.79	233.22	233.65	100
110	233.65	234.08	234.51	234.94	235.37	235.79	236.22	236.65	237.08	237.51	237.93	110
120	237.93	238.36	238.79	239.22	239.65	240.07	240.50	240.93	241.36	241.78	242.21	120
130	242.21	242.64	243.06	243.49	243.92	244.35	244.77	245.20	245.63	246.05	246.48	130
140	246.48	246.90	247.33	247.76	248.18	248.61	249.04	249.46	249.89	250.31	250.74	140
150	250.74	251.17	251.59	252.02	252.44	252.87	253.29	253.72	254.14	254.57	254.99	150
160	254.99	255.42	255.84	256.27	256.69	257.12	257.54	257.97	258.39	258.82	259.24	160
170	259.24	259.67	260.09	260.51	260.94	261.36	261.79	262.21	262.63	263.06	263.48	170
180	263.48	263.90	264.33	264.75	265.17	265.60	266.02	266.44	266.87	267.29	267.71	180
190	267.71	268.14	268.56	268.98	269.40	269.83	270.25	270.67	271.09	271.52	271.94	190
200	271.94	272.36	272.78	273.21	273.63	274.05	274.47	274.89	275.31	275.74	276.16	200
210	276.16	276.58	277.00	277.42	277.84	278.26	278.68	279.11	279.53	279.95	280.37	210
220	280.37	280.79	281.21	281.63	282.05	282.47	282.89	283.31	283.73	284.15	284.57	220
230	284.57	284.99	285.41	285.83	286.25	286.67	287.09	287.51	287.93	288.35	288.77	230
240	288.77	289.19	289.61	290.03	290.45	290.86	291.28	291.70	292.12	292.54	292.96	240
250	292.96	293.38	293.80	294.21	294.63	295.05	295.47	295.89	296.31	296.72	297.14	250
260	297.14	297.56	297.98	298.39	298.81	299.23	299.65	300.06	300.48	300.90	301.32	260

Appendix B

RTD Resistance vs Temperature (°F)
Platinum 200, Alpha = 0.00385

RESISTANCE IN OHMS

°F	0	1	2	3	4	5	6	7	8	9	10	°F
270	301.32	301.73	302.15	302.57	302.98	303.40	303.82	304.24	304.65	305.07	305.48	270
280	305.48	305.90	306.32	306.73	307.15	307.57	307.98	308.40	308.81	309.23	309.65	280
290	309.65	310.06	310.48	310.89	311.31	311.72	312.14	312.55	312.97	313.38	313.80	290
300	313.80	314.21	314.63	315.04	315.46	315.87	316.29	316.70	317.12	317.53	317.95	300
310	317.95	318.36	318.77	319.19	319.60	320.02	320.43	320.84	321.26	321.67	322.09	310
320	322.09	322.50	322.91	323.33	323.74	324.15	324.57	324.98	325.39	325.81	326.22	320
330	326.22	326.63	327.04	327.46	327.87	328.28	328.69	329.11	329.52	329.93	330.34	330
340	330.34	330.76	331.17	331.58	331.99	332.40	332.82	333.23	333.64	334.05	334.46	340
350	334.46	334.87	335.28	335.70	336.11	336.52	336.93	337.34	337.75	338.16	338.57	350
360	338.57	338.98	339.39	339.80	340.21	340.63	341.04	341.45	341.86	342.27	342.68	360
370	342.68	343.09	343.50	343.91	344.32	344.73	345.14	345.54	345.95	346.36	346.77	370
380	346.77	347.18	347.59	348.00	348.41	348.82	349.23	349.64	350.04	350.45	350.86	380
390	350.86	351.27	351.68	352.09	352.50	352.90	353.31	353.72	354.13	354.54	354.94	390
400	354.94	355.35	355.76	356.17	356.58	356.98	357.39	357.80	358.21	358.61	359.02	400
410	359.02	359.43	359.83	360.24	360.65	361.05	361.46	361.87	362.27	362.68	363.09	410
420	363.09	363.49	363.90	364.31	364.71	365.12	365.52	365.93	366.34	366.74	367.15	420
430	367.15	367.55	367.96	368.36	368.77	369.18	369.58	369.99	370.39	370.80	371.20	430
440	371.20	371.61	372.01	372.42	372.82	373.23	373.63	374.03	374.44	374.84	375.25	440
450	375.25	375.65	376.06	376.46	376.86	377.27	377.67	378.08	378.48	378.88	379.29	450
460	379.29	379.69	380.09	380.50	380.90	381.30	381.71	382.11	382.51	382.92	383.32	460
470	383.32	383.72	384.13	384.53	384.93	385.33	385.74	386.14	386.54	386.94	387.34	470
480	387.34	387.75	388.15	388.55	388.95	389.35	389.76	390.16	390.56	390.96	391.36	480
490	391.36	391.76	392.16	392.57	392.97	393.37	393.77	394.17	394.57	394.97	395.37	490
500	395.37	395.77	396.17	396.57	396.97	397.38	397.78	398.18	398.58	398.98	399.38	500
510	399.38	399.78	400.18	400.58	400.98	401.38	401.77	402.17	402.57	402.97	403.37	510
520	403.37	403.77	404.17	404.57	404.97	405.37	405.77	406.17	406.56	406.96	407.36	520
530	407.36	407.76	408.16	408.56	408.96	409.35	409.75	410.15	410.55	410.95	411.34	530
540	411.34	411.74	412.14	412.54	412.93	413.33	413.73	414.13	414.52	414.92	415.32	540
550	415.32	415.72	416.11	416.51	416.91	417.30	417.70	418.10	418.49	418.89	419.29	550
560	419.29	419.68	420.08	420.47	420.87	421.27	421.66	422.06	422.45	422.85	423.25	560
570	423.25	423.64	424.04	424.43	424.83	425.22	425.62	426.01	426.41	426.80	427.20	570
580	427.20	427.59	427.99	428.38	428.78	429.17	429.57	429.96	430.36	430.75	431.15	580
590	431.15	431.54	431.93	432.33	432.72	433.12	433.51	433.90	434.30	434.69	435.08	590
600	435.08	435.48	435.87	436.27	436.66	437.05	437.44	437.84	438.23	438.62	439.02	600
610	439.02	439.41	439.80	440.19	440.59	440.98	441.37	441.76	442.16	442.55	442.94	610
620	442.94	443.33	443.73	444.12	444.51	444.90	445.29	445.68	446.08	446.47	446.86	620
630	446.86	447.25	447.64	448.03	448.42	448.81	449.21	449.60	449.99	450.38	450.77	630
640	450.77	451.16	451.55	451.94	452.33	452.72	453.11	453.50	453.89	454.28	454.67	640
650	454.67	455.06	455.45	455.84	456.23	456.62	457.01	457.40	457.79	458.18	458.57	650
660	458.57	458.96	459.35	459.74	460.12	460.51	460.90	461.29	461.68	462.07	462.46	660
670	462.46	462.85	463.23	463.62	464.01	464.40	464.79	465.17	465.56	465.95	466.34	670
680	466.34	466.73	467.11	467.50	467.89	468.28	468.66	469.05	469.44	469.83	470.21	680
690	470.21	470.60	470.99	471.37	471.76	472.15	472.53	472.92	473.31	473.69	474.08	690
700	474.08	474.47	474.85	475.24	475.63	476.01	476.40	476.78	477.17	477.56	477.94	700
710	477.94	478.33	478.71	479.10	479.48	479.87	480.25	480.64	481.02	481.41	481.79	710
720	481.79	482.18	482.56	482.95	483.33	483.72	484.10	484.49	484.87	485.26	485.64	720
730	485.64	486.02	486.41	486.79	487.18	487.56	487.94	488.33	488.71	489.09	489.48	730
740	489.48	489.86	490.25	490.63	491.01	491.39	491.78	492.16	492.54	492.93	493.31	740
750	493.31	493.69	494.08	494.46	494.84	495.22	495.61	495.99	496.37	496.75	497.13	750
760	497.13	497.52	497.90	498.28	498.66	499.04	499.43	499.81	500.19	500.57	500.95	760
770	500.95	501.33	501.71	502.10	502.48	502.86	503.24	503.62	504.00	504.38	504.76	770
780	504.76	505.14	505.52	505.90	506.28	506.66	507.04	507.42	507.80	508.18	508.56	780
790	508.56	508.94	509.32	509.70	510.08	510.46	510.84	511.22	511.60	511.98	512.36	790
800	512.36	512.74	513.12	513.50	513.88	514.26	514.63	515.01	515.39	515.77	516.15	800
810	516.15	516.53	516.91	517.28	517.66	518.04	518.42	518.80	519.17	519.55	519.93	810
820	519.93	520.31	520.69	521.06	521.44	521.82	522.20	522.57	522.95	523.33	523.70	820
830	523.70	524.08	524.46	524.83	525.21	525.59	525.97	526.34	526.72	527.09	527.47	830
840	527.47	527.85	528.22	528.60	528.98	529.35	529.73	530.10	530.48	530.86	531.23	840
850	531.23	531.61	531.98	532.36	532.73	533.11	533.48	533.86	534.23	534.61	534.98	850
860	534.98	535.36	535.73	536.11	536.48	536.86	537.23	537.61	537.98	538.36	538.73	860
870	538.73	539.10	539.48	539.85	540.23	540.60	540.97	541.35	541.72	542.09	542.47	870
880	542.47	542.84	543.21	543.59	543.96	544.33	544.71	545.08	545.45	545.83	546.20	880
890	546.20	546.57	546.94	547.32	547.69	548.06	548.43	548.81	549.18	549.55	549.92	890
900	549.92	550.30	550.67	551.04	551.41	551.78	552.15	552.53	552.90	553.27	553.64	900
910	553.64	554.01	554.38	554.75	555.12	555.50	555.87	556.24	556.61	556.98	557.35	910

RTD Resistance vs Temperature (°F)
Platinum 200, Alpha = 0.00385

RESISTANCE IN OHMS

°F	0	1	2	3	4	5	6	7	8	9	10	°F
920	557.35	557.72	558.09	558.46	558.83	559.20	559.57	559.94	560.31	560.68	561.05	920
930	561.05	561.42	561.79	562.16	562.53	562.90	563.27	563.64	564.01	564.38	564.75	930
940	564.75	565.12	565.49	565.86	566.22	566.59	566.96	567.33	567.70	568.07	568.44	940
950	568.44	568.80	569.17	569.54	569.91	570.28	570.65	571.01	571.38	571.75	572.12	950
960	572.12	572.48	572.85	573.22	573.59	573.96	574.32	574.69	575.06	575.42	575.79	960
970	575.79	576.16	576.53	576.89	577.26	577.63	577.99	578.36	578.73	579.09	579.46	970
980	579.46	579.82	580.19	580.56	580.92	581.29	581.65	582.02	582.39	582.75	583.12	980
990	583.12	583.48	583.85	584.21	584.58	584.94	585.31	585.67	586.04	586.41	586.77	990
1000	586.77	587.13	587.50	587.86	588.23	588.59	588.96	589.32	589.69	590.05	590.42	1000
1010	590.42	590.78	591.14	591.51	591.87	592.24	592.60	592.96	593.33	593.69	594.05	1010
1020	594.05	594.42	594.78	595.14	595.51	595.87	596.23	596.60	596.96	597.32	597.68	1020
1030	597.68	598.05	598.41	598.77	599.13	599.50	599.86	600.22	600.58	600.95	601.31	1030
1040	601.31	601.67	602.03	602.39	602.76	603.12	603.48	603.84	604.20	604.56	604.92	1040
1050	604.92	605.29	605.65	606.01	606.37	606.73	607.09	607.45	607.81	608.17	608.53	1050
1060	608.53	608.89	609.26	609.62	609.98	610.34	610.70	611.06	611.42	611.78	612.14	1060
1070	612.14	612.50	612.86	613.22	613.58	613.94	614.29	614.65	615.01	615.37	615.73	1070
1080	615.73	616.09	616.45	616.81	617.17	617.53	617.89	618.24	618.60	618.96	619.32	1080
1090	619.32	619.68	620.04	620.39	620.75	621.11	621.47	621.83	622.18	622.54	622.90	1090
1100	622.90	623.26	623.62	623.97	624.33	624.69	625.05	625.40	625.76	626.12	626.47	1100
1110	626.47	626.83	627.19	627.54	627.90	628.26	628.61	628.97	629.33	629.68	630.04	1110
1120	630.04	630.40	630.75	631.11	631.47	631.82	632.18	632.53	632.89	633.24	633.60	1120
1130	633.60	633.96	634.31	634.67	635.02	635.38	635.73	636.09	636.44	636.80	637.15	1130
1140	637.15	637.51	637.86	638.22	638.57	638.93	639.28	639.63	639.99	640.34	640.70	1140
1150	640.70	641.05	641.41	641.76	642.11	642.47	642.82	643.17	643.53	643.88	644.24	1150
1160	644.24	644.59	644.94	645.30	645.65	646.00	646.35	646.71	647.06	647.41	647.77	1160
1170	647.77	648.12	648.47	648.82	649.18	649.53	649.88	650.23	650.59	650.94	651.29	1170
1180	651.29	651.64	651.99	652.34	652.70	653.05	653.40	653.75	654.10	654.45	654.81	1180
1190	654.81	655.16	655.51	655.86	656.21	656.56	656.91	657.26	657.61	657.96	658.31	1190
1200	658.31	658.67	659.02	659.37	659.72	660.07	660.42	660.77	661.12	661.47	661.82	1200
1210	661.82	662.17	662.52	662.87	663.22	663.57	663.91	664.26	664.61	664.96	665.31	1210
1220	665.31	665.66	666.01	666.36	666.71	667.06	667.41	667.75	668.10	668.45	668.80	1220
1230	668.80	669.15	669.50	669.84	670.19	670.54	670.89	671.24	671.58	671.93	672.28	1230
1240	672.28	672.63	672.98	673.32	673.67	674.02	674.36	674.71	675.06	675.41	675.75	1240
1250	675.75	676.10	676.45	676.79	677.14	677.49	677.83	678.18	678.53	678.87	679.22	1250
1260	679.22	679.57	679.91	680.26	680.60	680.95	681.30	681.64	681.99	682.33	682.68	1260
1270	682.68	683.02	683.37	683.71	684.06	684.41	684.75	685.10	685.44	685.79	686.13	1270
1280	686.13	686.48	686.82	687.16	687.51	687.85	688.20	688.54	688.89	689.23	689.58	1280
1290	689.58	689.92	690.26	690.61	690.95	691.29	691.64	691.98	692.33	692.67	693.01	1290
1300	693.01	693.36	693.70	694.04	694.39	694.73	695.07	695.41	695.76	696.10	696.44	1300
1310	696.44	696.79	697.13	697.47	697.81	698.16	698.50	698.84	699.18	699.52	699.87	1310
1320	699.87	700.21	700.55	700.89	701.23	701.58	701.92	702.26	702.60	702.94	703.28	1320
1330	703.28	703.62	703.96	704.31	704.65	704.99	705.33	705.67	706.01	706.35	706.69	1330
1340	706.69	707.03	707.37	707.71	708.05	708.39	708.73	709.07	709.41	709.75	710.09	1340
1350	710.09	710.43	710.77	711.11	711.45	711.79	712.13	712.47	712.81	713.15	713.49	1350
1360	713.49	713.83	714.17	714.50	714.84	715.18	715.52	715.86	716.20	716.54	716.88	1360
1370	716.88	717.21	717.55	717.89	718.23	718.57	718.90	719.24	719.58	719.92	720.26	1370
1380	720.26	720.59	720.93	721.27	721.61	721.94	722.28	722.62	722.95	723.29	723.63	1380
1390	723.63	723.97	724.30	724.64	724.98	725.31	725.65	725.99	726.32	726.66	726.99	1390
1400	726.99	727.33	727.67	728.00	728.34	728.67	729.01	729.35	729.68	730.02	730.35	1400
1410	730.35	730.69	731.02	731.36	731.69	732.03	732.37	732.70	733.04	733.37	733.70	1410
1420	733.70	734.04	734.37	734.71	735.04	735.38	735.71	736.05	736.38	736.72	737.05	1420
1430	737.05	737.38	737.72	738.05	738.39	738.72	739.05	739.39	739.72	740.05	740.39	1430
1440	740.39	740.72	741.05	741.39	741.72	742.05	742.39	742.72	743.05	743.38	743.72	1440
1450	743.72	744.05	744.38	744.71	745.05	745.38	745.71	746.04	746.38	746.71	747.04	1450
1460	747.04	747.37	747.70	748.04	748.37	748.70	749.03	749.36	749.69	750.02	750.36	1460
1470	750.36	750.69	751.02	751.35	751.68	752.01	752.34	752.67	753.00	753.33	753.66	1470
1480	753.66	753.99	754.32	754.66	754.99	755.32	755.65	755.98	756.31	756.64	756.97	1480
1490	756.97	757.30	757.62	757.95	758.28	758.61	758.94	759.27	759.60	759.93	760.26	1490
1500	760.26	760.59	760.92	761.25	761.58	761.90	762.23	762.56	762.89	763.22	763.55	1500
1510	763.55	763.88	764.20	764.53	764.86	765.19	765.52	765.84	766.17	766.50	766.83	1510
1520	766.83	767.15	767.48	767.81	768.14	768.46	768.79	769.12	769.45	769.77	770.10	1520
1530	770.10	770.43	770.75	771.08	771.41	771.73	772.06	772.39	772.71	773.04	773.37	1530
1540	773.37	773.69	774.02	774.34	774.67	775.00	775.32	775.65	775.97	776.30	776.62	1540
1550	776.62	776.95	777.27	777.60	777.93	778.25	778.58	778.90	779.23	779.55	779.88	1550
1560	779.88	780.20	780.52									1560

RTD Resistance vs Temperature (°C)
Platinum 500, Alpha = 0.00385

RESISTANCE IN OHMS

°C	0	1	2	3	4	5	6	7	8	9	10	°C
−200	92.47	94.63	96.79	98.95	101.11	103.26	105.42	107.57	109.72	111.87	114.02	−200
−190	114.02	116.16	118.30	120.45	122.59	124.72	126.86	129.00	131.13	133.26	135.39	−190
−180	135.39	137.52	139.64	141.77	143.89	146.01	148.13	150.25	152.37	154.49	156.60	−180
−170	156.60	158.71	160.82	162.93	165.04	167.15	169.25	171.35	173.46	175.56	177.66	−170
−160	177.66	179.75	181.85	183.94	186.04	188.13	190.22	192.31	194.40	196.48	198.57	−160
−150	198.57	200.65	202.73	204.82	206.89	208.97	211.05	213.13	215.20	217.27	219.35	−150
−140	219.35	221.42	223.49	225.55	227.62	229.69	231.75	233.81	235.88	237.94	240.00	−140
−130	240.00	242.05	244.11	246.17	248.22	250.28	252.33	254.38	256.43	258.48	260.53	−130
−120	260.53	262.58	264.62	266.67	268.71	270.75	272.80	274.84	276.88	278.91	280.95	−120
−110	280.95	282.99	285.02	287.06	289.09	291.12	293.15	295.19	297.21	299.24	301.27	−110
−100	301.27	303.30	305.32	307.35	309.37	311.39	313.42	315.44	317.46	319.48	321.49	−100
−90	321.49	323.51	325.53	327.54	329.56	331.57	333.58	335.60	337.61	339.62	341.63	−90
−80	341.63	343.63	345.64	347.65	349.65	351.66	353.66	355.67	357.67	359.67	361.67	−80
−70	361.67	363.67	365.67	367.67	369.67	371.67	373.66	375.66	377.65	379.65	381.64	−70
−60	381.64	383.63	385.63	387.62	389.61	391.60	393.59	395.57	397.56	399.55	401.53	−60
−50	401.53	403.52	405.50	407.49	409.47	411.45	413.44	415.42	417.40	419.38	421.36	−50
−40	421.36	423.33	425.31	427.29	429.27	431.24	433.22	435.19	437.17	439.14	441.11	−40
−30	441.11	443.08	445.05	447.02	448.99	450.96	452.93	454.90	456.87	458.84	460.80	−30
−20	460.80	462.77	464.73	466.70	468.66	470.62	472.59	474.55	476.51	478.47	480.43	−20
−10	480.43	482.39	484.35	486.31	488.27	490.22	492.18	494.14	496.09	498.05	500.00	−10
0	500.00	501.95	503.91	505.86	507.81	509.76	511.71	513.66	515.61	517.56	519.51	0
10	519.51	521.46	523.41	525.35	527.30	529.24	531.19	533.13	535.08	537.02	538.96	10
20	538.96	540.91	542.85	544.79	546.73	548.67	550.61	552.55	554.48	556.42	558.36	20
30	558.36	560.30	562.23	564.17	566.10	568.03	569.97	571.90	573.83	575.77	577.70	30
40	577.70	579.63	581.56	583.49	585.41	587.34	589.27	591.20	593.12	595.05	596.98	40
50	596.98	598.90	600.82	602.75	604.67	606.59	608.51	610.44	612.36	614.28	616.20	50
60	616.20	618.12	620.03	621.95	623.87	625.78	627.70	629.62	631.53	633.45	635.36	60
70	635.36	637.27	639.18	641.10	643.01	644.92	646.83	648.74	650.65	652.56	654.46	70
80	654.46	656.37	658.28	660.18	662.09	663.99	665.90	667.80	669.71	671.61	673.51	80
90	673.51	675.41	677.31	679.21	681.11	683.01	684.91	686.81	688.71	690.60	692.50	90
100	692.50	694.40	696.29	698.19	700.08	701.97	703.87	705.76	707.65	709.54	711.43	100
110	711.43	713.32	715.21	717.10	718.99	720.87	722.76	724.65	726.53	728.42	730.30	110
120	730.30	732.19	734.07	735.95	737.84	739.72	741.60	743.48	745.36	747.24	749.12	120
130	749.12	751.00	752.87	754.75	756.63	758.50	760.38	762.25	764.13	766.00	767.88	130
140	767.88	769.75	771.62	773.49	775.36	777.23	779.10	780.97	782.84	784.71	786.57	140
150	786.57	788.44	790.31	792.17	794.04	795.90	797.77	799.63	801.49	803.35	805.22	150
160	805.22	807.08	808.94	810.80	812.66	814.51	816.37	818.23	820.09	821.94	823.80	160
170	823.80	825.65	827.51	829.36	831.21	833.07	834.92	836.77	838.62	840.47	842.32	170
180	842.32	844.17	846.02	847.87	849.72	851.56	853.41	855.26	857.10	858.95	860.79	180
190	860.79	862.63	864.48	866.32	868.16	870.00	871.84	873.68	875.52	877.36	879.20	190
200	879.20	881.04	882.87	884.71	886.55	888.38	890.22	892.05	893.88	895.72	897.55	200
210	897.55	899.38	901.21	903.04	904.87	906.70	908.53	910.36	912.19	914.01	915.84	210
220	915.84	917.67	919.49	921.32	923.14	924.97	926.79	928.61	930.43	932.26	934.08	220
230	934.08	935.90	937.72	939.54	941.35	943.17	944.99	946.81	948.62	950.44	952.25	230
240	952.25	954.07	955.88	957.69	959.51	961.32	963.13	964.94	966.75	968.56	970.37	240
250	970.37	972.18	973.99	975.80	977.60	979.41	981.21	983.02	984.82	986.63	988.43	250
260	988.43	990.23	992.04	993.84	995.64	997.44	999.24	1001.04	1002.84	1004.64	1006.43	260
270	1006.43	1008.23	1010.03	1011.82	1013.62	1015.41	1017.21	1019.00	1020.79	1022.59	1024.38	270
280	1024.38	1026.17	1027.96	1029.75	1031.54	1033.33	1035.12	1036.91	1038.69	1040.48	1042.27	280
290	1042.27	1044.05	1045.84	1047.62	1049.40	1051.19	1052.97	1054.75	1056.53	1058.31	1060.09	290
300	1060.09	1061.87	1063.65	1065.43	1067.21	1068.99	1070.76	1072.54	1074.32	1076.09	1077.86	300
310	1077.86	1079.64	1081.41	1083.18	1084.96	1086.73	1088.50	1090.27	1092.04	1093.81	1095.58	310
320	1095.58	1097.35	1099.11	1100.88	1102.65	1104.41	1106.18	1107.94	1109.71	1111.47	1113.23	320
330	1113.23	1114.99	1116.76	1118.52	1120.28	1122.04	1123.80	1125.56	1127.31	1129.07	1130.83	330
340	1130.83	1132.58	1134.34	1136.10	1137.85	1139.60	1141.36	1143.11	1144.86	1146.62	1148.37	340
350	1148.37	1150.12	1151.87	1153.62	1155.37	1157.11	1158.86	1160.61	1162.36	1164.10	1165.85	350
360	1165.85	1167.59	1169.34	1171.08	1172.82	1174.57	1176.31	1178.05	1179.79	1181.53	1183.27	360
370	1183.27	1185.01	1186.75	1188.48	1190.22	1191.96	1193.69	1195.43	1197.17	1198.90	1200.63	370
380	1200.63	1202.37	1204.10	1205.83	1207.56	1209.29	1211.02	1212.75	1214.48	1216.21	1217.94	380
390	1217.94	1219.67	1221.39	1223.12	1224.85	1226.57	1228.30	1230.02	1231.74	1233.47	1235.19	390

RTD Resistance vs Temperature (°C)
Platinum 500, Alpha = 0.00385

RESISTANCE IN OHMS

°C	0	1	2	3	4	5	6	7	8	9	10	°C
400	1235.19	1236.91	1238.63	1240.35	1242.07	1243.79	1245.51	1247.23	1248.94	1250.66	1252.38	400
410	1252.38	1254.09	1255.81	1257.52	1259.24	1260.95	1262.66	1264.38	1266.09	1267.80	1269.51	410
420	1269.51	1271.22	1272.93	1274.64	1276.35	1278.06	1279.76	1281.47	1283.17	1284.88	1286.58	420
430	1286.58	1288.29	1289.99	1291.70	1293.40	1295.10	1296.80	1298.50	1300.20	1301.90	1303.60	430
440	1303.60	1305.30	1307.00	1308.69	1310.39	1312.09	1313.78	1315.48	1317.17	1318.87	1320.56	440
450	1320.56	1322.25	1323.94	1325.64	1327.33	1329.02	1330.71	1332.40	1334.08	1335.77	1337.46	450
460	1337.46	1339.15	1340.83	1342.52	1344.20	1345.89	1347.57	1349.26	1350.94	1352.62	1354.30	460
470	1354.30	1355.98	1357.66	1359.34	1361.02	1362.70	1364.38	1366.06	1367.73	1369.41	1371.09	470
480	1371.09	1372.76	1374.44	1376.11	1377.78	1379.46	1381.13	1382.80	1384.47	1386.14	1387.81	480
490	1387.81	1389.48	1391.15	1392.82	1394.49	1396.15	1397.82	1399.49	1401.15	1402.82	1404.48	490
500	1404.48	1406.14	1407.81	1409.47	1411.13	1412.79	1414.45	1416.11	1417.77	1419.43	1421.09	500
510	1421.09	1422.75	1424.41	1426.06	1427.72	1429.37	1431.03	1432.68	1434.34	1435.99	1437.64	510
520	1437.64	1439.29	1440.95	1442.60	1444.25	1445.90	1447.55	1449.19	1450.84	1452.49	1454.14	520
530	1454.14	1455.78	1457.43	1459.07	1460.72	1462.36	1464.00	1465.65	1467.29	1468.93	1470.57	530
540	1470.57	1472.21	1473.85	1475.49	1477.13	1478.77	1480.41	1482.04	1483.68	1485.32	1486.95	540
550	1486.95	1488.59	1490.22	1491.85	1493.49	1495.12	1496.75	1498.38	1500.01	1501.64	1503.27	550
560	1503.27	1504.90	1506.53	1508.16	1509.78	1511.41	1513.03	1514.66	1516.28	1517.91	1519.53	560
570	1519.53	1521.16	1522.78	1524.40	1526.02	1527.64	1529.26	1530.88	1532.50	1534.12	1535.74	570
580	1535.74	1537.35	1538.97	1540.59	1542.20	1543.82	1545.43	1547.04	1548.66	1550.27	1551.88	580
590	1551.88	1553.49	1555.10	1556.71	1558.32	1559.93	1561.54	1563.15	1564.76	1566.36	1567.97	590
600	1567.97	1569.58	1571.18	1572.79	1574.39	1575.99	1577.60	1579.20	1580.80	1582.40	1584.00	600
610	1584.00	1585.60	1587.20	1588.80	1590.40	1591.99	1593.59	1595.19	1596.78	1598.38	1599.97	610
620	1599.97	1601.57	1603.16	1604.75	1606.34	1607.94	1609.53	1611.12	1612.71	1614.30	1615.89	620
630	1615.89	1617.47	1619.06	1620.65	1622.24	1623.82	1625.41	1626.99	1628.58	1630.16	1631.74	630
640	1631.74	1633.32	1634.91	1636.49	1638.07	1639.65	1641.23	1642.81	1644.38	1645.96	1647.54	640
650	1647.54	1649.12	1650.69	1652.27	1653.84	1655.42	1656.99	1658.56	1660.14	1661.71	1663.28	650
660	1663.28	1664.85	1666.42	1667.99	1669.56	1671.13	1672.70	1674.26	1675.83	1677.40	1678.96	660
670	1678.96	1680.53	1682.09	1683.65	1685.22	1686.78	1688.34	1689.90	1691.46	1693.03	1694.58	670
680	1694.58	1696.14	1697.70	1699.26	1700.82	1702.37	1703.93	1705.49	1707.04	1708.60	1710.15	680
690	1710.15	1711.70	1713.26	1714.81	1716.36	1717.91	1719.46	1721.01	1722.56	1724.11	1725.66	690
700	1725.66	1727.21	1728.75	1730.30	1731.85	1733.39	1734.94	1736.48	1738.02	1739.57	1741.11	700
710	1741.11	1742.65	1744.19	1745.73	1747.27	1748.81	1750.35	1751.89	1753.43	1754.96	1756.50	710
720	1756.50	1758.04	1759.57	1761.11	1762.64	1764.17	1765.71	1767.24	1768.77	1770.30	1771.83	720
730	1771.83	1773.36	1774.89	1776.42	1777.95	1779.48	1781.01	1782.53	1784.06	1785.58	1787.11	730
740	1787.11	1788.63	1790.16	1791.68	1793.20	1794.72	1796.25	1797.77	1799.29	1800.81	1802.33	740
750	1802.33	1803.85	1805.36	1806.88	1808.40	1809.91	1811.43	1812.94	1814.46	1815.97	1817.49	750
760	1817.49	1819.00	1820.51	1822.02	1823.53	1825.04	1826.55	1828.06	1829.57	1831.08	1832.59	760
770	1832.59	1834.09	1835.60	1837.11	1838.61	1840.12	1841.62	1843.12	1844.63	1846.13	1847.63	770
780	1847.63	1849.13	1850.63	1852.13	1853.63	1855.13	1856.63	1858.13	1859.62	1861.12	1862.62	780
790	1862.62	1864.11	1865.61	1867.10	1868.59	1870.09	1871.58	1873.07	1874.56	1876.05	1877.54	790
800	1877.54	1879.03	1880.52	1882.01	1883.50	1884.99	1886.47	1887.96	1889.44	1890.93	1892.41	800
810	1892.41	1893.90	1895.38	1896.86	1898.35	1899.83	1901.31	1902.79	1904.27	1905.75	1907.23	810
820	1907.23	1908.70	1910.18	1911.66	1913.13	1914.61	1916.08	1917.56	1919.03	1920.51	1921.98	820
830	1921.98	1923.45	1924.92	1926.39	1927.86	1929.33	1930.80	1932.27	1933.74	1935.21	1936.67	830
840	1936.67	1938.14	1939.61	1941.07	1942.54	1944.00	1945.46	1946.93	1948.39	1949.85	1951.31	840

RTD Resistance vs Temperature (°F)
Platinum 500, Alpha = 0.00385

RESISTANCE IN OHMS

°F	0	1	2	3	4	5	6	7	8	9	10	°F
-330	90.06	91.26	92.47	93.67	94.87	96.07	97.27	98.47	99.67	100.87	102.07	-330
-320	102.07	103.26	104.46	105.66	106.85	108.05	109.24	110.44	111.63	112.82	114.02	-320
-310	114.02	115.21	116.40	117.59	118.78	119.97	121.16	122.35	123.54	124.72	125.91	-310
-300	125.91	127.10	128.28	129.47	130.65	131.84	133.02	134.21	135.39	136.57	137.75	-300
-290	137.75	138.94	140.12	141.30	142.48	143.66	144.84	146.01	147.19	148.37	149.55	-290
-280	149.55	150.72	151.90	153.08	154.25	155.43	156.60	157.77	158.95	160.12	161.29	-280
-270	161.29	162.46	163.64	164.81	165.98	167.15	168.32	169.49	170.65	171.82	172.99	-270
-260	172.99	174.16	175.32	176.49	177.66	178.82	179.99	181.15	182.32	183.48	184.64	-260
-250	184.64	185.81	186.97	188.13	189.29	190.45	191.61	192.77	193.93	195.09	196.25	-250
-240	196.25	197.41	198.57	199.73	200.88	202.04	203.20	204.35	205.51	206.66	207.82	-240
-230	207.82	208.97	210.13	211.28	212.43	213.59	214.74	215.89	217.04	218.19	219.35	-230
-220	219.35	220.50	221.65	222.80	223.95	225.09	226.24	227.39	228.54	229.69	230.83	-220
-210	230.83	231.98	233.13	234.27	235.42	236.56	237.71	238.85	240.00	241.14	242.28	-210
-200	242.28	243.43	244.57	245.71	246.85	247.99	249.14	250.28	251.42	252.56	253.70	-200
-190	253.70	254.84	255.98	257.11	258.25	259.39	260.53	261.67	262.80	263.94	265.08	-190
-180	265.08	266.21	267.35	268.48	269.62	270.75	271.89	273.02	274.16	275.29	276.42	-180
-170	276.42	277.56	278.69	279.82	280.95	282.08	283.21	284.34	285.48	286.61	287.74	-170
-160	287.74	288.87	289.99	291.12	292.25	293.38	294.51	295.64	296.76	297.89	299.02	-160
-150	299.02	300.14	301.27	302.40	303.52	304.65	305.77	306.90	308.02	309.15	310.27	-150
-140	310.27	311.39	312.52	313.64	314.76	315.89	317.01	318.13	319.25	320.37	321.49	-140
-130	321.49	322.61	323.73	324.85	325.97	327.09	328.21	329.33	330.45	331.57	332.69	-130
-120	332.69	333.81	334.92	336.04	337.16	338.28	339.39	340.51	341.63	342.74	343.86	-120
-110	343.86	344.97	346.09	347.20	348.32	349.43	350.55	351.66	352.77	353.89	355.00	-110
-100	355.00	356.11	357.23	358.34	359.45	360.56	361.67	362.78	363.90	365.01	366.12	-100
-90	366.12	367.23	368.34	369.45	370.56	371.67	372.78	373.89	374.99	376.10	377.21	-90
-80	377.21	378.32	379.43	380.53	381.64	382.75	383.86	384.96	386.07	387.17	388.28	-80
-70	388.28	389.39	390.49	391.60	392.70	393.81	394.91	396.02	397.12	398.22	399.33	-70
-60	399.33	400.43	401.53	402.64	403.74	404.84	405.95	407.05	408.15	409.25	410.35	-60
-50	410.35	411.45	412.55	413.66	414.76	415.86	416.96	418.06	419.16	420.26	421.36	-50
-40	421.36	422.46	423.55	424.65	425.75	426.85	427.95	429.05	430.14	431.24	432.34	-40
-30	432.34	433.44	434.53	435.63	436.73	437.82	438.92	440.02	441.11	442.21	443.30	-30
-20	443.30	444.40	445.49	446.59	447.68	448.78	449.87	450.96	452.06	453.15	454.25	-20
-10	454.25	455.34	456.43	457.52	458.62	459.71	460.80	461.89	462.99	464.08	465.17	-10
0	465.17	466.26	467.35	468.44	469.53	470.62	471.71	472.80	473.89	474.98	476.07	0
10	476.07	477.16	478.25	479.34	480.43	481.52	482.61	483.70	484.78	485.87	486.96	10
20	486.96	488.05	489.14	490.22	491.31	492.40	493.48	494.57	495.66	496.74	497.83	20
30	497.83	498.91	500.00	501.09	502.17	503.26	504.34	505.43	506.51	507.59	508.68	30
40	508.68	509.76	510.85	511.93	513.01	514.10	515.18	516.26	517.35	518.43	519.51	40
50	519.51	520.59	521.68	522.76	523.84	524.92	526.00	527.08	528.16	529.24	530.33	50
60	530.33	531.41	532.49	533.57	534.65	535.73	536.81	537.88	538.96	540.04	541.12	60
70	541.12	542.20	543.28	544.36	545.44	546.51	547.59	548.67	549.75	550.82	551.90	70
80	551.90	552.98	554.05	555.13	556.21	557.28	558.36	559.44	560.51	561.59	562.66	80
90	562.66	563.74	564.81	565.89	566.96	568.03	569.11	570.18	571.26	572.33	573.40	90
100	573.40	574.48	575.55	576.62	577.70	578.77	579.84	580.91	581.99	583.06	584.13	100
110	584.13	585.20	586.27	587.34	588.41	589.48	590.56	591.63	592.70	593.77	594.84	110
120	594.84	595.91	596.98	598.04	599.11	600.18	601.25	602.32	603.39	604.46	605.53	120
130	605.53	606.59	607.66	608.73	609.80	610.86	611.93	613.00	614.06	615.13	616.20	130
140	616.20	617.26	618.33	619.39	620.46	621.53	622.59	623.66	624.72	625.78	626.85	140
150	626.85	627.91	628.98	630.04	631.11	632.17	633.23	634.30	635.36	636.42	637.48	150
160	637.48	638.55	639.61	640.67	641.73	642.80	643.86	644.92	645.98	647.04	648.10	160
170	648.10	649.16	650.22	651.28	652.34	653.40	654.46	655.52	656.58	657.64	658.70	170
180	658.70	659.76	660.82	661.88	662.94	663.99	665.05	666.11	667.17	668.23	669.28	180
190	669.28	670.34	671.40	672.45	673.51	674.57	675.62	676.68	677.74	678.79	679.85	190
200	679.85	680.90	681.96	683.01	684.07	685.12	686.18	687.23	688.29	689.34	690.39	200
210	690.39	691.45	692.50	693.55	694.61	695.66	696.71	697.76	698.82	699.87	700.92	210
220	700.92	701.97	703.02	704.08	705.13	706.18	707.23	708.28	709.33	710.38	711.43	220
230	711.43	712.48	713.53	714.58	715.63	716.68	717.73	718.78	719.83	720.87	721.92	230
240	721.92	722.97	724.02	725.07	726.11	727.16	728.21	729.26	730.30	731.35	732.40	240
250	732.40	733.44	734.49	735.54	736.58	737.63	738.67	739.72	740.76	741.81	742.85	250
260	742.85	743.90	744.94	745.99	747.03	748.07	749.12	750.16	751.21	752.25	753.29	260

RTD Resistance vs Temperature (°F)
Platinum 500, Alpha = 0.00385

RESISTANCE IN OHMS

°F	0	1	2	3	4	5	6	7	8	9	10	°F
270	753.29	754.33	755.38	756.42	757.46	758.50	759.55	760.59	761.63	762.67	763.71	270
280	763.71	764.75	765.79	766.83	767.88	768.92	769.96	771.00	772.04	773.08	774.11	280
290	774.11	775.15	776.19	777.23	778.27	779.31	780.35	781.39	782.42	783.46	784.50	290
300	784.50	785.54	786.57	787.61	788.65	789.69	790.72	791.76	792.79	793.83	794.87	300
310	794.87	795.90	796.94	797.97	799.01	800.04	801.08	802.11	803.15	804.18	805.22	310
320	805.22	806.25	807.28	808.32	809.35	810.38	811.42	812.45	813.48	814.51	815.55	320
330	815.55	816.58	817.61	818.64	819.67	820.70	821.74	822.77	823.80	824.83	825.86	330
340	825.86	826.89	827.92	828.95	829.98	831.01	832.04	833.07	834.10	835.13	836.15	340
350	836.15	837.18	838.21	839.24	840.27	841.29	842.32	843.35	844.38	845.40	846.43	350
360	846.43	847.46	848.48	849.51	850.54	851.56	852.59	853.61	854.64	855.67	856.69	360
370	856.69	857.72	858.74	859.76	860.79	861.81	862.84	863.86	864.89	865.91	866.93	370
380	866.93	867.96	868.98	870.00	871.02	872.05	873.07	874.09	875.11	876.13	877.16	380
390	877.16	878.18	879.20	880.22	881.24	882.26	883.28	884.30	885.32	886.34	887.36	390
400	887.36	888.38	889.40	890.42	891.44	892.46	893.48	894.49	895.51	896.53	897.55	400
410	897.55	898.57	899.58	900.60	901.62	902.64	903.65	904.67	905.69	906.70	907.72	410
420	907.72	908.73	909.75	910.77	911.78	912.80	913.81	914.83	915.84	916.86	917.87	420
430	917.87	918.88	919.90	920.91	921.93	922.94	923.95	924.97	925.98	926.99	928.00	430
440	928.00	929.02	930.03	931.04	932.05	933.06	934.08	935.09	936.10	937.11	938.12	440
450	938.12	939.13	940.14	941.15	942.16	943.17	944.18	945.19	946.20	947.21	948.22	450
460	948.22	949.23	950.24	951.24	952.25	953.26	954.27	955.28	956.28	957.29	958.30	460
470	958.30	959.31	960.31	961.32	962.33	963.33	964.34	965.34	966.35	967.36	968.36	470
480	968.36	969.37	970.37	971.38	972.38	973.39	974.39	975.39	976.40	977.40	978.41	480
490	978.41	979.41	980.41	981.42	982.42	983.42	984.42	985.43	986.43	987.43	988.43	490
500	988.43	989.43	990.44	991.44	992.44	993.44	994.44	995.44	996.44	997.44	998.44	500
510	998.44	999.44	1000.44	1001.44	1002.44	1003.44	1004.44	1005.44	1006.43	1007.43	1008.43	510
520	1008.43	1009.43	1010.43	1011.42	1012.42	1013.42	1014.42	1015.41	1016.41	1017.41	1018.40	520
530	1018.40	1019.40	1020.40	1021.39	1022.39	1023.38	1024.38	1025.37	1026.37	1027.36	1028.36	530
540	1028.36	1029.35	1030.35	1031.34	1032.34	1033.33	1034.32	1035.32	1036.31	1037.30	1038.30	540
550	1038.30	1039.29	1040.28	1041.27	1042.27	1043.26	1044.25	1045.24	1046.23	1047.22	1048.21	550
560	1048.21	1049.21	1050.20	1051.19	1052.18	1053.17	1054.16	1055.15	1056.14	1057.13	1058.12	560
570	1058.12	1059.11	1060.09	1061.08	1062.07	1063.06	1064.05	1065.04	1066.02	1067.01	1068.00	570
580	1068.00	1068.99	1069.97	1070.96	1071.95	1072.93	1073.92	1074.91	1075.89	1076.88	1077.86	580
590	1077.86	1078.85	1079.84	1080.82	1081.81	1082.79	1083.78	1084.76	1085.74	1086.73	1087.71	590
600	1087.71	1088.70	1089.68	1090.66	1091.65	1092.63	1093.61	1094.59	1095.58	1096.56	1097.54	600
610	1097.54	1098.52	1099.51	1100.49	1101.47	1102.45	1103.43	1104.41	1105.39	1106.37	1107.35	610
620	1107.35	1108.33	1109.31	1110.29	1111.27	1112.25	1113.23	1114.21	1115.19	1116.17	1117.15	620
630	1117.15	1118.13	1119.10	1120.08	1121.06	1122.04	1123.01	1123.99	1124.97	1125.95	1126.92	630
640	1126.92	1127.90	1128.88	1129.85	1130.83	1131.80	1132.78	1133.75	1134.73	1135.71	1136.68	640
650	1136.68	1137.66	1138.63	1139.60	1140.58	1141.55	1142.53	1143.50	1144.47	1145.45	1146.42	650
660	1146.42	1147.39	1148.37	1149.34	1150.31	1151.28	1152.26	1153.23	1154.20	1155.17	1156.14	660
670	1156.14	1157.11	1158.08	1159.06	1160.03	1161.00	1161.97	1162.94	1163.91	1164.88	1165.85	670
680	1165.85	1166.82	1167.79	1168.75	1169.72	1170.69	1171.66	1172.63	1173.60	1174.57	1175.53	680
690	1175.53	1176.50	1177.47	1178.44	1179.40	1180.37	1181.34	1182.30	1183.27	1184.24	1185.20	690
700	1185.20	1186.17	1187.13	1188.10	1189.06	1190.03	1190.99	1191.96	1192.92	1193.89	1194.85	700
710	1194.85	1195.82	1196.78	1197.74	1198.71	1199.67	1200.63	1201.60	1202.56	1203.52	1204.48	710
720	1204.48	1205.45	1206.41	1207.37	1208.33	1209.29	1210.26	1211.22	1212.18	1213.14	1214.10	720
730	1214.10	1215.06	1216.02	1216.98	1217.94	1218.90	1219.86	1220.82	1221.78	1222.74	1223.70	730
740	1223.70	1224.65	1225.61	1226.57	1227.53	1228.49	1229.45	1230.40	1231.36	1232.32	1233.27	740
750	1233.27	1234.23	1235.19	1236.14	1237.10	1238.06	1239.01	1239.97	1240.92	1241.88	1242.84	750
760	1242.84	1243.79	1244.75	1245.70	1246.65	1247.61	1248.56	1249.52	1250.47	1251.42	1252.38	760
770	1252.38	1253.33	1254.28	1255.24	1256.19	1257.14	1258.10	1259.05	1260.00	1260.95	1261.90	770
780	1261.90	1262.85	1263.81	1264.76	1265.71	1266.66	1267.61	1268.56	1269.51	1270.46	1271.41	780
790	1271.41	1272.36	1273.31	1274.26	1275.21	1276.16	1277.11	1278.06	1279.00	1279.95	1280.90	790
800	1280.90	1281.85	1282.80	1283.74	1284.69	1285.64	1286.58	1287.53	1288.48	1289.42	1290.37	800
810	1290.37	1291.32	1292.26	1293.21	1294.15	1295.10	1296.05	1296.99	1297.94	1298.88	1299.82	810
820	1299.82	1300.77	1301.71	1302.66	1303.60	1304.54	1305.49	1306.43	1307.37	1308.32	1309.26	820
830	1309.26	1310.20	1311.15	1312.09	1313.03	1313.97	1314.91	1315.85	1316.80	1317.74	1318.68	830
840	1318.68	1319.62	1320.56	1321.50	1322.44	1323.38	1324.32	1325.26	1326.20	1327.14	1328.08	840
850	1328.08	1329.02	1329.96	1330.89	1331.83	1332.77	1333.71	1334.65	1335.58	1336.52	1337.46	850
860	1337.46	1338.40	1339.33	1340.27	1341.21	1342.14	1343.08	1344.02	1344.95	1345.89	1346.82	860
870	1346.82	1347.76	1348.69	1349.63	1350.56	1351.50	1352.43	1353.37	1354.30	1355.24	1356.17	870
880	1356.17	1357.10	1358.04	1358.97	1359.90	1360.84	1361.77	1362.70	1363.63	1364.57	1365.50	880
890	1365.50	1366.43	1367.36	1368.29	1369.22	1370.15	1371.09	1372.02	1372.95	1373.88	1374.81	890
900	1374.81	1375.74	1376.67	1377.60	1378.53	1379.46	1380.39	1381.31	1382.24	1383.17	1384.10	900
910	1384.10	1385.03	1385.96	1386.88	1387.81	1388.74	1389.67	1390.59	1391.52	1392.45	1393.37	910

RTD Resistance vs Temperature (°F)
Platinum 500, Alpha = 0.00385

RESISTANCE IN OHMS

°F	0	1	2	3	4	5	6	7	8	9	10	°F
920	1393.37	1394.30	1395.23	1396.15	1397.08	1398.00	1398.93	1399.86	1400.78	1401.71	1402.63	920
930	1402.63	1403.56	1404.48	1405.40	1406.33	1407.25	1408.18	1409.10	1410.02	1410.95	1411.87	930
940	1411.87	1412.79	1413.72	1414.64	1415.56	1416.48	1417.40	1418.33	1419.25	1420.17	1421.09	940
950	1421.09	1422.01	1422.93	1423.85	1424.77	1425.69	1426.61	1427.53	1428.45	1429.37	1430.29	950
960	1430.29	1431.21	1432.13	1433.05	1433.97	1434.89	1435.81	1436.72	1437.64	1438.56	1439.48	960
970	1439.48	1440.40	1441.31	1442.23	1443.15	1444.06	1444.98	1445.90	1446.81	1447.73	1448.64	970
980	1448.64	1449.56	1450.48	1451.39	1452.31	1453.22	1454.14	1455.05	1455.97	1456.88	1457.79	980
990	1457.79	1458.71	1459.62	1460.54	1461.45	1462.36	1463.27	1464.19	1465.10	1466.01	1466.92	990
1000	1466.92	1467.84	1468.75	1469.66	1470.57	1471.48	1472.40	1473.31	1474.22	1475.13	1476.04	1000
1010	1476.04	1476.95	1477.86	1478.77	1479.68	1480.59	1481.50	1482.41	1483.32	1484.22	1485.13	1010
1020	1485.13	1486.04	1486.95	1487.86	1488.77	1489.67	1490.58	1491.49	1492.40	1493.30	1494.21	1020
1030	1494.21	1495.12	1496.02	1496.93	1497.84	1498.74	1499.65	1500.55	1501.46	1502.37	1503.27	1030
1040	1503.27	1504.18	1505.08	1505.98	1506.89	1507.79	1508.70	1509.60	1510.51	1511.41	1512.31	1040
1050	1512.31	1513.22	1514.12	1515.02	1515.92	1516.83	1517.73	1518.63	1519.53	1520.43	1521.34	1050
1060	1521.34	1522.24	1523.14	1524.04	1524.94	1525.84	1526.74	1527.64	1528.54	1529.44	1530.34	1060
1070	1530.34	1531.24	1532.14	1533.04	1533.94	1534.84	1535.74	1536.63	1537.53	1538.43	1539.33	1070
1080	1539.33	1540.23	1541.12	1542.02	1542.92	1543.82	1544.71	1545.61	1546.51	1547.40	1548.30	1080
1090	1548.30	1549.20	1550.09	1550.99	1551.88	1552.78	1553.67	1554.57	1555.46	1556.36	1557.25	1090
1100	1557.25	1558.15	1559.04	1559.93	1560.83	1561.72	1562.61	1563.51	1564.40	1565.29	1566.19	1100
1110	1566.19	1567.08	1567.97	1568.86	1569.75	1570.65	1571.54	1572.43	1573.32	1574.21	1575.10	1110
1120	1575.10	1575.99	1576.88	1577.77	1578.66	1579.55	1580.44	1581.33	1582.22	1583.11	1584.00	1120
1130	1584.00	1584.89	1585.78	1586.67	1587.55	1588.44	1589.33	1590.22	1591.11	1591.99	1592.88	1130
1140	1592.88	1593.77	1594.65	1595.54	1596.43	1597.31	1598.20	1599.09	1599.97	1600.86	1601.74	1140
1150	1601.74	1602.63	1603.51	1604.40	1605.28	1606.17	1607.05	1607.94	1608.82	1609.70	1610.59	1150
1160	1610.59	1611.47	1612.35	1613.24	1614.12	1615.00	1615.89	1616.77	1617.65	1618.53	1619.41	1160
1170	1619.41	1620.30	1621.18	1622.06	1622.94	1623.82	1624.70	1625.58	1626.46	1627.34	1628.22	1170
1180	1628.22	1629.10	1629.98	1630.86	1631.74	1632.62	1633.50	1634.38	1635.26	1636.14	1637.01	1180
1190	1637.01	1637.89	1638.77	1639.65	1640.53	1641.40	1642.28	1643.16	1644.03	1644.91	1645.79	1190
1200	1645.79	1646.66	1647.54	1648.42	1649.29	1650.17	1651.04	1651.92	1652.79	1653.67	1654.54	1200
1210	1654.54	1655.42	1656.29	1657.17	1658.04	1658.91	1659.79	1660.66	1661.53	1662.41	1663.28	1210
1220	1663.28	1664.15	1665.02	1665.90	1666.77	1667.64	1668.51	1669.38	1670.26	1671.13	1672.00	1220
1230	1672.00	1672.87	1673.74	1674.61	1675.48	1676.35	1677.22	1678.09	1678.96	1679.83	1680.70	1230
1240	1680.70	1681.57	1682.44	1683.31	1684.18	1685.04	1685.91	1686.78	1687.65	1688.52	1689.38	1240
1250	1689.38	1690.25	1691.12	1691.98	1692.85	1693.72	1694.58	1695.45	1696.32	1697.18	1698.05	1250
1260	1698.05	1698.91	1699.78	1700.65	1701.51	1702.37	1703.24	1704.10	1704.97	1705.83	1706.70	1260
1270	1706.70	1707.56	1708.42	1709.29	1710.15	1711.01	1711.88	1712.74	1713.60	1714.46	1715.33	1270
1280	1715.33	1716.19	1717.05	1717.91	1718.77	1719.63	1720.50	1721.36	1722.22	1723.08	1723.94	1280
1290	1723.94	1724.80	1725.66	1726.52	1727.38	1728.24	1729.10	1729.96	1730.81	1731.67	1732.53	1290
1300	1732.53	1733.39	1734.25	1735.11	1735.96	1736.82	1737.68	1738.54	1739.39	1740.25	1741.11	1300
1310	1741.11	1741.96	1742.82	1743.68	1744.53	1745.39	1746.24	1747.10	1747.96	1748.81	1749.67	1310
1320	1749.67	1750.52	1751.38	1752.23	1753.08	1753.94	1754.79	1755.65	1756.50	1757.35	1758.21	1320
1330	1758.21	1759.06	1759.91	1760.76	1761.62	1762.47	1763.32	1764.17	1765.03	1765.88	1766.73	1330
1340	1766.73	1767.58	1768.43	1769.28	1770.13	1770.98	1771.83	1772.68	1773.53	1774.38	1775.23	1340
1350	1775.23	1776.08	1776.93	1777.78	1778.63	1779.48	1780.33	1781.18	1782.02	1782.87	1783.72	1350
1360	1783.72	1784.57	1785.41	1786.26	1787.11	1787.96	1788.80	1789.65	1790.50	1791.34	1792.19	1360
1370	1792.19	1793.03	1793.88	1794.72	1795.57	1796.42	1797.26	1798.11	1798.95	1799.79	1800.64	1370
1380	1800.64	1801.48	1802.33	1803.17	1804.01	1804.86	1805.70	1806.54	1807.39	1808.23	1809.07	1380
1390	1809.07	1809.91	1810.76	1811.60	1812.44	1813.28	1814.12	1814.96	1815.80	1816.65	1817.49	1390
1400	1817.49	1818.33	1819.17	1820.01	1820.85	1821.69	1822.53	1823.37	1824.21	1825.04	1825.88	1400
1410	1825.88	1826.72	1827.56	1828.40	1829.24	1830.07	1830.91	1831.75	1832.59	1833.42	1834.26	1410
1420	1834.26	1835.10	1835.94	1836.77	1837.61	1838.44	1839.28	1840.12	1840.95	1841.79	1842.62	1420
1430	1842.62	1843.46	1844.29	1845.13	1845.96	1846.80	1847.63	1848.47	1849.30	1850.13	1850.97	1430
1440	1850.97	1851.80	1852.63	1853.47	1854.30	1855.13	1855.96	1856.80	1857.63	1858.46	1859.29	1440
1450	1859.29	1860.12	1860.95	1861.79	1862.62	1863.45	1864.28	1865.11	1865.94	1866.77	1867.60	1450
1460	1867.60	1868.43	1869.26	1870.09	1870.92	1871.75	1872.57	1873.40	1874.23	1875.06	1875.89	1460
1470	1875.89	1876.72	1877.54	1878.37	1879.20	1880.03	1880.85	1881.68	1882.51	1883.33	1884.16	1470
1480	1884.16	1884.99	1885.81	1886.64	1887.46	1888.29	1889.11	1889.94	1890.76	1891.59	1892.41	1480
1490	1892.41	1893.24	1894.06	1894.89	1895.71	1896.53	1897.36	1898.18	1899.00	1899.83	1900.65	1490
1500	1900.65	1901.47	1902.29	1903.12	1903.94	1904.76	1905.58	1906.40	1907.23	1908.05	1908.87	1500
1510	1908.87	1909.69	1910.51	1911.33	1912.15	1912.97	1913.79	1914.61	1915.43	1916.25	1917.07	1510
1520	1917.07	1917.89	1918.71	1919.52	1920.34	1921.16	1921.98	1922.80	1923.61	1924.43	1925.25	1520
1530	1925.25	1926.07	1926.88	1927.70	1928.52	1929.33	1930.15	1930.97	1931.78	1932.60	1933.41	1530
1540	1933.41	1934.23	1935.04	1935.86	1936.67	1937.49	1938.30	1939.12	1939.93	1940.75	1941.56	1540
1550	1941.56	1942.37	1943.19	1944.00	1944.81	1945.63	1946.44	1947.25	1948.06	1948.88	1949.69	1550
1560	1949.69	1950.50	1951.31									1560

Appendix B

RTD Resistance vs Temperature (°C)
Nickel 120, Edison Curve 7

RESISTANCE IN OHMS

°C	0	1	2	3	4	5	6	7	8	9	10	°C
−70	73.10	73.75	74.40	75.05	75.71	76.36	77.01	77.66	78.31	78.97	79.62	−70
−60	79.62	80.27	80.93	81.58	82.24	82.89	83.55	84.20	84.86	85.51	86.17	−60
−50	86.17	86.83	87.48	88.14	88.80	89.46	90.12	90.78	91.44	92.10	92.76	−50
−40	92.76	93.42	94.08	94.75	95.41	96.08	96.74	97.41	98.08	98.74	99.41	−40
−30	99.41	100.08	100.75	101.42	102.09	102.77	103.44	104.12	104.79	105.47	106.15	−30
−20	106.15	106.83	107.51	108.19	108.87	109.56	110.24	110.93	111.62	112.31	113.00	−20
−10	113.00	113.69	114.39	115.08	115.78	116.48	117.18	117.88	118.59	119.29	120.00	−10
0	120.00	120.71	121.42	122.13	122.85	123.56	124.28	125.00	125.72	126.44	127.17	0
10	127.17	127.90	128.62	129.35	130.09	130.82	131.56	132.29	133.03	133.77	134.52	10
20	134.52	135.26	136.01	136.76	137.51	138.26	139.02	139.77	140.53	141.29	142.06	20
30	142.06	142.82	143.59	144.36	145.13	145.90	146.68	147.45	148.23	149.01	149.80	30
40	149.80	150.58	151.37	152.16	152.95	153.74	154.54	155.34	156.14	156.94	157.74	40
50	157.74	158.55	159.36	160.17	160.98	161.79	162.61	163.43	164.25	165.07	165.90	50
60	165.90	166.73	167.56	168.39	169.22	170.06	170.89	171.73	172.58	173.42	174.27	60
70	174.27	175.11	175.97	176.82	177.67	178.53	179.39	180.25	181.11	181.98	182.85	70
80	182.85	183.72	184.59	185.46	186.34	187.22	188.10	188.98	189.86	190.75	191.64	80
90	191.64	192.53	193.42	194.32	195.21	196.11	197.02	197.92	198.83	199.73	200.64	90
100	200.64	201.56	202.47	203.39	204.31	205.23	206.15	207.08	208.00	208.93	209.86	100
110	209.86	210.80	211.73	212.67	213.61	214.55	215.50	216.45	217.40	218.35	219.30	110
120	219.30	220.26	221.21	222.17	223.14	224.10	225.07	226.04	227.01	227.98	228.96	120
130	228.96	229.94	230.92	231.90	232.89	233.87	234.86	235.86	236.85	237.85	238.85	130
140	238.85	239.85	240.85	241.86	242.87	243.88	244.89	245.91	246.93	247.95	248.97	140
150	248.97	250.01	251.04	252.08	253.12	254.16	255.20	256.24	257.28	258.32	259.37	150
160	259.37	260.41	261.46	262.51	263.56	264.62	265.67	266.73	267.80	268.86	269.93	160
170	269.93	271.00	272.07	273.15	274.23	275.31	276.40	277.49	278.58	279.68	280.78	170
180	280.78	281.88	282.99	284.10	285.21	286.33	287.45	288.57	289.70	290.83	291.96	180
190	291.96	293.10	294.24	295.38	296.53	297.68	298.83	299.98	301.14	302.30	303.46	190
200	303.46	304.63	305.80	306.97	308.15	309.34	310.52	311.71	312.90	314.10	315.30	200
210	315.30	316.51	317.72	318.93	320.15	321.37	322.59	323.82	325.05	326.29	327.53	210
220	327.53	328.77	330.02	331.27	332.53	333.79	335.05	336.32	337.59	338.86	340.14	220
230	340.14	341.43	342.71	344.00	345.30	346.59	347.90	349.20	350.51	351.82	353.14	230
240	353.14	354.46	355.79	357.12	358.45	359.79	361.13	362.47	363.82	365.17	366.53	240
250	366.53	367.89	369.25	370.62	371.99	373.37	374.75	376.13	377.52	378.91	380.31	250
260	380.31	381.71	383.11	384.52	385.93	387.35	388.77	390.19	391.62	393.05	394.49	260
270	394.49	395.93	397.37	398.82	400.27	401.73	403.19	404.65	406.12	407.59	409.07	270
280	409.07	410.55	412.03	413.52	415.01	416.51	418.01	419.51	421.02	422.53	424.05	280
290	424.05	425.57	427.09	428.62	430.15	431.69	433.23	434.77	436.32	437.87	439.43	290

RTD Resistance vs Temperature (°F)
Nickel 120, Edison Curve 7

RESISTANCE IN OHMS

°F	0	1	2	3	4	5	6	7	8	9	10	°F
−90	74.55	74.91	75.27	75.63	76.00	76.36	76.72	77.08	77.44	77.81	78.17	−90
−80	78.17	78.53	78.90	79.26	79.62	79.98	80.35	80.71	81.07	81.44	81.80	−80
−70	81.80	82.16	82.53	82.89	83.25	83.62	83.98	84.35	84.71	85.08	85.44	−70
−60	85.44	85.80	86.17	86.53	86.90	87.26	87.63	87.99	88.36	88.73	89.09	−60
−50	89.09	89.46	89.82	90.19	90.56	90.92	91.29	91.66	92.02	92.39	92.76	−50
−40	92.76	93.13	93.50	93.86	94.23	94.60	94.97	95.34	95.71	96.08	96.45	−40
−30	96.45	96.82	97.19	97.56	97.93	98.30	98.67	99.04	99.41	99.78	100.16	−30
−20	100.16	100.53	100.90	101.27	101.65	102.02	102.39	102.77	103.14	103.52	103.89	−20
−10	103.89	104.27	104.64	105.02	105.40	105.77	106.15	106.53	106.90	107.28	107.66	−10
0	107.66	108.04	108.42	108.80	109.18	109.56	109.94	110.32	110.70	111.08	111.47	0

RTD Resistance vs Temperature (°F)
Nickel 120, Edison Curve 7

RESISTANCE IN OHMS

°F	0	1	2	3	4	5	6	7	8	9	10	°F
10	111.47	111.85	112.23	112.62	113.00	113.39	113.77	114.16	114.54	114.93	115.32	10
20	115.32	115.70	116.09	116.48	116.87	117.26	117.65	118.04	118.43	118.82	119.21	20
30	119.21	119.61	120.00	120.39	120.79	121.18	121.58	121.98	122.37	122.77	123.17	30
40	123.17	123.56	123.96	124.36	124.76	125.16	125.56	125.96	126.36	126.77	127.17	40
50	127.17	127.57	127.98	128.38	128.79	129.19	129.60	130.00	130.41	130.82	131.23	50
60	131.23	131.64	132.05	132.46	132.87	133.28	133.69	134.10	134.52	134.93	135.35	60
70	135.35	135.76	136.18	136.59	137.01	137.43	137.84	138.26	138.68	139.10	139.52	70
80	139.52	139.94	140.36	140.79	141.21	141.63	142.06	142.48	142.91	143.33	143.76	80
90	143.76	144.19	144.61	145.04	145.47	145.90	146.33	146.76	147.19	147.63	148.06	90
100	148.06	148.49	148.93	149.36	149.80	150.23	150.67	151.11	151.54	151.98	152.42	100
110	152.42	152.86	153.30	153.74	154.19	154.63	155.07	155.52	155.96	156.40	156.85	110
120	156.85	157.30	157.74	158.19	158.64	159.09	159.54	159.99	160.44	160.89	161.34	120
130	161.34	161.79	162.25	162.70	163.16	163.61	164.07	164.53	164.98	165.44	165.90	130
140	165.90	166.36	166.82	167.28	167.74	168.20	168.66	169.13	169.59	170.06	170.52	140
150	170.52	170.99	171.45	171.92	172.39	172.86	173.33	173.80	174.27	174.74	175.21	150
160	175.21	175.68	176.15	176.63	177.10	177.58	178.05	178.53	179.01	179.48	179.96	160
170	179.96	180.44	180.92	181.40	181.88	182.36	182.85	183.33	183.81	184.30	184.78	170
180	184.78	185.27	185.75	186.24	186.73	187.22	187.70	188.19	188.68	189.17	189.67	180
190	189.67	190.16	190.65	191.14	191.64	192.13	192.63	193.12	193.62	194.12	194.62	190
200	194.62	195.12	195.61	196.11	196.62	197.12	197.62	198.12	198.62	199.13	199.63	200
210	199.63	200.14	200.64	201.15	201.66	202.17	202.67	203.18	203.69	204.20	204.71	210
220	204.71	205.23	205.74	206.25	206.77	207.28	207.80	208.31	208.83	209.35	209.86	220
230	209.86	210.38	210.90	211.42	211.94	212.46	212.98	213.51	214.03	214.55	215.08	230
240	215.08	215.60	216.13	216.66	217.18	217.71	218.24	218.77	219.30	219.83	220.36	240
250	220.36	220.89	221.43	221.96	222.50	223.03	223.57	224.10	224.64	225.18	225.71	250
260	225.71	226.25	226.79	227.33	227.87	228.42	228.96	229.50	230.05	230.59	231.14	260
270	231.14	231.68	232.23	232.78	233.32	233.87	234.42	234.97	235.52	236.08	236.63	270
280	236.63	237.18	237.74	238.29	238.85	239.40	239.96	240.52	241.08	241.63	242.19	280
290	242.19	242.75	243.32	243.88	244.44	245.01	245.57	246.13	246.70	247.27	247.83	290
300	247.83	248.40	248.97	249.55	250.12	250.70	251.27	251.85	252.43	253.00	253.58	300
310	253.58	254.16	254.73	255.31	255.89	256.47	257.05	257.63	258.20	258.78	259.37	310
320	259.37	259.95	260.53	261.11	261.69	262.28	262.86	263.45	264.03	264.62	265.20	320
330	265.20	265.79	266.38	266.97	267.56	268.15	268.74	269.34	269.93	270.52	271.12	330
340	271.12	271.72	272.31	272.91	273.51	274.11	274.71	275.31	275.92	276.52	277.13	340
350	277.13	277.73	278.34	278.95	279.56	280.17	280.78	281.39	282.01	282.62	283.24	350
360	283.24	283.85	284.47	285.09	285.71	286.33	286.95	287.58	288.20	288.82	289.45	360
370	289.45	290.08	290.71	291.33	291.96	292.60	293.23	293.86	294.49	295.13	295.76	370
380	295.76	296.40	297.04	297.68	298.32	298.96	299.60	300.24	300.88	301.52	302.17	380
390	302.17	302.81	303.46	304.11	304.76	305.41	306.06	306.71	307.37	308.02	308.68	390
400	308.68	309.34	309.99	310.65	311.31	311.98	312.64	313.30	313.97	314.64	315.30	400
410	315.30	315.97	316.64	317.31	317.99	318.66	319.34	320.01	320.69	321.37	322.05	410
420	322.05	322.73	323.41	324.10	324.78	325.47	326.15	326.84	327.53	328.22	328.91	420
430	328.91	329.61	330.30	330.99	331.69	332.39	333.09	333.79	334.49	335.19	335.90	430
440	335.90	336.60	337.31	338.01	338.72	339.43	340.14	340.85	341.57	342.28	343.00	440
450	343.00	343.72	344.43	345.15	345.87	346.59	347.32	348.04	348.77	349.49	350.22	450
460	350.22	350.95	351.68	352.41	353.14	353.88	354.61	355.35	356.08	356.82	357.56	460
470	357.56	358.30	359.04	359.79	360.53	361.28	362.02	362.77	363.52	364.27	365.02	470
480	365.02	365.77	366.53	367.29	368.04	368.80	369.56	370.32	371.08	371.84	372.61	480
490	372.61	373.37	374.14	374.90	375.67	376.44	377.21	377.99	378.76	379.53	380.31	490
500	380.31	381.09	381.87	382.65	383.43	384.21	384.99	385.78	386.56	387.35	388.14	500
510	388.14	388.93	389.72	390.51	391.30	392.10	392.89	393.69	394.49	395.29	396.09	510
520	396.09	396.89	397.70	398.50	399.31	400.11	400.92	401.73	402.54	403.35	404.17	520
530	404.17	404.98	405.80	406.61	407.43	408.25	409.07	409.89	410.71	411.54	412.36	530
540	412.36	413.19	414.02	414.85	415.68	416.51	417.34	418.18	419.01	419.85	420.69	540
550	420.69	421.53	422.37	423.21	424.05	424.89	425.74	426.59	427.43	428.28	429.13	550
560	429.13	429.98	430.84	431.69	432.55	433.40	434.26	435.12	435.98	436.84	437.70	560
570	437.70	438.57	439.43									570

RTD Resistance vs Temperature (°C)
Copper 10, SAMA RC21-4-1966

RESISTANCE IN OHMS

°C	0	1	2	3	4	5	6	7	8	9	10	°C
−70	6.33	6.37	6.41	6.45	6.49	6.53	6.57	6.60	6.64	6.68	6.72	−70
−60	6.72	6.76	6.80	6.84	6.88	6.92	6.96	6.99	7.03	7.07	7.11	−60
−50	7.11	7.15	7.19	7.23	7.27	7.31	7.34	7.38	7.42	7.46	7.50	−50
−40	7.50	7.54	7.58	7.62	7.65	7.69	7.73	7.77	7.81	7.85	7.89	−40
−30	7.89	7.92	7.96	8.00	8.04	8.08	8.12	8.16	8.19	8.23	8.27	−30
−20	8.27	8.31	8.35	8.39	8.43	8.46	8.50	8.54	8.58	8.62	8.66	−20
−10	8.66	8.70	8.73	8.77	8.81	8.85	8.89	8.93	8.97	9.00	9.04	−10
0	9.04	9.08	9.12	9.16	9.20	9.23	9.27	9.31	9.35	9.39	9.43	0
10	9.43	9.47	9.50	9.54	9.58	9.62	9.66	9.70	9.74	9.77	9.81	10
20	9.81	9.85	9.89	9.93	9.97	10.01	10.04	10.08	10.12	10.16	10.20	20
30	10.20	10.24	10.27	10.31	10.35	10.39	10.43	10.47	10.51	10.54	10.58	30
40	10.58	10.62	10.66	10.70	10.74	10.78	10.81	10.85	10.89	10.93	10.97	40
50	10.97	11.01	11.05	11.08	11.12	11.16	11.20	11.24	11.28	11.31	11.35	50
60	11.35	11.39	11.43	11.47	11.51	11.55	11.58	11.62	11.66	11.70	11.74	60
70	11.74	11.78	11.82	11.85	11.89	11.93	11.97	12.01	12.05	12.09	12.12	70
80	12.12	12.16	12.20	12.24	12.28	12.32	12.35	12.39	12.43	12.47	12.51	80
90	12.51	12.55	12.59	12.62	12.66	12.70	12.74	12.78	12.82	12.86	12.89	90
100	12.89	12.93	12.97	13.01	13.05	13.09	13.13	13.16	13.20	13.24	13.28	100
110	13.28	13.32	13.36	13.39	13.43	13.47	13.51	13.55	13.59	13.63	13.66	110
120	13.66	13.70	13.74	13.78	13.82	13.86	13.90	13.93	13.97	14.01	14.05	120
130	14.05	14.09	14.13	14.17	14.20	14.24	14.28	14.32	14.36	14.40	14.43	130
140	14.43	14.47	14.51	14.55	14.59	14.63	14.67	14.70	14.74	14.78	14.82	140
150	14.82	14.86	14.90	14.94	14.97	15.01	15.05	15.09	15.13	15.17	15.21	150
160	15.21	15.24	15.28	15.32	15.36	15.40	15.44	15.47	15.51	15.55	15.59	160
170	15.59	15.63	15.67	15.71	15.74	15.78	15.82	15.86	15.90	15.94	15.98	170
180	15.98	16.01	16.05	16.09	16.13	16.17	16.21	16.25	16.28	16.32	16.36	180
190	16.36	16.40	16.44	16.48	16.51	16.55	16.59	16.63	16.67	16.71	16.75	190
200	16.75	16.78	16.82	16.86	16.90	16.94	16.98	17.02	17.05	17.09	17.13	200
210	17.13	17.17	17.21	17.25	17.29	17.32	17.36	17.40	17.44	17.48	17.52	210
220	17.52	17.55	17.59	17.63	17.67	17.71	17.75	17.79	17.82	17.86	17.90	220
230	17.90	17.94	17.98	18.02	18.06	18.09	18.13	18.17	18.21	18.25	18.29	230
240	18.29	18.33	18.36	18.40	18.44	18.48	18.52	18.56	18.59	18.63	18.67	240

RTD Resistance vs Temperature (°F)
Copper 10, SAMA RC21-4-1966

RESISTANCE IN OHMS

°F	0	1	2	3	4	5	6	7	8	9	10	°F
-90	6.42	6.44	6.46	6.48	6.51	6.53	6.55	6.57	6.59	6.61	6.64	-90
-80	6.64	6.66	6.68	6.70	6.72	6.74	6.77	6.79	6.81	6.83	6.85	-80
-70	6.85	6.87	6.89	6.92	6.94	6.96	6.98	7.00	7.02	7.05	7.07	-70
-60	7.07	7.09	7.11	7.13	7.15	7.18	7.20	7.22	7.24	7.26	7.28	-60
-50	7.28	7.31	7.33	7.35	7.37	7.39	7.41	7.43	7.46	7.48	7.50	-50
-40	7.50	7.52	7.54	7.56	7.59	7.61	7.63	7.65	7.67	7.69	7.71	-40
-30	7.71	7.74	7.76	7.78	7.80	7.82	7.84	7.86	7.89	7.91	7.93	-30
-20	7.93	7.95	7.97	7.99	8.01	8.04	8.06	8.08	8.10	8.12	8.14	-20
-10	8.14	8.16	8.19	8.21	8.23	8.25	8.27	8.29	8.31	8.34	8.36	-10
0	8.36	8.38	8.40	8.42	8.44	8.46	8.49	8.51	8.53	8.55	8.57	0
10	8.57	8.59	8.61	8.64	8.66	8.68	8.70	8.72	8.74	8.76	8.79	10
20	8.79	8.81	8.83	8.85	8.87	8.89	8.91	8.94	8.96	8.98	9.00	20
30	9.00	9.02	9.04	9.06	9.08	9.11	9.13	9.15	9.17	9.19	9.21	30
40	9.21	9.23	9.26	9.28	9.30	9.32	9.34	9.36	9.38	9.41	9.43	40
50	9.43	9.45	9.47	9.49	9.51	9.53	9.56	9.58	9.60	9.62	9.64	50
60	9.64	9.66	9.68	9.71	9.73	9.75	9.77	9.79	9.81	9.83	9.86	60
70	9.86	9.88	9.90	9.92	9.94	9.96	9.98	10.01	10.03	10.05	10.07	70
80	10.07	10.09	10.11	10.13	10.15	10.18	10.20	10.22	10.24	10.26	10.28	80
90	10.28	10.30	10.33	10.35	10.37	10.39	10.41	10.43	10.45	10.48	10.50	90
100	10.50	10.52	10.54	10.56	10.58	10.60	10.63	10.65	10.67	10.69	10.71	100
110	10.71	10.73	10.75	10.78	10.80	10.82	10.84	10.86	10.88	10.90	10.93	110
120	10.93	10.95	10.97	10.99	11.01	11.03	11.05	11.08	11.10	11.12	11.14	120
130	11.14	11.16	11.18	11.20	11.22	11.25	11.27	11.29	11.31	11.33	11.35	130
140	11.35	11.37	11.40	11.42	11.44	11.46	11.48	11.50	11.52	11.55	11.57	140
150	11.57	11.59	11.61	11.63	11.65	11.67	11.70	11.72	11.74	11.76	11.78	150
160	11.78	11.80	11.82	11.85	11.87	11.89	11.91	11.93	11.95	11.97	12.00	160
170	12.00	12.02	12.04	12.06	12.08	12.10	12.12	12.15	12.17	12.19	12.21	170
180	12.21	12.23	12.25	12.27	12.29	12.32	12.34	12.36	12.38	12.40	12.42	180
190	12.42	12.44	12.47	12.49	12.51	12.53	12.55	12.57	12.59	12.62	12.64	190
200	12.64	12.66	12.68	12.70	12.72	12.74	12.77	12.79	12.81	12.83	12.85	200
210	12.85	12.87	12.89	12.92	12.94	12.96	12.98	13.00	13.02	13.04	13.07	210
220	13.07	13.09	13.11	13.13	13.15	13.17	13.19	13.22	13.24	13.26	13.28	220
230	13.28	13.30	13.32	13.34	13.36	13.39	13.41	13.43	13.45	13.47	13.49	230
240	13.49	13.51	13.54	13.56	13.58	13.60	13.62	13.64	13.66	13.69	13.71	240
250	13.71	13.73	13.75	13.77	13.79	13.81	13.84	13.86	13.88	13.90	13.92	250
260	13.92	13.94	13.96	13.99	14.01	14.03	14.05	14.07	14.09	14.11	14.14	260
270	14.14	14.16	14.18	14.20	14.22	14.24	14.26	14.29	14.31	14.33	14.35	270
280	14.35	14.37	14.39	14.41	14.43	14.46	14.48	14.50	14.52	14.54	14.56	280
290	14.56	14.58	14.61	14.63	14.65	14.67	14.69	14.71	14.73	14.76	14.78	290
300	14.78	14.80	14.82	14.84	14.86	14.88	14.91	14.93	14.95	14.97	14.99	300
310	14.99	15.01	15.03	15.06	15.08	15.10	15.12	15.14	15.16	15.18	15.21	310
320	15.21	15.23	15.25	15.27	15.29	15.31	15.33	15.36	15.38	15.40	15.42	320
330	15.42	15.44	15.46	15.48	15.50	15.53	15.55	15.57	15.59	15.61	15.63	330
340	15.63	15.65	15.68	15.70	15.72	15.74	15.76	15.78	15.80	15.83	15.85	340
350	15.85	15.87	15.89	15.91	15.93	15.95	15.98	16.00	16.02	16.04	16.06	350
360	16.06	16.08	16.10	16.13	16.15	16.17	16.19	16.21	16.23	16.25	16.28	360
370	16.28	16.30	16.32	16.34	16.36	16.38	16.40	16.43	16.45	16.47	16.49	370
380	16.49	16.51	16.53	16.55	16.57	16.60	16.62	16.64	16.66	16.68	16.70	380
390	16.70	16.72	16.75	16.77	16.79	16.81	16.83	16.85	16.87	16.90	16.92	390
400	16.92	16.94	16.96	16.98	17.00	17.02	17.05	17.07	17.09	17.11	17.13	400
410	17.13	17.15	17.17	17.20	17.22	17.24	17.26	17.28	17.30	17.32	17.35	410
420	17.35	17.37	17.39	17.41	17.43	17.45	17.47	17.50	17.52	17.54	17.56	420
430	17.56	17.58	17.60	17.62	17.64	17.67	17.69	17.71	17.73	17.75	17.77	430
440	17.77	17.79	17.82	17.84	17.86	17.88	17.90	17.92	17.94	17.97	17.99	440
450	17.99	18.01	18.03	18.05	18.07	18.09	18.12	18.14	18.16	18.18	18.20	450
460	18.20	18.22	18.24	18.27	18.29	18.31	18.33	18.35	18.37	18.39	18.42	460
470	18.42	18.44	18.46	18.48	18.50	18.52	18.54	18.57	18.59	18.61	18.63	470
480	18.63	18.65	18.67									480

Glossary

A

absolute humidity *n*: the amount of moisture present in the air. It may be expressed in milligrams of water per cubic metre of air. Compare *relative humidity*.

absolute pressure *n*: total pressure measured from an absolute vacuum. It equals the sum of the gauge pressure and the atmospheric pressure. Expressed in pounds per square inch.

absolute temperature scale *n*: a scale of temperature measurement in which zero degrees is absolute zero. On the Rankine absolute temperature scale, which is based on degrees Fahrenheit, water freezes at 492° and boils at 672°. On the Kelvin absolute temperature scale, which is based on degrees Celsius, water freezes at 273° and boils at 373°. See *absolute zero temperature*.

absolute viscosity *n*: the property by which a fluid in motion offers resistance to shear and flow. Usually expressed as newton-seconds/metre or poise.

absolute zero temperature *n*: a hypothetical temperature at which there is a total absence of heat. Since heat is a result of energy caused by molecular motion, there is no motion of molecules with respect to each other at absolute zero.

actuator *n*: an actuator is a device that provides the force to vary an orifice area through which a control agent flows. It accomplishes this by positioning the valve stem into the valve or other control element. Actuators are classified according to the form of input signal and output power used. Thus, actuators can be mechanical, pneumatic, electric, hydraulic, or a combination of these types.

adjustable choke *n*: a choke in which the position of a conical needle, sleeve, or plate may be changed with respect to their seat to vary the rate of flow; may be manual or automatic. See *choke*.

air bubble *n*: air from the regulated supply of a positive flow indicator bubbles to the surface of liquid from a tube that extends nearly to the bottom of the tank or container. The accumulation of air at the surface then flows out to the bubble tube and to the indicator or controller.

air-purge system *n*: in an air-purge system, a pipe is placed in the liquid with the outlet end connected above the vessel where it represents the zero reference level. Regulated air pressure is applied to the bubble tube through a positive flow indicator.

air-trap method *n*: a method, similar to a diaphragm box, that overcomes objections of internal mounting in corrosive liquids and effects of high temperature. A box is placed in a position to trap air under the box and liquid pressure is transmitted to an indicator-recorder.

analog *adj*: of or pertaining to an instrument that measures a continuous variable that is proportional to another variable over a given range. For example, temperature can be represented or measured as voltage, its analog.

analog input module *n*: a functional module of a programmable logic controller (PLC) that is used to accept an analog signal from a proportional transmitter and convert the analog signal to a digital signal for processing by the system processor.

analog output module *n*: a functional module of a programmable logic controller (PLC) that is used to deliver an analog current signal of, typically, 4-20 mA, to an analog transducer such as a current to pressure (I/P) transducer.

API gravity *n*: the measure of the density or gravity of liquid petroleum products on the North American continent, derived from relative density in accordance with the following equation:

API gravity at 60°F = 141.5/specific density − 131.5

API gravity is expressed in degrees, a specific gravity of 1.0 being equivalent to 10°API. See *gravity*.

API scale *n*: a scale of liquid gravity measurement units called degrees API, devised and adopted by the American Petroleum Institute. Although the scale is very different from an ordinary specific gravity scale, it bears a definite relation to it as follows:

$$°API = \frac{140}{G} - 130$$

where G is specific gravity of the petroleum with reference to water, both at 60°F.

The API scale has particular advantages: it provides finer graduations between whole number units and lends itself to schemes for correcting to a temperature standard of 60°F.

ASCII code *abbr*: American Standard Code for Information Interchange.

atmospheric pressure *n*: the pressure exerted by the weight of the atmosphere. At sea level, the pressure is approximately 14.7 pounds per square inch (101.325 kilopascals), often referred to as 1 atmosphere. Also called barometric pressure.

automatic control *n*: a device that regulates various factors (such as flow rate, pressure, or temperature) of a system without supervision or operation by personnel. See *instrumentation*.

B

back-pressure *n*: 1. the pressure maintained on equipment or systems through which a fluid flows. 2. in reference to engines, a term used to describe the resistance to the flow of exhaust gas through the exhaust pipe. 3. the operating pressure level measured downstream from a measuring device.

back-pressure control *n*: a type of control in which a pneumatic controller maintains a fixed pressure by controlling the amount of upstream pressure in a vessel such as a fractionating tower or an oil and gas separator.

back-pressure regulator *n*: a device used to reduce or increase and maintain a certain amount of back-pressure, or upstream pressure, existing in a pneumatic line.

barometric pressure *n*: see *atmospheric pressure*.

base quantity *n*: one of a small number of physical quantities in a system of measurement that is defined, independent of other physical quantities, by means of a physical standard and by procedures for comparing the quantity to be measured with the standard. Also known as fundamental quantity.

baud *n*: in digital communication terminology, the transmission rate of binary digits, or bits, per second.

Baumé scale *n*: either of two arbitrary scales—one for liquids lighter than water and the other for liquids heavier than water—that indicates specific gravity in degrees.

BCD *abbr*: binary coded decimal.

bellows *n pl*: a pressure-sensing element of cylindrical shape whose walls contain convolutions that cause the length of the bellows to change when pressure is applied.

beta factor *n*: used in fluid flow measurements that use orifice plates, the beta factor is the ratio of the orifice diameter to the pipe diameter.

bimetallic *n*: two metals with different thermal coefficients of expansion—one with a low coefficient and the other with high coefficient of expansion—held together through brazing, welding, or riveting, for the purpose of bending into an arc. When subjected to temperature variations, the bimetallic elements can be used for actuation, indication, and control.

bimetal thermometers *n pl*: use two strips of metal in their operation, one with an extremely low coefficient of thermal expansion and the other a rather high coefficient.

binary *n*: a device or signal that has two discrete states or conditions; an on-off switch has two states and is considered a binary device; a voltage or current switching on and off over time is considered a binary signal. A binary number uses two symbols, 1 and 0, that represent two discrete states or conditions; in specific combinations, the 1s and 0s can represent equivalent decimal numbers.

binary coded decimal (BCD) system *n*: a special combination of four binary digits arranged with its 1s and 0s in a specific order to minimize changes in the position of the binary 1 as decimal counting occurs from zero to nine.

bottomhole choke *n*: a device with a restricted opening placed in the lower end of the tubing to control the rate of flow. See *choke*.

Bourdon tube *n*: a pressure-sensing element consisting of a twisted or curved tube of noncircular cross section, which tends to straighten when pressure is applied internally. By the movements of an indicator over a circular scale, a Bourdon tube indicates the pressure applied.

bubble point *n*: 1. the temperature and pressure at which part of a liquid begins to convert to gas. For example, if a certain volume of liquid is held at constant pressure, but its temperature is increased, a point is reached when bubbles of gas begin to form in the liquid. That is the bubble point. Similarly, if a certain volume of liquid is held at a constant temperature but the pressure is reduced, the point at which gas begins to form is the bubble point. Compare *dew point*. 2. the temperature and pressure at which gas, held in solution in crude oil, breaks out of solution as free gas.

buoyancy *n*: the apparent loss of weight of an object immersed in a fluid. If the object is floating, the immersed portion displaces a volume of fluid the weight of which is equal to the weight of the object.

C

CAO *abbr*: computer assisted operations.

capacitance (C) *n*: the ratio of an impressed electrical charge on a conductor to the change in electrical potential; measured in farads. See *capacitor*.

capacitor *n*: an electrical device that, when wired in an electrical circuit, stores a charge of electricity and returns the charge to the line when certain electrical conditions occur. Physically consists of two parallel conductors separated with an insulating material called the dielectric. Its surface area, plate separation, and dielectric constant determine the amount of capacitance in farads. Also called a condenser.

capacity *n*: 1. volume capability of an electrical capacitor in relation to its ability to store electrical charges. 2. volume capability of a container. See *capacitance, storage capacity*.

capsule *n*: 1. a small container that holds liquid or gas. 2. a lightweight metal unit to which is attached a single metallic diaphragm.

Celsius scale *n*: the metric scale of temperature measurement used universally by scientists. On this scale, 0° represents the freezing point of water and 100° its boiling point at a barometric pressure of 760 mm. Degrees Celsius are converted to degrees Fahrenheit by using the following equation:

$$°F = \tfrac{9}{5}(°C) + 32.$$

The Celsius scale was formerly called the centigrade scale; now, however, the term "Celsius" is preferred in the International System of Units (SI).

centigrade scale *n*: see *Celsius scale*.

choke *n*: a device with an orifice installed in a line to restrict the flow of fluids. Surface chokes are part of the Christmas tree on a well and contain a choke nipple, or bean, with a small-diameter bore that serves to restrict the flow. Chokes are also used to control the rate of flow of the drilling mud out of the hole when the well is closed in with the blowout preventer and a kick is being circulated out of the hole. See *adjustable choke, bottomhole choke, positive choke*.

clear *n*: to restore a storage device, memory device, or binary stage to a prescribed state, usually that denoting zero. Also known as reset.

closed loop *n*: a system of components that utilizes feedback from its output for comparison with the desired set point; any difference results in automatic correction of the desired output. Used in automatic control of processes and equipment without human intervention.

coefficient of expansion *n*: the increment in volume of a unit volume of solid, liquid, or gas for a rise of temperature of 1° at constant pressure. Also called coefficient of cubical expansion, coefficient of thermal expansion, expansion coefficient, expansivity.

computer assisted operations (CAO) *n pl*: a method of data acquisition and control that uses computer technology such as programmable logic controllers or digital controllers.

continuous bleed relay *n*: a pneumatic relay whose operating air pressure escapes on a gradual basis.

control *n*: 1. manual or automatic regulation of a process. A means or device to direct and regulate a process or sequence of events. 2. the section of a digital computer that carries out instructions in proper sequence, interprets each coded instruction, and applies the proper signals to the arithmetic unit and other parts in accordance with this interpretation.

3. a mathematical check used in some computer operations. 4. a test made to determine the extent of error in experimental observations or measurements. 5. a procedure carried out to give a standard of comparison in an experiment.

control agents *n pl*: devices, equipment, or fluid used to bring about corrective action to achieve the desired values of level, temperature, or speed.

controlled variable (CV) *n*: process variables, such as temperature, speed, pressure, and level, can be controlled or regulated by varying other processes or control agents.

controller pressure *n*: pressure used for control of devices in a process; for example, a current-to-pressure transducer's output pressure controls a proportional valve position.

controlling means *n*: elements of a controller that are involved in producing a corrective action. Information is sent back from the process and compared to the desired set point and corrective action of the controlled variable is taken if there is any deviation.

conventional system *n*: a system meeting industry standards.

counting rate meter *n*: an instrument that indicates the time rate of occurrence of input pulses to a radiation counter, averaged over a time interval. Also known as rate meter.

current transmitter *n*: electronic circuitry and components that process a sensor signal created by a process variable from the primary element and produce an mA signal output. Within this circuitry are devices known as operational amplifiers (op amps) that perform functions such as amplification, comparison, addition and subtraction, integration or summing, differentiation or rate, and others. Often referred to as 2-wire transmitters, since the power supply is in series with the signal wires.

cycling *n*: the process by which effluent gas from a gas reservoir is passed through a gas-processing plant or separation system and the remaining residue gas returned to the reservoir. The word recycling is sometimes used for this function, but it is not the preferred term.

D

Dall tube *n*: a Dall tube is similar to a venturi tube but its efficiency is greater. It is designed as follows, 1) a short uniform section followed by an abrupt shoulder that begins an inlet cone; and, 2) a low-pressure tap at the throat leads to an annular groove that encircles the throat. Although the efficiency of a Dall tube is difficult to account for, data indicates it produces about twice the differential pressure of good venturi design, and its pressure recovery is superior to any other form of head meter primary element.

dampener *n*: an air- or inert gas-filled device that minimizes pressure surges in the output line of a mud pump. Sometimes called a surge dampener.

dead band *n*: the difference in the input of a device for a fixed output when operated in ascending and descending directions; for example, a level switch on rising liquid may operate at level x but operate at a different level y when the liquid level decreases. The difference between level x and y is the dead band.

dead time *n*: 1. the time interval between a change in the input signal to a process control system and the response to the signal. 2. the time interval, after a response to one signal or event, during which a system is unable to respond to another. Also known as insensitive time.

dead zone *n*: see *dead band*.

degree API *n*: a unit of measurement of the American Petroleum Institute that indicates the weight, or density, of oil. See *API gravity*.

density *n*: the mass or weight of a substance per unit volume. For instance, the density of a drilling mud may be 10 pounds per gallon, 74.8 pounds per cubic foot, or 1,198.2 kilograms per cubic metre. Specific gravity, relative density, and API gravity are other units of density.

derivative *adj*: a closed-loop control system function usually referred to as rate control; circuitry or components used to change the response of a system to changes in its output or input. Obtains its name from calculus where the derivative of a function is its incremental change.

derivative control *n*: a method of control that adjusts an output based on the rate of change of the process under control.

dew point *n*: the temperature and pressure at which a liquid begins to condense out of a gas. For example, if a constant pressure is held on a certain volume of gas but the temperature is reduced, a point is then reached at which droplets of liquid condense out of the gas. That point is the dew point of the gas at that pressure. Similarly, if a constant temperature is maintained on a volume of gas but the pressure is increased, the point at which liquid begins to condense out is the dew point at that temperature. Compare *bubble point*.

diaphragm *n*: a sensing element consisting of a thin, usually circular, plate that is deformed by pressure applied across the plate.

diaphragm actuator *n*: a pneumatic operator that provides the mechanical force to vary the orifice area through which the control agent flows.

diaphragm box *n*: the physical chamber that contains a flexible diaphragm.

differential *n*: 1. the difference in quantity or degree between two measurements or units. For example, the pressure differential across a choke is the variation between the pressure on one side and that on the other. 2. the value or volume payment accompanying an exchange of oil for oil. The payment serves as compensation for quality, location, or gravity differences between the oils being exchanged.

differential pressure *n*: the pressure existing across an orifice plate in a fluid flow application; pressure is converted into flow units. The difference between two fluid pressures; for example, the difference between the pressure in a reservoir and in a wellbore drilled in the reservoir, or between atmospheric pressure at sea level and at 10,000 feet (3,048 metres). Also called pressure differential.

differential-pressure gauge *n*: a pressure-measuring device actuated by two or more pressure-sensitive elements that act in opposition to produce an indication of the difference between two pressures.

differential-pressure transmitter *n*: 1. a pressure measuring device that senses the difference between two pressures and produces a proportional output, usually 4-20 mA, that covers the calibrated range of the transmitter. 2. an electrical device that senses very small bellows or diaphragm movement in a bellows meter. 3. allows a computer to calculate flow rates directly by producing an electrical output that is proportional to the square root of the differerential pressure that is measured by the transmitter.

digital *adj*: pertaining to data in the form of digits, especially electronic data stored in the form of a binary code.

dimensional analysis *n*: a technique that involves the study of dimensions of physical quantities, used primarily as a tool for obtaining information about physical systems too complicated for full mathematical solutions to be feasible.

direct-reading viscometer *n*: a viscosity measuring device that operates on the principle of relative motion between concentric cylinders separated by annular volumes of the liquid being measured and whose reading is taken visually through a graduated scale.

discrete *n*: digital; a device that is either on or off; an on-off controller that has been replaced by newer PLCs.

displacer *n*: a spherical or cylindrical object that is a component part of a pipe prover that moves through the prover pipe. The displacer has an elastic seal that contacts the inner pipe wall of a prover to prevent leakage. The displacer is made to move through the prover pipe by the flowing fluid and displaces a known measured volume of fluid between two fixed detecting devices.

distance-velocity lag *n*: the delay caused by the amount of time required to transport material or propagate a signal or condition from one point to another. Also known as transportation lag, transport lag.

draft gauge *n*: a modified U-tube manometer used to measure draft of low head gas heads, such as draft pressure in a furnace, or small differential pressures; for example, less than 2 inches (5 centimetres) of water.

E

electrical resistance *n*: the restriction to current flow in an electrical circuit with an impressed voltage; characteristics of conductive material. Measured in units of ohms.

electrolyte *n*: 1. a chemical that, when dissolved in water, dissociates into positive and negative ions, thus increasing its electrical conductivity. 2. the electrically conductive solution that must be present for a corrosion cell to exist.

elevate zero *v*: a procedure used in measuring liquid level to provide a starting point (zero) when the instrument reading is physically below the actual level before calibration.

energy *n*: the capability of a body for doing work. Potential energy is this capability due to the position or state of the body. Kinetic energy is the capability due to the motion of the body.

ephemeris second *n*: the fundamental unit of time of the International System of Units of 1960, equal to 1/31556925.9747 of the tropical year defined by the mean motion of the sun in longitude at the epoch 1900 January 0 day 12 hours.

EPROM *abbr*: erasable programmable read-only memory.

erasable programmable read-only memory (EPROM) *n*: in a computer, a type of read-only memory (ROM) that can be erased and thus be reprogrammed for continuous use.

ergonomics *n pl*: the science that deals with the methods, procedures, and environment involved between humans and machines.

F

Fahrenheit scale *n*: a temperature scale devised by Gabriel Fahrenheit, in which 32° represents the freezing point and 212° the boiling point of water at standard sea-level pressure. Fahrenheit degrees may be converted to Celsius degrees by using the following formula:

$$°C = \frac{5}{9}(°F-32).$$

farm tap regulator *n*: a first-stage pressure regulator that decreases the gas pressure from a high-pressure line to a pressure suitable for use in rural homes or farms.

feedback *n*: automatic control of a process requires the use of feedback. Information is sent back from the process and compared to the desired set point and corrective action of the controlled variable is taken if there is any deviation. If a difference exists between the set point (reference) and the actual condition (feedback), the controlling means (such as a valve) is activated to respond for correction. This arrangement of components is commonly called a closed-loop control system.

filled system *n*: filled systems may consist of the following elements: 1) a metallic bulb containing a fluid—liquid, gas, vapor, or mercury—whose volume or pressure responds to temperature changes. 2) a capillary tube that provides a means of transmission between the bulb and the indicating device. 3) an indicating device using a spiral Bourdon tube to drive a pointer or recording pen. 4) compensating elements to offset the effects of a varying ambient temperature.

Filled systems are relatively trouble-free and accurate, although failure of most components requires complete replacement of the mechanism.

filled-system thermometers *n*: used as direct-reading devices or with compensating elements to offset the effects of ambient temperature; used in process control to provide remote indication of temperature or the recording of temperatures; and, are simple in design and inexpensive to manufacture. Gas-filled thermometers have the greatest measurement capabilities, operate over wide temperature ranges, and have fast response to temperature changes.

Types of filled systems include 1) liquid- and mercury-filled thermometers; completely filled systems that operate on the principle of volumetric expansion. 2) vapor pressure or partially filled thermometers that operate on the pressure principle. 3) gas-filled systems; contain nitrogen or helium gas under pressure that are capable of measuring extremely low temperatures as well as reasonably high values.

final control element *n*: that portion of the controlling means that directly changes the value of the manipulated variable.

flapper valve *n*: a hinged closure mechanism operating in a pivot manner, used to shut off tubing flow. Also called a kick-stand valve.

floating control *n*: control device in which the speed of correction of the control element (such as a piston in a hydraulic relay) is proportional to the error signal. Also known as proportional-speed control.

flow *n*: a current or stream of fluid.

flow bean *n*: a plug with a small hole drilled through it, placed in the flow line at a wellhead to restrict flow if it is too high. Compare *choke*.

flow controller *n*: a device used to control the flow of gases, vapors, liquids, slurries, pastes, or solid particles.

flow nozzle *n*: a restriction installed in a line in which fluid is flowing that produces a pressure differential. The volume of fluid can be determined by measurement of the differential. Flow nozzles can handle dirty and abrasive gases better than orifices.

flow rate *n*: 1. time required for a given quantity of flowable material to flow a measured distance. 2. weight or volume of flowable material flowing per unit time. Also know as rate of flow.

flow regulator *n*: a device used to control the flow of a fluid in a system; typically used in a closed-loop system consisting of a flow set point, flow-rate feedback, and flow valve actuator.

flow sensor *n*: a tool inserted into a pipeline or other container that can sense the flow of fluid within the container.

fluidity *n*: the reciprocal of viscosity. The measure of rate with which a fluid is continuously deformed by a shearing stress; ease of flowing.

force *n*: that which causes, changes, or stops the motion of a body.

force per unit area *n*: an expression of pressure measurement in which a force, such as a pound, bears down on a given area, such as a square inch.

four-wire voltage transmitter *n*: these transmitters get their name based on the number of electrical wires used to power and operate them. Two of the wires to the transmitter provide the DC power for operation (typically 24 volts DC) and the other two wires are used for the signal loop or voltage output.

frequency *n*: cyclic variation of electrical signals or mechanical motion; electrical frequency is expressed in cycles per second or hertz.

fundamental quantity *n*: see *base quantity*.

G

gas *n*: a compressible fluid that completely fills any container in which it is confined. Technically, a gas will not condense when it is compressed and cooled, because a gas can exist only above the critical temperature for its particular composition. Below the critical temperature, this form of matter is known as a vapor, because liquid can exist and condensation can occur. Sometimes the terms "gas" and "vapor" are used interchangeably. The latter, however, should be used for those streams in which condensation can occur and that originate from, or are in equilibrium with, a liquid phase.

gas meter *n*: the gas meter is probably most evident to consumers of natural gas who use it to monitor their gas usage and form the basis for monthly gas bills. It is an accurate instrument that measures volumetric quantities over a wide range of flow values and does not require any electrical power to operate. Where the gas pressure is not unreasonably low, this meter is quite adequate to perform its function.

gauge cock *n*: a valve that activates or isolates a pressure measuring gauge.

gauge glass *n*: a glass tube or metal housing with a glass window that is connected to a vessel to indicate the level of the liquid contents.

gauge pressure *n*: 1. the amount of pressure exerted on the interior walls of a vessel by the fluid contained in it (as indicated by a pressure gauge). It is expressed in pounds per square inch gauge or in kilopascals. Gauge pressure plus atmospheric pressure equals absolute pressure. 2. gauge pressure measured relative to atmospheric pressure considered as zero.

gravitational force *n*: the force acting upon objects on earth, where the force is the product of the object's mass, and the acceleration of gravity is at 32 ft/sec/sec. Measured in units of pounds, grams, etc.

gravity *n*: 1. the attraction exerted by the earth's mass on objects at its surface. 2. the weight of a body.

H

HART protocol *n*: refers to the full name of "highway addressable remote transducer," a form of digital communication produced by a transducer that uses frequency bursts and lapses to simulate the logic 1 or 0.

The frequency can be superimposed on a DC bus or circuit by the transducer or transmitter for data transmission to a remote device for the purposes of programming or modification.

HART protocol programming language *n*: a definite arrangement of digital data produced by the HART transducer that is an established protocol recognized by the receiving unit.

head *n*: 1. the height of a column of liquid required to produce a specific pressure. See *hydraulic head*. 2. for centrifugal pumps, the velocity of flowing fluid converted into pressure expressed in feet or metres of flowing fluid. Also called velocity head. 3. that part of a machine (such as a pump or an engine) that is on the end of the cylinder opposite the crankshaft. Also called cylinder head.

head pressure *n*: head pressure is based on a column of water that exerts 0.433 psi for each vertical foot of height and converting this into pressure is a simple matter of multiplying height with this constant; i.e., 300 ft × 0.433 = 130 psi.

hexadecimal number *n*: a number consisting of 4 binary digits (bits) whose 16 individual counts or values are expressed by decimal numbers 0–9 and alphabetical characters A–F. A hexadecimal number is a simplified representation of a binary number since it produces one symbol to replace four binary digits. The base number of a hexadecimal number is 16.

high-vacuum range *n*: ranges of vacuum pressure measurement that are in the range of 1 micrometre of mercury to 1 millimetre of mercury.

HMI *abbr*: human machine interface.

horsepower (hp) *n*: an expression of power that is equal to 550 foot-pounds per second; 1 hp equals approximately 745.7 watts.

hp *abbr*: horsepower.

human-machine-interface (HMI) *n*: equipment that allows an operator to interface with a system. Formerly referred to as man machine interface (MMI), HMI is more precise and has broader meaning. The science that deals with the methods, procedures, and environment involved between humans and machines is called ergonomics and works to assure that operators perform their assignments efficiently, with ease, and with a minimum of error.

humidity *n*: a measure of moisture content in dry air; an expression for the weight of water dispersed as vapor in a unit weight of dry air or other gas, usually expressed as grains of water per pound of dry air. Relative humidity is expressed as a percentage of the total moisture that the air can support at a given temperature. Relative humidity of 100% indicates the air is saturated with moisture and water vapor will begin to condense and fall out as liquid.

hunting *n*: 1. undesirable oscillation of an automatic control system, wherein the controlled variable swings on both sides of the desired value. 2. operation of a selector in moving from terminal to terminal until one is found which is idle. 3. irregular engine speed resulting from instability of the governing device.

hydrate *n*: a hydrocarbon and water compound that is formed under reduced temperature and pressure in gathering, compression, and transmission facilities for gas. Hydrates often accumulate in troublesome amounts and impede fluid flow. They resemble snow or ice and decompose at atmospheric pressure. *v*: to enlarge by taking water on or in.

hydraulic head *n*: the force exerted by a column of liquid expressed by the height of the liquid above the point at which the pressure is measured. Although "head" refers to distance or height, it is used to express pressure, since the force of the liquid column is directly proportional to its height. Also called head or hydrostatic head. Compare *hydrostatic pressure*.

hydrometer *n*: an instrument with a graduated stem, used to determine the gravity of liquids. The liquid to be measured is placed in a cylinder, and the hydrometer lowered into it. It floats at a certain level in the liquid (high if the liquid is light, low if it is heavy), and the stem markings indicate the gravity of the liquid.

hydrostatic pressure *n*: the force exerted by a body of fluid at rest. It increases directly with the density and the depth of the fluid and is expressed in many different units, including pounds per square inch or kilopascals. The hydrostatic pressure of fresh water is 0.433 pounds per square inch per foot (9.792 kilopascals per metre) of depth. In drilling, the term refers to the pressure exerted by the drilling fluid in the wellbore. In a water drive field, the term refers to the pressure that may furnish the primary energy for production. In level measurement of liquid in a vessel, it is the pressure exerted by the liquid column at the zero reference point.

hygroscopic *adj*: absorbing or attracting moisture from the air.

I

inches of mercury *n*: a measure of atmospheric pressure based on the height, in inches, atmospheric pressure at a given location raises a column of mercury, which is usually confined in a tube.

inches of water *n*: a measure of hydrostatic, or head, pressure based on the height, in inches, liquid pressure in contact with a column of water in a tube raises the column of water.

input PLC module *n*: a functional module of a programmable logic controller that accepts discrete or analog signals from a process.

instrument *n*: a device for measuring, and sometimes recording and controlling, the value of a quantity under observation.

instrumentation *n*: designing, manufacturing, and utilizing physical instruments or instrument systems for detection, observation, measurement, automatic control, automatic computation, communication, or data processing.

integral *n*: 1. a solution to a differential equation. 2. the sum of variables in a control. 3. the damping effect placed in a control system.

integral control *n*: in an automatic process control system utilizing feedback, integral control is a method used to improve performance in regulating the process by gaining control at a faster rate. Another term that describes this activity is proportional plus reset.

integral gain *n*: in automatic control, the amount of error is multiplied by the amount of time the error has been detected, which determines how severe the problem is. Integral gain (Ki) is used as a multiplier to give the operator control over how aggressive he wants this control to correct the problem.

integrating network *n*: a circuit or network whose output waveform is the time integral of its input waveform. Also known as integrator.

integrator *n*: a computer device that approximates the mathematical process of integration. See *integrating network*.

interface unit *n*: a device used to buffer, isolate, or change the character of control signals in an automatic controller.

invar *n*: a proprietary low-expansion metal alloy consisting of 36 percent nickel, 0.35 percent manganese, and 63.65 percent iron that contains a minimum of carbon and other elements; used for watch parts and the measuring guides of accurate instruments.

K

Kelvin (K) *n*: the fundamental unit of thermodynamic temperature in the metric system. See *Kelvin temperature scale*.

Kelvin temperature scale *n*: a temperature scale with the degree interval of the Celsius scale and the zero point at absolute zero. On the Kelvin scale, water freezes at 273.16° and boils at 373.16°. See *absolute temperature scale*.

kinematic viscosity *n*: the absolute viscosity of a fluid divided by the density of the fluid at the temperature of the viscosity measurement. Usually expressed in square metres/second.

kinetic energy *n*: the energy which a body possesses because of its motion; in classical mechanics, equal to one-half of the body's mass times the square of its speed.

L

ladder logic *n*: a method used to describe the arrangement of electromechanical relays in a control system and consists of vertical lines called rails and horizontal lines called rungs. The general sequence of operation of the control is from left to right, top to bottom, of the ladder logic. In PLCs, ladder logic is also used to identify input and output device loca-tion, their logic relationship, and is part of the programming language.

length *n*: extension in space.

level *n*: 1. the physical position of a liquid relative to a lower reference, measured in inches, feet, millimetres, metres, etc. 2. a value of pressure or temperature compared to a reference amount.

level control *n*: maintaining level within specified limits in a tank or vessel can be accomplished with on-off control or with proportional-plus-reset control. On-off control can be achieved with level switches that control either the inflow or outflow of the liquid in the tank.

line pressure *n*: the amount of pressure, or the force per unit area, exerted on a surface by the fluid parallel to a pipe wall.

liquid level *n*: the uppermost surface of liquid in a tank or vessel; liquid level is controlled by level controllers, measured with liquid-level gauges, and viewed with liquid-level indicators.

lower range value (LRV) *n*: in instrumentation, the measure of a process variable, such as pressure or temperature, over the desired range of measurement at its minimum point or lower range value (LRV) and its maximum point or upper range value (URV).

LRV *abbr*: lower range value.

M

magnetic flowmeter *n*: a device that senses and indicates the amount of fluid flowing in a line. Operates on the principle that a magnetic field induces voltage in an electrical conductor moving through the magnetic field. The conductor is the fluid flowing in the line. Field coils mounted diametrically on the sides of a special section of line create the magnetic field.

magnetic meter *n*: a meter used to measure the electrical conductivity of liquids.

manipulated variable *n*: action items, such as rate of flow, used to change the control variables. Flow control of fluids has become an important part of controlling other process variables such as pressure, temperature, and liquid level. When used in this manner, flow is referred to as the manipulated variable.

manometer *n*: a U-shaped piece of glass tubing containing a liquid (usually water or mercury) that is used to measure the pressure of gases or liquids. When pressure is applied, the liquid level in one arm rises while the level in the other drops. A set of calibrated markings beside one of the arms permits a pressure reading to be taken, usually in inches or millimetres.

manual control *n*: 1. an operator-controlled process whereby adjustments to maintain the desired value is performed manually by the operator without any automatic feedback. 2. where a measuring device indicates pressure, flow, level, or temperature of a process variable, or when operators use, in combination, a measuring instrument and a control device in process control to maintain the desired value.

mass *n*: the quantity of matter a substance contains, independent of such external conditions as the buoyancy of the atmosphere or the acceleration caused by gravity.

mass flowmeter *n*: an instrument used to measure the mass rate of flow of a fluid in a system. Expressed as mass per unit time where mass is volume times density and expressed as a gram in the metric system. Used to measure volume per unit time (i.e., gallons/minute, barrels/hour, etc.) by converting the fluid from mass/density/time.

McLeod gauge *n*: a form of a column instrument capable of measuring pressures down to 1 micrometre of mercury column but finds no application in automatic controls since it cannot transmit its information to a remote point. It is primarily used as a standard of measurement to check the accuracy of other instruments.

measuring means *n pl*: elements of a controller that are used to measure and communicate to the controlling means either the value of the controlled variable or its deviation.

mechanical actuator *n*: a device that exerts a force to produce motion or movement to a different position. A process valve contains a mechanical actuator used to open, close, or position the valve's opening.

mercury bell *n*: a device that uses a direct method of measuring differential pressure between two ports. An inverted bell housing is floated in a pool of mercury located in a sealed chamber. Any differential

pressure will cause the bell to move upward in the mercury and operate a pressure indicator.

metering *n*: instruments or recorders used to monitor the quantity of process variables used in a process, including temperature, flow, level, pressure, etc.

minutes per repeat *n*: function of a proportional-plus-reset control system that determines how often the final control element is changed to correct for differences in the desired result. A method of control for constantly monitoring a system for an error against the set point and performing corrections in discrete time intervals to maintain the set-point level.

mixing valves *n pl*: valves used to properly regulate the mixing of two fluids, one hot and one cold, for the purpose of regulating temperature of the resulting fluid.

mode *n*: one of several alternative conditions or methods of operation of a device.

multiple-capacity system *n*: a system containing more than one capacity, such as heat energy storage, that enters the process of control.

multivariable flow metering *n*: the regulation and control of a flowing fluid based on two or more process variables such as level, temperature, or pressure. The result is a desired fluid product corrected by monitoring the process variables.

multivariable transmitter *n*: a process transmitter that monitors several process parameters at one time at different input ports and produces a calculated result—for example, a gas flow multivariable transmitter accepts inputs of temperature, differential pressure, and line pressure to compute the gas flow rate.

N

negative feedback *n*: feedback in which a portion of the output of a circuit, device, or machine is fed back 180° out of phase with the input signal, resulting in a decrease of amplification so as to stabilize the amplification with respect to time or frequency, and a reduction in distortion and noise. Also known as inverse feedback; reverse feedback; stabilized feedback. Feedback that tends to reduce the output in a system.

nonbleed relay *n*: a pneumatic relay whose operating air pressure is retained and sealed in its chamber.

nozzle *n*: 1. in drilling, a passageway through which fluid flows and through which the fluid is restricted. Used in flow measurement by creating a differential pressure or used to increase fluid velocity through smaller restrictions in the pipe. 2. a passageway through jet bits that causes the drilling fluid to be ejected from the bit at high velocity. The jets of mud clear the bottom of the hole. Nozzles come in different sizes that can be interchanged on the bit to adjust the velocity with which the mud exits the bit. 3. the part of the fuel system of an engine that has small holes in it to permit fuel to enter the cylinder. Properly known as a fuel-injection nozzle, but also called a spray valve. The needle valve is directly above the nozzle.

O

octal number *n*: the representation of a three-digit binary number by using the decimal numbers 0 through 7. The decimal numbers 8 and 9 do not exist in the octal numbering system, which has a base of 8.

on-off control *n*: a method of control that uses discrete devices to turn operating devices on or off. These discrete devices may include switches, level switches, pressure switches, temperature switches, and other items that only have two distinct states or conditions.

open loop *n*: a control system that does not incorporate automatic feedback but uses manual adjustments to maintain the desired function.

optical pyrometer *n*: an instrument that determines the temperature of a very hot surface from its incandescent brightness; the image of the surface is focused in the plane of an electrically heated wire, and current through the wire is adjusted until the wire blends into the image of the surface. Also known as disappearing filament pyrometer.

orifice *n*: an opening of a measured diameter that is used for measuring the flow of fluid through a pipe or for delivering a given amount of fluid through a fuel nozzle. In measuring the flow of fluid through a pipe, the orifice must be of smaller diameter than the pipe diameter. It is drilled into an orifice plate held by an orifice fitting.

orifice meter *n*: an instrument used to measure the flow of fluid through a pipe. The orifice meter is an inferential device that measures and records the pressure differential created by the passage of a fluid through an orifice of critical diameter placed in the line. The rate of flow is calculated from the differential pressure and the static, or line, pressure and other factors such as the temperature and density of the fluid, the size of the pipe, and the size of the orifice.

orifice plate *n*: a sheet of metal, usually circular, in which a hole of specific size is made for use in an orifice fitting for the measurement of fluids.

output PLC module *n*: equipment used as part of a PLC system that is used to send commands in the form of current, voltage, or switch closure to operate other equipment. The output module receives its commands from the PLC processor in the form of digital signals.

P

pH *abbr*: an indicator of the acidity or alkalinity of a substance or solution, represented on a scale of 0 to 14, with 0 to 6.9 being acidic, 7 being neither acidic nor basic (i.e., neutral), and 7.1 to 14 being basic. These values are based on hydrogen ion content and activity.

pH factor *n*: the degree of acidity or alkalinity of a substance expressed on a scale of 0 to 14, with 7 being neutral, 7.1 and higher being basic, and 6.9 and lower being acidic.

PID *abbr*: proportional, integral, and derivative.

PID input module *n*: a module that accepts an input reference signal from within a PLC that establishes the desired operating point, compares it to a sensor from the process, and then produces an output signal from the same module to control an output device such as an actuator or proportional valve.

PID loop control *n*: a method of control that utilizes proportional, integral, and derivative techniques to maximize performance and response of the process.

piezoelectricity *n*: electricity or electric polarity created by pressure in a crystalline substance, such as quartz.

piezometer ring *n*: a device that converts pressure into an electrical signal through the piezoelectric crystal principle. See *piezoelectricity*.

Pirani gauge *n*: a gauge that measures inferential pressure and is concerned with certain electrical effects that are observed in an environment of rarefied air or other gas. The gauge consists of a battery, ammeter, and a pressure-controlled glass bulb containing a quantity of resistance wire whose resistance is a function of temperature. Current flowing in the resistance element causes it to heat up, thus changing its resistance.

PLC *abbr*: programmable logic controller.

plug *n*: any object or device that blocks a hole or passageway (such as a cement plug in a borehole) or, the movable part of a valve used to block or restrict the passageway of fluids in the valve.

pneumatic *adj*: operated by air pressure.

poise *n*: the viscosity of a liquid in which a force of 1 dyne (a unit of measurement of small amounts of force) exerted tangentially on a surface of 1 square centimetre of either of two parallel planes 1 centimetre apart will move one plane at the rate of 1 centimetre per second in reference to the other plane, with the space between the two planes filled with the liquid.

poppet valve *n*: a valve that rises perpendicularly to or from its seat. The exhaust valves in a conventional internal combustion automobile engine are poppet valves.

positioner *n*: a device used where friction between a valve stem and its packing causes unsatisfactory response. Where friction becomes troublesome, a positioner and an actuator ensure that the valve is positively positioned at the setting required by the process under control.

positive choke *n*: a choke in which the orifice size must be changed to change the rate of flow through the choke.

positive displacement *n*: a type of flowmeter that accumulates a fixed quantity of fluid for measurement.

positive-displacement meter *n*: a flowmeter that utilizes the energy from the flowing fluid to operate indicating or totalizing meters. It typically rotates a fixed quantity of fluid for each segment of rotation and operates an electrical switch during each rotation. Such a meter is calibrated in terms of quantity of fluid, not rate.

potential energy *n*: the capacity to do work that a body or system has by virtue of its position or configuration.

potentiometer *n*: 1. an instrument to measure electromotive forces by comparison with a known potential difference. 2. a resistor used chiefly as a voltage divider.

power *n*: 1. the time rate of doing work; watts, horsepower, ft-lbs/second. 2. the value that is assigned to a mathematical expression and its exponent.

pressure *n*: stress or force that is exerted uniformly in all directions; it is usually measured in terms of force exerted per unit area, such as pounds per square inch (psi) or newtons per square metre (N/m^2).

pressure controller *n*: an electronic or pneumatic device that maintains a constant pressure at a specific point in a process, such as a pressure-operated valve.

pressure regulator *n*: a device for maintaining pressure in a line, downstream from the device.

pressure-relief valve *n*: a valve that opens at a preset pressure to relieve excessive pressures within a vessel or line. Also called a pop valve, relief valve, safety valve, or safety relief valve.

pressure sensor *n*: a device that senses the process variable of pressure and produces a proportional signal, usually in electrical form. Typical pressure sensors are capacitors and resistive strain gauges.

pressure taps *n pl*: in an orifice fitting, the threaded holes on each side of the orifice plate. Small pipes are screwed into the holes to connect the fitting with a flow recorder. The taps are used so that pressure differential on either side of an orifice plate can be recorded. See *orifice meter*.

primary element *n*: the device in an instrumentation system that produces a desired result that can be measured. An example is an orifice plate and associated hardware used for creating differential pressure from flowing fluids.

processor *n*: in PLC systems, the primary device that contains the microprocessor, memory, logic functions, and timing and scanning elements that create software programs, read input conditions, and issue output commands.

process reaction rate *n*: the speed with which a chemical or industrial process produces a change in the original composition of components.

process variables *n pl*: quantities such as pressure, temperature, flow, and level in a process system.

programmable logic controller (PLC) *n*: a device used to manage, or control, another device or devices that govern the operation of a system or process. An operator, using an attached computer, can program the controller to maintain a given set of desirable circumstances and to respond to changes or upsets in the system or process using ladder logic, which is a logic system that operates much like the rungs on a ladder—that is, before the next rung on the ladder can be scaled, the controller must determine that certain conditions are met on the current rung.

programming *n, adj*: software instructions delivered from a personal computer into a PLC processor in a particular format; i.e., the method used for converting ladder logic diagrams of control systems into input data for PLCs. Programming instructions to the PLC processor can also be in the form of Boolean logic.

proportional *adj*: the closeness of agreement between the desired set point and the actual process level in a control system. In a proportional control system, the proportional band determines how much change in the controlled variable is required to operate the final control element (valve) from fully open to fully closed.

proportional band *n*: the range of values of the controlled variable that cause a controller to operate over its full range.

proportional (P) control *n*: control in which the amount of corrective action is proportional to the amount of error; often controls pressure, flow rate, or temperature in a process system.

proportional (P) controller *n*: a controller whose output is proportional to the error signal.

proportional gain *n*: the ratio of the full range of final control element to the range of the controlled variable.

proportional, integral, and derivative (PID) *n*: mathematical terms applied to automated control system theory that relate to the more practical terms of proportional, proportional-plus-reset, and proportional-plus-reset-plus-rate. Certain closed-loop controls use proportional, integral, and derivative functions as feedback.

proportional-plus-reset controller *n*: the combined proportional and integral (PI) elements in a control system where the reset, or integral, function returns the controlled variable to the desired set point at regular intervals to eliminate any error in the system.

proportional-plus-reset-plus-rate *n*: a combination of proportional, integral, and derivative control (PID) in a control system where rate (derivative) is added to the PI system to improve speed of correction and response and improve stability.

proportional-speed floating control *n*: see *floating control*.

psychrometer *n*: a device used to measure the amount of water vapor, or relative humidity, of the air. See *sling psychrometer*.

pulse module *n*: an input PLC module that accepts electrical pulses at a particular rate to measure flow rates of fluids from a pulsing transmitter.

pump *n*: a device that increases the pressure on a fluid or raises it to a higher level. Various types of pumps include the bottomhole pump, centrifugal pump, hydraulic pump, jet pump, mud pump, reciprocating pump, rotary pump, sucker rod pump, and submersible pump.

pyrometer *n*: an instrument for measuring temperatures, especially those above the range of mercury thermometers.

Q

quantity meter *n*: a fluid meter that measures the volume of flow.

R

ranging *n*: the calibration of process measuring instruments between its lower range value (zero) and its upper range value (span).

Rankine temperature scale *n*: a temperature scale with the zero point at absolute zero. On the Rankine scale, water freezes at 491.60° and boils at 671.69°. See *absolute temperature scale, absolute zero temperature*.

rate meter *n*: see *counting rate meter*.

rate response *n*: the speed with which a controlled variable changes in value.

reflux *n*: in a distillation process, that part of the condensed overhead stream that is returned to the fractionating column as a source of cooling. *v*: in distillation extraction of fluids from a core, to use a solvent to flow over a core sample a second time to clean it.

regulated mixing valves *n pl*: the blending of two fluids for the purpose of regulating temperature at the output.

regulator pressure *n*: the desired set point of pressure controlled by a pressure regulator.

relative density *n*: 1. the ratio of the weight of a given volume of a substance at a given temperature to the weight of an equal volume of a standard substance at the same temperature. For example, if 1 cubic inch of water at 39°F (3.9°C) weighs 1 unit and 1 cubic inch of another solid or liquid at 39°F weighs 0.95 unit, then the relative density of the substance is 0.95. In determining the relative density of gases, the comparison is made with the standard of air or hydrogen. 2. the ratio of the mass of a given volume of a substance to the mass of a like volume of a standard substance, such as water or air.

relative humidity *n*: often used in weather reports. The amount of water that can exist as a vapor in the air varies with temperature—the higher the temperature, the more water the air will support. Relative humidity is expressed as a percentage of the total moisture that the air can support at a given temperature. Thus, as the temperature rises, the relative humidity decreases, and vice versa, assuming the absolute moisture content remains constant.

relative viscosity *n*: the ratio of absolute viscosity of the fluid being measured to the viscosity of water at 68°F.

relay *n*: a device used to open or close electrical circuits or to perform other control operations automatically.

relief valve *n*: see *pressure-relief valve*.

repeats per minute *n*: the time required for a controller to adjust itself back to its set point when a change in its input or output occurs.

reset *n*: in process control, the process of maintaining a prescribed output through reset action. *v*: to restore a setting or input to its original value.

resistance *n*: in electrical circuits, the opposition or restriction to current flow.

resistance temperature detector (RTD) *n*: an electronic device that senses temperature by means of changes in electrical resistance caused by changes in temperature.

response lag *n*: in process control, the difference in time between a command from the set point to the actual change in the controlled variable.

response time *n*: the time for the final element in a control system to reach 63.2 percent of its final value when an input or output disturbance occurs.

RTD *abbr*: resistance temperature detector.

rungs *n*: in a relay ladder logic diagram, the horizontal lines containing input and output devices.

S

Saybolt Second Universal (SSU) *n*: a unit for measuring the viscosity of lighter petroleum products and lubricating oils. See *Saybolt viscometer*.

Saybolt viscometer *n*: an instrument used to measure the viscosity of fluids, consisting basically of a container with a hole or jet of a standard size in the bottom. The time required for the flow of a specific volume of fluids recorded in seconds at three temperatures (100°F–37.8°C, 130°F–54.4°C, and 210°F–98.9°C). The time measurement unit is referred to as the Saybolt Second Universal (SSU).

set point *n*: the desired value or output in a control system. Establishes the reference in a closed-loop control system.

sight glass *n*: a glass tube or a glass-faced section on a process line or vessel; used for visual reading of liquid levels or of manometer pressures.

single-capacity system *n*: a process liquid level tank system where the only stored energy is the liquid weight and height, or level.

sling psychrometer *n*: a hygrometer, or psychrometer, attached to a handle and chain cord, which is whirled rapidly through the air. It also has two thermometers. The bulb of one is kept wet so that the cooling that results from evaporation as the thermometers are whirled makes it register a lower temperature than the dry one. The difference between the dry and wet readings is a measure of the relative humidity of the atmosphere. See *psychrometer*.

smart mass-flow transmitter *n*: an electronic mass flow transmitter that contains a microprocessor and memory for the purpose of producing an analog signal proportional to the mass rate of flow of a fluid.

smart transmitter *n*: a process instrument that contains the capability of manipulating changes in its calibration, ranging, or other data by using microprocessors and memory devices. When used in conjunction with a programming terminal or interface device, a smart transmitter can have its characteristics modified through software communication.

solenoid *n*: an electromagnet that operates a valve when electrical voltage is applied to its coil. The solenoid can either close or open the valve, depending on its arrangement.

span adjustment *n*: the adjustment or calibration of a process transmitter to acquire or set its upper range value (URV).

specific gravity *n*: the ratio of the mass of a given volume of substance to the mass of an equal volume of water. It is a dimensionless ratio that provides the relative weight of a liquid to that of water such as oil. See *relative density*.

specific viscosity *n*: the ratio of the absolute viscosity of a substance to that of a standard fluid, like water, with the viscosity of both fluids being measured at the same temperature.

state of equilibrium *n*: a state where a process variable is maintained under fixed operating conditions without change occurring.

static pressure *n*: 1. the stationary or line pressure existing in a vessel or pipe. 2. the pressure exerted by a fluid upon a surface that is at rest in relation to the fluid. 3. the pressure exhibited at the surface or point downhole during the time the well is shut in. 4. surface or bottomhole pressure after sufficient time has elapsed for the pressure to become stable.

step change *n*: the change of a variable from one value to another in a single process, taking a negligible amount of time; in automatic controls, the step change can occur as an input change from the operator or process, or the step change can occur from the feedback sensing element.

storage capacity *n*: the quantity of data that can be retained simultaneously in a storage device; usually measured in bits, digits, characters, bytes, or words. Also known as capacity memory capacity.

straightening vanes *n pl*: bundles of small-diameter tubing tack-welded together in a concentric pattern and placed in the upstream section of an orifice-meter run for the purpose of reducing considerably the amount of straight pipe required upstream of the orifice. They eliminate swirls and crosscurrents set up by the pipe fittings and valves preceding the meter tube.

stuffing box *n*: a device that prevents leakage along a piston, rod, propeller shaft, or other moving part that passes through a hole in a cylinder or vessel. It consists of a box or chamber made by enlarging the hole and a gland containing compressed packing.

system resistance *n*: the resistance encountered when transferring energy from one source to another, such as that encountered in a hot-water system. Any component, or part, of the system that opposes the free transfer of energy between two capacities is called a resistance.

T

temperature *n*: a measure of heat or the absence of heat, expressed in degrees Fahrenheit or Celsius. The latter is the standard used in countries on the metric system.

temperature sensor *n*: a sensing element and its housing, if any, and defined as the part of a temperature device that is positioned in a liquid the temperature of which is being measured.

temperature transmitters *n pl*: devices that sense the process variable of temperature and convert it to a proportional electrical signal such as 4-20 milliamperes. Transmitters can be reranged or calibrated over a specific temperature range for better resolution around the temperature of interest.

thermistor *n*: a temperature-sensitive element with variable resistivity; often installed in the power circuit of an electric motor to shut down the motor should it overheat.

thermocouple *n*: a device consisting of two dissimilar metals bonded together, with electrical connections to each. When the device is exposed to heat, an electrical current is generated, the magnitude of which varies with the temperature. It is used to measure temperatures higher than those that can be measured by an ordinary thermometer, such as those in engine exhaust.

thermometer *n*: an instrument that measures temperature. Thermometers provide a way to estimate temperature from its effect on a substance with known characteristics (such as a gas that expands when heated). Various types of thermometers measure temperature by measuring the change in pressure of a gas kept at a constant volume, the change in electrical resistance of metals, or the galvanic effect of dissimilar metals in contact. The most common thermometer is the mercury-filled glass tube, which indicates temperature by the expansion of the liquid mercury.

thermoresistive elements *n pl*: devices that change their electrical property of resistance when exposed to varying temperatures. Typical of these devices in instrumentation is a resistance temperature detector (RTD) that is constructed of platinum, nickel, or copper.

throttling range *n*: the full range of control over which the final control element in a process can be operated. See *proportional band*.

time *n*: 1. the dimension of the physical universe which, at a given place, orders the sequence of events. 2. a designated instant in this sequence, as the time of day. Also known as epoch.

transfer lag *n*: the amount of time delay involved in receiving an electrical signal by an electronic circuit and producing a modified form of this signal at its output.

transmitter *n*: an instrument or transducer that converts a process variable such as temperature, pressure, level, or flow into a proportional electrical current or voltage.

transportation lag *n*: see *distance-velocity lag*.

trim *n*: 1. the difference between fore and aft draft readings on a marine vessel or an offshore drilling rig. 2. the use of special corrosion-resistant metals in blowout preventers, valves, and other oilfield devices to minimize the effects of hydrogen sulfide. *v*: 1. to minimize the difference between fore and aft readings. 2. to install H_2S corrosion-resistant materials in oilfield equipment used in H_2S environments.

turbine flowmeter *n*: a device placed in a flow line consisting of a turbine wheel and magnetic pickup to measure the fluid flow. The flow rate is proportional to the electrical frequency produced by the magnetic pickup.

two-wire current transmitter *n*: a common term for a process transmitter that converts a process variable into a proportional 4-20 mA current signal into a two-wire control circuit.

U

ultra-high vacuum range *n*: a range of extremely low pressures wherein the pressure is only 10^{-10} millimetres of mercury or less.

unstable *adj*: an undesirable oscillation or variation of a system's output at its final control element.

upper range value (URV) *n*: the upper value of calibration of a process transmitter where its lower value is the lower range value (LRV).

URV *abbr*: upper range value.

V

vacuum *n*: 1. a space that is theoretically devoid of all matter and that exerts zero pressure. 2. a condition that exists in a system when pressure is reduced below atmospheric pressure.

vacuum pressure *n*: any pressure below atmospheric pressure.

valve *n*: a device used to control the rate of flow in a line to open or shut off a line completely, or to serve as an automatic or semiautomatic safety device. Those used extensively include the check valve, gate valve, globe valve, needle valve, plug valve, and pressure-relief valve.

valve positioner *n*: an attachment that fits on a valve actuator; assures that valve stem movement follows the demands of a controller with great accuracy by acting as a force amplifier between the controller and the actuator. Used where friction occurs between valve stem and packing and when the force of flowing fluid on the valve plug causes unsatisfactory response.

vapor *n*: a substance in the gaseous state that can be liquefied by compression or cooling.

variable quantities *n pl*: quantities of a process variable that can change their chemical or physical composition due to changes in pressure, temperature, or volume.

variable speed *n*: the resulting variable speed output of an electrical motor when subjected to variable voltage or frequency.

velocity of flow equation *n*: a mathematical representation of fluid flow based on predetermined conditions.

venturi effect *n*: the drop in pressure resulting from the increased velocity of a fluid as it flows through a constricted section of a pipe.

venturi section *n*: consists of either a venturi tube or a Dall tube and is one of the most efficient of the primary restrictive elements. It is very difficult and expensive to manufacture to uniform quality standards and is not easy nor quick to replace.

venturi tube *n*: a device inserted into the flow line that places a restriction to the flow for the purpose of developing a differential pressure. When compared with other primary restrictive elements such as orifice plates, it is the most efficient but more costly and difficult to manufacture.

viscometer *n*: a device used to determine the viscosity of a substance. Also called a viscosimeter.

viscosimeter *n*: see *viscometer*.

viscosity *n*: a measure of the resistance of a fluid to flow. Resistance is brought about by the internal friction resulting from the combined effects of cohesion and adhesion. The viscosity of petroleum products is commonly expressed in terms of the time required for a specific volume of the liquid to flow through an orifice of a specific size at a given temperature.

viscosity index *n*: an index used to establish the tendency of an oil to become less viscous at increasing temperatures.

voltage transmitter *n*: four-wire process instruments that convert process variables such as temperature, pressure, level, and flow to a proportional signal, such as 1 to 5 volts DC. These devices are not as accurate as current transmitters since their signal accuracy is affected by wire resistance and electrical interference.

volume booster relay *n*: a device that uses pneumatic pressure in a valve to improve the valve's response time.

vortex flowmeter *n*: a meter that measures the rate of flow with a piezoelectric crystal inside the meter. The piezoelectric crystal detects forces produced by vortices and produces AC voltage whose frequency is directly proportional to flow rate.

W

weight *n*: 1. in mud terminology, refers to the density of a drilling fluid. 2. of a measurement, expresses degree of confidence in result of measurement of a certain quantity compared with result of another measurement of the same quantity.

weight-loaded regulator *n*: a pressure-regulator valve for pressure vessels or flow systems; the regulator is preloaded by counterbalancing weights to open (or close) at the upper (or lower) limit of a preset pressure range.

wet-bulb thermometer *n*: a thermometer having the bulb covered with a cloth, usually muslin or cambric, saturated with water.

Wheatstone bridge *n*: an electrical comparison circuit used to convert variable resistance from a sensing element into a proportional voltage. Output voltage can be calibrated in units of pressure, temperature, or other process variables, or modified to produce a proportional current output for control or indication.

working pressure *n*: 1. the maximum pressure at which an item is to be used at a specified temperature. 2. the current pressure functioning in a process.

Z

zero adjustment *n*: a procedure used in process transmitters to establish the minimum level of measurement of a process variable, or zero. Zero is established first when calibrating transmitters followed by the span adjustment.

Ziegler-Nichols tuning *n:* a method used to establish the optimum performance of a closed-loop automatic control system. This method uses an experimental procedure to establish system gain and time constants within proportional (P), proportional-plus-reset (PI), and proportional-plus-reset-plus-rate (PID) systems.

Index

12-bit A/D converter, 200
135-ohm resistors, 46

A

abscissa, 32
absolute pressure, 85, 86
absolute viscosity, 27, 177–78
acceleration, 20
acid, 185
actuators, 39–48
   air-loaded diaphragm, 41–42
   combination, 48
   diaphragm, 40
   electric, 43–46
   electric-motor-operated, 44–46
   electrohydraulic, 48
   electropneumatic, 48
   hydraulic, 46–48
   mechanical, 40
   overview, 39–40
   piston, 40, 42, 102–3
   pneumatic, 40–43
   reverse-acting diaphragm, 40
   solenoid, 43–44
   spring-loaded diaphragm, 40–41
A/D converter, 200
adjustable flow beans, 157–58
air-bubble (air-purge) system, 133, 172–75
air compressor controllers, PLC, 104, *105*
air compressors, pressure-controlled, 97–98
air-loaded diaphragm actuators, 41–42
air-operated injection pumps, 48–49
air-purge (air-bubble) system, 133, 172–75
air relays, 54–55, 62, 68
air-to-open valves, 67
Allen-Bradley Panelview, 83
Allen-Bradley PLC-5, 196–203
ambient temperature, 110
American Petroleum Institute (API) scale, 171–72
American Standard Code for Information Interchange (ASCII), 82, 195, 209–12
amount of substance, *17*
analog circuits and equipment, 71–73
analog modules, in PLCs, 201–2
angle-body valves, 31
API (American Petroleum Institute) scale, 171–72
area measurement, 16, *17*

ASCII (American Standard Code for Information Interchange), 82, 195, 209–12
asynchronous transmission, 82
atmospheric pressure, 85, 86
automatic control, 5–6
automatic reset, for pneumatic controllers, 58

B

back-pressure regulators, 103–4
baffle plates, 66, 67–68
baffles, 53–57
Bakelite sliding valves, 156
balancing relay, *47*
barometric pressure, 86
bauds, 82
Baumé scale, 172
BCD (binary coded decimal), 195, 209
bellows, 90–91
   and rate of change, 61
   in relief valves, 122
   and valve positioners, 66–68
bellows orifice meters, 150
bellows-spring assembly, 56
bell-type gauges, 91
benefits of instrumentation, 1–2
beta factors, 147
bimetal thermometers, 112
binary coded decimal (BCD), 195, 209
binary numbering system, 193, 194, *195*, 207
BISYNC standard, 82
boilers, pressure-controlled, 97–98
bonnets, 38
Boolean symbols, 190–91
booster relays, 64–65
bottom product discharge rate, 168
Bourdon gauges, 95
Bourdon tubes/springs, 54, 85, 87–88, 130
B-type thermocouple, 215–22
bubble tube, 133
bulbs, rubber, 92
buoyancy instruments, 126–28
Bureau of Mines dew-point tester, 184–85
butterfly valve bodies, 32

C

cabling, parallel vs. serial, 82
CAOs (computer assisted operations), 203

capacitance level measurement and controls, 135–36
capacitor plates, 135
capacitors, 135
capacity, 7
capsules, 89
Celsius, Anders, 109
Celsius scale, 16, 18, 109
centimetre-gram-second (cgs), 27
centrifugal pumps, 50, 164, *165*
cgs (centimetre-gram-second), 27
characterized V-port valve plugs, 34
Charles, Jacques, 109
chokes, 157
closed-loop control system, 6, 48–49, *80*
closed-loop sight glasses, 125–26
closed-tank liquid-level indication, 132
coefficient of expansion, 109
coil CR1, 191, 192
combination actuators, 48
commercial pneumatic controllers, 61–64
common buses, 189
computer assisted operations (CAOs), 203
concentric orifices, 147
continuous bleed air relays, 62
control agents, 6
control, defined, 1
controlled variables, 2, 4
controlled-volume pumps, 48–50
controller set-point regulation by vapor pressure differential, 168–69
controlling means, 6
control of processes
    methods or modes of, 9–14
        floating mode control, 10–11
        on-off, or two-position mode, 9–10
        PID controls, 13–14
        proportional control, 11–12
        proportional plus-reset mode control, 12–13
        proportional plus-reset plus rate, 13
    need for, 1–2
    types of controls, 4–8
control variables, 6
conventional system of measurement, 15, *17*
cooling, evaporative, 182
copper 295–96
copper-constantan thermocouple, 114
critically damped responses, 204
C-tube, 88
current transmitters, two-wire, 93–94, 118–20
C_v (flow coefficient), 34
cycling, 6–7

D

Dall tubes, 145–46
dampeners, 91–92
data code, 82
data communication equipment (DCE), 82
Data Highway Plus, 83
data terminal equipment (DTE), 82
data transfer protocols, 82
data transmission rate, 82
DCE (data communication equipment), 82
dead band, 7
dead time, 9
decimal numbering system, 192–93, *195*
density, 171–77
    measuring devices, 172–77
    measuring scales, 171–72
    SI units of measurement, *17*
derivative control, 14
dew point, 184–85
diaphragm actuators, 40
diaphragms
    in differential-pressure devices, 158
    in gas meters, 155–56
    for level measurement in open tanks, 132–33
    in liquid-level gauges, 90
    metallic and non-metallic, 88–90
    in piston pneumatic actuators, 68
    slack, 89–90
    and valve positioners, 66–67
    why not satisfactory for large differential pressures, 102
dielectric, 135
differential pressure, 85, 86, 88, 144
differential-pressure devices, 158–59
differential-pressure gauges, *90*, *92*
differential-pressure transmitters, 80, 137
dimensions, flow measurement, 143
dimensions of various quantities, 24–27
direct-acting ported valve, *30*
direct measurement, 2–3
direct-reading instruments, 125–26
discharge rate, 167–68
displacer floats, 175–77
displacer instruments, 128–30
distributed control systems, 81–83
double-ported valves, *29*, 30
draft gauges, 90
dry-bulb thermometers, 183
DTE (data terminal equipment), 82
D valves, 156
dynamic viscosity, 177–78
dyne, 27

Index

E

eccentric orifices, 147
EEPROM memory, 121, 190
EIA (Electrical Industries Association), 82
electric actuators, 43–46
electrical current, units of measurement for, *17*
Electrical Industries Association (EIA), 82
electrical level measuring devices, 134–38
electrical noise, 93–94
electric fields, 20
electric liquid-level controllers, 134, *135*
electric-motor-operated actuators, 44–46
electric variable-speed drive, 50
electrodes, 135
electrohydraulic actuators, 48
electrolytes, 186
electronic automatic controls, 71–84
    analog circuits and equipment, 71–73
    distributed control systems, 81–83
    human-machine-interface (HMI), 83–84
    modes of control and control loops, 73–78
        overview, 73–74
        proportional control mode, 74–76
        proportional-plus-integral control (PI), or proportional-plus-reset mode, 76–77
        proportional-plus-integral-plus-derivative (PID) control, or proportional-plus-reset-plus-rate mode, 77–78
    programmable logic controllers (PLC) control systems, 79–81
    specialized flow computers, 81
    system stability and loop tuning, 78–79
electronic differential-pressure flowmeters, 151
electronic flow controllers, 159–62
electronic flow sensors and meters, 150–56
    electronic differential-pressure flowmeters, 151
    magnetic flowmeters, 151–52
    mass flowmeters, 152–53
    positive displacement meters, 154–56
    turbine flowmeters, 153
    vortex flowmeters, 153–54
electronic pressure measurement, 92–95
electronic temperature sensors, 112–17
electronic temperature transmitters, 117–21
electronic transmitter configurations, 94
electropneumatic actuators, 48
elevation of zero, 137, 138
end connections, for valves, 38–39
energy
    kinetic, 16, 23, 24
    potential, 23, 24
    units of measurement for, *17*, 22–24

Engler degree system, 179
Engler seconds system, 178
English system of measurement, 15
ephemeris second, 18
equal percentage valve plugs, 35
ergonomics, 83
error control, 82–83
E-type thermocouple, 223–28
evaporative cooling, 182
expansion, coefficient of, 109

F

Fahrenheit, Gabriel, 108
Fahrenheit scale, 18–19, 108
feedback, 5–6
feed-rate control, 166–67
filled temperature systems, 110–12
filters, for pneumatic actuators, 43
final control elements, 29–50
    actuators, 39–48
        combination, 48
        electric, 43–46
        hydraulic, 46–48
        mechanical, 40
        overview, 39–40
        pneumatic, 40–43
    controlled-volume pumps, 48–50
    overview, 6
    sizing and piping arrangements, 39
    valves, 29–39
        characteristics of, 32–34
        design details, 37–39
        guides and seats for, 36
        plugs for, 34–36
        trim of, 36–37
        valve bodies, 29–32
    variable-volume pumps, 50
fixed flow beans, 157
flappers, 53–57
    in differential-pressure devices, 158
    in displacer instruments, 130
    in Foxboro Model 40 pneumatic controller, *64*
    and rate of change, 61
flexure tube, 130
floating control, 10–11, *45*
floats, 126–28, 129–30, 175–77
flow beans, 157–58
flow characteristics of valves, 32–34
flow coefficient (C_v), 34
flow control, 157–70
    electronic flow controllers, 159–62
    in fractionating columns, 166

integral flow controllers, 162–69
    control of fraction withdrawal rate, 166–69
    flow control in fractionating columns, 166
    gas and steam flow control, 162–63
    liquid flow control, 163–65
  mechanical flow control elements, 157–59

flow measurement, 143–56
  defining, 143
  electronic flow sensors and meters, 150–56
    electronic differential-pressure flowmeters, 151
    magnetic flowmeters, 151–52
    mass flowmeters, 152–53
    positive-displacement meters, 154–56
    turbine flowmeters, 153
    vortex flowmeters, 153–54
  flow rate, 2, 26
  mechanical flow sensors and meters, 144–50
    bellows orifice meters, 150
    calculating flow velocity, 148–49
    installation arrangements for primary elements, 148
    mercury manometer orifice meters, 149
    restrictive elements, 144–47
    variable-area meters, 150

flow nozzles, 144, 146
flow regulators, 123
flow velocity, calculating, 148–49
fluidity, 178
fluid-straightening vanes, 148
force, units of measurement for, *17*, 19–22
force-balance sensing devices, 158
four-wire voltage transmitters, 72, 92–93
Foxboro Model 40 pneumatic controller, *63, 64*
fractionating columns, 166
frequency, units of measurement for, *17*
frequency-counter-to-binary converter, 201–2
friction, and valve positioners, 65–66

G

gain, 12
gallons, 16
gas and steam flow control, 162–63
gases, measuring electrical effects occurring in, 95–96
gas-filled systems, 111–12
gas lines, pressure relief valves in, 99–100
gas meters, 154–56
gas-operated injection pumps, 48–49
gas thermal conduction, 95
gate valve bodies, 32
gauge cocks, 125
gauge glasses, 125–26
gauge pressure, 85, 86

gauges
  bell-type, 91
  Bourdon, 95
  differential-pressure, *90, 91, 92*
  draft, 90
  liquid-level, diaphragms used in, 90
  McLeod, 87, 95
  Pirani, 95–96
  thermocouple vacuum, 96–97
GENET system, 83
globe valve bodies, 29–30
gold-leaf grids, 184
gram, 20
gravitational force, 19
gravitational force, determining mass by balancing, 21
gravity, specific, 26, 171–77
Gray binary code, 195, 212
guides and seats, for valves, 36

H

Handbook of Chemistry and Physics, 182
HART (highway addressable remote transducer), 160
HAT pressure switches, 191, 192
head meters, 144
head pressure, 87, 125, 144
heat, units of measurement for, *17*
helical Bourdon tube, *88*
helical elements, 112
hexadecimal numbering system, 194, *195*, 208–9
H-H pressure switches, 191
high high-level switch (hi-hi *LS*), 138–39
high-pressure regulators, 102–3
high-vacuum range, 95
highway addressable remote transducer (HART), 160
hi-hi *LS* (high high-level switch), 138–39
HMI (human-machine-interface), 83–84, 195
horsepower, 24
hot buses, 189
hot-water temperature, 4–5
human-machine-interface (HMI), 83–84, 195
humidity
  measuring, 180–85
  overview, 107
hunting. *See* cycling
hydrates, 184
hydraulic actuators, 46–48
hydrogen ions, 185
hydrometers, 172
hydrostatic level measurements, 134
hydrostatic pressure, 3, 86, 98
hydrostatic pressure instruments, 131
hygroscopic materials, 182